ADVANCED ENGINEERING ANALYSIS

ADVANCED ENGINEERING ANALYSIS

J. N. REDDY
Virginia Polytechnic Institute and State University

M. L. RASMUSSEN
University of Oklahoma

175 YEARS OF PUBLISHING
1807 1982

A WILEY-INTERSCIENCE PUBLICATION

JOHN WILEY & SONS
New York • Chichester • Brisbane • Toronto • Singapore

Library of Congress Cataloging in Publication Data:

Reddy, J. N.
Advanced engineering analysis.

"A Wiley-Interscience Publication."
Bibliography: p.
Includes index.
1. Engineering mathematics. I. Rasmussen, M. L.
II. Title.

TA330.R43 515'.63'02462 81-14730
ISBN 0-471-09349-1 AACR2

Printed in the United States of America

10 9 8 7 6 5 4 3 2 1

To our teachers, John Tinsley Oden and Krishnamurty Karamcheti

PREFACE

The objective of this book is to present for engineers and applied scientists the basic mathematical concepts of vector and tensor analysis, the extension of these concepts into abstract function spaces (functional analysis), and the unification of these subjects with the variational calculus and associated methods of numerical approximation. Vector and tensor analysis is fundamental to understanding and dealing with a vast range of physical problems and disciplines, and is an indispensable tool for engineering analysis as a subject in itself. In addition, the classical notion of vectors and tensors in Euclidean space, with its physical applications, leads naturally to the modern abstract notion of vectors in function spaces, and thus to the subject of functional analysis. These abstract notions of vector and function spaces provide powerful new concepts and tools of analysis. In particular, they lend themselves directly to approximation methods stemming from the calculus of variations. The variational calculus in turn is related intimately to vector analysis in its complementary representation and interpretation of physical phenomena. Thus the three subjects of this book, vector and tensor analysis, functional analysis, and variational calculus, are mutually related and form a fundamental foundation for modern engineering analysis.

This book is the outgrowth of class notes which the authors have developed and taught over a decade at four major universities. The book is intended for undergraduate seniors and first-year graduate students in engineering and the applied sciences. Senior standing in engineering or a course in differential equations is a prerequisite for the understanding of the material in this book. The subject matter should serve as text for a two quarter course, or a one-semester course in any two of the three chapters on engineering analysis.

The text is divided into three parts: 1. Elements of Vector and Tensor Analysis, 2. Elements of Functional Analysis, and 3. Calculus of Variations and Variational Methods. Numerous examples, most of which are applications of the concepts to problems in various fields of engineering, are provided throughout the book. Many exercise problems are included to test and extend the understanding of the subject matter. A number of these exercise problems are intended to explore related ideas and applications of the concepts covered.

The conclusions of proofs and examples are indicated by the symbol ■.

There are several sections that can be skipped in a first reading of the book (or, if required as prerequisite material, omitted in the syllabus of the course). The material is intended for a semester or two quarter courses, although the material is better suited for a two quarter sequence (Elements of Vectors and Tensors at the undergraduate senior level and Elements of Functional Analysis and Calculus of Variations and Variational Methods at the first-year graduate level).

The authors wish to acknowledge with great pleasure and appreciation the skillful typing of Mrs. Jo Ann Christina, Mrs. Rose Benda, Mrs. Marlene Taylor, and Mrs. Vanessa McCoy.

J. N. Reddy
M. L. Rasmussen

Blacksburg, Virginia
Norman, Oklahoma
December 1981

CONTENTS

ADVANCED ENGINEERING ANALYSIS

Chapter One

ELEMENTS OF VECTOR AND TENSOR ANALYSIS

1.1 INTRODUCTION

1.1.1 Opening Comments

Our approach in this volume is evolutionary; that is, we wish to begin with concepts that are simple and intuitive, and then generalize these ideas step by step into a broader and more abstract body of analysis. This is a natural, inductive approach, more or less in accord with the historical development of the subject. At every stage of development, the subject is useful to the engineer, physicist, and practitioner of applied mechanics. In this chapter we study vector and tensor analysis, which is extremely helpful for a deeper understanding and insight into physical problems.

The term *vector* is used to imply a *physical* vector that has "magnitude and direction" and obeys certain rules of addition and multiplication. In Chapter 2 we consider more general, abstract objects (than physical vectors), which are also called vectors. It transpires that the physical vector is a special case of what is known as a "vector from a linear vector space." Then the notion of vectors in modern mathematical analysis is an abstraction of the elementary notion of a physical vector. While the definition of a vector in abstract analysis does not require the vector to have a magnitude, in nearly all cases of practical interest the vector is endowed with a magnitude (in which case the vector is said to belong to a normed vector space). Thus the present chapter is concerned with vectors from a special normed vector space, that is, physical vectors. In Chapter 2 we consider the more general abstract notion of a vector from a linear vector space.

1.1.2 Concept of Ordinary Vector

In the analysis of physical phenomena we are concerned with quantities that may be classified according to the information needed to specify them completely. Consider two groups:

Scalars	**Nonscalar quantities**
Mass	Force
Temperature	Moment
Density	Stress
Volume	Acceleration
Time	Displacement

After units have been selected, the scalars are given by a single number. Nonscalar quantities need not only a magnitude specified, but also additional information, such as direction. Nonscalar quantities that obey certain rules (such as the parallelogram law of addition) are called *vectors*. Not all nonscalar quantities are vectors. Some quantities require more information than vectors, for instance, a stress. A stress requires the specification of not only a force, but also an area upon which the force acts. A stress is a *tensor*.

In analyzing physical phenomena we set up relations between various quantities that characterize the phenomena (such as Newton's laws, energy conservation, etc.). As a means of expressing a natural law, a coordinate system in a chosen frame of reference can be introduced, and the various physical quantities involved can be expressed in terms of measurements made in that system. The law thus depends upon the chosen coordinate system and may appear different in another type of coordinate system. The laws of nature, however, should be independent of the artificial choice of a coordinate system, and we may seek to represent the law in a manner independent of a particular coordinate system. A way of doing this is provided by vector and tensor analysis. When vector notation is used, a particular coordinate system need not be introduced. Consequently use of vector notation in formulating natural laws leaves them *invariant* to coordinate transformations. A study of physical phenomena by means of vector equations often leads to a deeper understanding of the problem in addition to bringing simplicity and versatility into the analysis.

1.2 VECTOR ALGEBRA

1.2.1 Representation of a Vector

If P and Q are any two points in space, the directed line segment from P to Q locates the position of Q with respect to P. Such a directed line segment is called a *position vector* (see Fig. 1.1). The length of the arrow gives the magnitude of the distance from P to Q, while the *sense* of the arrow indicates the direction. In a similar fashion, we can represent any vector by an arrow as above.

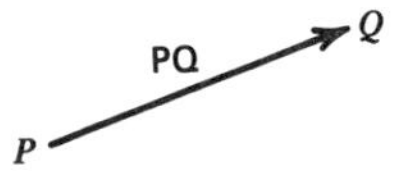

Figure 1.1 Representation of a position vector.

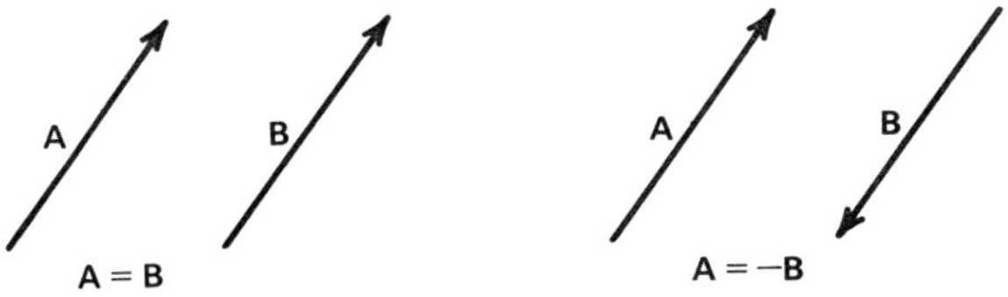

Figure 1.2 Parallel vectors.

Notation In written or typed material it is customary to place an arrow or a bar over the letter denoting the vector, such as $\vec{r}$ or $\vec{v}$. Sometimes the typesetter's mark of a tilde under the letter is used, such as $\underset{\sim}{r}$ or $\underset{\sim}{v}$. In printed material the vector letter is denoted by a boldface letter, such as used in this volume.

For the directed line segment we write **PQ**. The magnitude of the vector **A** is given by $|\mathbf{A}|$ or just A.

Two vectors **A** and **B** are equal if their magnitudes are equal, $|\mathbf{A}|=|\mathbf{B}|$, and if their directions and sense are equal (see Fig. 1.2). Consequently a vector is not changed if it is moved parallel to itself. This means that the position of a vector in space may be chosen arbitrarily. In certain applications, however, the actual point of location of a vector may be important (for instance, a moment or a force acting on a body). A vector associated with a given point is known as a *localized* or *bound* vector. Otherwise it is a free vector. Also one might conceive of a *sliding vector*.

When two or more vectors are parallel to the same line, they are said to be *collinear*; parallel to the same plane, they are *coplanar*. Any two vectors are coplanar.

1.2.2 Addition and Subtraction

Let P and Q be two points in space and let **OP** and **OQ** be the position vectors from some reference point O, as shown in Fig. 1.3. We define **OQ** to be the sum of the vectors **OP** and **PQ**:

$$\mathbf{OQ}=\mathbf{OP}+\mathbf{PQ}. \tag{1.1}$$

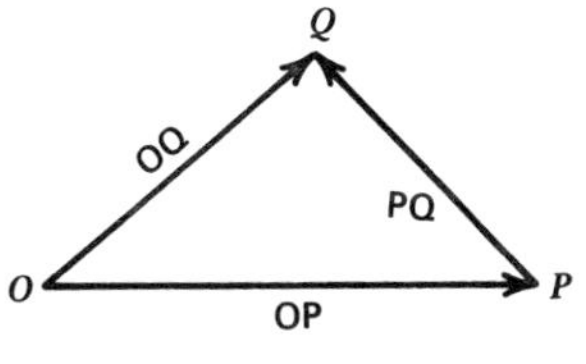

Figure 1.3 Triangle law of addition.

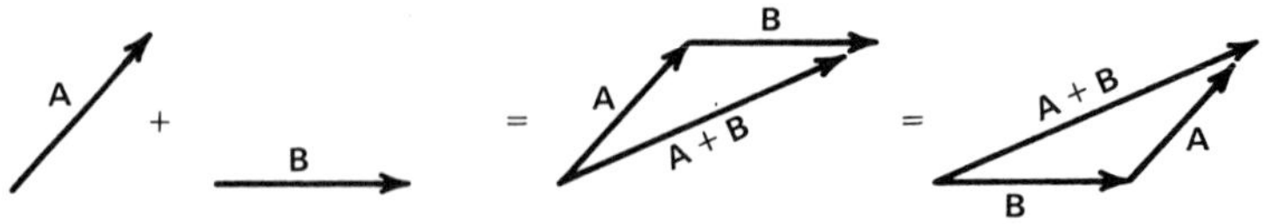

Figure 1.4 Addition of vectors.

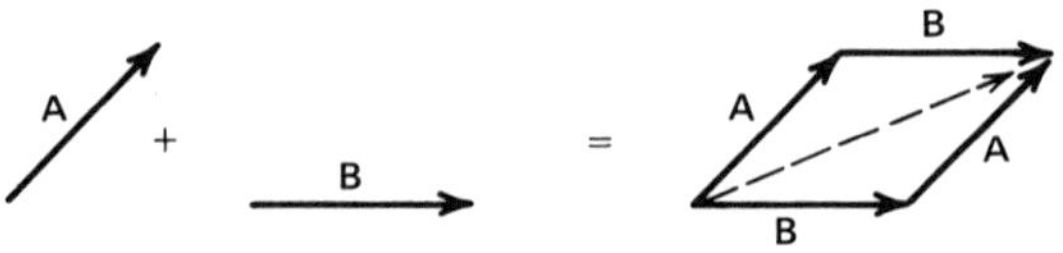

Figure 1.5 Parallelogram law of addition.

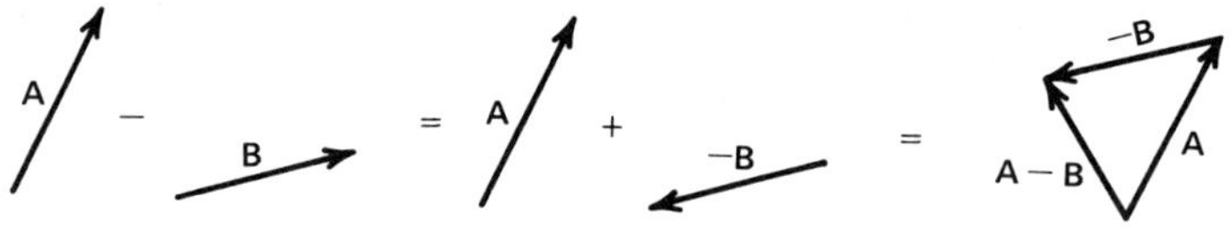

Figure 1.6 Subtraction of vectors.

This notion of addition is applicable to vector quantities other than position vectors.

Let **A** and **B** be any two vectors. Then we can add them as shown in Fig. 1.4. The combination of the two diagrams in Fig. 1.4 gives the parallelogram shown in Fig. 1.5.

Thus we say the vectors add according to the *parallelogram law* of addition so that

$$\mathbf{C}=\mathbf{A}+\mathbf{B}=\mathbf{B}+\mathbf{A}. \tag{1.2}$$

We thus see that vector addition is *commutative.*

Subtraction of vectors is carried out along the same lines. To form the difference $\mathbf{A}-\mathbf{B}$, we write

$$\mathbf{A}-\mathbf{B}=\mathbf{A}+(-\mathbf{B}), \tag{1.3}$$

and subtraction reduces to the operation of addition (see Fig. 1.6). The negative vector $-\mathbf{B}$ has the same magnitude as **B**, but has the opposite *sense.*

1.2.3 Definition of a Vector (Geometric)

A vector is a quantity that possesses both magnitude and direction and obeys the parallelogram law of addition.

Obeying the law is important because there are quantities having both magnitude and direction that do not obey this law. A finite rotation of a rigid body is not a vector, although infinitesimal rotations are. The above is a *geometrical* definition. That vectors can be represented graphically is an *incidental* rather than a fundamental feature of the vector concept.

1.2.4 Multiplication by a Scalar

If a vector **A** is multiplied by a scalar m (or number), one obtains another vector which has m times the magnitude of **A**:

$$m\mathbf{A}=\mathbf{A}m \quad \text{and} \quad \begin{aligned} |m\mathbf{A}| &= m|\mathbf{A}|, \quad m>0 \\ &= -m|\mathbf{A}|, \quad m<0. \end{aligned} \tag{1.4}$$

Multiplication by a scalar also commutes.

1.2.5 Unit Vector

A vector of unit length is called a *unit vector*. The unit vector may be defined as follows:

$$\hat{\mathbf{e}}_A=\frac{\mathbf{A}}{A}. \tag{1.5}$$

We may now write

$$\mathbf{A}=A\hat{\mathbf{e}}_A. \tag{1.6}$$

Thus *any vector may be represented as a product of its magnitude and a unit vector*.

A unit vector is used to designate direction. It does not have any physical dimensions. We denote a unit vector by a "hat" (caret) above the boldface letter.

1.2.6 Zero Vector

A vector of zero magnitude is called a *zero vector* or a *null vector*. All null vectors are considered equal to each other without consideration as to direction:

$$\mathbf{A}+\mathbf{0}=\mathbf{A} \quad \text{and} \quad 0\mathbf{A}=\mathbf{0}. \tag{1.7}$$

Summary The laws that govern addition, subtraction, and scalar multiplication of vectors are identical with those governing the operations of scalar algebra.

1.2.7 Linear Dependence

The concepts of collinear and coplanar vectors can be stated in algebraic terms. A set of n vectors is said to be linearly dependent if a set of n numbers $\beta_1, \beta_2, \ldots, \beta_n$ can be found such that

$$\beta_1 \mathbf{A}_1 + \beta_2 \mathbf{A}_2 + \cdots + \beta_n \mathbf{A}_n = \mathbf{0}, \tag{1.8}$$

where $\beta_1, \beta_2, \ldots, \beta_n$ cannot all be zero. If this expression cannot be satisfied, the vectors are said to be *linearly independent*.

If two vectors are linearly dependent, then they are *collinear*. If three vectors are linearly dependent, then they are *coplanar*. Four or more vectors in three-dimensional space are always linearly dependent.

We can state linear dependence by saying that one of the n vectors can be represented by a linear combination of the others (for $\beta_n \neq 0$, say):

$$\mathbf{A}_n = -\frac{1}{\beta_n}(\beta_1 \mathbf{A}_1 + \beta_2 \mathbf{A}_2 + \cdots + \beta_{n-1} \mathbf{A}_{n-1}). \tag{1.9}$$

1.2.8 Scalar Product of Two Vectors

Besides addition, subtraction, and multiplication by a scalar, we must consider multiplication of two vectors. There are several ways the product of two vectors can be defined. We consider first the so-called scalar product.

Let us recall the concept of work. When a force **F** acts on a mass point and moves through an infinitesimal displacement **ds**, the work done by the force is defined by the *projection* of the force in the direction of the displacement times the magnitude of the displacement (see Fig. 1.7). Such an operation may be defined for any two vectors. Since the result of the product is a scalar, it is called the *scalar product*. We denote this product as follows:

$$\mathbf{F} \cdot \mathbf{ds} = F\,ds \cos\theta, \qquad 0 \leq \theta \leq \pi. \tag{1.10}$$

The scalar product is also known as the *dot product* and *inner product*.

A few simple results follow from the above definition:

1 Since $\mathbf{A} \cdot \mathbf{B} = \mathbf{B} \cdot \mathbf{A}$, the scalar product is commutative.

2 If the vectors **A** and **B** are perpendicular to each other, then $\mathbf{A} \cdot \mathbf{B} = AB\cos(\pi/2) = 0$. Conversely, if $\mathbf{A} \cdot \mathbf{B} = 0$, then either **A** or **B** is zero or **A** is perpendicular, or *orthogonal*, to **B**.

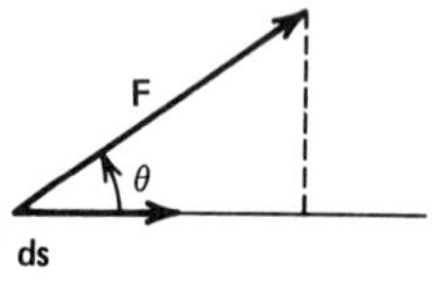

Figure 1.7 Representation of work.

3 If two vectors **A** and **B** are parallel and in the same direction, then $\mathbf{A}\cdot\mathbf{B}=AB\cos 0=AB$, since $\cos 0=1$. Thus the scalar product of a vector with itself is equal to the square of its magnitude:

$$\mathbf{A}\cdot\mathbf{A}=AA=A^2. \tag{1.11}$$

The product $\mathbf{A}\cdot\mathbf{A}$ is sometimes denoted by $\mathbf{A}^2$.

4 The orthogonal projection of a vector **A** in any direction $\hat{\mathbf{e}}$ is given by $\mathbf{A}\cdot\hat{\mathbf{e}}$.

5 The scalar product follows the distributive law also:

$$\mathbf{A}\cdot(\mathbf{B}+\mathbf{C})=(\mathbf{A}\cdot\mathbf{B})+(\mathbf{A}\cdot\mathbf{C}). \tag{1.12}$$

1.2.9 Vector Product of Two Vectors

To introduce this product, consider the concept of a *moment* due to a force. Let us describe the moment about a point O of a force **F** acting at a point P, such as shown in Fig. 1.8. By definition the magnitude of the moment is given by

$$M=Fl, \tag{1.13}$$

where l is the lever arm for the force about the point O. If **r** denotes the vector **OP** and θ the angle between **r** and **F** as shown, such that $0\leq\theta\leq\pi$, we have $l=r\sin\theta$, and thus

$$M=Fr\sin\theta. \tag{1.14}$$

A direction can now be assigned to the moment. Drawing the vectors **F** and **r** from the common origin O, we note that the rotation due to **F** tends to bring **r** into **F** (see Fig. 1.9).

We now set up an axis of rotation perpendicular to the plane formed by **F** and **r**. Along this axis of rotation we set up a preferred direction as that in which a right-handed screw would advance when turned in the direction of rotation due to the moment (see Fig. 1.10). Along this axis of rotation we draw a unit vector $\hat{\mathbf{e}}_M$ and agree that it represents the direction of the moment **M**. Thus we have

$$\mathbf{M}=Fr\sin\theta\,\hat{\mathbf{e}}_M$$

$$\equiv\mathbf{r}\times\mathbf{F}. \tag{1.15}$$

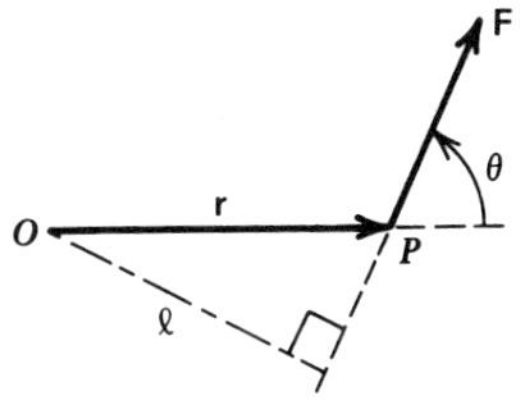

Figure 1.8 Representation of a moment.

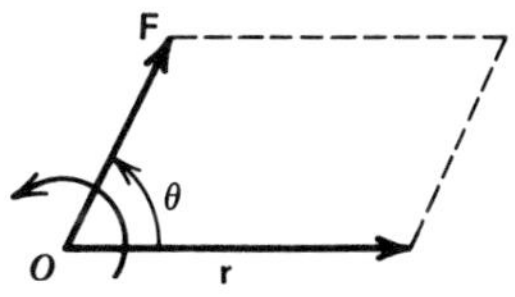

Figure 1.9 Direction of rotation.

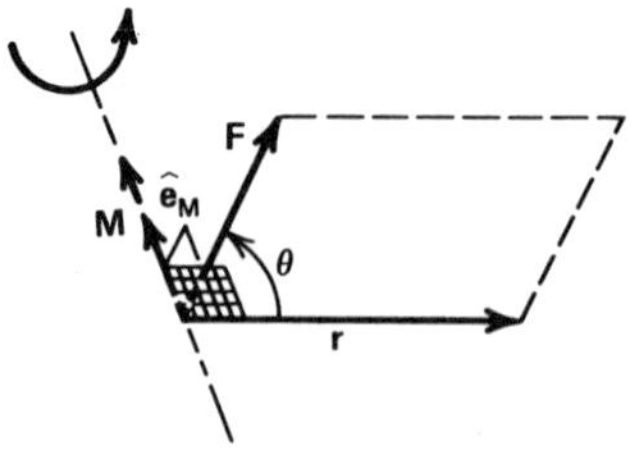

Figure 1.10 Axis of rotation.

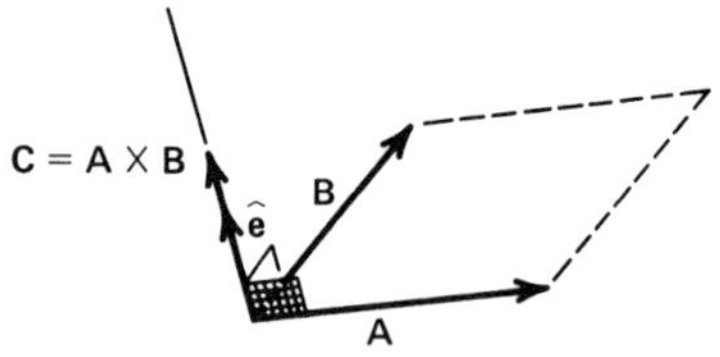

Figure 1.11 Representation of the vector product.

According to this expression, **M** may be looked upon as resulting from a special operation between the two vectors **F** and **r**. It is thus the basis for defining a product between any two vectors. Since the result of such a product is a vector, it may be called the *vector product*.

The vector product of the vectors **A** and **B** is a vector **C** whose magnitude is equal to the product of the magnitude of **A** and **B** times the sine of the angle measured from A to B such that $0 \leq \theta \leq \pi$, and whose direction is specified by the condition that **C** be perpendicular to the plane of the vectors **A** and **B** and points in the direction in which a right-handed screw advances when turned so as to bring **A** into **B**.

The vector product is usually denoted by

$$\mathbf{C} = \mathbf{A} \times \mathbf{B} = AB \sin(A, B)\hat{\mathbf{e}}. \tag{1.16}$$

This product is called the *cross product*, *skew product*, and also *outer product*, as well as vector product (see Fig. 1.11).

From the above definition a few simple results follow.

1 The products $\mathbf{A}\times\mathbf{B}$ and $\mathbf{B}\times\mathbf{A}$ are not equal. In fact, we have

$$\mathbf{A}\times\mathbf{B} = -\mathbf{B}\times\mathbf{A}. \tag{1.17}$$

Thus the vector product *does not commute*. We must therefore preserve the order of the vectors when vector products are involved.

2 If two vectors **A** and **B** are parallel to each other, then $\theta = \pi$, 0 and $\sin\theta = 0$. Thus

$$\mathbf{A}\times\mathbf{B} = \mathbf{0}.$$

Conversely, if $\mathbf{A}\times\mathbf{B} = \mathbf{0}$, then either **A** or **B** is zero, or they are parallel vectors. It follows that the vector product of a vector with itself is zero; that is, $\mathbf{A}\times\mathbf{A} = \mathbf{0}$.

3 The distributive law still holds, but the order of the factors must be maintained:

$$(\mathbf{A}+\mathbf{B})\times\mathbf{C} = (\mathbf{A}\times\mathbf{C}) + (\mathbf{B}\times\mathbf{C}). \tag{1.18}$$

1.2.10 Plane Area as a Vector

The magnitude of the vector $\mathbf{C}=\mathbf{A}\times\mathbf{B}$ is equal to the area of the parallelogram formed by the vectors **A** and **B**, as shown in Fig. 1.12. In fact, the vector **C** may be considered to represent *both* the magnitude and the direction of the product **A** and **B**. Thus a plane area may be looked upon as possessing a direction in addition to a magnitude, the directional character arising out of the need to specify an orientation of the plane in space.

It is customary to denote the direction of a plane area by means of a unit vector drawn normal to that plane. To fix the direction of the normal, we assign a *sense of travel* along the contour of the boundary of the plane area in question. The direction of the normal is taken by convention as that in which a right-handed screw advances as it is rotated according to the sense of travel along the boundary curve or contour (see Fig. 1.13). Let the unit normal vector be given by $\hat{\mathbf{n}}$. Then the area can be denoted by $\mathbf{S}=S\hat{\mathbf{n}}$.

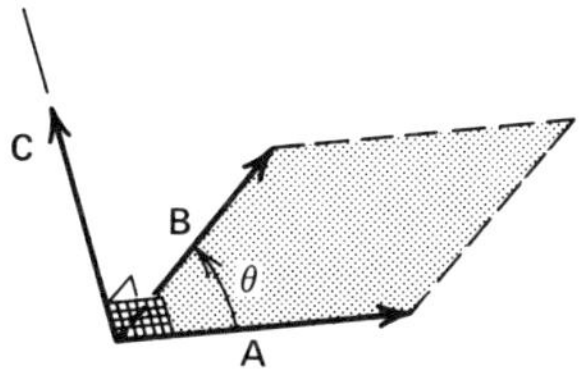

Figure 1.12 Plane area as a vector.

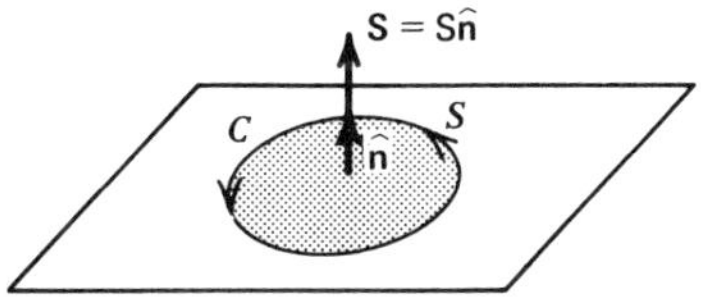

Figure 1.13 Unit normal vector and sense of travel.

Example 1.1 Representation of a plane as a vector has many uses. The vector can be used to determine the area of an inclined plane in terms of its projected area.

1 Consider a cylinder (height of the cylinder is not important). We wish to determine the plane area of the surface obtained by cutting the cylinder with an inclined plane whose normal is $\hat{\mathbf{n}}$. Let the plane area of the inclined surface be S, and that of the base of the cylinder be S' (see Fig. 1.14*a*). We have

$$\mathbf{S}'=S'\hat{\mathbf{n}}' \qquad \text{and} \qquad \mathbf{S}=S\hat{\mathbf{n}}.$$

Since S' is the projection of **S** along $\hat{\mathbf{n}}'$ (if the angle between $\hat{\mathbf{n}}$ and $\hat{\mathbf{n}}'$ is acute; otherwise the negative of it),

$$S'=\mathbf{S}\cdot\hat{\mathbf{n}}'=S\hat{\mathbf{n}}\cdot\hat{\mathbf{n}}'. \tag{1.19}$$

2 Similar ideas apply to the case of a cube (or a prism) cut by an inclined plane whose normal is $\hat{\mathbf{n}}$. We can express the areas of the sides of the resulting tetrahedron in terms of the area of the inclined surface. For reference purposes we label the sides of the cube by 1, 2, and 3 and the

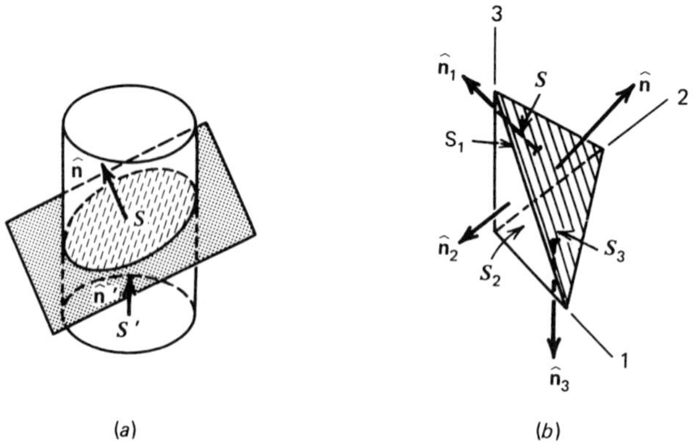

Figure 1.14 Representation of inclined plane areas by unit normal vectors.

normals and surface areas by $(\hat{\mathbf{n}}_1, S_1)$, $(\hat{\mathbf{n}}_2, S_2)$, and $(\hat{\mathbf{n}}_3, S_3)$, respectively (i.e., S_i is the surface area of the plane perpendicular to the ith line or $\hat{\mathbf{n}}_i$ vector), as shown in Fig. 1.14*b*. Then we have

$$S_1 = S\hat{\mathbf{n}}\cdot\hat{\mathbf{n}}_1, \qquad S_2 = S\hat{\mathbf{n}}\cdot\hat{\mathbf{n}}_2, \qquad S_3 = S\hat{\mathbf{n}}\cdot\hat{\mathbf{n}}_3. \quad \blacksquare \tag{1.20}$$

1.2.11 Velocity of a Point of a Rotating Rigid Body

The description of the velocity of a point of a rotating rigid body is an important example of geometrical and physical applications of vectors. Suppose a rigid body is rotating with an angular velocity ω about an axis, and we wish to describe the velocity of some point P of the body (see Fig. 1.15). Let $\mathbf{V}$ denote the velocity at the point. Each point of the body describes a circle that lies in a plane perpendicular to the axis with its center on the axis. The radius of the circle is the perpendicular distance from the axis to the point of interest. The magnitude of the velocity is equal to ωa. The direction of $\mathbf{V}$ is perpendicular to a and to the axis of rotation. We denote the direction of the velocity by the unit vector $\hat{\mathbf{e}}$. Thus we have

$$\mathbf{V} = \omega a\hat{\mathbf{e}}. \tag{1.21}$$

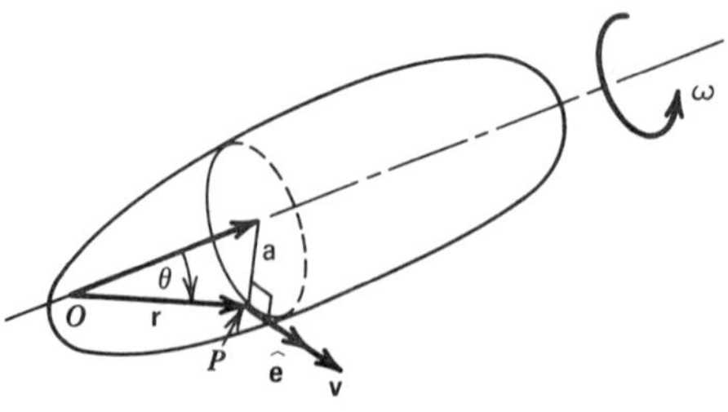

Figure 1.15 Velocity at a point in a rotating rigid body.

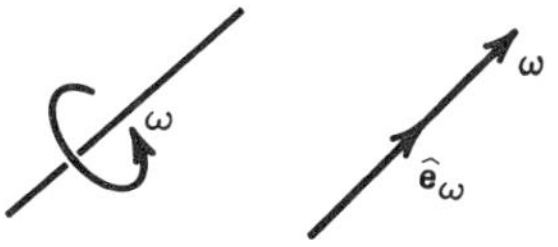

Figure 1.16 Angular velocity as a vector.

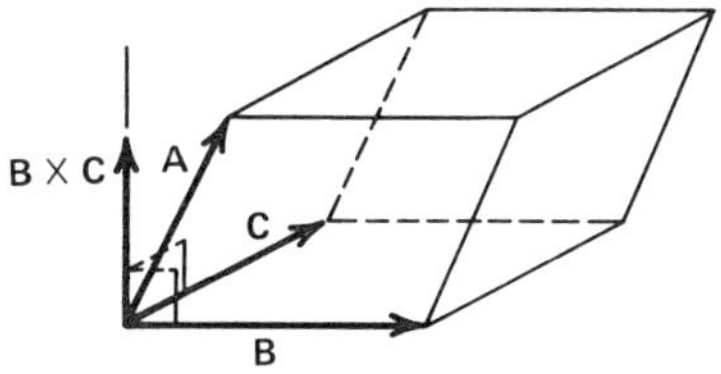

Figure 1.17 Scalar triple product as the volume of a parallelepiped.

Let O be a reference point on the axis of revolution, and let $\mathbf{OP}=\mathbf{r}$. We then have $a=r\sin\theta$, so that

$$\mathbf{V}=\omega r\sin\theta\,\hat{\mathbf{e}}. \tag{1.22}$$

The angular velocity is a vector since it has an assigned direction, magnitude, and obeys the parallelogram law of addition. We denote it by $\boldsymbol{\omega}$ and represent its direction in the sense of a right-handed screw (see Fig. 1.16). If we further let $\hat{\mathbf{e}}_r$ be a unit vector in the direction of $\mathbf{r}$, we see that

$$\hat{\mathbf{e}}_\omega\times\hat{\mathbf{e}}_r=\hat{\mathbf{e}}\sin\theta. \tag{1.23}$$

With these relations we have

$$\mathbf{V}=\boldsymbol{\omega}\times\mathbf{r}. \tag{1.24}$$

Thus the velocity of a point of a rigid body rotating about an axis is given by the vector product of $\boldsymbol{\omega}$ and a position vector $\mathbf{r}$ drawn from any reference point on the axis of revolution.

1.2.12 Multiple Products

Consider the various products of three vectors:

$$\mathbf{A}(\mathbf{B}\cdot\mathbf{C}),\qquad \mathbf{A}\cdot(\mathbf{B}\times\mathbf{C}),\qquad \mathbf{A}\times(\mathbf{B}\times\mathbf{C}). \tag{1.25}$$

The product $\mathbf{A}(\mathbf{B}\cdot\mathbf{C})$ is merely a multiplication of the vector $\mathbf{A}$ by the scalar $(\mathbf{B}\cdot\mathbf{C})$.

Scalar Triple Product

The product $\mathbf{A}\cdot(\mathbf{B}\times\mathbf{C})$ is a scalar. It can be seen that the product $\mathbf{A}\cdot(\mathbf{B}\times\mathbf{C})$, except for the algebraic sign, is the volume of the parallelepiped formed by the vectors $\mathbf{A}$, $\mathbf{B}$, and $\mathbf{C}$, as shown in Fig. 1.17.

We also note the following properties:

1 The dot and cross can be interchanged without changing the value:

$$\mathbf{A}\cdot\mathbf{B}\times\mathbf{C}=\mathbf{A}\times\mathbf{B}\cdot\mathbf{C}\equiv[\mathbf{ABC}]. \tag{1.26}$$

2 A cyclical permutation of the order of the vectors leaves the result unchanged:

$$\mathbf{A}\cdot\mathbf{B}\times\mathbf{C}=\mathbf{C}\cdot\mathbf{A}\times\mathbf{B}=\mathbf{B}\cdot\mathbf{C}\times\mathbf{A}\equiv[\mathbf{ABC}]. \tag{1.27}$$

3 If the cyclic order is changed, the sign changes:

$$\mathbf{A}\cdot\mathbf{B}\times\mathbf{C}=-\mathbf{A}\cdot\mathbf{C}\times\mathbf{B}=-\mathbf{C}\cdot\mathbf{B}\times\mathbf{A}=-\mathbf{B}\cdot\mathbf{A}\times\mathbf{C}. \tag{1.28}$$

4 A necessary and sufficient condition for any three vectors, **A**, **B**, **C** to be coplanar is that $\mathbf{A}\cdot(\mathbf{B}\times\mathbf{C})=0$. Note also that the scalar triple product is zero when any two vectors are the same.

Vector Triple Product

This result is a vector. The vector $\mathbf{A}\times(\mathbf{B}\times\mathbf{C})$ is normal to the plane formed by **A** and $(\mathbf{B}\times\mathbf{C})$. The vector $(\mathbf{B}\times\mathbf{C})$, however, is perpendicular to the plane formed by **B** and **C**. This means that $\mathbf{A}\times(\mathbf{B}\times\mathbf{C})$ lies in the plane formed by **B** and **C** and is perpendicular to **A** (see Fig. 1.18). Thus $\mathbf{A}\times(\mathbf{B}\times\mathbf{C})$ can be expressed as a linear combination of **B** and **C**:

$$\mathbf{A}\times(\mathbf{B}\times\mathbf{C})=m_1\mathbf{B}+n_1\mathbf{C}. \tag{1.29}$$

Likewise, we would find that

$$(\mathbf{A}\times\mathbf{B})\times\mathbf{C}=m_2\mathbf{A}+n_2\mathbf{B}. \tag{1.30}$$

Thus the parentheses *cannot* be interchanged or removed. It can be shown that

$$m_1=\mathbf{A}\cdot\mathbf{C}, \qquad n_1=-\mathbf{A}\cdot\mathbf{B},$$

and hence that

$$\mathbf{A}\times(\mathbf{B}\times\mathbf{C})\equiv(\mathbf{A}\cdot\mathbf{C})\mathbf{B}-(\mathbf{A}\cdot\mathbf{B})\mathbf{C}. \tag{1.31}$$

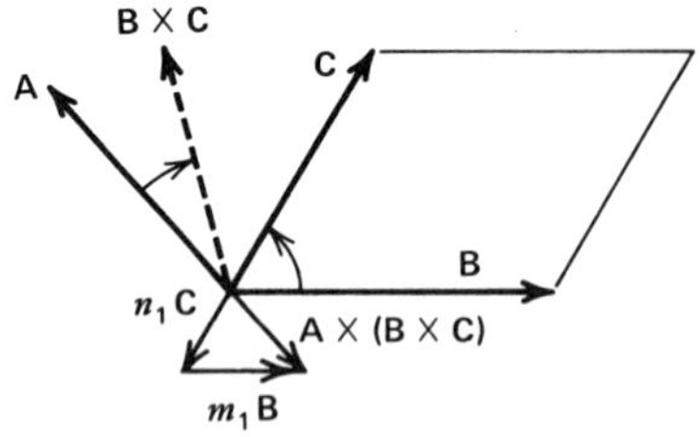

Figure 1.18 The vector triple product.

Example 1.2 The equation of a plane perpendicular to a vector **A** and passing through the terminal point of vector **B** can be obtained without the use of any coordinate system (see Fig. 1.19). Let *O* be the origin and *B* the terminal point

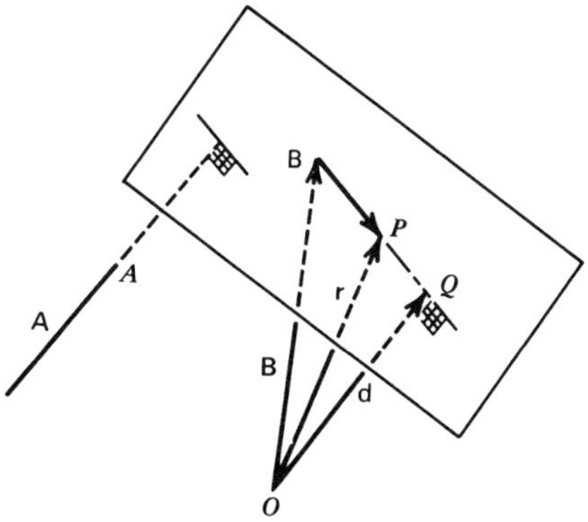

Figure 1.19 Plane perpendicular to **A**, and passing through the terminal point of **B**.

of vector **B**. Draw a directed line segment from O to Q, such that **OQ** is parallel to **A** and Q is in the plane. Then $\mathbf{OQ}=\alpha\mathbf{A}$, where α is a scalar. Let P be an arbitrary point on the line BQ. If the position vector of the point P is **r**, then

$$\mathbf{BP}=\mathbf{OP}-\mathbf{OB}=\mathbf{r}-\mathbf{B}.$$

Since **BP** is perpendicular to $\mathbf{OQ}=\alpha\mathbf{A}$, we must have

$$\mathbf{BP}\cdot\mathbf{OQ}=0 \qquad \text{or} \qquad (\mathbf{r}-\mathbf{B})\cdot\mathbf{A}=0,$$

which is the equation of the plane in question.

The perpendicular distance from O to the plane is the magnitude of **OQ**. However, we do not know its magnitude (or, α is not known). The distance is also given by the projection of vector **B** along **OQ**:

$$d=\mathbf{B}\cdot\frac{\mathbf{OQ}}{|\mathbf{OQ}|}=\mathbf{B}\cdot\hat{\mathbf{e}}_A,$$

where $\hat{\mathbf{e}}_A$ is the unit vector along **A**, $\hat{\mathbf{e}}_A=\mathbf{A}/A$. ■

Example 1.3 Let **A** and **B** be any two vectors in space. Then vector **A** can be expressed in terms of components along (i.e., parallel) and perpendicular to **B**: the component of **A** along **B** is given by $(\mathbf{A}\cdot\hat{\mathbf{e}}_B)$, where $\hat{\mathbf{e}}_B=\mathbf{B}/B$. The component of **A** perpendicular to **B** and in the plane of **A** and **B** is given by the vector triple product $\hat{\mathbf{e}}_B\times(\mathbf{A}\times\hat{\mathbf{e}}_B)$. Thus,

$$\mathbf{A}=(\mathbf{A}\cdot\hat{\mathbf{e}}_B)\hat{\mathbf{e}}_B+\hat{\mathbf{e}}_B\times(\mathbf{A}\times\hat{\mathbf{e}}_B). \tag{1.32}$$

Alternately, using Eq. (1.31) with $\mathbf{A}=\mathbf{C}=\hat{\mathbf{e}}_B$ and $\mathbf{B}=\mathbf{A}$, we obtain

$$\hat{\mathbf{e}}_B\times(\mathbf{A}\times\hat{\mathbf{e}}_B)=\mathbf{A}-(\hat{\mathbf{e}}_B\cdot\mathbf{A})\hat{\mathbf{e}}_B. \quad ■$$

Exercises 1.1

1 Find the equation of a line (or a set of lines) passing through the terminal point of a vector **A** and in the direction of vector **B**.

2 Consider the tetrahedron of Example 1.1. Suppose that the lengths of the sides along the three axes are dx_1, dx_2, and dx_3. Then the areas are given

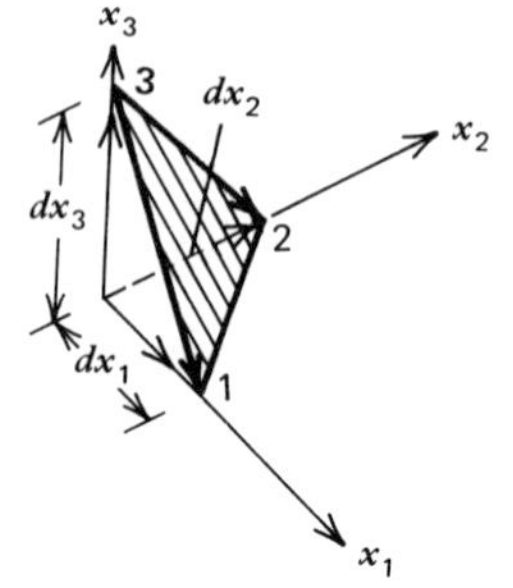

by

$$S_i = \tfrac{1}{2} dx_j dx_k, \qquad i \neq j \neq k, \qquad i, j, k = 1,2,3.$$

Using the cross product of the vectors connecting points 3 and 1 and points 3 and 2, derive the equations (1.20).

3 Find the equation of a plane connecting the terminal points of vectors **A**, **B**, and **C**. Assume that all three vectors are referred to a common origin.

4 Prove with the help of vectors that the diagonals of a parallelogram bisect each other.

5 Prove the vector identity in Eq. (1.31) without the use of a coordinate system.

6 Let P and Q denote two points in space, and let these points be represented by two vectors **A** and **B** with a common origin, as shown in the sketch. Deduce that the straight line through points P and Q can be

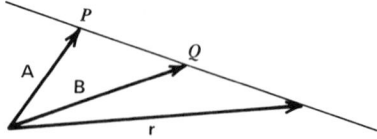

represented by the vector equation

$$(\mathbf{r} - \mathbf{A}) \times (\mathbf{B} - \mathbf{A}) = \mathbf{0}.$$

7 Show that the position vector that divides a line PQ in the ratio $k:l$ is given by

$$\mathbf{r}=\frac{l}{k+l}\mathbf{A}+\frac{k}{k+l}\mathbf{B},$$

where **A** and **B** are the vectors that designate points P and Q.

8 Deduce that the vector equation for a sphere with its center located at A and with a radius C is given by

$$(\mathbf{r}-\mathbf{A})\cdot(\mathbf{r}-\mathbf{A})=C^2.$$

Draw a sketch showing vectors **r** and **A** and radius C.

9 Represent a tetrahedron by the three noncoplanar vectors **A**, **B**, and **C** as shown in the sketch. Show that the vectorial sum of the areas of the tetrahedron sides is zero.

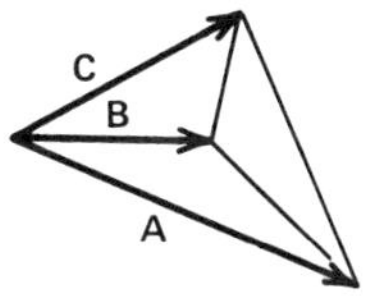

10 Verify that the following identity is true:

$$(\mathbf{A}\cdot\mathbf{B})^2+(\mathbf{A}\times\mathbf{B})^2=(AB)^2,$$

where **A** and **B** are any two vectors.

11 If **A**, **B**, and **C** are noncoplanar vectors, determine that the vectors

$$\mathbf{r}_1=\mathbf{A}-3\mathbf{B}+2\mathbf{C}$$

$$\mathbf{r}_2=2\mathbf{A}-5\mathbf{B}+3\mathbf{C}$$

$$\mathbf{r}_3=\mathbf{A}-5\mathbf{B}+4\mathbf{C}$$

are linearly dependent.

1.2.13 Components of a Vector

So far we have proceeded on a geometrical description of a vector as a directed line segment. We now embark on an analytical description of a vector and some of the operations associated with this description. Such a description yields a connection between vectors and ordinary numbers and relates opera-

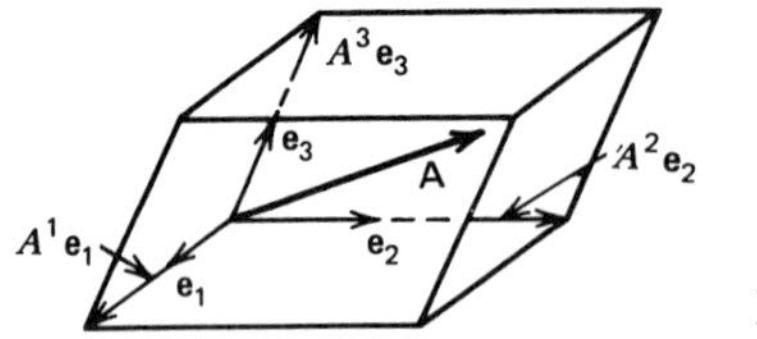

Figure 1.20 Components of a vector.

tions on vectors with those on numbers. The analytical description is based on the notion of components of a vector.

If there exists in a space (i.e., a manifold) a set of n linearly independent vectors, and a set of $n+1$ linearly independent vectors cannot be found, then that space is said to be *n-dimensional.* In what follows, we shall consider a three-dimensional space, and the extensions to n dimensions will be evident (except for a few exceptions). In a three-space a set of four linearly independent vectors cannot be found. On the other hand many sets of three linearly independent vectors can be found. Let us choose any one set and denote it as follows:

$$\mathbf{e}_1, \mathbf{e}_2, \mathbf{e}_3.$$

This set is called a *basis* (or base system).

It is clear from the concept of linear dependence that we can represent any vector in three-space as a linear combination of the basis vectors (see Fig. 1.20):

$$\mathbf{A} = A^1\mathbf{e}_1 + A^2\mathbf{e}_2 + A^3\mathbf{e}_3. \tag{1.33}$$

The vectors $A^1\mathbf{e}_1$, $A^2\mathbf{e}_2$, and $A^3\mathbf{e}_3$ are called the *vector components* of **A**, and A^1, A^2, and A^3 are called the *scalar components* or *measure numbers* of **A** associated with the basis $\mathbf{e}_1, \mathbf{e}_2, \mathbf{e}_3$. The reason for putting the superscripts on the components and subscripts on the basis will become apparent later.

1.2.14 Dual or Reciprocal Basis

Associated with any arbitrary basis is another basis that can be derived from it. We can construct this basis in the following way: Taking the scalar product of the vector **A** with the cross product $\mathbf{e}_1 \times \mathbf{e}_2$, we obtain

$$\mathbf{A} \cdot (\mathbf{e}_1 \times \mathbf{e}_2) = A^3\mathbf{e}_3 \cdot (\mathbf{e}_1 \times \mathbf{e}_2)$$

since $\mathbf{e}_1 \times \mathbf{e}_2$ is perpendicular to both $\mathbf{e}_1$ and $\mathbf{e}_2$. Solving for A^3 gives

$$A^3 = \mathbf{A} \cdot \frac{\mathbf{e}_1 \times \mathbf{e}_2}{\mathbf{e}_3 \cdot (\mathbf{e}_1 \times \mathbf{e}_2)} = \mathbf{A} \cdot \frac{\mathbf{e}_1 \times \mathbf{e}_2}{[\mathbf{e}_1\mathbf{e}_2\mathbf{e}_3]}. \tag{1.34a}$$

In similar fashion we can obtain the following expressions for A^1 and A^2:

$$A^1 = \mathbf{A} \cdot \frac{\mathbf{e}_2 \times \mathbf{e}_3}{[\mathbf{e}_1 \mathbf{e}_2 \mathbf{e}_3]}, \qquad A^2 = \mathbf{A} \cdot \frac{\mathbf{e}_3 \times \mathbf{e}_1}{[\mathbf{e}_1 \mathbf{e}_2 \mathbf{e}_3]}. \tag{1.34b}$$

We thus observe that we can obtain the components A^1, A^2, and A^3 by taking the scalar product of the vector **A** with special vectors, which we denote as follows:

$$\mathbf{e}^1 = \frac{\mathbf{e}_2 \times \mathbf{e}_3}{[\mathbf{e}_1 \mathbf{e}_2 \mathbf{e}_3]}, \qquad \mathbf{e}^2 = \frac{\mathbf{e}_3 \times \mathbf{e}_1}{[\mathbf{e}_1 \mathbf{e}_2 \mathbf{e}_3]}, \qquad \mathbf{e}^3 = \frac{\mathbf{e}_1 \times \mathbf{e}_2}{[\mathbf{e}_1 \mathbf{e}_2 \mathbf{e}_3]}. \tag{1.35}$$

The set of vectors $\mathbf{e}^1, \mathbf{e}^2, \mathbf{e}^3$ constitutes the *dual* or *reciprocal* basis.

Notice from the basic definitions that we have the following relations:

$$\mathbf{e}^1 \cdot \mathbf{e}_1 = \mathbf{e}^2 \cdot \mathbf{e}_2 = \mathbf{e}^3 \cdot \mathbf{e}_3 = 1, \tag{1.36a}$$

and in general

$$\mathbf{e}^i \cdot \mathbf{e}_j = \delta^i_j \begin{cases} 1, & i=j \\ 0, & i \neq j \end{cases}. \tag{1.36b}$$

The symbol δ^i_j is called the *Kronecker delta*.

The dual basis is linearly independent because in the linear relation

$$\alpha_1 \mathbf{e}^1 + \alpha_2 \mathbf{e}^2 + \alpha_3 \mathbf{e}^3 = \mathbf{0}$$

all of the scalars α_1, α_2, and α_3 are zero. Indeed, by taking the scalar product of the above relation with $\mathbf{e}_i$ we obtain $\alpha_i = 0$ for $i = 1$, 2, and 3.

It is possible, since the dual basis is linearly independent, to express a vector in terms of the dual basis:

$$\mathbf{A} = A_1 \mathbf{e}^1 + A_2 \mathbf{e}^2 + A_3 \mathbf{e}^3. \tag{1.37}$$

Notice now that the components associated with the dual basis have subscripts.

By an analogous process as that above we can show that the original basis can be expressed in terms of the dual basis in the following way:

$$\mathbf{e}_1 = \frac{\mathbf{e}^2 \times \mathbf{e}^3}{[\mathbf{e}^1 \mathbf{e}^2 \mathbf{e}^3]}, \qquad \mathbf{e}_2 = \frac{\mathbf{e}^3 \times \mathbf{e}^1}{[\mathbf{e}^1 \mathbf{e}^2 \mathbf{e}^3]}, \qquad \mathbf{e}_3 = \frac{\mathbf{e}^1 \times \mathbf{e}^2}{[\mathbf{e}^1 \mathbf{e}^2 \mathbf{e}^3]}. \tag{1.38}$$

Of course in the evaluation of the cross products we shall always use the right-hand rule.

It follows from the above expressions that

$$A^1 = \mathbf{A}\cdot\mathbf{e}^1, \qquad A^2 = \mathbf{A}\cdot\mathbf{e}^2, \qquad A^3 = \mathbf{A}\cdot\mathbf{e}^3, \qquad \text{or} \qquad A^i = \mathbf{A}\cdot\mathbf{e}^i$$

$$A_1 = \mathbf{A}\cdot\mathbf{e}_1, \qquad A_2 = \mathbf{A}\cdot\mathbf{e}_2, \qquad A_3 = \mathbf{A}\cdot\mathbf{e}_3, \qquad \text{or} \qquad A_i = \mathbf{A}\cdot\mathbf{e}_i. \quad (1.39)$$

Note that we really have only one basis since the dual basis was derived from the given basis. However, we have two ways of expressing the same vector for a given basis. This gives rise, as we shall see, to the descriptions *cogredient* (A_1, A_2, A_3) and *contragredient* (A^1, A^2, A^3) components of a vector. This terminology is based on the way the components transform as a given coordinate system is transformed to another coordinate system. The components (A_1, A_2, A_3) transform in the same way as the basis $\{\mathbf{e}_1, \mathbf{e}_2, \mathbf{e}_3\}$, whereas the components (A^1, A^2, A^3) transform by another rule (contrary-wise!) in the same way as $\{\mathbf{e}^1, \mathbf{e}^2, \mathbf{e}^3\}$. As we shall find, a particular basis, called a *unitary basis*, can be defined in terms of a related coordinate system. For a unitary basis, the term cogredient is called *covariant* and contragredient is called *contravariant*. The utility of a covariant and contravariant system will become apparent shortly.

1.2.15 Summation Convention

It is useful to abbreviate a summation of terms by understanding that a repeated index means summation over all values of that index. Thus the summation

$$\mathbf{A} = \sum_{i=1}^{3} A^i \mathbf{e}_i$$

can be shortened to

$$\mathbf{A} = A^i \mathbf{e}_i. \qquad (1.40)$$

The repeated index is a *dummy index* and thus can be replaced by *any other symbol that has not already been used*. Thus we can also write

$$\mathbf{A} = A^i \mathbf{e}_i = A^m \mathbf{e}_m$$

and so on.

As an example, we can record the following useful expressions for resolving an arbitrary vector into components for an arbitrary basis system:

$$\mathbf{A} = (\mathbf{A}\cdot\mathbf{e}^i)\mathbf{e}_i$$
$$\mathbf{A} = (\mathbf{A}\cdot\mathbf{e}_j)\mathbf{e}^j. \qquad (1.41)$$

1.2.16 Orthonormal Basis Systems

When a basis is unit and orthogonal, that is, orthonormal, we have

$$[\mathbf{e}_1\mathbf{e}_2\mathbf{e}_3]=1, \tag{1.42}$$

and we find that

$$\mathbf{e}^1=\mathbf{e}_1, \qquad \mathbf{e}^2=\mathbf{e}_2, \qquad \mathbf{e}^3=\mathbf{e}_3. \tag{1.43}$$

Hence for an orthonormal system, there is no distinction between cogredient and contragredient components. In many situations an *orthonormal basis* simplifies calculations.

For an orthonormal basis the vectors **A** and **B** can be written as

$$\begin{aligned}\mathbf{A}&=\hat{A}_1\hat{\mathbf{e}}_1+\hat{A}_2\hat{\mathbf{e}}_2+\hat{A}_3\hat{\mathbf{e}}_3=\hat{A}_i\hat{\mathbf{e}}_i\\ \mathbf{B}&=\hat{B}_1\hat{\mathbf{e}}_1+\hat{B}_2\hat{\mathbf{e}}_2+\hat{B}_3\hat{\mathbf{e}}_3=\hat{B}_i\hat{\mathbf{e}}_i,\end{aligned} \tag{1.44}$$

where $\mathbf{e}^i=\mathbf{e}_i\equiv\hat{\mathbf{e}}_i$ $(i=1,2,3)$ is the orthonormal basis, and $\hat{A}_i$ and $\hat{B}_i$ are the corresponding *physical components* (i.e., the components have the same physical dimensions as the vector).

If $(\mathbf{e}_1,\mathbf{e}_2,\mathbf{e}_3)$ is a linearly independent set, then the Gram-Schmidt orthonormalization process can be used to convert the set to an orthonormal set (see Exercises 1.2.12 and 1.2.13).

It is convenient at this time to introduce the alternating symbol ε_{ijk} for representing the cross product of two orthonormal vectors in a right-handed basis system. We define the cross product $\hat{\mathbf{e}}_i\times\hat{\mathbf{e}}_j$ for a right-handed system as

$$\hat{\mathbf{e}}_i\times\hat{\mathbf{e}}_j\equiv\varepsilon_{ijk}\hat{\mathbf{e}}_k, \tag{1.45}$$

where

$$\varepsilon_{ijk}=\begin{cases}1, & \text{if } ijk \text{ are in cyclic order and not repeated } (i\neq j\neq k)\\ -1, & \text{if } ijk \text{ are not in cyclic order and not repeated } (i\neq j\neq k)\\ 0, & \text{if any of } ijk \text{ are repeated.}\end{cases} \tag{1.46}$$

The symbol ε_{ijk} is called the *alternating symbol* (also *permutation symbol*) or *alternating tensor*, since it is a Cartesian component of a third-order tensor.

In an orthonormal basis the scalar and vector products can be expressed in the index form using the Kronecker delta $\delta_{ij}\equiv\hat{\mathbf{e}}_i\cdot\hat{\mathbf{e}}_j$, and the alternating symbols:

$$\mathbf{A}\cdot\mathbf{B}=(\hat{A}_i\hat{\mathbf{e}}_i)\cdot(\hat{B}_j\hat{\mathbf{e}}_j)=\hat{A}_i\hat{B}_j\delta_{ij}=\hat{A}_i\hat{B}_i$$
$$\mathbf{A}\times\mathbf{B}=(\hat{A}_i\hat{\mathbf{e}}_i)\times(\hat{B}_j\hat{\mathbf{e}}_j)=\hat{A}_i\hat{B}_j\varepsilon_{ijk}\hat{\mathbf{e}}_k. \tag{1.47}$$

Further, the Kronecker delta and the permutation symbol are related by the identity, known as the *ε–δ identity*,

$$\varepsilon_{ijk}\varepsilon_{imn}=\delta_{jm}\delta_{kn}-\delta_{jn}\delta_{km}. \tag{1.48}$$

The permutation symbol and the Kronecker delta prove to be very useful in proving vector identities. Since a vector form of any identity is invariant (i.e., valid in any coordinate system), it suffices to prove it in one coordinate system. In particular, an orthonormal system is very convenient because of the permutation symbol and the Kronecker delta. The following example illustrates some of the uses of δ_{ij} and ε_{ijk}.

Example 1.4 We wish to express the vector operation $(\mathbf{A}\times\mathbf{B})\cdot(\mathbf{C}\times\mathbf{D})$ in an alternate vector form (to establish a vector identity):

$$\begin{aligned}(\mathbf{A}\times\mathbf{B})\cdot(\mathbf{C}\times\mathbf{D})&=\hat{A}_i\hat{B}_j\varepsilon_{ijk}\hat{\mathbf{e}}_k\cdot\hat{C}_m\hat{D}_n\varepsilon_{mnp}\hat{\mathbf{e}}_p\\
&=\hat{A}_i\hat{B}_j\hat{C}_m\hat{D}_n\varepsilon_{ijk}\varepsilon_{mnp}\delta_{kp}\\
&=\hat{A}_i\hat{B}_j\hat{C}_m\hat{D}_n\varepsilon_{ijk}\varepsilon_{mnk}\\
&=\hat{A}_i\hat{B}_j\hat{C}_m\hat{D}_n(\delta_{im}\delta_{jn}-\delta_{in}\delta_{jm})\\
&=\hat{A}_i\hat{B}_j\hat{C}_m\hat{D}_n\delta_{im}\delta_{jn}-\hat{A}_i\hat{B}_j\hat{C}_m\hat{D}_n\delta_{in}\delta_{jm},\end{aligned}$$

where we have used the ε–δ identity. Since $\hat{C}_m\delta_{im}=\hat{C}_i$ (or $\hat{A}_i\delta_{im}=\hat{A}_m$, etc.), we have

$$\begin{aligned}(\mathbf{A}\times\mathbf{B})\cdot(\mathbf{C}\times\mathbf{D})&=\hat{A}_i\hat{B}_j\hat{C}_i\hat{D}_j-\hat{A}_i\hat{B}_j\hat{C}_j\hat{D}_i\\
&=\hat{A}_i\hat{C}_i\hat{B}_j\hat{D}_j-\hat{A}_i\hat{D}_i\hat{B}_j\hat{C}_j\\
&=(\mathbf{A}\cdot\mathbf{C})(\mathbf{B}\cdot\mathbf{D})-(\mathbf{A}\cdot\mathbf{D})(\mathbf{B}\cdot\mathbf{C}).\quad\blacksquare\end{aligned} \tag{1.49}$$

Although the above vector identity is established in an orthonormal coordinate system, it holds in a general coordinate system. That is, the vector identity is invariant.

1.2.17 Specification of a Vector

We have seen that a vector can be specified by three components in either a cogredient or a contragredient system. Thus a vector can be uniquely determined by an ordered set of three numbers which form the components of the vector.

Quantities other than components may be used to specify a vector. To illustrate this, consider a vector **A** with contragredient components (A^1, A^2, A^3) associated with a cogredient basis $\{\mathbf{e}_1, \mathbf{e}_2, \mathbf{e}_3\}$. Let α, β, and γ denote the angles **A** makes with the basis $\{\mathbf{e}_1, \mathbf{e}_2, \mathbf{e}_3\}$:

$$\cos\alpha = \frac{\mathbf{A}\cdot\mathbf{e}_1}{|\mathbf{A}|\,|\mathbf{e}_1|} \tag{1.50a}$$

$$\cos\beta = \frac{\mathbf{A}\cdot\mathbf{e}_2}{|\mathbf{A}|\,|\mathbf{e}_2|} \tag{1.50b}$$

$$\cos\gamma = \frac{\mathbf{A}\cdot\mathbf{e}_3}{|\mathbf{A}|\,|\mathbf{e}_3|}. \tag{1.50c}$$

Now each of the sets (A^1, A^2, A^3), (A^1, β, γ), (α, A^2, γ), and (α, β, A^3), specifies the magnitude and direction of **A**.

We may thus describe a vector *analytically* as an ordered set of three numbers that is related in some way to the basis $\{\mathbf{e}_1, \mathbf{e}_2, \mathbf{e}_3\}$. The set is ordered in the sense that the first number corresponds to $\mathbf{e}_1$, the second number to $\mathbf{e}_2$, and the third number to $\mathbf{e}_3$. Furthermore, the three numbers must obey certain rules because not every ordered set specifies a vector. To see how such rules are set up, we could consider any vector and specify it in two independent basis systems. While the various components are different in the two different systems, the vector as a whole must be *invariant*. Thus a rule can be set up that expresses one set of numbers in terms of a corresponding set in the other system. *We thus define a vector analytically as an ordered set of three numbers that obey certain specific rules.* These rules replace, in effect, the so-called parallelogram law of addition. We have yet to establish these rules.

1.2.18 Transformation Law for Different Bases

Suppose that we introduce a second (barred) basis:

$$\{\bar{\mathbf{e}}_1, \bar{\mathbf{e}}_2, \bar{\mathbf{e}}_3\}$$

and its dual

$$\{\bar{\mathbf{e}}^1, \bar{\mathbf{e}}^2, \bar{\mathbf{e}}^3\}.$$

Now we can express the same vector in four ways:

$$\left.\begin{aligned}\mathbf{A}&=A^i\mathbf{e}_i\\&=A_j\mathbf{e}^j\end{aligned}\right\},\qquad \text{unbarred basis}$$

$$\left.\begin{aligned}&=\bar{A}^m\bar{\mathbf{e}}_m\\&=\bar{A}_n\bar{\mathbf{e}}^n\end{aligned}\right\},\qquad \text{barred basis.} \tag{1.51}$$

From Eq. (1.39) we have

$$\bar{A}^s=\mathbf{A}\cdot\bar{\mathbf{e}}^s, \tag{1.52a}$$

and it follows by substitution of Eq. (1.51) for **A** that

$$\bar{A}^s=(\mathbf{e}_i\cdot\bar{\mathbf{e}}^s)A^i=(\mathbf{e}^j\cdot\bar{\mathbf{e}}^s)A_j. \tag{1.52b}$$

Similarly we obtain

$$\bar{A}_s=(\mathbf{e}^j\cdot\bar{\mathbf{e}}_s)A_j=(\mathbf{e}_i\cdot\bar{\mathbf{e}}_s)A^i. \tag{1.52c}$$

The first two members of Eqs. (1.52b) and (1.52c) give the transformation rules between the contragredient and the cogredient components in the two basis systems.

By means of Eqs. (1.41) we find that the basis systems are related by

$$\bar{\mathbf{e}}^s=(\bar{\mathbf{e}}^s\cdot\mathbf{e}_j)\mathbf{e}^j=(\bar{\mathbf{e}}^s\cdot\mathbf{e}^j)\mathbf{e}_j, \tag{1.53a}$$

or because j is a dummy index, we have also

$$\bar{\mathbf{e}}^s=(\mathbf{e}_i\cdot\bar{\mathbf{e}}^s)\mathbf{e}^i=(\mathbf{e}^i\cdot\bar{\mathbf{e}}^s)\mathbf{e}_i. \tag{1.53b}$$

Likewise we have

$$\bar{\mathbf{e}}_s=(\mathbf{e}^j\cdot\bar{\mathbf{e}}_s)\mathbf{e}_j=(\mathbf{e}_j\cdot\bar{\mathbf{e}}_s)\mathbf{e}^j. \tag{1.54}$$

If we now write

$$a_s^j\equiv\mathbf{e}^j\cdot\bar{\mathbf{e}}_s \tag{1.55a}$$

$$b_i^s\equiv\bar{\mathbf{e}}^s\cdot\mathbf{e}_i, \tag{1.55b}$$

then we have in summary

$$\bar{\mathbf{e}}_s=a_s^j\mathbf{e}_j,\qquad \bar{A}_s=a_s^jA_j,\qquad \text{cogredient law} \tag{1.55c}$$

$$\bar{\mathbf{e}}^s=b_i^s\mathbf{e}^i,\qquad \bar{A}^s=b_i^sA^i,\qquad \text{contragredient law.} \tag{1.55d}$$

Thus there are two transformation laws, and the subscripts and superscripts are assigned according to which law is satisfied. The subscripted basis vectors and the subscripted components transform according to the same law, the cogredient law, and the superscripted basis vectors and superscripted components transform according to another law, the contragredient law.

There are also relations between the subscripted and superscripted basis vectors and components in the two systems. These relations are called *mixed* laws. Let

$$c_{js} = \mathbf{e}_j \cdot \bar{\mathbf{e}}_s \tag{1.56a}$$

$$d^{is} \equiv \mathbf{e}^i \cdot \bar{\mathbf{e}}^s. \tag{1.56b}$$

Then we have

$$\left.\begin{array}{ll} \bar{\mathbf{e}}_s = c_{js}\mathbf{e}^j, & \bar{A}_s = c_{js}A^j \\ \bar{\mathbf{e}}^s = d^{is}\bar{\mathbf{e}}_i, & \bar{A}^s = d^{is}A_i \end{array}\right\}, \quad \text{mixed laws.} \tag{1.56c}$$

In an orthonormal system the transformation Eqs. (1.55) and (1.56) reduce to

$$\hat{\bar{\mathbf{e}}}_i = (\hat{\bar{\mathbf{e}}}_i \cdot \hat{\mathbf{e}}_j)\hat{\mathbf{e}}_j = \beta_{ij}\hat{\mathbf{e}}_j. \tag{1.57}$$

The coefficients β_{ij} can be interpreted as the direction cosines of the barred coordinate system with respect to the unbarred coordinates:

$$\beta_{ij} = \text{cosine of the angle between } \hat{\bar{\mathbf{e}}}_i \text{ and } \hat{\mathbf{e}}_j.$$

Obviously, β_{ij} is not symmetric (i.e., $\beta_{ij} \neq \beta_{ji}$).

Example 1.5 Let $\hat{\mathbf{e}}_i$ $(i=1,2,3)$ be a set of orthonormal base vectors, and define new right-handed coordinate base vectors by

$$\hat{\bar{\mathbf{e}}}_1 = \frac{2\hat{\mathbf{e}}_1 + 2\hat{\mathbf{e}}_2 + \hat{\mathbf{e}}_3}{3}, \qquad \hat{\bar{\mathbf{e}}}_2 = \frac{\hat{\mathbf{e}}_1 - \hat{\mathbf{e}}_2}{\sqrt{2}}.$$

The third base vector is uniquely determined by

$$\hat{\bar{\mathbf{e}}}_3 = \hat{\bar{\mathbf{e}}}_1 \times \hat{\bar{\mathbf{e}}}_2 = \frac{(\hat{\mathbf{e}}_1 + \hat{\mathbf{e}}_2 - 4\hat{\mathbf{e}}_3)}{3\sqrt{2}}.$$

The relations above between the barred and unbarred systems can be written with the help of the array (see Section 1.3)

$$[\beta]=\frac{1}{3\sqrt{2}}\begin{bmatrix} 2\sqrt{2} & 2\sqrt{2} & \sqrt{2} \\ 3 & -3 & 0 \\ 1 & 1 & -4 \end{bmatrix}$$

as $\{\hat{\mathbf{e}}\}=[\beta]\{\hat{\mathbf{e}}\}$. ■

Exercises 1.2

1 Verify the following identities:

(a) $\delta_{ii}=3$.

(b) $\delta_{ij}\delta_{ij}=\delta_{ii}$.

(c) $\delta_{ij}\delta_{jk}=\delta_{ik}$.

(d) $\varepsilon_{ijk}\varepsilon_{ijk}=6$.

(e) $A_i A_j \varepsilon_{ijk}=0$.

2 Prove the ε–δ identity using the vector identity in Eq. (1.31).

3 Using the index notation prove the identity

$$(\mathbf{A}\times\mathbf{B})\cdot(\mathbf{B}\times\mathbf{C})\times(\mathbf{C}\times\mathbf{A})=(\mathbf{A}\cdot(\mathbf{B}\times\mathbf{C}))^2.$$

4 Prove the following vector identity in an orthonormal system using index-summation notation:

$$(\mathbf{A}\times\mathbf{B})\times(\mathbf{C}\times\mathbf{D})=[\mathbf{A}\cdot(\mathbf{C}\times\mathbf{D})]\mathbf{B}-[\mathbf{B}\cdot(\mathbf{C}\times\mathbf{D})]\mathbf{A}.$$

5 Determine whether the following set of vectors is linearly independent:

$$\mathbf{A}=\hat{\mathbf{e}}_1+\hat{\mathbf{e}}_2,\qquad \mathbf{B}=\hat{\mathbf{e}}_2+\hat{\mathbf{e}}_4,\qquad \mathbf{C}=\hat{\mathbf{e}}_3+\hat{\mathbf{e}}_4,\qquad \mathbf{D}=\hat{\mathbf{e}}_1+\hat{\mathbf{e}}_2+\hat{\mathbf{e}}_3+\hat{\mathbf{e}}_4.$$

Here $\hat{\mathbf{e}}_i$ are orthonormal unit base vectors in a four-space.

6 The angles between the barred and unbarred coordinate lines are given by

	$\hat{\bar{\mathbf{e}}}_1$	$\hat{\bar{\mathbf{e}}}_2$	$\hat{\bar{\mathbf{e}}}_3$
$\hat{\mathbf{e}}_1$	60°	30°	90°
$\hat{\mathbf{e}}_2$	150°	60°	90°
$\hat{\mathbf{e}}_3$	90°	90°	0°

Determine the direction cosines of the transformation.

7 In a rectangular Cartesian coordinate system find the length and direction cosines of a vector **A** that extends from the point $(1,-1,3)$ to the midpoint of the line segment from the origin to the point $(6,-6,4)$.

8 The vectors **A** and **B** are defined as follows:

$$\mathbf{A}=3\hat{\mathbf{i}}-4\hat{\mathbf{k}}$$

$$\mathbf{B}=2\hat{\mathbf{i}}-2\hat{\mathbf{j}}+\hat{\mathbf{k}},$$

where $\hat{\mathbf{i}}$, $\hat{\mathbf{j}}$, and $\hat{\mathbf{k}}$ are an orthonormal basis.

(a) Find the orthogonal projection of **A** on **B**.

(b) Find the angle between the positive directions of the vectors.

9 A rigid body rotates with the angular speed ω about the line $x=y=z$ in a rectangular Cartesian coordinate system. Find the velocity and speed of the particle at the point $(1,2,2)$.

10 Let the vectors $\{\hat{\mathbf{i}},\hat{\mathbf{j}},\hat{\mathbf{k}}\}$ constitute an orthonormal basis. In terms of this basis, define a cogredient basis by

$$\mathbf{e}_1=\frac{\sqrt{3}}{4}\hat{\mathbf{i}}+\frac{1}{4}\hat{\mathbf{j}}$$

$$\mathbf{e}_2=\frac{1}{2}\hat{\mathbf{i}}+\frac{3}{2}\hat{\mathbf{j}}$$

$$\mathbf{e}_3=\hat{\mathbf{k}}.$$

(a) Determine the dual or reciprocal (contragredient) basis $\{\mathbf{e}^1,\mathbf{e}^2,\mathbf{e}^3\}$ in terms of the orthonormal basis $\{\hat{\mathbf{i}},\hat{\mathbf{j}},\hat{\mathbf{k}}\}$.

(b) Determine the magnitudes (or norms) $|\mathbf{e}_1|$, $|\mathbf{e}_2|$, $|\mathbf{e}_3|$, $|\mathbf{e}^1|$, $|\mathbf{e}^2|$, and $|\mathbf{e}^3|$.

(c) Sketch the basis vectors $\mathbf{e}_1$, $\mathbf{e}_2$, $\mathbf{e}^1$, and $\mathbf{e}^2$ on a graph (to scale approximately).

11 Corresponding to the cogredient basis defined in Exercise 10, let the contragredient components of a vector **A** be given by $A^1=1$, $A^2=2$, $A^3=3$. What are the values of the cogredient components A_1, A_2, and A_3?

12 (*Gram-Schmidt orthonormalization.*) Let $(\mathbf{e}_1,\mathbf{e}_2,\ldots,\mathbf{e}_n)$ be any linearly independent set of vectors. Construct an orthonormal set $(\hat{\mathbf{e}}_1,\hat{\mathbf{e}}_2,\ldots,\hat{\mathbf{e}}_n)$ using the following procedure: Since $\mathbf{e}_1$ is an element of a linearly independent set, $\mathbf{e}_1\neq 0$, and therefore $|\mathbf{e}_1|>0$, let $\hat{\mathbf{e}}_1\equiv\mathbf{e}_1/|\mathbf{e}_1|$. Clearly, $|\hat{\mathbf{e}}_1|=1$. Next choose the second vector $\mathbf{e}_2$ from the original set and require that the vector $\mathbf{e}_2'=\mathbf{e}_2-\alpha\hat{\mathbf{e}}_1$ is orthogonal to $\hat{\mathbf{e}}_1$:

$$0=\hat{\mathbf{e}}_1\cdot(\mathbf{e}_2-\alpha\hat{\mathbf{e}}_1)=(\hat{\mathbf{e}}_1\cdot\mathbf{e}_2)-\alpha|\hat{\mathbf{e}}_1|^2,$$

or $\alpha=(\hat{\mathbf{e}}_1\cdot\mathbf{e}_2)$. The second element of the desired set is then obtained by $\hat{\mathbf{e}}_2=[\mathbf{e}_2-(\hat{\mathbf{e}}_1\cdot\mathbf{e}_2)\hat{\mathbf{e}}_1]/|\mathbf{e}_2'|$. Continue the procedure to obtain the $(r+1)$th element

$$\mathbf{e}'_{r+1}=\mathbf{e}_{r+1}-(\hat{\mathbf{e}}_1\cdot\mathbf{e}_{r+1})\hat{\mathbf{e}}_1-(\hat{\mathbf{e}}_2\cdot\mathbf{e}_{r+1})\hat{\mathbf{e}}_2-\cdots-(\hat{\mathbf{e}}_r\cdot\mathbf{e}_{r+1})\hat{\mathbf{e}}_r$$

$$\hat{\mathbf{e}}_{r+1}=\frac{\mathbf{e}'_{r+1}}{|\mathbf{e}'_{r+1}|}.$$

13 Construct an orthonormal basis from the set of cogredient vectors in Exercise 10.

Using the Gram-Schmidt orthonormalization process, construct the orthonormal sets associated with the sets of vectors in 14 and 15:

14 $\mathbf{e}_1=\hat{\mathbf{e}}_1+\hat{\mathbf{e}}_3, \quad \mathbf{e}_2=\hat{\mathbf{e}}_1+2\hat{\mathbf{e}}_2-2\hat{\mathbf{e}}_3, \quad \mathbf{e}_3=2\hat{\mathbf{e}}_1-\hat{\mathbf{e}}_2+\hat{\mathbf{e}}_3.$

15 $\mathbf{e}_1=2\hat{\mathbf{e}}_1+\hat{\mathbf{e}}_2+\hat{\mathbf{e}}_3, \quad \mathbf{e}_2=\hat{\mathbf{e}}_1-2\hat{\mathbf{e}}_2+\hat{\mathbf{e}}_3, \quad \mathbf{e}_3=-2\hat{\mathbf{e}}_1+\hat{\mathbf{e}}_2+\hat{\mathbf{e}}_3.$

16 Use the following definition of inner product for functions (which can be regarded as "vectors" in a generalized sense, as will be seen in Chapter 2) to construct the orthonormal set associated with the set $\{1, x, x^2\}$: $(f(x), g(x))=\int_{-1}^{1} f(x)g(x)\,dx$.

1.3 MATRICES AND LINEAR EQUATIONS

1.3.1 Introductory Comments

In the preceding sections we studied the algebra of ordinary vectors and the transformation of vector components from one coordinate system to another. For example, the cogredient law (1.55c) relates the components of a vector in the barred coordinate system to the unbarred coordinate system. Writing Eq. (1.55c) in expanded form,

$$\bar{A}_1=a_1^1A_1+a_1^2A_2+a_1^3A_3$$

$$\bar{A}_2=a_2^1A_1+a_2^2A_2+a_2^3A_3$$

$$\bar{A}_3=a_3^1A_1+a_3^2A_2+a_3^3A_3,$$

we see that there are nine coefficients relating the components A_i to $\bar{A}_i$. The form of these linear equations suggests writing down the scalars of a_i^j (jth component in the ith equation) in the rectangular array:

$$[A]=\begin{bmatrix} a_1^1 & a_1^2 & a_1^3 \\ a_2^1 & a_2^2 & a_2^3 \\ a_3^1 & a_3^2 & a_3^3 \end{bmatrix}.$$

Now let us agree to write the above linear equations in the form

$$\begin{Bmatrix} \bar{A}_1 \\ \bar{A}_2 \\ \bar{A}_3 \end{Bmatrix} = \begin{bmatrix} a_1^1 & a_1^2 & a_1^3 \\ a_2^1 & a_2^2 & a_2^3 \\ a_3^1 & a_3^2 & a_3^3 \end{bmatrix} \begin{Bmatrix} A_1 \\ A_2 \\ A_3 \end{Bmatrix}. \tag{1.58}$$

This notation enables us, as will be seen shortly, to calculate A_i in terms of $\bar{A}_i$, that is, to solve the linear Eqs. (1.58). This serves as motivation to study the properties of rectangular arrays, called *matrices*, when they satisfy certain properties.

The reader who has had prior exposure to matrices could skip this section and go to Section 1.4 directly.

1.3.2 Definition of a Matrix

A *matrix* is a rectangular array of elements. Two matrices are equal if and only if they are identical.

If a matrix has m rows and n columns, we will say that it is m by n $(m\times n)$, the number of rows always being listed first. The element in the ith row and jth column of a matrix $[A]$ is generally denoted by a_{ij}, and we will sometimes write $[A]=[a_{ij}]$ to denote this. A square matrix is one that has the same number of rows as columns. An $n\times n$ matrix is said to be of *order n*. The elements of a square matrix for which the row number and the column number are the same (i.e., a_{ii}) are called *diagonal elements*, or simply the *diagonal*. A square matrix is said to be a *diagonal matrix* if all of the off-diagonal elements are zero. An *identity matrix*, denoted by $[I]$, is a diagonal matrix whose elements are all 1's. Examples of a diagonal and an identity matrix are given below:

$$\begin{bmatrix} 5 & 0 & 0 & 0 \\ 0 & 2 & 0 & 0 \\ 0 & 0 & 1 & 0 \\ 0 & 0 & 0 & 3 \end{bmatrix}, \qquad [I]=\begin{bmatrix} 1 & 0 & 0 & 0 \\ 0 & 1 & 0 & 0 \\ 0 & 0 & 1 & 0 \\ 0 & 0 & 0 & 1 \end{bmatrix}.$$

The sum of the diagonal elements is called the *trace* of the matrix.

If the matrix has only one row or one column, we will normally use only a single subscript to designate its elements. For example,

$$\{X\}=\begin{Bmatrix} x_1 \\ x_2 \\ x_3 \end{Bmatrix}, \qquad \{Y\}=\{y_1, y_2, y_3\}$$

denote a column matrix and a row matrix, respectively. Row and column matrices can be used to denote the components of a first-order tensor (i.e., a vector). For example, consider a vector $\mathbf{A}=a_1\hat{\mathbf{e}}_1+a_2\hat{\mathbf{e}}_2+a_3\hat{\mathbf{e}}_3$ in a Cartesian system. We can represent **A** as a *product* of a row matrix with a column matrix,

$$\mathbf{A}=\{a_1 \quad a_2 \quad a_3\}\begin{Bmatrix}\hat{\mathbf{e}}_1\\ \hat{\mathbf{e}}_2\\ \hat{\mathbf{e}}_3\end{Bmatrix}.$$

Note that the vector **A** is obtained by multiplying the ith element in the row matrix with the ith element in the column matrix and adding them. This gives us a strong motivation for defining the product of two matrices.

1.3.3 Matrix Multiplication

Let $\{R\}=\{r_1, r_2,\ldots, r_m\}$ be a row matrix of $1\times m$ and let

$$\{C\}=\begin{Bmatrix}c_1\\ c_2\\ \vdots\\ c_m\end{Bmatrix}$$

be a column matrix of $m\times 1$. We define the product $\{R\}\{C\}$ to be the scalar

$$\{R\}\{C\}=\{r_1, r_2,\ldots, r_m\}\begin{Bmatrix}c_1\\ c_2\\ \vdots\\ c_m\end{Bmatrix}=r_1c_1+r_2c_2+\cdots+r_mc_m$$

$$=\sum_{i=1}^{m} r_ic_i. \tag{1.59}$$

More generally, let $[A]=[a_{ij}]$ be $m\times n$ and $[B]=[b_{ij}]$ be $n\times p$ matrices. The product $[A][B]$ is defined to be the $m\times p$ matrix $[C]=[c_{ij}]$ with

$$c_{ij}=\{i\text{th row of }[A]\}\begin{Bmatrix}j\text{th}\\ \text{column}\\ \text{of }[B]\end{Bmatrix}=\{a_{i1}, a_{i2},\ldots, a_{in}\}\begin{Bmatrix}b_{1j}\\ b_{2j}\\ \vdots\\ b_{nj}\end{Bmatrix}$$

$$=a_{i1}b_{1j}+a_{i2}b_{2j}+\cdots+a_{in}b_{nj}=\sum_{k=1}^{n} a_{ik}b_{kj}. \tag{1.60}$$

Several comments are in order on the matrix multiplication. If $[A]$ is $m\times n$ and $[B]$ is $p\times q$, then:

1 The product $[A][B]$ is defined only if the number of columns n in A is equal to the number of rows p in B:

$$\underset{m\times n}{[A]}\;\underset{n\times q}{[B]}=\underset{m\times q}{[AB]}.$$

2 Similarly, the product $[B][A]$ is defined only if $q=m$.
3 If $[A][B]$ is defined, $[B][A]$ may or may not be defined. If both $[A][B]$ and $[B][A]$ are defined, it is not necessary that they be of the same size.
4 The products $[A][B]$ and $[B][A]$ are of the same size if and only if both $[A]$ and $[B]$ are square matrices of the same size.
5 The products $[A][B]$ and $[B][A]$ are in general not equal (even if they are of equal size), that is, the matrix multiplication is not *commutative*.
6 A constant multiple of a matrix is equal to the matrix obtained by multiplying all of the elements by the constant.
7 If $[A]$ is a square matrix, the powers of $[A]$ are defined by $[A]^2=[A][A]$, $[A]^3=[A][A]^2=[A]^2[A]$, and so on.
8 Matrix multiplication is associative: $([A][B])[C]=[A]([B][C])$.
9 The product of any square matrix with the identity matrix is the matrix itself.

Some of the above statements should be proved as exercises by the reader.

1.3.4 Inverse of a Matrix

If $[A]$ is an $n\times n$ matrix and $[B]$ is any $n\times n$ matrix such that $[A][B]=[B][A]=[I]$, then $[B]$ is called an *inverse* of $[A]$. If it exists, the inverse of a matrix is unique (a consequence of the associative law). If both $[B]$ and $[C]$ are inverses for $[A]$, then by definition,

$$[A][B]=[B][A]=[A][C]=[C][A]=[I]$$

Since matrix multiplication is associative, we have

$$[B][A][C]=([B][A])[C]=[I][C]=[C]$$

$$=[B]([A][C])=[B][I]=[B].$$

This shows that $[B]=[C]$, and the inverse is unique. The inverse of $[A]$ is denoted by $[A]^{-1}$.

A matrix is said to be *singular* if it does not have an inverse.

1.3.5 Matrix Addition

The *sum* of two matrices of the same size is defined to be a matrix of the same size obtained by simply adding the corresponding elements. If $[A]=[a_{ij}]$ is an $m\times n$ matrix and $[B]$ is an $m\times n$ matrix, their sum is an $m\times n$ matrix, $[C]=[c_{ij}]$, with

$$c_{ij}=a_{ij}+b_{ij} \qquad \text{for all } i,\, j. \tag{1.61}$$

Matrix addition has the following properties:

1. Addition is commutative: $[A]+[B]=[B]+[A]$.
2. Addition is associative: $[A]+([B]+[C])=([A]+[B])+[C]$.
3. There exists a unique matrix [0], such that $[A]+[0]=[0]+[A]=[A]$. The matrix [0] is called *zero matrix*; it contains all zeros.
4. For each matrix $[A]$ there exists a unique matrix $-[A]$ such that $[A]+(-[A])=[0]$.
5. Addition is distributive with respect to scalar multiplication: $\alpha([A]+[B])=\alpha[A]+\alpha[B]$.
6. Addition is distributive with respect to matrix multiplication:

$$([A]+[B])[C]=[A][C]+[B][C].$$

1.3.6 Transpose of a Matrix

If $[A]$ is an $m\times n$ matrix, then the $n\times m$ matrix obtained by interchanging its rows and columns is called the *transpose* of $[A]$ and is denoted by $[A]^{\mathrm{T}}$. The following basic properties of a transpose should be noted.

1. $([A]^{\mathrm{T}})^{\mathrm{T}}=[A]$.
2. $([A]+[B])^{\mathrm{T}}=[A]^{\mathrm{T}}+[B]^{\mathrm{T}}$.
3. $([A][B])^{\mathrm{T}}=[B]^{\mathrm{T}}[A]^{\mathrm{T}}$ (note the order).
4. If $[A]$ is nonsingular, $([A]^{-1})^{\mathrm{T}}=([A]^{\mathrm{T}})^{-1}$.

1.3.7 Symmetric, Skew Symmetric, and Triangular Matrices

A square matrix $[A]$ of real numbers is said to be *symmetric* if $[A]^{\mathrm{T}}=[A]$. It is said to be *skew symmetric* if $[A]^{\mathrm{T}}=-[A]$.

In terms of the elements of $[A]$, these definitions imply that $[A]$ is symmetric if and only if $a_{ij}=a_{ji}$, and it is skew symmetric if and only if $a_{ij}=-a_{ji}$. Note that the diagonal elements of a skew symmetric matrix are always zero since $a_{ij}=-a_{ij}$ implies $a_{ij}=0$ for $i=j$.

A matrix $[A]$ is said to be *upper triangular* if $a_{ij}=0$ for $i>j$ (all elements of $[A]$ which lie below the diagonal are zero). Similarly $[A]$ is said to be *lower triangular* if $a_{ij}=0$ for $i<j$.

1.3.8 Elementary Matrix Operations

An *elementary* row (column) *operation* on a matrix is an operation of one of the following types:

1 Two rows (columns) are interchanged.
2 A row (column) equation is multiplied by a nonzero scalar.
3 A row (column) is replaced by itself plus a scalar multiple of another row (column).

Two matrices $[A]$ and $[B]$ are said to be *row* (column) *equivalent* if $[B]$ is obtainable from $[A]$ by a finite sequence of elementary row (column) operations. Matrix $[A]$ is *equivalent* to $[B]$ if $[B]$ is obtainable from $[A]$ by a finite sequence of elementary operations.

A matrix $[E]$ is said to be in *row-reduced echelon form* if:

1 The first nonzero element of each row is 1, and the column in which it appears is a column of the identity matrix.
2 The 0 rows, if any, come last.
3 The leading 1's in the nonzero rows occur in positions $(1, c_1)$, $(2, c_2), \ldots, (r, c_r)$, then $c_1 < c_2 < \cdots < c_r$, where c_1, c_2, etc. denote columns.

An example of a matrix in echelon form is given below:

$$\begin{bmatrix} 1 & 2 & 0 & 3 & 0 & 0 & 5 \\ 0 & 0 & 1 & 0 & 2 & 0 & -1 \\ 0 & 0 & 0 & 0 & 1 & 2 & 3 \\ 0 & 0 & 0 & 0 & 0 & 1 & 2 \\ 0 & 0 & 0 & 0 & 0 & 0 & 0 \end{bmatrix}.$$

The leading 1's in the nonzero rows occur in positions (1,1), (2,3), (3,5), and (4,6).

A *submatrix* of a matrix $[A]$ is a matrix obtained from $[A]$ by deleting certain rows and/or columns of $[A]$. One can also think of *partitioning* $[A]$ into submatrices by certain horizontal and vertical lines. One can multiply partitioned matrices, treating the submatrices like the elements of the matrix. However, we must make sure that all the sums and products of the submatrices are well defined. An example of partitioning of a matrix $[A]$ is shown below:

$$[A] = \left[\begin{array}{ccc|c} a_{11} & a_{12} & a_{13} & a_{14} \\ a_{21} & a_{22} & a_{23} & a_{24} \\ a_{31} & a_{32} & a_{33} & a_{34} \\ \hline a_{41} & a_{42} & a_{43} & a_{44} \end{array}\right] = \left[\begin{array}{c|c} \underset{3\times 3}{[A_{11}]} & \underset{3\times 1}{[A_{12}]} \\ \hline \underset{1\times 3}{[A_{21}]} & \underset{1\times 1}{[A_{22}]} \end{array}\right].$$

A matrix $[A]$ is said to be *augmented* by another matrix $[B]$ having the same number of rows as $[A]$ if $[A]$ and $[B]$ are placed columnwise in the same matrix: $[A \,\vdots\, B]$. One can perform elementary operations on the *augmented* matrix $[A \,\vdots\, B]$.

The *rank* of any matrix $[A]$ is the order of the largest square submatrix of $[A]$ whose determinant does not vanish.

1.3.9 Determinant of a Matrix

Let $[A]=[a_{ij}]$ be an $n \times n$ matrix. We wish to associate with $[A]$ a scalar that in some sense measures the "size" of $[A]$ and indicates whether or not $[A]$ is nonsingular.

The *determinant* of the matrix $[A]=[a_{ij}]$ is defined to be the scalar $\det[A]=|A|$ computed according to the rule

$$\det[A]=|a_{ij}|=\sum_{i=1}^{n}(-1)^{i+1}a_{i1}|A_{i1}|,$$

where $|A_{i1}|$ is the determinant of the $(n-1)\times(n-1)$ matrix that remains on deleting out the ith row and the first column of $[A]$. For 1×1 matrices the determinant is defined according to $|a_{11}|=a_{11}$. For convenience we define the determinant of a zeroth-order matrix to be unity. In the above definition special attention is given to the first column of the matrix $[A]$. We call it the expansion of $|A|$ according to the first column of $[A]$. One can expand $|A|$ according to any column or row:

$$|A|=\sum_{i=1}^{n}(-1)^{i+j}a_{ij}|A_{ij}|, \tag{1.62}$$

where $|A_{ij}|$ is the determinant of the matrix obtained by deleting the ith row and jth column of matrix $[A]$.

Example 1.6 Let us find the determinant of the matrix

$$[A]=\begin{bmatrix} 2 & 5 & -1 \\ 1 & 4 & 3 \\ 2 & -3 & 5 \end{bmatrix}.$$

Using the definition (1.62), we have

$$\begin{aligned}
|A| &= \sum_{i=1}^{3}(-1)^{i+1}a_{i1}|A_{i1}| \\
&=(-1)^2 a_{11}\begin{vmatrix} 4 & 3 \\ -3 & 5 \end{vmatrix}+(-1)^3 a_{21}\begin{vmatrix} 5 & -1 \\ -3 & 5 \end{vmatrix}+(-1)^4 a_{31}\begin{vmatrix} 5 & -1 \\ 4 & 3 \end{vmatrix} \\
&=2\big((-1)^2(4)(5)+(-1)^3(-3)3\big)-1\big((-1)^2(5)(5)+(-1)^3(-3)(-1)\big) \\
&\quad +2\big((-1)^2(5)(3)+(-1)^3(4)(-1)\big) \\
&=2(20+9)-(25-3)+2(15+4)=74. \quad \blacksquare
\end{aligned}$$

The cross product of two vectors **A** and **B** can be expressed as the value of the determinant,

$$\mathbf{A}\times\mathbf{B}\equiv\begin{vmatrix} \hat{\mathbf{e}}_1 & \hat{\mathbf{e}}_2 & \hat{\mathbf{e}}_3 \\ \hat{A}_1 & \hat{A}_2 & \hat{A}_3 \\ \hat{B}_1 & \hat{B}_2 & \hat{B}_3 \end{vmatrix}, \tag{1.63a}$$

and the scalar triple product can be expressed as the value of a determinant:

$$\mathbf{A}\cdot(\mathbf{B}\times\mathbf{C})\equiv\begin{vmatrix} \hat{A}_1 & \hat{A}_2 & \hat{A}_3 \\ \hat{B}_1 & \hat{B}_2 & \hat{B}_3 \\ \hat{C}_1 & \hat{C}_2 & \hat{C}_3 \end{vmatrix}. \tag{1.63b}$$

The verification of these results is left as an exercise for the reader.

Example 1.7 We wish to prove that the determinant of a 3×3 matrix A can be expressed in the form

$$|A|=\varepsilon_{ijk}a_{1i}a_{2j}a_{3k},$$

where a_{ij} is the element occupying the ith row and the jth column of the matrix. In view of Eqs. (1.47) and (1.63b) we have

$$\mathbf{A}\cdot\mathbf{B}\times\mathbf{C}=\varepsilon_{ijk}\hat{A}_i\hat{B}_j\hat{C}_k=\begin{vmatrix} \hat{A}_1 & \hat{A}_2 & \hat{A}_3 \\ \hat{B}_1 & \hat{B}_2 & \hat{B}_3 \\ \hat{C}_1 & \hat{C}_2 & \hat{C}_3 \end{vmatrix}.$$

Now let $\hat{A}_i=a_{1i}$, $\hat{B}_i=a_{2i}$, and $\hat{C}_i=a_{3i}$ in the above equation to obtain the identity:

$$\begin{vmatrix} a_{11} & a_{12} & a_{13} \\ a_{21} & a_{22} & a_{23} \\ a_{31} & a_{32} & a_{33} \end{vmatrix}=\varepsilon_{ijk}a_{1i}a_{2j}a_{3k}. \quad \blacksquare$$

We note the following properties of determinants:

1. $\det([A][B])=\det[A]\cdot\det[B]$.
2. $\det[A]^{\mathrm{T}}=\det[A]$.
3. $\det(\alpha[A])=\alpha^n\det[A]$, where α is a scalar and n is the order of $[A]$.
4. If $[A']$ is a matrix obtained from $[A]$ by multiplying a row (or column) of $[A]$ by a scalar α, then $\det[A']=\alpha\det[A]$.
5. If $[A']$ is the matrix obtained from $[A]$ by interchanging any two rows (or columns) of $[A]$, then $\det[A']=-\det[A]$.

6 If $[A]$ has two rows (or columns) one of which is a scalar multiple of another (i.e., linearly dependent), $\det[A]=0$.

7 If $[A']$ is the matrix obtained from $[A]$ by adding a multiple of one row (or column) to another, then $\det[A']=\det[A]$.

We redefine (in fact, the definition given earlier is an indirect definition) singular matrices in terms of their determinants. A matrix is said to be *singular* if and only if its determinant is zero. By property 6 above the determinant of a matrix is zero if it has linearly dependent rows (or columns).

1.3.10 Minor, Cofactor, and Adjunct of a Matrix

For an $n\times n$ matrix $[A]$ the determinant of the $(n-1)\times(n-1)$ submatrix of $[A]$ obtained by deleting row i and column j of $[A]$ is called the *minor* of a_{ij} and is denoted by $M_{ij}(A)$. The quantity $\mathrm{cof}_{ij}(A)\equiv(-1)^{i+j}M_{ij}(A)$ is called the *cofactor* of a_{ij}. The determinant of $[A]$ can be cast in terms of the minor and cofactor of a_{ij}:

$$\det[A]=\sum_{i=1}^{n} a_{ij}\mathrm{cof}_{ij}(A) \tag{1.64}$$

for any value of j.

The *adjunct* (also called *adjoint*) of a matrix $[A]$ is the transpose of the matrix obtained from $[A]$ by replacing each element by its cofactor. The adjunct of $[A]$ is denoted by $\mathrm{Adj}(A)$.

Now we have the essential tools to compute the inverse of a matrix. If $[A]$ is nonsingular (i.e., $\det[A]\neq0$), the inverse $[A]^{-1}$ of $[A]$ can be computed according to

$$[A]^{-1}=\frac{1}{\det[A]}\mathrm{Adj}(A). \tag{1.65}$$

Example 1.8 Consider the matrix of Example 1.6. We have, for example,

$$M_{11}(A)=\begin{vmatrix} 4 & 3 \\ -3 & 5 \end{vmatrix}, \quad M_{12}(A)=\begin{vmatrix} 1 & 3 \\ 2 & 5 \end{vmatrix}, \quad M_{13}(A)=\begin{vmatrix} 1 & 4 \\ 2 & -3 \end{vmatrix}$$

$$\mathrm{cof}_{11}(A)=(-1)^2M_{11}(A)=4\times5-(-3)3=29$$

$$\mathrm{cof}_{12}(A)=(-1)^3M_{12}(A)=-(1\times5-3\times2)=1$$

$$\mathrm{cof}_{13}(A)=(-1)^4M_{13}(A)=1\times(-3)-2\times4=-11.$$

The Adj(A) is given by

$$\text{Adj}(A)=\begin{bmatrix} \text{cof}_{11}(A) & \text{cof}_{12}(A) & \text{cof}_{13}(A) \\ \text{cof}_{21}(A) & \text{cof}_{22}(A) & \text{cof}_{23}(A) \\ \text{cof}_{31}(A) & \text{cof}_{32}(\text{A}) & \text{cof}_{33}(\text{A}) \end{bmatrix}^{\text{T}}$$

$$=\begin{bmatrix} 29 & -22 & 19 \\ 1 & 12 & -7 \\ -11 & 16 & 3 \end{bmatrix}.$$

The determinant is given by (expanding by the first row)

$$|A|=2(29)+5(1)+(-1)(-11)=74.$$

The inverse of $[A]$ can be now computed using Eq. (1.65),

$$[A]^{-1}=\frac{1}{74}\begin{bmatrix} 29 & -22 & 19 \\ 1 & 12 & -7 \\ -11 & 16 & 3 \end{bmatrix}.$$

It can be easily verified that $[A][A]^{-1}=[I]$. ■

1.3.11 Solution of Linear Equations

Consider the system of linear equations of the form

$$[A]\{X\}=\{B\}, \tag{1.66}$$

where $[A]=[a_{ij}]$ is an $m\times n$ matrix, $\{X\}=\{x_i\}$ is an $n\times 1$ column matrix and $\{B\}=\{b_i\}$ is an $m\times 1$ column matrix. We wish to find answers to the following questions concerning the solution of Eq. (1.66):

1 Does the system possess a solution?
2 If a solution exists, is it unique?
3 How are the solution(s) to be determined?

If $[A]$ is a square nonsingular matrix, then the answers to the first two questions are clear since we can multiply Eq. (1.66) from the left by $[A]^{-1}$ to obtain $\{X\}=[A]^{-1}\{B\}$ as the unique solution of Eq. (1.66). The solutions can be determined in a number of ways: by computing the inverse of $[A]$ according to Eq. (1.65) or by the Gauss-Jordan elimination method. The latter involves reducing the augmented matrix $[A \mid B]$ to a row-reduced (upper) echelon form by means of elementary operations.

If $[A]$ is not square or if $[A]$ is square and singular, the nature of the solution(s) is not clear. We investigate this case in the following. The nature of the solutions of the system (1.66) depends on the nature of the homogeneous system associated with Eq. (1.66):

$$[A]\{X\}=\{0\}. \tag{1.67}$$

If $\{X_p\}$ is a fixed solution of the system (1.66), then any other solution of Eq. (1.66) can be written as $\{X_h+X_p\}$, where $\{X_h\}$ is the solution of the homogeneous system (1.67). This follows from the fact that

$$\begin{aligned}[A]\{X_h+X_p\}&=[A]\{X_h\}+[A]\{X_p\}\\ &=\{0\}+\{B\}.\end{aligned}$$

We state without proof the fact that every matrix $[A]$ is row equivalent to a matrix in row-reduced echelon form (see [3, 13, 14]):

Theorem 1.1 Every $m\times n$ matrix $[A]=[a_{ij}]$ is row equivalent to a matrix in row-reduced echelon form. ■

The basic method of solving the system $[A]\{X\}=\{B\}$ is to find an equivalent system $[E]\{X\}=\{C\}$ for which the solutions are readily obtainable. Two systems of equations are equivalent if and only if they have precisely the same solutions. The systems $[A]\{X\}=\{B\}$ and $[E]\{X\}=\{C\}$ are equivalent if the augmented matrices $[A \mid B]$ and $[E \mid C]$ are row equivalent. We close this section with the following important observations and an illustrative example.

Let the systems $[A]\{X\}=\{B\}$ and $[E]\{X\}=\{C\}$ be equivalent and let $[E]$ be in row-reduced echelon form. Then the linear system (1.66) has:

1 No solution if $[E \mid C]$ has more nonzero rows than $[E]$.
2 A unique solution $\{X\}=\{C\}$ if $[E]=[I]$.
3 Many solutions if both $[E]$ and $[E \mid C]$ have $r<n$ nonzero rows; r of the unknowns can be expressed in terms of the remaining $n-r$ unknowns.

Example 1.9 Consider the system of linear equations

$$4x_1+3x_2+2x_3-x_4=4$$

$$5x_1+4x_2+3x_3-x_4=4$$

$$2x_1+2x_2+x_3-2x_4=3$$

$$11x_1+6x_2+4x_3+x_4=11.$$

The augmented matrix $[A \mid B]$ is given by

$$\begin{bmatrix} 4 & 3 & 2 & -1 & \vert & 4 \\ 5 & 4 & 3 & -1 & \vert & 4 \\ 2 & 2 & 1 & -2 & \vert & 3 \\ 11 & 6 & 4 & 1 & \vert & 11 \end{bmatrix}.$$

We use elementary row operations to reduce the augmented matrix to row-reduced echelon form. To obtain the 1 in position (1, 1), we multiply row 1 by -1 and add row 2 to row 1. This can be written compactly as $R_2+(-1)R_1$:

$$\begin{bmatrix} 1 & 1 & 1 & 0 & \vert & 0 \\ 5 & 4 & 3 & -1 & \vert & 4 \\ 2 & 2 & 1 & -2 & \vert & 3 \\ 11 & 6 & 4 & 1 & \vert & 11 \end{bmatrix}.$$

Now multiples of row 1 can be added to obtain zeros in the first column $(-5R_1+R_2;\ -2R_1+R_3;\ -11R_1+R_4)$. We have

$$\begin{bmatrix} 1 & 1 & 1 & 0 & \vert & 0 \\ 0 & -1 & -2 & -1 & \vert & 4 \\ 0 & 0 & -1 & -2 & \vert & 3 \\ 0 & -5 & -7 & 1 & \vert & 11 \end{bmatrix},$$

$$(-5)R_2+R_4 \rightarrow \begin{bmatrix} 1 & 1 & 1 & 0 & \vert & 0 \\ 0 & -1 & -2 & -1 & \vert & 4 \\ 0 & 0 & -1 & -2 & \vert & 3 \\ 0 & 0 & 3 & 6 & \vert & -9 \end{bmatrix}.$$

Finally the operations $3R_3+R_4$, R_3+R_1, $-R_3+R_1$, $-2R_3+R_2$ yield the desired form:

$$\begin{bmatrix} 1 & 0 & 0 & 1 & \vert & 1 \\ 0 & 1 & 0 & -3 & \vert & 2 \\ 0 & 0 & 1 & 2 & \vert & -3 \\ 0 & 0 & 0 & 0 & \vert & 0 \end{bmatrix}.$$

The rank of this matrix is 3. The system of equations corresponding to the above augmented matrix is

$$x_1+x_4=1$$

$$x_2-3x_4=2 \qquad \text{or} \qquad [E]\{X\}=\{C\}.$$

$$x_3+2x_4=-3$$

The system falls into case 3 above, many solutions: $r=3$, $n=m=4$. Hence $r=3$

of the unknowns can be expressed in terms of the remaining $n-r=1$ unknown. For example, x_1, x_2, and x_3 can be solved in terms of x_4:

$$x_1=1-x_4, \qquad x_2=2+3x_4, \qquad x_3=-3-2x_4. \quad \blacksquare$$

Exercises 1.3

1 Express the following equations in matrix form:

(a) $3x_1+5x_2+x_3-x_4=2$
$-4x_1+x_2-2x_3+x_4=-3$
$2x_1+x_2+5x_3+x_4=5$
$2x_1-x_2+x_3=4.$

(b) $3x_1+2x_2+16x_3+5x_4=0$
$2x_1+x_2+5x_3+x_4=5$
$x_1-x_2-x_3-x_4=-2$
$-5x_1+11x_2+13x_3+3x_4=6.$

2 Find the transformation matrix for the following orthogonal coordinate transformation: Rotate 90° about the $\hat{\mathbf{e}}_3$-axis and then rotate 45° about the position assumed by the $\hat{\mathbf{e}}_1$-axis after the first rotation.

3 Evaluate the following matrix products:

(a) $$\begin{bmatrix} 4 & -1 & 2 \\ -1 & 0 & 1 \\ 2 & 1 & 0 \end{bmatrix}\begin{bmatrix} -3 & 5 & 6 \\ -1 & 2 & 2 \\ 1 & -1 & -1 \end{bmatrix}.$$

(b) $$\begin{bmatrix} -3 & 6 & 3 \\ -1 & 2 & -1 \\ 7 & -8 & -5 \end{bmatrix}\begin{bmatrix} 3 & -1 & 2 \\ 2 & 1 & 1 \\ 1 & -3 & 0 \end{bmatrix}.$$

(c) $$\begin{bmatrix} 5 & 2 & 10 & 14 \\ 0 & -2 & 2 & 12 \\ 17 & 18 & 3 & -4 \end{bmatrix}\begin{bmatrix} 2 & 0 \\ 4 & 1 \\ 6 & 3 \\ 8 & 5 \end{bmatrix}.$$

4 If $[A]$ is a symmetric $n\times n$ matrix, and $[B]$ is any $n\times n$ matrix, show that $[B]^{\mathrm{T}}[A][B]$ is symmetric.

5 If $[A]$ and $[B]$ are nonsingular $n\times n$ matrices, prove that:

(a) $[A][B]$ is nonsingular.

(b) $([A][B])^{-1}=[B]^{-1}[A]^{-1}$.

6 If $f(x)=a_0+a_1x+a_2x^2+\cdots+a_nx^n$, and $[A]$ is any square matrix, we define the polynomial in $[A]$ by

$$f(A)=a_0[I]+a_1[A]+a_2[A]^2+\cdots+a_n[A]^n.$$

If $[A]=\begin{bmatrix} 1 & -1 \\ -1 & 1 \end{bmatrix}$ and $f(x)=x^2-2x+1$, compute $f(A)$.

7 *Matrix condensation*. Given an $n\times n$ partitioned matrix equation

$$\left[\begin{array}{c|c}[K_{11}] & [K_{12}] \\ \hline [K_{21}] & [K_{22}]\end{array}\right]\left\{\begin{array}{c}\{\Delta_1\} \\ \hline \{\Delta_2\}\end{array}\right\}=\left\{\begin{array}{c}\{F_1\} \\ \hline \{F_2\}\end{array}\right\},$$

express it in the form (i.e., eliminate $\{\Delta_2\}$)

$$[\hat{K}]\{\Delta_1\}=\{\hat{F}\}$$

by assuming that $[K_{22}]$ is nonsingular. Here $\{\Delta_1\}$ is an $(n-r)\times 1$ vector and $\{\Delta_2\}$ an $r\times 1$ vector.

8 Consider the quadratic form in n variables

$$Q(x_1,x_2,\ldots,x_n)=\sum_{i,j=1}^{n} a_{ij}x_ix_j, \qquad \text{with } a_{ij}=a_{ji}.$$

The quadratic form can be expressed in matrix form as

$$\sum_{i=1}^{n} x_i\left(\sum_{j=1}^{n} a_{ij}x_j\right)=\{X\}^{\mathrm{T}}[A]\{X\}.$$

Find the symmetric matrix associated with the quadratic form

$$Q(x_1,x_2,x_3)=x_1^2+2x_1x_2+4x_1x_3+3x_2^2+x_2x_3+7x_3^2.$$

9 Reduce the matrices of Exercise 1 to row-reduced echelon form and determine their ranks.

10 Show that the matrix

$$\begin{bmatrix} 1 & 2 & 3 & 3 & 10 & 6 \\ 2 & 1 & 0 & 0 & 2 & 3 \\ 2 & 2 & 2 & 1 & 5 & 5 \\ -1 & 1 & 3 & 2 & 5 & 2 \end{bmatrix}$$

can be reduced to the echelon form

$$\begin{bmatrix} 1 & 0 & -1 & 0 & 1 & 1 \\ 0 & 1 & 2 & 0 & 0 & 1 \\ 0 & 0 & 0 & 1 & 3 & 1 \\ 0 & 0 & 0 & 0 & 0 & 0 \end{bmatrix}.$$

11 Find the determinants of the square matrices in Exercise 1.3.3.

12 Prove the following identity by actual calculation:

$$\varepsilon_{ijk}=\begin{vmatrix}\delta_{i1} & \delta_{i2} & \delta_{i3}\\ \delta_{j1} & \delta_{j2} & \delta_{j3}\\ \delta_{k1} & \delta_{k2} & \delta_{k3}\end{vmatrix}.$$

13 Prove that the determinant of a 3×3 matrix $[A]$ can be expressed in the form

$$\det A=\tfrac{1}{6}\varepsilon_{ijk}\varepsilon_{rst}a_{ir}a_{js}a_{kt},$$

where a_{ij} is the element in the ith row and jth column of the matrix $[A]$.

14 Prove that

$$[\mathbf{ABC}][\mathbf{CEF}]=\begin{vmatrix}\mathbf{A}\cdot\mathbf{D} & \mathbf{A}\cdot\mathbf{E} & \mathbf{A}\cdot\mathbf{F}\\ \mathbf{B}\cdot\mathbf{D} & \mathbf{B}\cdot\mathbf{E} & \mathbf{B}\cdot\mathbf{F}\\ \mathbf{C}\cdot\mathbf{D} & \mathbf{C}\cdot\mathbf{E} & \mathbf{C}\cdot\mathbf{F}\end{vmatrix},$$

and from there show that $[\mathbf{e}_1\mathbf{e}_2\mathbf{e}_3][\mathbf{e}^1\mathbf{e}^2\mathbf{e}^3]=1$, and

$$\varepsilon_{ijk}\varepsilon_{rst}=\begin{vmatrix}\delta_{ir} & \delta_{is} & \delta_{it}\\ \delta_{jr} & \delta_{js} & \delta_{jt}\\ \delta_{kr} & \delta_{ks} & \delta_{kt}\end{vmatrix}.$$

15 Show that $\mathrm{Adj}([A][B])=\mathrm{Adj}[A]\cdot\mathrm{Adj}[B]$.

16 Find the inverse of the following matrices:

(a) $\begin{bmatrix}4 & 1 & 2\\ 1 & 4 & 1\\ 2 & 1 & 4\end{bmatrix}.$

(b) $\begin{bmatrix}1 & 2 & 3\\ 2 & 3 & 4\\ 3 & 4 & 5\end{bmatrix}.$

(c) $\begin{bmatrix}0 & 1 & 2\\ 1 & 0 & 3\\ 2 & 3 & 0\end{bmatrix}.$

(d) $\begin{bmatrix}1 & -2 & 2\\ -2 & 1 & 2\\ -2 & -2 & -1\end{bmatrix}.$

17 Let $[A]$ be an $n\times n$ square matrix with elements a_{ij}. Recall that the *cofactor* of a_{ij}, denoted here by A_{ij}, is related to the *minor* M_{ij} of a_{ij} by

$$A_{ij}=(-1)^{i+j}M_{ij}.$$

By definition, the determinant $|A|$ is given by

$$|A| = \sum_{k=1}^{n} a_{ik}A_{ik} = \sum_{k=1}^{n} a_{kj}A_{kj} \text{ (no sum on } i \text{ and } j)$$

for any value of i and j $(i, j \leq n)$. Show that

$$\sum_{k=1}^{n} a_{rk}A_{ik} = |A|\delta_{ri}, \qquad \sum_{k=1}^{n} a_{ks}A_{kj} = |A|\delta_{sj}.$$

18 (*Cramer's Rule*) Using the result of the above exercise, show that the solution to a set of n linear equations in n unknown quantities

$$\sum_{k=1}^{n} a_{ij}x_k = b_i, \qquad i = 1, 2, \ldots, n,$$

in the case when the determinant of the matrix of coefficients is not zero ($|a_{ij}| \neq 0$), is given by

$$x_k = \frac{1}{|A|} \sum_{i=1}^{n} A_{ik}b_i, \qquad k = 1, 2, \ldots, n.$$

19 Using Cramer's rule (result of the above exercise) determine the solution to the following equations:

(a) $2x_1 - x_2 - x_3 = 2$
$x_1 + 2x_2 + x_3 = 2$
$4x_1 - 8x_2 - 5x_3 = 2.$

(b) $2x_1 - x_2 = 1$
$-x_1 + 4x_2 - x_3 = 2$
$-x_2 + 2x_3 = 1.$

20 Find the solution(s), if they exist, of the following systems of equations:

(a) $x_1 + x_2 + x_3 + x_4 = 1$
$x_1 + 2x_2 + 3x_3 - 4x_4 = 0$
$2x_1 + 3x_2 + 5x_3 - 5x_4 = 0$
$3x_1 - 4x_2 - 5x_3 + 8x_4 = 0.$

(b) $4x_1 + 6x_2 + 8x_3 = c_1$
$6x_1 + 12x_2 + 18x_3 = c_2$
$8x_1 + 18x_2 + \frac{144}{5}x_3 = c_3$
where c_1, c_2, and c_3 are constants.

(c) $2x_1 + x_2 + 5x_3 + x_4 = 5$
$x_1 + x_2 + 7x_3 + 3x_4 = 0$
$3x_1 + 6x_2 - 2x_3 + x_4 = 8$
$2x_2 + 10x_3 + 8x_4 = 0.$

21 (*Eigenvalues of a Matrix*) The *eigenvalue problem* associated with a square matrix $[A]$ consists of finding the scalars λ, called the *eigenvalues* of $[A]$, such that there exists a nonzero vector $\{X\}$, called the *eigenvector*, satisfying $[A]\{X\}=\lambda\{X\}$, equivalently, $[A-\lambda I]\{X\}=\{0\}$.

(a) Show that this homogeneous equation has a nontrivial solution only if $\det[A-\lambda I]$ is zero. (This problem will be studied in more detail in Section 1.5.19.)

(b) If the eigenvalues $\lambda_1, \lambda_2, \ldots, \lambda_n$ are all different, show that the eigenvectors associated with them are linearly independent.

22 Find the eigenvalues (using the result of Exercise 1.3.21a) of the matrices in Exercise 16.

23 (*Caley-Hamilton Theorem*) Let $[A]$ be an $n\times n$ matrix, and denote by $p(\lambda)$ the determinant of $[A-\lambda I]$, called the *characteristic polynomial.* Show that $p(A)=0$. Here $p(A)$ is defined as in Exercise 1.3.6.

24 (*Matrices of Functions*) Let $[A(t)]$ be an $n\times n$ matrix whose elements are all functions of t. We define the derivative and integral of $[A(t)]$ in terms of its elements:

$$\frac{d}{dt}[A(t)]=\left[\frac{da_{ij}(t)}{dt}\right], \qquad \int_a^b [A(t)]\,dt=\left[\int_0^b a_{ij}(t)\,dt\right].$$

show that

(a) $\dfrac{d}{dt}([A(t)]\cdot[B(t)])=\dfrac{d}{dt}[A(t)]\cdot[B(t)]+[A(t)]\cdot\dfrac{d}{dt}[B(t)].$

(b) $\dfrac{d}{dt}(e^{t[A]})=[A]e^{t[A]},$

where e=exponential and $[A]$ is an $n\times n$ matrix.

(c) $\dfrac{d}{dt}(\det[A])=\sum_{i=1}^{n}\det(A_i(t)),$

where $[A]$ is an $n\times n$ matrix and $A_i(t)$ is obtained from $[A]$ by differentiating the ith row only.

1.4 COORDINATE SYSTEMS AND VECTOR CALCULUS

1.4.1 Differentiation with Respect to a Scalar

Suppose that a vector is given as a function of a scalar, say $\mathbf{A}=\mathbf{A}(t)$. In general the vector $\mathbf{A}$ will have different magnitudes and different directions for different values of t, which we can picture schematically as shown in Fig. 1.21. With the tails of the vector being at the same point for different values of the scalar t, the tip of the arrow draws out a trajectory, as shown in Fig. 1.21.

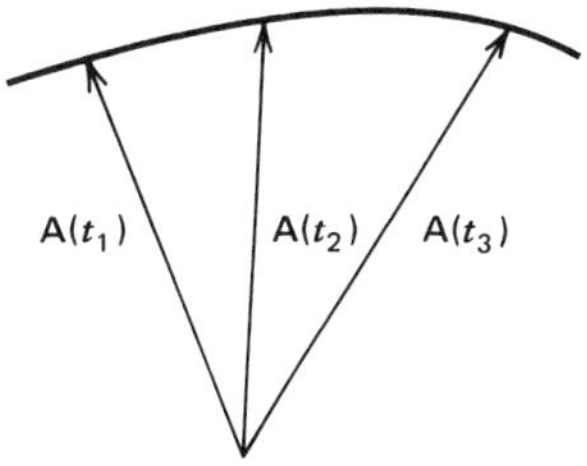

1.21 Variation of a vector as a function of a scalar.

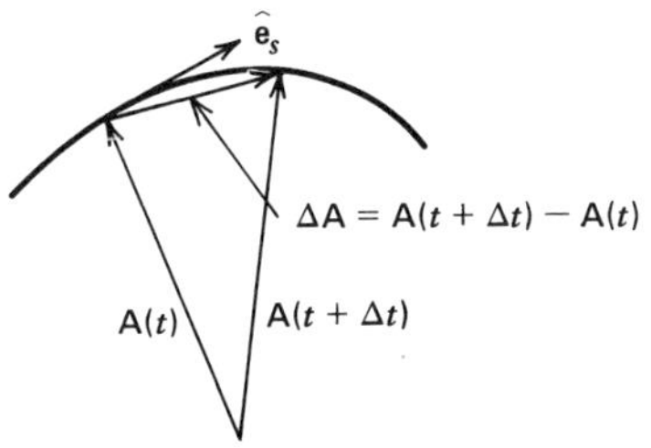

Figure 1.22 Differential change in a vector.

Consider now two values of t differing by an infinitesimal amount, say t and $t+\Delta t$. Then the variation in Fig. 1.21 becomes as shown in Fig. 1.22.

With this picture in mind it is easy to visualize the definition of the derivative of a vector with respect to t:

$$\frac{d\mathbf{A}}{dt}=\lim_{\Delta t\to 0}\frac{\mathbf{A}(t+\Delta t)-\mathbf{A}(t)}{\Delta t}. \tag{1.68}$$

Now let $\Delta s=|\Delta\mathbf{A}|$ so that s is the distance measured along the trajectory. Then we have

$$\frac{d\mathbf{A}}{dt}=\lim_{\Delta t\to 0}\frac{\Delta\mathbf{A}}{\Delta s}\frac{\Delta s}{\Delta t}.$$

In the limit, $\Delta\mathbf{A}/\Delta s$ is a unit vector that is tangent to the trajectory, say $\hat{\mathbf{e}}_s$. In the limit, therefore, we have

$$\frac{d\mathbf{A}}{dt}=\frac{ds}{dt}\hat{\mathbf{e}}_s. \tag{1.69}$$

It is clear that the derivative of the vector has a different direction and magnitude than the vector itself. The magnitude of the derivative ds/dt is the rate of change of distance s with respect to t along the trajectory. Observe that the distance s has the same dimensions as the vector $\mathbf{A}$ itself.

An important special case occurs when the vector has a constant length. In general we note that

$$\mathbf{A}\cdot\mathbf{A}=A^2(t). \tag{1.70a}$$

From differentiation of both sides of this equation it follows that

$$\mathbf{A}\cdot\frac{d\mathbf{A}}{dt}=A\frac{dA}{dt}. \tag{1.70b}$$

When **A** has a constant length, then $dA/dt=0$, and we have

$$\mathbf{A}\cdot\frac{d\mathbf{A}}{dt}=0. \tag{1.70c}$$

We deduce from this result that either $\mathbf{A}=\mathbf{0}$, or $d\mathbf{A}/dt=\mathbf{0}$, or **A** is perpendicular to $d\mathbf{A}/dt$. If $d\mathbf{A}/dt=\mathbf{0}$, then the vector not only has a constant length, but also a constant direction. If the direction of **A** varies, but its length is fixed, then the *derivative* dA/dt *is perpendicular to* **A**.

Consider now the derivative of a vector in terms of its components. If we write

$$\mathbf{A}=A^1(t)\mathbf{e}_1+A^2(t)\mathbf{e}_2+A^3(t)\mathbf{e}_3,$$

we must remember that the basis vectors are also functions of the scalar t. Thus we have

$$\frac{d\mathbf{A}}{dt}=\frac{dA^1}{dt}\mathbf{e}_1+\frac{dA^2}{dt}\mathbf{e}_2+\frac{dA^3}{dt}\mathbf{e}_3+A^1\frac{d\mathbf{e}_1}{dt}+A^2\frac{d\mathbf{e}_2}{dt}+A^3\frac{d\mathbf{e}_3}{dt}. \tag{1.71}$$

Basis vectors that are *constant* are associated with *Cartesian* systems. Only in these systems do the derivatives $d\mathbf{e}_i/dt$ vanish.

Example 1.10 Consider a particle constrained to move in a circular orbit of radius a at a constant speed v. Then the velocity vector can be expressed as [see Eq. (1.24)],

$$\frac{d\mathbf{r}}{dt}=\mathbf{v}=\boldsymbol{\omega}\times\mathbf{r},\qquad \boldsymbol{\omega}=\frac{v}{a}\hat{\mathbf{e}}_z,\qquad \mathbf{r}=a\hat{\mathbf{e}}_r,$$

where $\hat{\mathbf{e}}_r$ is the unit vector along $\mathbf{r}$, $\hat{\mathbf{e}}_r=\mathbf{r}/a$, and $\hat{\mathbf{e}}_z$ is the unit vector perpendicular to the plane of the orbit. The acceleration of the particle is given by (since $\hat{\mathbf{e}}_z$ does not change its direction with time)

$$\begin{aligned}\mathbf{a}=\frac{d\mathbf{v}}{dt}&=\frac{d\boldsymbol{\omega}}{dt}\times\mathbf{r}+\boldsymbol{\omega}\times\frac{d\mathbf{r}}{dt}\\&=\frac{v}{a}\frac{d\hat{\mathbf{e}}_z}{dt}\times\mathbf{r}+\boldsymbol{\omega}\times\mathbf{v}=\boldsymbol{\omega}\times\mathbf{v}\\&=\left(\frac{v}{a}\right)\hat{\mathbf{e}}_z\times(\boldsymbol{\omega}\times\mathbf{r})=\frac{v^2}{a}[\hat{\mathbf{e}}_z\times(\hat{\mathbf{e}}_z\times\hat{\mathbf{e}}_r)]\\&=\frac{v^2}{a}[(\hat{\mathbf{e}}_z\cdot\hat{\mathbf{e}}_r)\hat{\mathbf{e}}_z-(\hat{\mathbf{e}}_z\cdot\hat{\mathbf{e}}_z)\hat{\mathbf{e}}_r]\\&=-\frac{v^2}{a}\hat{\mathbf{e}}_r.\end{aligned}$$

Alternately, let $\mathbf{v}=v\hat{\mathbf{e}}_t$. Then $\mathbf{v}=v\hat{\mathbf{e}}_t=\boldsymbol{\omega}\times\mathbf{r}=v(\hat{\mathbf{e}}_z\times\hat{\mathbf{e}}_r)$ or $\hat{\mathbf{e}}_t=\hat{\mathbf{e}}_z\times\hat{\mathbf{e}}_r$. That is, $\hat{\mathbf{e}}_t$ is the unit vector normal to $\hat{\mathbf{e}}_z$ and $\hat{\mathbf{e}}_r$. The vectors $(\hat{\mathbf{e}}_r,\hat{\mathbf{e}}_t,\hat{\mathbf{e}}_z)$ form an orthonormal basis. Further, $a\,d\hat{\mathbf{e}}_r/dt=\mathbf{v}=v\hat{\mathbf{e}}_t$ implies that

$$\frac{d\hat{\mathbf{e}}_r}{dt}=\omega\hat{\mathbf{e}}_t=\omega(\hat{\mathbf{e}}_z\times\hat{\mathbf{e}}_r)=\boldsymbol{\omega}\times\hat{\mathbf{e}}_r. \quad \blacksquare$$

1.4.2 Cartesian Coordinates

When the basis vectors are constant, that is, with fixed lengths and directions, the basis is called *Cartesian*. The general Cartesian system is oblique. When the basis vectors are unit and orthogonal (orthonormal), the basis system is called *rectangular Cartesian*, or simply *Cartesian*.

Let us denote an orthonormal Cartesian basis by

$$\{\hat{\mathbf{e}}_x,\hat{\mathbf{e}}_y,\hat{\mathbf{e}}_z\} \qquad \text{or} \qquad \{\hat{\mathbf{i}}_1,\hat{\mathbf{i}}_2,\hat{\mathbf{i}}_3\}.$$

It is sometimes convenient to distinguish orthonormal Cartesian bases by the letter $\hat{\mathbf{i}}$. The coordinate curves tangent to these vectors are straight lines. For an arbitrary fixed origin, these curves are described by Cartesian coordinates, denoted by (x, y, z) or (x^1, x^2, x^3). The familiar rectangular Cartesian coordinate system is shown in Fig. 1.23. We shall always use right-handed coordinate systems.

A position vector to an arbitrary point (x, y, z) or (x^1, x^2, x^3), measured from the origin, is given by

$$\begin{aligned}\mathbf{r}&=x\hat{\mathbf{e}}_x+y\hat{\mathbf{e}}_y+z\hat{\mathbf{e}}_z\\ &=x^1\hat{\mathbf{i}}_1+x^2\hat{\mathbf{i}}_2+x^3\hat{\mathbf{i}}_3\end{aligned} \tag{1.72a}$$

or, in summation notation, by

$$\mathbf{r}=x^j\hat{\mathbf{i}}_j. \tag{1.72b}$$

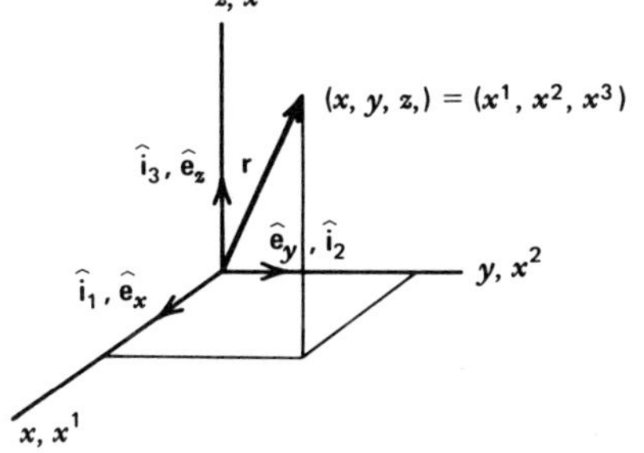

Figure 1.23 Rectangular Cartesian coordinates.

The distance between two infinitesimally removed points is given by

$$d\mathbf{r}\cdot d\mathbf{r}=(ds)^2=dx^j dx^j$$

$$=(dx)^2+(dy)^2+(dx)^2. \tag{1.73}$$

1.4.3 Curvilinear Coordinates

Consider now a transformation to a new set of coordinates denoted by (q^1, q^2, q^3). The transformation equations are

$$\begin{aligned} q^1&=q^1(x^1,x^2,x^3)=q^1(x,y,z)\\ q^2&=q^2(x^1,x^2,x^3)=q^2(x,y,z)\\ q^3&=q^3(x^1,x^2,x^3)=q^3(x,y,z). \end{aligned} \tag{1.74a}$$

The inverse transformation* is

$$\begin{aligned} x^1&=x=x(q^1,q^2,q^3)\\ x^2&=y=y(q^1,q^2,q^3)\\ x^3&=z=z(q^1,q^2,q^3). \end{aligned} \tag{1.74b}$$

The functions $q^1(x, y, z)=\text{constant}$, $q^2(x, y, z)=\text{constant}$, and $q^3(x, y, z)=$ constant denote three surfaces in space. The intersection of any two of these surfaces defines a *coordinate curve*, as shown in Fig. 1.24.

When the above transformation is *nonlinear*, the coordinate curves denoted by q^1, q^2, and q^3 are curved lines, and the (q^1, q^2, q^3) system is called *curvilinear*. When the transformation is *linear*, the coordinate curves will be *straight lines*, but not necessarily parallel to the original (x, y, z) system, and a new Cartesian system will be defined.

Because of the transformations defined, the position vector $\mathbf{r}$ will be a function of the new coordinates (q^1, q^2, q^3), that is,

$$\mathbf{r}=\mathbf{r}(q^1,q^2,q^3). \tag{1.75}$$

The differential $d\mathbf{r}$ can now be written

$$d\mathbf{r}=\frac{\partial\mathbf{r}}{\partial q^1}dq^1+\frac{\partial\mathbf{r}}{\partial q^2}dq^2+\frac{\partial\mathbf{r}}{\partial q^3}dq^3=\frac{\partial\mathbf{r}}{\partial q^i}dq^i. \tag{1.76}$$

*The inverse exists if and only if the Jacobian does not vanish, that is, when $|\partial x^j/\partial q^i|\neq 0$.

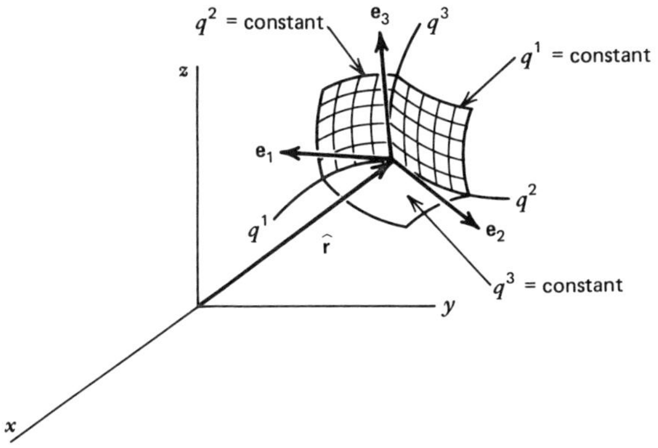

Figure 1.24 Curvilinear coordinates.

The partial derivatives

$$\frac{\partial \mathbf{r}}{\partial q^1}, \qquad \frac{\partial \mathbf{r}}{\partial q^2}, \qquad \frac{\partial \mathbf{r}}{\partial q^3}$$

are vectors that are tangent to the coordinate curves as shown in Fig. 1.24. These vectors can be taken as a basis system associated with the coordinates (q^1, q^2, q^3). This basis is referred to as the *unitary system*. Note that these vectors in general are not unit nor orthogonal. We now denote the unitary basis by $(\mathbf{e}_1, \mathbf{e}_2, \mathbf{e}_3)$ as follows:

$$\mathbf{e}_1 \equiv \frac{\partial \mathbf{r}}{\partial q^1}, \qquad \mathbf{e}_2 \equiv \frac{\partial \mathbf{r}}{\partial q^2}, \qquad \mathbf{e}_3 \equiv \frac{\partial \mathbf{r}}{\partial q^3}. \tag{1.77}$$

A differential distance is denoted by

$$d\mathbf{r} = dq^1\mathbf{e}_1 + dq^2\mathbf{e}_2 + dq^3\mathbf{e}_3 = dq^i\mathbf{e}_i. \tag{1.78}$$

Observe that the q's have superscripts whereas the unitary basis has subscripts. The dq's thus are referred to as the *contravariant components* of the differential vector $d\mathbf{r}$. The unitary basis will satisfy the *covariant* transformation law and thus is a *covariant basis*.

The unitary basis can be described in terms of the original Cartesian basis as follows:

$$\begin{aligned}
\mathbf{e}_1 &= \frac{\partial \mathbf{r}}{\partial q^1} = \frac{\partial x}{\partial q^1}\hat{\mathbf{e}}_x + \frac{\partial y}{\partial q^1}\hat{\mathbf{e}}_y + \frac{\partial z}{\partial q^1}\hat{\mathbf{e}}_z \\
\mathbf{e}_2 &= \frac{\partial \mathbf{r}}{\partial q^2} = \frac{\partial x}{\partial q^2}\hat{\mathbf{e}}_x + \frac{\partial y}{\partial q^2}\hat{\mathbf{e}}_y + \frac{\partial z}{\partial q^2}\hat{\mathbf{e}}_z \\
\mathbf{e}_3 &= \frac{\partial \mathbf{r}}{\partial q^3} = \frac{\partial x}{\partial q^3}\hat{\mathbf{e}}_x + \frac{\partial y}{\partial q^3}\hat{\mathbf{e}}_y + \frac{\partial z}{\partial q^3}\hat{\mathbf{e}}_z.
\end{aligned} \tag{1.79}$$

In the summation convention we have

$$\mathbf{e}_i \equiv \frac{\partial \mathbf{r}}{\partial q^i} = \frac{\partial x^j}{\partial q^i}\hat{\mathbf{i}}_j, \qquad i=1,2,3. \tag{1.80}$$

We can associate with the covariant base vectors $\mathbf{e}_i$, a dual or reciprocal basis, defined by Eq. (1.35).

1.4.4 The Fundamental Metric

The square of the infinitesimal distance between two points can now be written in the unitary basis as

$$\begin{aligned}(ds)^2 &= d\mathbf{r}\cdot d\mathbf{r} = (\mathbf{e}_i\cdot\mathbf{e}_j)\,dq^i dq^j \\ &= g_{ij}\,dq^i dq^j,\end{aligned} \tag{1.81}$$

where

$$g_{ij} \equiv \mathbf{e}_i \cdot \mathbf{e}_j. \tag{1.82}$$

The terms denoted by the symbol g_{ij} are the components of the *fundamental metric tensor*. According to the above definition, g_{ij} is symmetric, that is, the indices can be interchanged without changing the value of the metric, $g_{ij}=g_{ji}$.

The metric tensor plays a key role in the fundamental absolute tensor calculus. Whereas we have "derived" a form for g_{ij} by essentially starting with things we "know," that is, a Cartesian system, the fundamental tensor calculus uses the metric tensor as a starting point from which all else follows. The fundamental metric describes the nature of a space, or manifold. A manifold may be either "curved" or "flat." A flat space is said to be *Euclidean*. If it is possible to find a transformation to a coordinate system such that all the g_{ij}'s are constant, the space is Euclidean. If it is not possible, it is said to be non-Euclidean, or *Riemannian*.

The amazing property of the fundamental metric is that it is possible to make measurements within the space in such a way as to determine whether that space is flat or curved. As an example, consider a two-dimensional space, which is a surface. If the surface is flat, we can mark out a triangle and measure the sum of the included angles to be 180°. This is a well-known result from Euclidean geometry. On the other hand, consider a curved surface—the surface of a sphere. If a triangle is marked out from three points by "straight lines," which are geodesics or segments of great circles, the sum of the included angles turns out to be greater than 180°. This information leads to the conclusion that the surface is not flat, and further, it reveals information about the curvature of the surface. Other surfaces may be found for which the sum of the included angles of a triangle is less than 180°. These surfaces are also curved, but of a different nature than the spherical type surface. The fundamental tensor calculus allows these notions to be generalized to n-space. A mathematical apparatus such as this was the ticket for Einstein and his general theory of relativity. For most engineering applications we will be concerned with Euclidean spaces.

Example 1.11 Consider the nonlinear transformation

$$x^1 = a - q^2 \sin\left(\frac{q^1}{a}\right)$$

$$x^2 = a - (a - q^2)\cos\left(\frac{q^1}{a}\right)$$

$$x^3 = q^3,$$

where a is a constant. The unitary base vectors are given by

$$\mathbf{e}_1 = \frac{\partial \mathbf{r}}{\partial q^1} = \frac{\partial x^i}{\partial q^1}\hat{\mathbf{i}}_i = -\frac{q^2}{a}\cos\left(\frac{q^1}{a}\right)\hat{\mathbf{i}}_1 + \left(1 - \frac{q^2}{a}\right)\sin\left(\frac{q^1}{a}\right)\hat{\mathbf{i}}_2$$

$$\mathbf{e}_2 = \frac{\partial \mathbf{r}}{\partial q^2} = \frac{\partial x^i}{\partial q^2}\hat{\mathbf{i}}_i = -\sin\left(\frac{q^1}{a}\right)\hat{\mathbf{i}}_1 + \cos\left(\frac{q^1}{a}\right)\hat{\mathbf{i}}_2$$

$$\mathbf{e}_3 = \frac{\partial \mathbf{r}}{\partial q^3} = \frac{\partial x^i}{\partial q^3}\hat{\mathbf{i}}_i = \hat{\mathbf{i}}_3.$$

The components of the fundamental metric are

$$g_{11} = \mathbf{e}_1 \cdot \mathbf{e}_1 = \left(1 - \frac{2q^2}{a}\right)\sin^2\left(\frac{q^1}{a}\right) + \left(\frac{q^2}{a}\right)^2$$

$$g_{12} = \mathbf{e}_1 \cdot \mathbf{e}_2 = \sin\left(\frac{q^1}{a}\right)\cos\left(\frac{q^1}{a}\right)$$

$$g_{22} = 1, \qquad g_{13} = 0, \qquad g_{23} = 0, \qquad g_{33} = 1. \quad \blacksquare$$

1.4.5 The Norm of a Unitary Vector

The length, magnitude, or *norm* of a unitary vector is defined as

$$|\mathbf{e}_i| = \|\mathbf{e}_i\|. \tag{1.83}$$

It is given in terms of the metric coefficient as follows:

$$|\mathbf{e}_i| = [\mathbf{e}_i \cdot \mathbf{e}_i]^{1/2}$$

$$= \sqrt{g_{ii}} \qquad \text{(no summation)}. \tag{1.84}$$

Then we have

$$|\mathbf{e}_1| = \sqrt{g_{11}}, \qquad |\mathbf{e}_2| = \sqrt{g_{22}}, \qquad |\mathbf{e}_3| = \sqrt{g_{33}}. \tag{1.85}$$

1.4.6 Relation Between Covariant and Contravariant Components and Bases

We now recall from Eq. (1.41) that any vector can be expressed in terms of a given basis as

$$\mathbf{A} = (\mathbf{A} \cdot \mathbf{e}^i)\mathbf{e}_i, \tag{1.86}$$

or in terms of its dual basis by

$$\mathbf{A} = (\mathbf{A} \cdot \mathbf{e}_i)\mathbf{e}^i. \tag{1.87}$$

The dual of the unitary basis can be derived in terms of the unitary basis by setting $\mathbf{A} = \mathbf{e}^j$ in Eq. (1.86). Thus

$$\mathbf{e}^j = (\mathbf{e}^j \cdot \mathbf{e}^i)\mathbf{e}_i. \tag{1.88}$$

The product $\mathbf{e}^j \cdot \mathbf{e}^i$ is denoted by g^{ij}. Thus we have

$$\mathbf{e}^j = g^{ij}\mathbf{e}_i. \tag{1.89}$$

The terms denoted by g^{ij} are also components of the metric tensor, but they are the *contravariant components*, whereas the terms g_{ij} are the *covariant components*. We shall amplify on this shortly.

In similar fashion we can express the dual unitary basis as a linear combination of its dual basis. To do this set $\mathbf{A} = \mathbf{e}_j$ in Eq. (1.87). We obtain

$$\begin{aligned} \mathbf{e}_j &= (\mathbf{e}_j \cdot \mathbf{e}_i)\mathbf{e}^i \\ &= g_{ij}\mathbf{e}^i. \end{aligned} \tag{1.90}$$

Similar relations exist between the contravariant and covariant components of a vector. We can write either

$$\mathbf{A} = A^i\mathbf{e}_i$$

or

$$\mathbf{A} = A_i\mathbf{e}^i. \tag{1.91}$$

Scalar multiply the first of these equations by $\mathbf{e}_j$ and obtain, since $\mathbf{e}_j \cdot \mathbf{A} = A_j$,

$$A_j = (\mathbf{e}_i \cdot \mathbf{e}_j) A^i$$
$$= g_{ij} A^i = g_{ji} A^i. \tag{1.92}$$

Scalar multiply the second equation by $\mathbf{e}^j$ and obtain, since $\mathbf{e}^j \cdot \mathbf{A} = A^j$,

$$A^j = (\mathbf{e}^i \cdot \mathbf{e}^j) A_i$$
$$= g^{ij} A_i = g^{ji} A_i. \tag{1.93}$$

These expressions are analogous to the corresponding expressions for the basis vectors.

The above equations are sometimes known as the process of the *raising and lowering of indices*. The covariant or contravariant components or basis systems can be obtained in terms of each other if the tensor components g_{ij} and g^{ij} are known. If we assume that the unitary basis (which is the covariant basis) is known, then g_{ij} can be computed. Further, the dual basis (contravariant) can be computed from the cross-product expressions given by Eqs. (1.35). When these are known, then g^{ij} can be determined.

A further relation between g_{ij} and g^{ij} can be obtained by scalar multiplying the equation $\mathbf{e}_j = g_{ij} \mathbf{e}^i$ by $\mathbf{e}^k$. We then get

$$\delta_j^k = g_{ij} g^{ik}. \tag{1.94}$$

Notice that only the index i is summed on the right-hand side.

1.4.7 Matrix Notation for the Metric Tensor

It is very useful by matrix methods to obtain the relation between g_{ij} and g^{ij} instead by means of the cross-product relations, Eqs. (1.35). Among other reasons, the generalization to n-space is easily facilitated. The matrix notation of $A_i = g_{ij} A^j$ is

$$\begin{Bmatrix} A_1 \\ A_2 \\ A_3 \end{Bmatrix} = \begin{bmatrix} g_{11} & g_{12} & g_{13} \\ g_{21} & g_{22} & g_{23} \\ g_{31} & g_{32} & g_{33} \end{bmatrix} \begin{Bmatrix} A^1 \\ A^2 \\ A^3 \end{Bmatrix}. \tag{1.95}$$

We wish now to invert this relation and solve for A^i in terms of A_i. First we define g as the determinant of the matrix $[g_{ij}]$, that is,

$$g \equiv \det[g_{ij}]. \tag{1.96}$$

Further, let M_{ij} be the minor of g_{ij}. Let $\tilde{M}_{ij}$ be the cofactor:

$$\tilde{M}_{ij} \equiv (-1)^{i+j} M_{ij} \qquad \text{(no summation)}. \tag{1.97}$$

According to Cramer's rule, we can solve for A^1 as follows (see Exercise 1.3.18):

$$A^1 = \frac{1}{g}\begin{vmatrix} A_1 & g_{12} & g_{13} \\ A_2 & g_{22} & g_{23} \\ A_3 & g_{32} & g_{33} \end{vmatrix} = \frac{A_1 M_{11} - A_2 M_{21} + A_3 M_{31}}{g}$$

$$= \frac{A_1 \tilde{M}_{11} + A_2 \tilde{M}_{21} + A_3 \tilde{M}_{31}}{g} = \frac{\tilde{M}_{i1}}{g} A_i.$$

Now define

$$G^{ij} = \frac{\tilde{M}_{ij}}{g}. \tag{1.98}$$

Then we have

$$A^1 = G^{i1} A_i.$$

The general result is

$$A^j = G^{ij} A_i. \tag{1.99}$$

From our previous result that $A^j = g^{ij} A_i$ we deduce that

$$g^{ij} = G^{ij}.$$

Thus when the metric tensor g_{ij} is known, it is not necessary to use the cross-product formulas to determine g^{ij}. We will find that the determinant g also plays an important role.

Analogous to Eq. (1.45), we can define the permutation symbol in a general curvilinear (right-handed) system by

$$\mathbf{e}_i \times \mathbf{e}_j = \mathcal{E}_{ijk} \mathbf{e}^k, \tag{1.100}$$

where $\mathcal{E}_{ijk}$ is the permutation symbol in the general curvilinear system,

$$\mathcal{E}_{ijk} \equiv \mathbf{e}_i \times \mathbf{e}_j \cdot \mathbf{e}_k, \tag{1.101a}$$

and is related to ε_{ijk} of Eq. (1.46) by

$$\mathcal{E}_{ijk} = \sqrt{g}\, \varepsilon_{ijk}. \tag{1.101b}$$

Similarly, we define

$$\mathcal{E}^{ijk} \equiv \mathbf{e}^i \times \mathbf{e}^j \cdot \mathbf{e}^k = \frac{1}{\sqrt{g}} \varepsilon_{ijk}. \tag{1.102}$$

1.4.8 Physical Component of a Vector

The physical component $\hat{A}^i$ of a vector is defined as

$$\hat{A}^i \hat{\mathbf{e}}_i = A^i \mathbf{e}_i \qquad \text{(no summation)}, \tag{1.103}$$

where $\hat{\mathbf{e}}_i$ is the unit vector,

$$\hat{\mathbf{e}}_i = \frac{\mathbf{e}_i}{|\mathbf{e}_i|} = \frac{\mathbf{e}_i}{\sqrt{g_{ii}}} \qquad \text{(no summation)}. \tag{1.104}$$

Thus we have

$$\hat{A}^i = A^i \mathbf{e}_i \cdot \hat{\mathbf{e}}_i$$

$$= A^i \sqrt{g_{ii}} \qquad \text{(no summation)}. \tag{1.105}$$

Because the unitary basis vectors may have physical dimensions, the contravariant component A^i may not have the same physical dimensions that **A** has. On the other hand, the physical component $\hat{A}^i$ does have the same dimensions as **A**. The physical component is essentially the length of the vector component $A^i \mathbf{e}_i$.

By the same token, the physical components corresponding to the covariant components are given by

$$\hat{A}_i = A_i \sqrt{g^{ii}} \qquad \text{(no summation)}. \tag{1.106}$$

1.4.9 Orthogonal Curvilinear Systems

When the transformation given in Section 1.4.3 is orthogonal, the basis vectors are mutually perpendicular. However, they still are not of unit length. It is useful to introduce a unit orthogonal basis so that covariant and contravariant aspects need not be distinguished. We do this by means of *scale factors* defined as follows:

$$h_1 = |\mathbf{e}_1| = \sqrt{g_{11}}, \qquad h_2 = |\mathbf{e}_2| = \sqrt{g_{22}}, \qquad h_3 = |\mathbf{e}_3| = \sqrt{g_{33}}. \tag{1.107}$$

Thus we have

$$\mathbf{e}_1 = h_1 \hat{\mathbf{e}}_1, \qquad \mathbf{e}_2 = h_2 \hat{\mathbf{e}}_2, \qquad \mathbf{e}_3 = h_3 \hat{\mathbf{e}}_3. \tag{1.108}$$

It follows that

$$d\mathbf{r}=(h_1\,dq^1)\hat{\mathbf{e}}_1+(h_2\,dq^2)\hat{\mathbf{e}}_2+(h_3\,dq^3)\hat{\mathbf{e}}_3 \tag{1.109a}$$

and

$$(ds)^2=(h_1\,dq^1)^2+(h_2\,dq^2)^2+(h_3\,dq^3)^2. \tag{1.109b}$$

Thus the differential *distances* measured along coordinate lines are

$$ds_1=h_1\,dq^1, \qquad ds_2=h_2\,dq^2, \qquad ds_3=h_3\,dq^3. \tag{1.110}$$

The name *scale factors* for h_1, h_2, h_3 is appropriate since dq^1, dq^2, and dq^3 may not have the dimensions of distance (they may be angles, for instance).

The scale factors may be computed from the orthogonal transformation as follows:

$$\frac{\partial\mathbf{r}}{\partial q^i}=\frac{\partial x}{\partial q^i}\hat{\mathbf{e}}_x+\frac{\partial y}{\partial q^i}\hat{\mathbf{e}}_y+\frac{\partial z}{\partial q^i}\hat{\mathbf{e}}_z \tag{1.111}$$

$$\begin{aligned} h_1&\equiv\left|\frac{\partial\mathbf{r}}{\partial q^1}\right|=\left[\left(\frac{\partial x}{\partial q^1}\right)^2+\left(\frac{\partial y}{\partial q^1}\right)^2+\left(\frac{\partial z}{\partial q^1}\right)^2\right]^{1/2}\\ h_2&\equiv\left|\frac{\partial\mathbf{r}}{\partial q^2}\right|=\left[\left(\frac{\partial x}{\partial q^2}\right)^2+\left(\frac{\partial y}{\partial q^2}\right)^2+\left(\frac{\partial z}{\partial q^2}\right)^2\right]^{1/2}\\ h_3&\equiv\left|\frac{\partial\mathbf{r}}{\partial q^3}\right|=\left[\left(\frac{\partial x}{\partial q^3}\right)^2+\left(\frac{\partial y}{\partial q^3}\right)^2+\left(\frac{\partial z}{\partial q^3}\right)^2\right]^{1/2}. \end{aligned} \tag{1.112}$$

When $\mathbf{e}_1$, $\mathbf{e}_2$, and $\mathbf{e}_3$ are orthogonal, we have

$$g_{ij}=0, \qquad i\neq j. \tag{1.113}$$

The determinant of the metric tensor is thus

$$g=\begin{vmatrix} g_{11} & 0 & 0\\ 0 & g_{22} & 0\\ 0 & 0 & g_{33}\end{vmatrix} \tag{1.114}$$

$$=g_{11}g_{22}g_{33}=(h_1h_2h_3)^2. \tag{1.115}$$

It is further easy to show that the cofactor $\tilde{M}_{ij}$ has the values

$$\tilde{M}_{11}=g_{22}g_{33}, \qquad \tilde{M}_{22}=g_{11}g_{33}, \qquad \tilde{M}_{33}=g_{11}g_{22}, \qquad \tilde{M}_{ij}=0, \qquad i\neq j. \tag{1.116}$$

The contravariant components of the metric tensor now can be computed as

$$\begin{aligned} g^{11}&=G^{11}=\frac{\tilde{M}_{11}}{g}=\frac{g_{22}g_{33}}{g_{11}g_{22}g_{33}}=\frac{1}{g_{11}}=\frac{1}{h_1^2} \\ g^{22}&=\frac{1}{g_{22}}=\frac{1}{h_2^2} \\ g^{33}&=\frac{1}{g_{33}}=\frac{1}{h_3^2} \\ g^{ij}&=0, \qquad i\neq j. \end{aligned} \tag{1.117}$$

1.4.10 Examples of Orthogonal Curvilinear Coordinate Systems

Two commonly used orthogonal curvilinear coordinate systems are *spherical coordinates* and *cylindrical coordinates.*

Spherical Coordinates

$$\begin{aligned} q^1&=r \\ q^2&=\theta \\ q^3&=\phi. \end{aligned} \tag{1.118}$$

$$\begin{aligned} x&=r\sin\theta\cos\phi \\ y&=r\sin\theta\sin\phi \\ z&=r\cos\theta. \end{aligned} \tag{1.119}$$

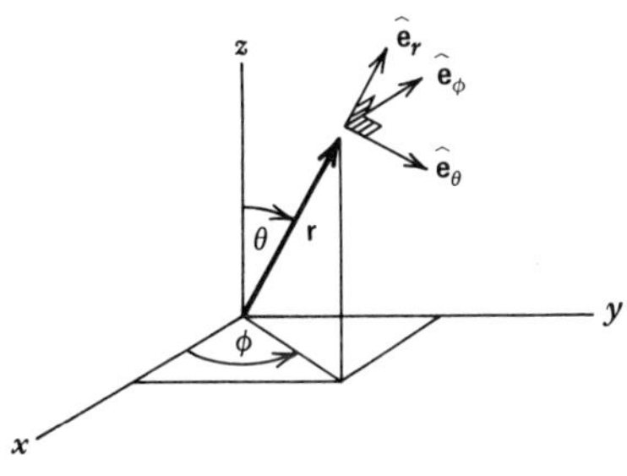

Figure 1.25 Spherical coordinates.

$$r=(x^2+y^2+z^2)^{1/2}$$

$$\theta=\cos^{-1}\frac{z}{(x^2+y^2+z^2)^{1/2}} \tag{1.120}$$

$$\phi=\tan^{-1}\left(\frac{y}{x}\right).$$

$$h_1=h_r=1$$

$$h_2=h_\theta=r \tag{1.121}$$

$$h_3=h_\phi=r\sin\theta.$$

Thus we also find:

$$g_{11}=1, \qquad g_{22}=r^2, \qquad g_{33}=(r\sin\theta)^2$$

$$g^{11}=1, \qquad g^{22}=r^{-2}, \qquad g^{33}=(r\sin\theta)^{-2}, \tag{1.122}$$

$$g_{ij}=g^{ij}=0, \qquad i\neq j.$$

Also, we have

$$\mathbf{e}_r=\hat{\mathbf{e}}_r, \qquad \mathbf{e}_\theta=r\hat{\mathbf{e}}_\theta, \qquad \mathbf{e}_\phi=r\sin\theta\,\hat{\mathbf{e}}_\phi \tag{1.123}$$

$$\mathbf{e}^r=\hat{\mathbf{e}}_r, \qquad \mathbf{e}^\theta=\frac{1}{r}\hat{\mathbf{e}}_\theta, \qquad \mathbf{e}^\phi=\frac{1}{r\sin\theta}\hat{\mathbf{e}}_\phi.$$

It is clear that the unitary basis and its dual are not unit vectors in general.

In an orthogonal basis system the physical components $\hat{A}^i$ and $\hat{A}_i$ of a vector $\mathbf{A}$ are equal. Thus for spherical coordinates, we have

$$\hat{A}^r=\hat{A}_r=A^r=A_r$$

$$\hat{A}^\theta=\hat{A}_\theta=rA^\theta=\frac{A_\theta}{r} \tag{1.124}$$

$$\hat{A}^\phi=\hat{A}_\phi=r\sin\theta\,A^\phi=\frac{A_\phi}{r\sin\theta}.$$

In terms of the Cartesian basis we have

$$\hat{\mathbf{e}}_i = \frac{1}{h_i}\frac{\partial x^j}{\partial q^i}\hat{\mathbf{i}}_j \qquad \text{(no summation on } i\text{)}$$

$$\begin{aligned}
\hat{\mathbf{e}}_r &= \sin\theta\cos\phi\,\hat{\mathbf{e}}_x + \sin\theta\sin\phi\,\hat{\mathbf{e}}_y + \cos\theta\,\hat{\mathbf{e}}_z \\
\hat{\mathbf{e}}_\theta &= \cos\theta\cos\phi\,\hat{\mathbf{e}}_x + \cos\theta\sin\phi\,\hat{\mathbf{e}}_y - \sin\theta\,\hat{\mathbf{e}}_z \\
\hat{\mathbf{e}}_\phi &= -\sin\phi\,\hat{\mathbf{e}}_x + \cos\phi\,\hat{\mathbf{e}}_y .
\end{aligned} \tag{1.125}$$

On the other hand, the Cartesian basis in terms of the spherical orthonormal basis is

$$\begin{aligned}
\hat{\mathbf{i}}_i &= (\hat{\mathbf{i}}_i \cdot \hat{\mathbf{e}}_j)\hat{\mathbf{e}}_j \\
\hat{\mathbf{e}}_x &= \sin\theta\cos\phi\,\hat{\mathbf{e}}_r + \cos\theta\cos\phi\,\hat{\mathbf{e}}_\theta - \sin\phi\,\hat{\mathbf{e}}_\phi \\
\hat{\mathbf{e}}_y &= \sin\theta\sin\phi\,\hat{\mathbf{e}}_r + \cos\theta\sin\phi\,\hat{\mathbf{e}}_\theta + \cos\phi\,\hat{\mathbf{e}}_\phi \\
\hat{\mathbf{e}}_z &= \cos\theta\,\hat{\mathbf{e}}_r - \sin\theta\,\hat{\mathbf{e}}_\theta .
\end{aligned} \tag{1.126}$$

The position vector is given by

$$\mathbf{r} = r\hat{\mathbf{e}}_r .$$

These results are very useful in many problems.

Cylindrical Coordinates

$$\begin{aligned}
q^1 &= R \\
q^2 &= \phi \\
q^3 &= Z .
\end{aligned} \tag{1.127}$$

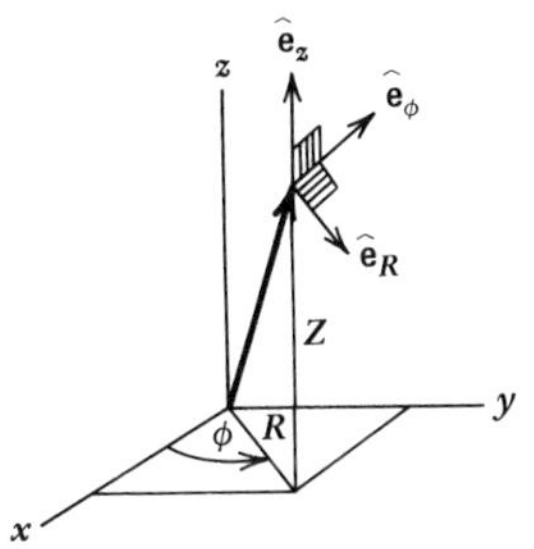

Figure 1.26 Cylindrical coordinates.

$$x=R\cos\phi$$
$$y=R\sin\phi \tag{1.128}$$
$$z=Z=q^3.$$

$$R=(x^2+y^2)^{1/2}$$
$$\phi=\tan^{-1}\left(\frac{y}{x}\right) \tag{1.129}$$
$$z=Z.$$

$$h_1=h_R=1$$
$$h_2=h_\phi=R \tag{1.130}$$
$$h_3=h_z=1.$$

The metric coefficients are

$$g_{11}=1, \quad g_{22}=R^2, \quad g_{33}=1$$
$$g^{11}=1, \quad g^{22}=R^{-2}, \quad g^{33}=1$$

$$g_{ij}=g^{ij}=0, \qquad i\neq j. \tag{1.131}$$

In addition we have

$$\mathbf{e}_R=\hat{\mathbf{e}}_R, \quad \mathbf{e}_\phi=R\hat{\mathbf{e}}_\phi, \quad \mathbf{e}_z=\hat{\mathbf{e}}_z$$
$$\mathbf{e}^R=\hat{\mathbf{e}}_R, \quad \mathbf{e}^\phi=\frac{1}{R}\hat{\mathbf{e}}_\phi, \quad \mathbf{e}^z=\hat{\mathbf{e}}_z. \tag{1.132}$$

In terms of the Cartesian basis we have

$$\hat{\mathbf{e}}_R=\cos\phi\,\hat{\mathbf{e}}_x+\sin\phi\,\hat{\mathbf{e}}_y$$
$$\hat{\mathbf{e}}_\phi=-\sin\phi\,\hat{\mathbf{e}}_x+\cos\phi\,\hat{\mathbf{e}}_y \tag{1.133}$$
$$\hat{\mathbf{e}}_z=\hat{\mathbf{e}}_z.$$

The Cartesian basis in terms of the cylindrical orthonormal basis is

$$\hat{\mathbf{e}}_x=\cos\phi\,\hat{\mathbf{e}}_R-\sin\phi\,\hat{\mathbf{e}}_\phi$$
$$\hat{\mathbf{e}}_y=\sin\phi\,\hat{\mathbf{e}}_R+\cos\phi\,\hat{\mathbf{e}}_\phi \tag{1.134}$$
$$\hat{\mathbf{e}}_z=\hat{\mathbf{e}}_z.$$

The physical components of a vector **A** are related to the covariant and contravariant components by

$$\hat{A}^R=\hat{A}_R=A^R=A_R$$

$$\hat{A}^\phi=\hat{A}_\phi=RA^\phi=\frac{A_\phi}{R} \tag{1.135}$$

$$\hat{A}^z=\hat{A}_z=A^z=A_z.$$

The position vector is given by

$$\mathbf{r}=R\hat{\mathbf{e}}_R+Z\hat{\mathbf{e}}_z. \tag{1.136}$$

1.4.11 Relation Between Two Curvilinear Coordinate Systems

Consider now barred and unbarred coordinate systems that are related by the following transformations:

$$q^1=q^1(\bar{q}^1,\bar{q}^2,\bar{q}^3)$$

$$q^2=q^2(\bar{q}^1,\bar{q}^2,\bar{q}^3) \tag{1.137}$$

$$q^3=q^3(\bar{q}^1,\bar{q}^2,\bar{q}^3).$$

The inverse transformation is

$$\bar{q}^1=\bar{q}^1(q^1,q^2,q^3)$$

$$\bar{q}^2=\bar{q}^2(q^1,q^2,q^3) \tag{1.138}$$

$$\bar{q}^3=\bar{q}^3(q^1,q^2,q^3).$$

The unitary bases in the the two systems are $\{\mathbf{e}_1,\mathbf{e}_2,\mathbf{e}_3\}$ and $\{\bar{\mathbf{e}}_1,\bar{\mathbf{e}}_2,\bar{\mathbf{e}}_3\}$, and the differential distance $d\mathbf{r}$ is denoted by

$$d\mathbf{r}=dq^i\mathbf{e}_i, \qquad d\mathbf{r}=d\bar{q}^j\bar{\mathbf{e}}_j. \tag{1.139}$$

Dotting these equations with $\mathbf{e}^s$ and solving for dq^s gives

$$dq^s=(\mathbf{e}^s\cdot\bar{\mathbf{e}}_j)\,d\bar{q}^j, \qquad s=1,2,3. \tag{1.140}$$

Dotting with $\bar{\mathbf{e}}^s$ and solving for $d\bar{q}^s$ gives

$$d\bar{q}^s=(\bar{\mathbf{e}}^s\cdot\mathbf{e}_i)\,dq^i, \qquad s=1,2,3. \tag{1.141}$$

We can obtain similar relations between the dq's for each system directly from the transformations themselves, that is,

$$dq^s = \frac{\partial q^s}{\partial \bar{q}^j} d\bar{q}^j, \qquad s=1,2,3$$

$$d\bar{q}^s = \frac{\partial \bar{q}^s}{\partial q^i} dq^i, \qquad s=1,2,3. \tag{1.142}$$

Comparison of these two sets of relations leads to the following identification:

$$\mathbf{e}^s \cdot \bar{\mathbf{e}}_j = \frac{\partial q^s}{\partial \bar{q}^j}$$

$$\bar{\mathbf{e}}^s \cdot \mathbf{e}_i = \frac{\partial \bar{q}^s}{\partial q^i}. \tag{1.143}$$

From Section 1.2.18 we also deduce that

$$a_s^j = \mathbf{e}^j \cdot \bar{\mathbf{e}}_s = \frac{\partial q^j}{\partial \bar{q}^s}$$

$$b_i^s = \bar{\mathbf{e}}^s \cdot \mathbf{e}_i = \frac{\partial \bar{q}^s}{\partial q^i}. \tag{1.144}$$

Thus the relationships between the two basis systems are given in terms of the transformation equations themselves.

The cogredient and contragredient laws given in Section 1.2.18 can now be written

$$\bar{\mathbf{e}}_s = \frac{\partial q^j}{\partial \bar{q}^s} \mathbf{e}_j, \qquad \bar{A}_s = \frac{\partial q^j}{\partial \bar{q}^s} A_j, \qquad \text{covariant law}$$

$$\bar{\mathbf{e}}^s = \frac{\partial \bar{q}^s}{\partial q^i} \mathbf{e}^i, \qquad \bar{A}^s = \frac{\partial \bar{q}^s}{\partial q^i} A^i, \qquad \text{contravariant law.} \tag{1.145}$$

The relations between the covariant and contravariant components of a vector and the unitary bases and the dual bases are thus given in terms of the transformation equations themselves.

Note that scalar multiplying the covariant law by $\bar{\mathbf{e}}^i$ gives

$$\delta_s^i = \frac{\partial q^j}{\partial \bar{q}^s} \frac{\partial \bar{q}^i}{\partial q^j}. \tag{1.146}$$

1.4.12 Definition of a Vector (Analytical)

The covariant and contravariant transformation laws given above now provide a rule that replaces our so-called parallelogram law of addition given in our original geometric definition of a vector. We now can define a vector in the following way:

Covariant Vector

If an ordered triple (A_1, A_2, A_3) associated with the coordinates (q^1, q^2, q^3) in one system is related to the ordered triple $(\bar{A}_1, \bar{A}_2, \bar{A}_3)$ in another coordinate system $(\bar{q}^1, \bar{q}^2, \bar{q}^3)$ by means of the transformation

$$\bar{A}_j = \frac{\partial q^i}{\partial \bar{q}^j} A_i, \qquad j = 1,2,3, \tag{1.147}$$

the elements of the ordered triple are called the *covariant components of a vector*.

Contravariant Vector

If an ordered triple (A^1, A^2, A^3) associated with the coordinates (q^1, q^2, q^3) in one system is related to an ordered triple $(\bar{A}^1, \bar{A}^2, \bar{A}^3)$ in another coordinate system $(\bar{q}^1, \bar{q}^2, \bar{q}^3)$ by means of the transformations

$$\bar{A}^j = \frac{\partial \bar{q}^j}{\partial q^i} A^i, \qquad j = 1,2,3, \tag{1.148}$$

the elements of the ordered triple are called the *contravariant components of a vector*.

Observe that a *vector is a vector*, that is, it is invariant with coordinate transformations. On the other hand, the *individual components* are not invariant. The terminology *covariant vector* and *contravariant vector* does not mean different vectors, but two different ways of describing the same vector.

If the coordinate systems involved in a transformation are all *rectangular Cartesian* coordinates, then the transformations will all be linear and orthogonal. It is therefore not necessary to distinguish between covariant and contravariant. Generally only subscripts are then used. The two transformation laws above become identical, and vectors under these conditions are called *Cartesian vectors*.

An entity may be a vector under one group of transformations, but not a vector under all groups of transformations. As an example consider a set of translations of Cartesian systems and the so-called *position vector* associated with each system.

The displacement of the system is given by

$$\mathbf{r}=\bar{\mathbf{r}}+\mathbf{r}_0$$

$$x^i\hat{\mathbf{i}}_i=\bar{x}^i\hat{\bar{\mathbf{i}}}_i+x_0^i\hat{\mathbf{i}}_i, \qquad \hat{\mathbf{i}}_i=\hat{\bar{\mathbf{i}}}_i \tag{1.149}$$

or

$$x^i=\bar{x}^i+x_0^i, \qquad i=1,2,3.$$

From this expression we deduce that

$$\frac{\partial x^i}{\partial \bar{x}^j}=\frac{\partial \bar{x}^i}{\partial \bar{x}^j}=\delta_j^i$$

$$\frac{\partial \bar{x}^i}{\partial x^j}=\frac{\partial x^i}{\partial x^j}=\delta_j^i. \tag{1.150}$$

The covariant and contravariant laws reduce to the same form, since we are dealing with Cartesian systems. Hence we have

$$\bar{A}^j=\delta_j^i A^i=A^j$$

$$\bar{A}_j=\delta_j^i A_i=A_j. \tag{1.151}$$

Thus the components of a vector remain invariant under the translation transformation. Yet the position vectors in the two systems are related by

$$x^j=\bar{x}^j+x_0^j. \tag{1.152}$$

Consequently the position vector is not really a vector under a translation transformation (this can be seen from Fig. 1.27, since $\mathbf{r}$ and $\bar{\mathbf{r}}$ are not identical). Thus it is more appropriate to refer to the position vector as a *position arrow*.

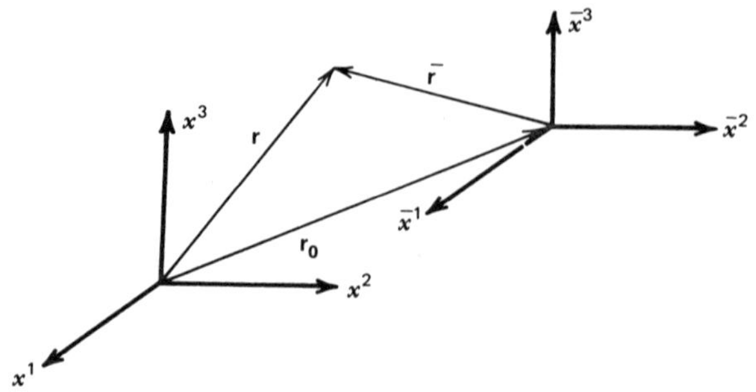

Figure 1.27 Translation of Cartesian systems.

1.4.13 Derivatives of Orthonormal Basis Vectors

The basis vectors associated with a curvilinear coordinate system vary, in general, in length and orientation from point to point in space. For orthonormal basis systems the basis vectors are always unit length, but their orientation may vary from point to point. Thus an orthonormal basis system can be conceived as a rigid triad that rotates from point to point in space. The change in an orthonormal basis vector can thus be brought about by a rigid-body rotation.

In a rotating rigid body the velocity at point **r** measured from the axis of rotation is related to the angular rotation vector by the familiar formula

$$\mathbf{V}=\boldsymbol{\omega}\times\mathbf{r}. \tag{1.153}$$

If we now set $\mathbf{V}=d\mathbf{r}/dt$ and $\boldsymbol{\omega}=d\boldsymbol{\Phi}/dt$ and eliminate the differential time dt, we have

$$d\mathbf{r}=d\boldsymbol{\Phi}\times\mathbf{r}, \tag{1.154}$$

where $d\boldsymbol{\Phi}$ is the differential angle of rotation, which can be regarded as a vector. In a rigid rotation **r** does not change in length, but only in orientation. Thus we can set $\mathbf{r}=\hat{\mathbf{e}}_i$ as one of the unit basis vectors in an orthonormal triad. We then obtain

$$d\hat{\mathbf{e}}_i=d\boldsymbol{\Phi}\times\hat{\mathbf{e}}_i, \qquad i=1,2,3 \tag{1.155}$$

as the formula that gives the change in a unit vector brought about by an infinitesimal rotation $d\boldsymbol{\Phi}$. The differential rotation can be obtained geometrically in many cases.

Consider first a *cylindrical* coordinate system, as shown in Fig. 1.28. In this system $\hat{\mathbf{e}}_z$ is a constant, but $\hat{\mathbf{e}}_R$ and $\hat{\mathbf{e}}_\phi$ vary in orientation with the angle ϕ. It is not difficult to see that a change $d\phi$ produces a different rotation given by

$$d\boldsymbol{\Phi}=d\phi\hat{\mathbf{e}}_z. \tag{1.156}$$

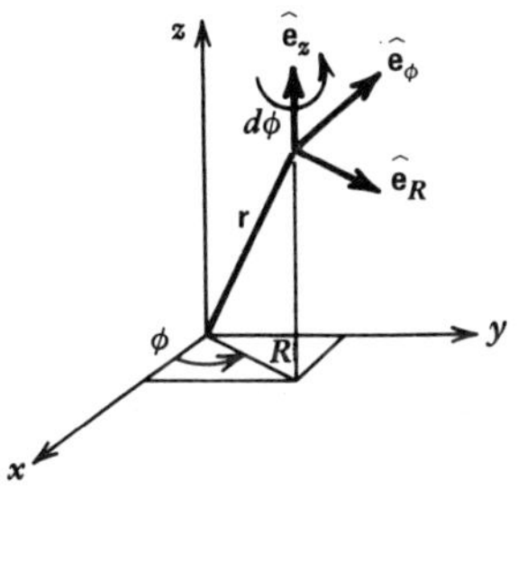

Figure 1.28 Cylindrical coordinate system.

Thus we have

$$d\hat{\mathbf{e}}_i = d\phi(\hat{\mathbf{e}}_z \times \hat{\mathbf{e}}_i), \tag{1.157}$$

and it follows that

$$\begin{aligned} d\hat{\mathbf{e}}_R &= d\phi\hat{\mathbf{e}}_\phi \\ d\hat{\mathbf{e}}_\phi &= -d\phi\hat{\mathbf{e}}_R \\ d\hat{\mathbf{e}}_z &= 0. \end{aligned} \tag{1.158}$$

It is easy to see that the changes in $\hat{\mathbf{e}}_R$ and $\hat{\mathbf{e}}_\phi$ are perpendicular to themselves.

The differential rotation $d\boldsymbol{\Phi}$ is also easy to obtain geometrically in a *spherical coordinate system* (see Fig. 1.29). In this system there are two angles that cause rotation, and we have

$$d\boldsymbol{\Phi} = d\phi\hat{\mathbf{e}}_z + d\theta\hat{\mathbf{e}}_\phi. \tag{1.159}$$

Representing $\hat{\mathbf{e}}_z$ in terms of the spherical orthonormal basis, we get

$$\hat{\mathbf{e}}_z = (\hat{\mathbf{e}}_z \cdot \hat{\mathbf{e}}_r)\hat{\mathbf{e}}_r + (\hat{\mathbf{e}}_z \cdot \hat{\mathbf{e}}_\theta)\hat{\mathbf{e}}_\theta + (\hat{\mathbf{e}}_z \cdot \hat{\mathbf{e}}_\phi)\hat{\mathbf{e}}_\phi. \tag{1.160}$$

From geometrical considerations we find

$$\begin{aligned} \hat{\mathbf{e}}_z \cdot \hat{\mathbf{e}}_r &= \cos\theta \\ \hat{\mathbf{e}}_z \cdot \hat{\mathbf{e}}_\theta &= -\sin\theta \\ \hat{\mathbf{e}}_z \cdot \hat{\mathbf{e}}_\phi &= 0. \end{aligned} \tag{1.161}$$

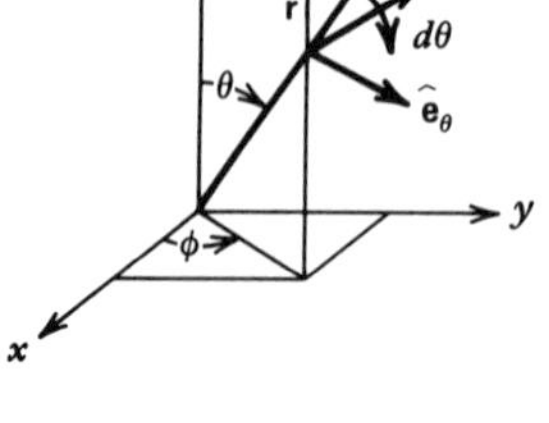

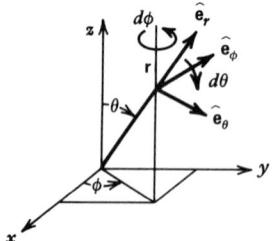

Figure 1.29 Spherical coordinate system.

Thus we obtain

$$d\mathbf{\Phi}=(d\phi\cos\theta)\hat{\mathbf{e}}_r-(d\phi\sin\theta)\hat{\mathbf{e}}_\theta+(d\theta)\hat{\mathbf{e}}_\phi. \tag{1.162}$$

It follows that

$$\begin{aligned} d\hat{\mathbf{e}}_r&=(d\theta)\hat{\mathbf{e}}_\theta+(d\phi\sin\theta)\hat{\mathbf{e}}_\phi\\ d\hat{\mathbf{e}}_\theta&=-(d\theta)\hat{\mathbf{e}}_r+(d\phi\cos\theta)\hat{\mathbf{e}}_\phi\\ d\hat{\mathbf{e}}_\phi&=-(d\phi\sin\theta)\hat{\mathbf{e}}_r-(d\phi\cos\theta)\hat{\mathbf{e}}_\theta. \end{aligned} \tag{1.163}$$

The changes in the unit vectors are again perpendicular to themselves.

This procedure can be set up for a general orthogonal curvilinear coordinate system. Here we make use of the curl of a vector to be discussed in Section 1.4.20. By taking the curl of the equation $\mathbf{V}=\boldsymbol{\omega}\times\mathbf{r}$, we can obtain

$$d\mathbf{\Phi}=\tfrac{1}{2}\,\mathrm{curl}\,(d\mathbf{r}). \tag{1.164}$$

In general orthogonal curvilinear coordinates, we have

$$d\mathbf{r}=h_1\,dq^1\hat{\mathbf{e}}_1+h_2\,dq^2\hat{\mathbf{e}}_2+h_3\,dq^3\hat{\mathbf{e}}_3 \tag{1.165}$$

and

$$\begin{aligned} \mathrm{curl}\,(d\mathbf{r})&=\frac{1}{h_1h_2h_3}\begin{vmatrix} h_1\hat{\mathbf{e}}_1 & h_2\hat{\mathbf{e}}_2 & h_3\hat{\mathbf{e}}_3\\ \dfrac{\partial}{\partial q^1} & \dfrac{\partial}{\partial q^2} & \dfrac{\partial}{\partial q^3}\\ h_1(h_1\,dq^1) & h_2(h_2\,dq^2) & h_3(h_3\,dq^3) \end{vmatrix}\\ &=\frac{\hat{\mathbf{e}}_1}{h_2h_3}\left[dq^3\frac{\partial}{\partial q^2}(h_3^2)-dq^2\frac{\partial}{\partial q^3}(h_2^2)\right]\\ &\quad+\frac{\hat{\mathbf{e}}_2}{h_1h_3}\left[dq^1\frac{\partial}{\partial q^3}(h_1^2)-dq^3\frac{\partial}{\partial q^1}(h_3^2)\right]\\ &\quad+\frac{\hat{\mathbf{e}}_3}{h_1h_2}\left[dq^2\frac{\partial}{\partial q^1}(h_2^2)-dq^1\frac{\partial}{\partial q^2}(h_1^2)\right]. \end{aligned} \tag{1.166}$$

Substituting into the expression $d\hat{\mathbf{e}}_i = \mathbf{d\Phi} \times \hat{\mathbf{e}}_i$ now gives

$$
\begin{aligned}
d\hat{\mathbf{e}}_1 &= \frac{\hat{\mathbf{e}}_2}{2h_1h_2}\left[dq^2\frac{\partial}{\partial q^1}(h_2^2) - dq^1\frac{\partial}{\partial q^2}(h_1^2)\right] \\
&\quad + \frac{\hat{\mathbf{e}}_3}{2h_1h_3}\left[dq^3\frac{\partial}{\partial q^1}(h_3^2) - dq^1\frac{\partial}{\partial q^3}(h_1^2)\right] \\
d\hat{\mathbf{e}}_2 &= \frac{\hat{\mathbf{e}}_1}{2h_1h_2}\left[dq^1\frac{\partial}{\partial q^2}(h_1^2) - dq^2\frac{\partial}{\partial q^1}(h_2^2)\right] \\
&\quad + \frac{\hat{\mathbf{e}}_3}{2h_2h_3}\left[dq^3\frac{\partial}{\partial q^2}(h_3^2) - dq^2\frac{\partial}{\partial q^3}(h_2^2)\right] \qquad (1.167) \\
d\hat{\mathbf{e}}_3 &= \frac{\hat{\mathbf{e}}_1}{2h_1h_3}\left[dq^1\frac{\partial}{\partial q^3}(h_1^2) - dq^3\frac{\partial}{\partial q^1}(h_3^2)\right] \\
&\quad + \frac{\hat{\mathbf{e}}_2}{2h_2h_3}\left[dq^2\frac{\partial}{\partial q^3}(h_2^2) - dq^3\frac{\partial}{\partial q^2}(h_3^2)\right].
\end{aligned}
$$

These results reduce to those previously obtained for cylindrical and spherical systems.

Since we can write in general

$$d\hat{\mathbf{e}}_i = \frac{\partial \hat{\mathbf{e}}_i}{\partial q^j}dq^j, \qquad (1.168)$$

comparison of this expression with Eqs. (1.167) allows the partial derivatives of the unit basis vectors to be identified:

$$
\begin{aligned}
&\frac{\partial \hat{\mathbf{e}}_1}{\partial q^1} = -\frac{\hat{\mathbf{e}}_2}{h_2}\frac{\partial h_1}{\partial q^2} - \frac{\hat{\mathbf{e}}_3}{h_3}\frac{\partial h_1}{\partial q^3}, \quad \frac{\partial \hat{\mathbf{e}}_1}{\partial q^2} = \frac{\hat{\mathbf{e}}_2}{h_1}\frac{\partial h_2}{\partial q^1}, \quad \frac{\partial \hat{\mathbf{e}}_1}{\partial q^3} = \frac{\hat{\mathbf{e}}_3}{h_1}\frac{\partial h_3}{\partial q^1} \\
&\frac{\partial \hat{\mathbf{e}}_2}{\partial q^1} = \frac{\hat{\mathbf{e}}_1}{h_2}\frac{\partial h_1}{\partial q^2}, \quad \frac{\partial \hat{\mathbf{e}}_2}{\partial q^2} = -\frac{\hat{\mathbf{e}}_1}{h_1}\frac{\partial h_2}{\partial q^1} - \frac{\hat{\mathbf{e}}_3}{h_3}\frac{\partial h_2}{\partial q^3}, \quad \frac{\partial \hat{\mathbf{e}}_2}{\partial q^3} = \frac{\hat{\mathbf{e}}_3}{h_2}\frac{\partial h_3}{\partial q^2} \qquad (1.169) \\
&\frac{\partial \hat{\mathbf{e}}_3}{\partial q^1} = \frac{\hat{\mathbf{e}}_1}{h_3}\frac{\partial h_1}{\partial q^3}, \quad \frac{\partial \hat{\mathbf{e}}_3}{\partial q^2} = \frac{\hat{\mathbf{e}}_2}{h^3}\frac{\partial h_2}{\partial q^3}, \quad \frac{\partial \hat{\mathbf{e}}_3}{\partial q^3} = -\frac{\hat{\mathbf{e}}_1}{h_1}\frac{\partial h_3}{\partial q^1} - \frac{\hat{\mathbf{e}}_2}{h_2}\frac{\partial h_3}{\partial q^2}.
\end{aligned}
$$

The generalization of these results to general curvilinear coordinate systems leads to a more abstract representation involving what is known as the *Christoffel symbol of the second kind* (see Section 1.4.15).

Exercises 1.4

1 Prove the following identities:

(a) $\dfrac{d}{dt}[\mathbf{ABC}]=\left[\dfrac{d\mathbf{A}}{dt}\mathbf{BC}\right]+\left[\mathbf{A}\dfrac{d\mathbf{B}}{dt}\mathbf{C}\right]+\left[\mathbf{AB}\dfrac{d\mathbf{C}}{dt}\right].$

(b) $\dfrac{d}{dt}\left[\mathbf{A}\dfrac{d\mathbf{A}}{dt}\dfrac{d^2\mathbf{A}}{dt^2}\right]=\left[\mathbf{A}\dfrac{d\mathbf{A}}{dt}\dfrac{d^3\mathbf{A}}{dt^3}\right].$

2 If x_i are rectangular Cartesian coordinates, and θ^i are the parabolic curvilinear coordinates defined by the transformation

$$x_1=\theta^1\theta^2\cos\theta^3$$

$$x_2=\theta^1\theta^2\sin\theta^3$$

$$x_3=\frac{1}{2}\left[(\theta^1)^2-(\theta^2)^2\right],$$

determine:

(a) The covariant base vectors $\mathbf{e}_i$.

(b) The components of the fundamental metric g_{ij}.

(c) The reciprocal base vectors.

(d) The arc length ds.

(e) The physical components of a vector.

(f) The scale factors.

3 Let x_i be the rectangular Cartesian coordinates, and ξ^i the elliptic cylindrical coordinates defined by

$$x_1=a\cosh\xi^1\cos\xi^2$$

$$x_2=a\sinh\xi^1\sin\xi^2$$

$$x_3=\xi^3,\qquad a=\text{constant}.$$

Determine:

(a) The covariant base vectors.

(b) The components of the fundamental metric g_{ij}.

(c) The reciprocal base vectors.

(d) The arc length.

4 If **A** and **B** are two vectors in a curvilinear coordinate system, find the cosine of the angle between **A** and **B** in terms of the components of **A** and **B**, and the fundamental metric g_{ij}.

5 Show that $\det(g_{ij})\equiv g=[\mathbf{e}_1\mathbf{e}_2\mathbf{e}_3]^2$.

6 Determine the transformation matrix between x_i and θ^i of Exercise 2.

7 Consider a curvilinear tetrahedron in the curvilinear coordinates q^1, q^2, and q^3, with sides $\mathbf{e}_1$, $\mathbf{e}_2$, and $\mathbf{e}_3$ along q^1, q^2, and q^3, respectively. Using the procedure of Exercise 1.1.2, show that the relation between the slant area S and the areas S_i bounded by the coordinate lines is given by

$$S\hat{\mathbf{n}}=\sqrt{g}\left(\mathbf{e}^3\,dq^1\,dq^2+\mathbf{e}^1\,dq^2\,dq^3+\mathbf{e}^2\,dq^3\,dq^1\right)$$

$$=\frac{\hat{S}_3}{\sqrt{g^{33}}}\mathbf{e}^3+\frac{\hat{S}_2}{\sqrt{g^{22}}}\mathbf{e}^2+\frac{\hat{S}_1}{\sqrt{g^{11}}}\mathbf{e}^1,$$

where

$$\hat{S}_i=S_i\sqrt{g^{ii}} \qquad \text{(no summation)}.$$

8 Consider two rectangular Cartesian coordinate systems that are rotated with respect to each other and have a common origin. Let one system be denoted as a barred system, so that a position vector can be written in each of the systems as

$$\mathbf{r}=x_i\hat{\mathbf{e}}_i$$

$$=\bar{x}_j\hat{\bar{\mathbf{e}}}_j,$$

where $\{\hat{\mathbf{e}}_i\}$ and $\{\hat{\bar{\mathbf{e}}}_j\}$ are the respective orthonormal Cartesian bases in the unbarred and barred systems. By requiring that the position vector $\mathbf{r}$ be invariant under a rotation of the coordinate systems, deduce that the transformation between the coordinates is given by

$$\bar{x}_1=a_{11}x_1+a_{12}x_2+a_{13}x_3$$

$$\bar{x}_2=a_{21}x_1+a_{22}x_2+a_{23}x_3$$

$$\bar{x}_3=a_{31}x_1+a_{32}x_2+a_{33}x_3$$

or, more compactly,

$$\bar{x}_i=a_{ij}x_j, \qquad i=1,2,3,$$

where the terms a_{ij} can be identified as the direction cosines

$$a_{ij}\equiv\hat{\bar{\mathbf{e}}}_i\cdot\hat{\mathbf{e}}_j=\cos\left(\hat{\bar{\mathbf{e}}}_i,\hat{\mathbf{e}}_j\right).$$

Deduce further that the basis vectors obey the same transformation

$$\hat{\bar{\mathbf{e}}}_i=a_{ij}\hat{\mathbf{e}}_j$$

and that the following orthogonality conditions hold:

$$a_{ij}a_{kj}=\delta_{ik}.$$

9 Suppose that the position vector $\mathbf{r}$ of a material point in a body is given by

$$\mathbf{r}=x_i\hat{\mathbf{e}}_i+u_i\hat{\mathbf{e}}_i,$$

where $\hat{\mathbf{e}}_i$ are the orthonormal Cartesian basis vectors, x_i are the Cartesian components, and u_i are called the displacement components of the point. Compute the components of the metric tensor g_{ij}, and $\gamma_{ij}\equiv\frac{1}{2}(g_{ij}-\delta_{ij})$.

10 (*Continuation of Exercise* 9) The *principal invariants* of g_{ij} are defined by

$$I_1=g_{ii},\qquad I_2=\frac{1}{2}(g_{ii}g_{jj}-g_{ij}g_{ij}),\qquad I_3=g\equiv\det(g_{ij}).$$

Express I_1, I_2, and I_3 in terms of γ_{ij} defined in Exercise 9.

1.4.14 Derivatives of Vectors in Rotating Reference Frames

In setting up physical problems, the motion of a frame of reference is of importance. In particular we wish to examine the rate of change of a vector as it is perceived by two observers, each in a different reference frame rotating with respect to one another.

Let us denote the nonrotating reference system by unbarred quantities and the rotating reference frame by barred quantities. Let $\boldsymbol{\omega}$ denote the angular rotation of the barred system with respect to the unbarred system (see Fig. 1.30).

In the nonrotating system the basis vectors $\{\mathbf{e}_1,\mathbf{e}_2,\mathbf{e}_3\}$ do not change with time. In the rotating system the basis vectors $\{\bar{\mathbf{e}}_1,\bar{\mathbf{e}}_2,\bar{\mathbf{e}}_3\}$ are rotating with the angular velocity $\boldsymbol{\omega}$.

The time derivatives now can be written as

$$\left.\begin{aligned}\mathbf{A}&=A^i\mathbf{e}_i\\ \frac{d\mathbf{A}}{dt}&=\frac{dA^i}{dt}\mathbf{e}_i\end{aligned}\right\},\qquad \text{nonrotating system}$$
$$\left.\begin{aligned}\mathbf{A}&=\bar{A}^i\bar{\mathbf{e}}_i\\ \frac{d\mathbf{A}}{dt}&=\frac{d\bar{A}^i}{dt}\bar{\mathbf{e}}_i+\bar{A}^i\frac{d\bar{\mathbf{e}}_i}{dt}\end{aligned}\right\},\qquad \text{rotating system.} \tag{1.170}$$

The rate of change $d\bar{\mathbf{e}}_i/dt$ is given by

$$\frac{d\bar{\mathbf{e}}_i}{dt}=\boldsymbol{\omega}\times\bar{\mathbf{e}}_i \tag{1.171}$$

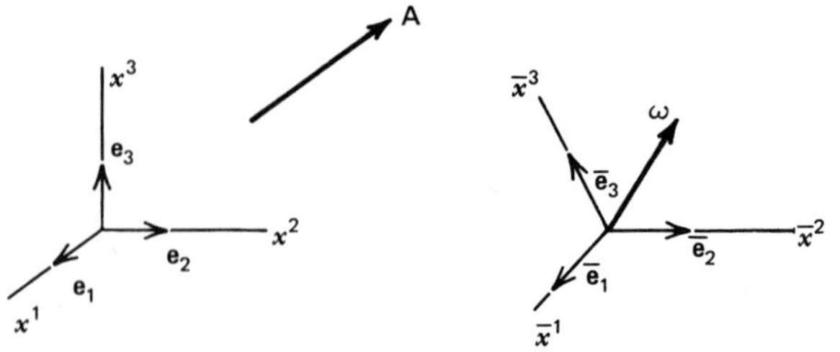

Figure 1.30 Rotating and nonrotating reference systems.

because the change is brought about by a rigid-body rotation. To an observer in the rotating frame, however, the basis vectors appear to be constant. Thus we write

$$\frac{d\bar{A}^i}{dt}\bar{\mathbf{e}}_i \equiv \left(\frac{d\mathbf{A}}{dt}\right)_{\text{rot}}. \tag{1.172}$$

The relationship of the time derivatives in the two frames is thus given by

$$\begin{aligned}\left(\frac{d\mathbf{A}}{dt}\right)_{\text{nonrot}} &= \left(\frac{d\mathbf{A}}{dt}\right)_{\text{rot}} + \bar{A}^i(\boldsymbol{\omega}\times\bar{\mathbf{e}}_i)\\ &= \left(\frac{d\mathbf{A}}{dt}\right)_{\text{rot}} + \boldsymbol{\omega}\times\mathbf{A}.\end{aligned} \tag{1.173}$$

Consider now the velocity vector as an example. Let the two frames of reference have the same origin. Then the position vector is given by $\mathbf{r}=\bar{\mathbf{r}}$. It follows that

$$\left(\frac{d\mathbf{r}}{dt}\right)_{\text{nonrot}} = \left(\frac{d\mathbf{r}}{dt}\right)_{\text{rot}} + \boldsymbol{\omega}\times\mathbf{r}. \tag{1.174}$$

The term $(d\mathbf{r}/dt)_{\text{rot}}$ is the velocity of a point as it is perceived by an observer in the rotating system. Thus we write

$$(\mathbf{V})_{\text{nonrot}} = (\mathbf{V})_{\text{rot}} + \boldsymbol{\omega}\times\mathbf{r}. \tag{1.175}$$

If $(\mathbf{V})_{\text{rot}}$ is zero, then the point in question is fixed with respect to the rotating system, and the above expression reduces to that of a rotating rigid body.

The accelerations in the two systems can be obtained by applying the operator (which holds for time derivatives on vector quantities)

$$\left(\frac{d}{dt}\right)_{\text{nonrot}} = \left(\frac{d}{dt}\right)_{\text{rot}} + \boldsymbol{\omega}\times \tag{1.176}$$

to the last equation above. We get

$$\left(\frac{d^2}{dt^2}\mathbf{r}\right)_{\text{nonrot}}=\frac{d}{dt}\left(\frac{d\mathbf{r}}{dt}\right)_{\text{rot}}+\frac{d}{dt}(\boldsymbol{\omega}\times\mathbf{r})$$

$$=\frac{d}{dt}\left(\frac{d\mathbf{r}}{dt}\right)_{\text{rot}}+\boldsymbol{\omega}\times\frac{d\mathbf{r}}{dt}+\frac{d\boldsymbol{\omega}}{dt}\times\mathbf{r}$$

$$=\left(\frac{d^2\mathbf{r}}{dt^2}\right)_{\text{rot}}+\boldsymbol{\omega}\times\left(\frac{d\mathbf{r}}{dt}\right)_{\text{rot}}+\boldsymbol{\omega}\times\left[\left(\frac{d\mathbf{r}}{dt}\right)_{\text{rot}}+\boldsymbol{\omega}\times\mathbf{r}\right]+\frac{d\boldsymbol{\omega}}{dt}\times\mathbf{r},$$

or finally,

$$\left(\frac{d^2\mathbf{r}}{dt^2}\right)_{\text{nonrot}}=\left(\frac{d^2\mathbf{r}}{dt^2}\right)_{\text{rot}}+2\boldsymbol{\omega}\times\left(\frac{d\mathbf{r}}{dt}\right)_{\text{rot}}+\boldsymbol{\omega}\times(\boldsymbol{\omega}\times\mathbf{r})+\frac{d\boldsymbol{\omega}}{dt}\times\mathbf{r}. \tag{1.177a}$$

Thus the acceleration in the nonrotating frame is made up of four parts:

$\left(\frac{d^2\mathbf{r}}{dt^2}\right)_{\text{rot}}$,	acceleration relative to rotating reference frame
$2\boldsymbol{\omega}\times\left(\frac{d\mathbf{r}}{dt}\right)_{\text{rot}}$,	Coriolis acceleration
$\boldsymbol{\omega}\times(\boldsymbol{\omega}\times\mathbf{r})$,	centripetal (or negative centrifugal) acceleration
$\frac{d\boldsymbol{\omega}}{dt}\times\mathbf{r}$,	angular acceleration.

If the origin of the rotating frame of reference is different from that of the fixed frame of reference, we have

$$\left(\frac{d^2\mathbf{r}}{dt^2}\right)_{\text{nonrot}}=\frac{d^2\mathbf{R}}{dt^2}+\left(\frac{d^2\mathbf{r}}{dt^2}\right)_{\text{rot}}+2\boldsymbol{\omega}\times\left(\frac{d\mathbf{r}}{dt}\right)_{\text{rot}}+\boldsymbol{\omega}\times(\boldsymbol{\omega}\times\mathbf{r})+\frac{d\boldsymbol{\omega}}{dt}\times\mathbf{r}, \tag{1.177b}$$

where $\mathbf{R}$ is the position vector of the origin of the rotating frame of reference with respect to the nonrotating frame of reference, and $d^2\mathbf{R}/dt^2$ is the absolute acceleration of the origin of the rotating frame of reference.

If we regard this result in the context of Newton's second law, which must be written as $\mathbf{F}=m\mathbf{a}$ in an inertial or nonrotating reference frame, we have

$$\mathbf{F}=m\left(\frac{d^2\mathbf{r}}{dt^2}\right)_{\text{nonrot}}$$

$$=m\left(\frac{d^2\mathbf{r}}{dt^2}\right)_{\text{rot}}+2m\boldsymbol{\omega}\times\left(\frac{d\mathbf{r}}{dt}\right)_{\text{rot}}+m\boldsymbol{\omega}\times(\boldsymbol{\omega}\times\mathbf{r})+m\frac{d\boldsymbol{\omega}}{dt}\times\mathbf{r}$$

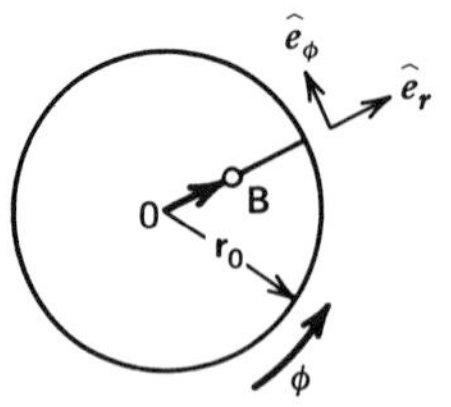

Figure 1.31 A ball with radial harmonic motion relative to a harmonically oscillating disk.

or

$$\mathbf{F}-2m\boldsymbol{\omega}\times\left(\frac{d\mathbf{r}}{dt}\right)_{\text{rot}}-m\boldsymbol{\omega}\times(\boldsymbol{\omega}\times\mathbf{r})-m\frac{d\boldsymbol{\omega}}{dt}\times\mathbf{r}=m\left(\frac{d^2\mathbf{r}}{dt^2}\right)_{\text{rot}}. \tag{1.178}$$

To an observer in the rotating reference frame it appears that the left-hand side causes changes in velocity. Thus the terms on the left, besides **F**, are called *inertial* or *effective forces*. In particular, $[-2m\boldsymbol{\omega}\times(d\mathbf{r}/dt)_{\text{rot}}]$ is called the *Coriolis force*, and $[-m\boldsymbol{\omega}\times(\boldsymbol{\omega}\times\mathbf{r})]$ is called the *centrifugal force*.

Example 1.12 A ball B moves along a straight radial groove in a circular disk of radius r_0 which is pivoted about a perpendicular axis through its center O (see Fig. 1.31). The ball moves relative to the disk such that

$$r=\frac{r_0}{2}(1+\sin\omega t),$$

and the disk rotates according to

$$\phi=\phi_0\sin\omega t.$$

We wish to find the expression for the absolute acceleration of B.

Let the rotating coordinate system be fixed at O in the disk such that the origins O and O' coincide. We have

$$\mathbf{r}=r\hat{\mathbf{e}}_r=\frac{r_0}{2}(1+\sin\omega t)\hat{\mathbf{e}}_r$$

$$\boldsymbol{\omega}=\dot{\phi}\hat{\mathbf{e}}_z=\omega\phi_0\cos\omega t\,\hat{\mathbf{e}}_z.$$

Now using Eq. (1.177b) we have

$$\begin{aligned}\mathbf{a}=0&+\left(-\frac{r_0\omega^2}{2}\sin\omega t\hat{\mathbf{e}}_r\right)+r_0\omega^2\phi_0\cos^2\omega t\hat{\mathbf{e}}_\phi\\&+\left[-\frac{r_0\omega^2}{2}\phi_0^2(1+\sin\omega t)\cos^2\omega t\hat{\mathbf{e}}_r\right]+\left[-\frac{r_0\omega^2}{2}\phi_0(1+\sin\omega t)\sin\omega t\hat{\mathbf{e}}_\phi\right]\\&=-\frac{r_0\omega^2}{2}\left[\sin\omega t+\phi_0^2(1+\sin\omega t)\cos^2\omega t\right]\hat{\mathbf{e}}_r\\&+r_0\omega^2\phi_0\left[\cos^2\omega t-\frac{1}{2}(1+\sin\omega t)\sin\omega t\right]\hat{\mathbf{e}}_\phi.\end{aligned}$$

■

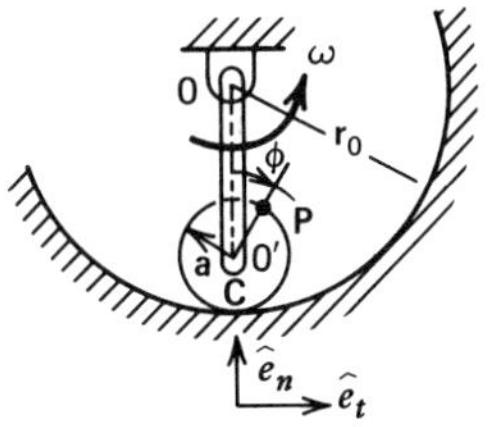

Figure 1.32 A wheel rolling on a fixed circular track.

Example 1.13 Consider a wheel of radius a rolling on the inside of a fixed circular track of radius r_0. An arm connecting the fixed point O and the wheel hub O' moves at a constant angular velocity ω. Assuming that there is no slipping between wheel and track, find the acceleration of a point P on the circumference of the wheel. The position of P relative to the arm is given by the angle ϕ.

We choose the origin of the fixed system at O, and that of the rotating system at O'. The normal and tangential unit vectors are taken as shown in Fig. 1.32.

Since there is no slipping of the wheel on the track, the velocity of point C (instantaneous center) with respect to the fixed system is given by

$$V_C = r_0\omega.$$

The velocity of the same point in the moving system is given by

$$V_C = \dot{\phi}a.$$

Hence we have a relation between $\dot{\phi}$ and ω:

$$\dot{\phi} = \frac{\omega r_0}{a}.$$

The position vector from O to O' is given by

$$\mathbf{R} = -(r_0 - a)\hat{\mathbf{e}}_n,$$

and the position vector of point P from O' is

$$\mathbf{r} = a(\sin\phi\,\hat{\mathbf{e}}_t + \cos\phi\,\hat{\mathbf{e}}_n).$$

We now proceed to calculate the acceleration of P using Eq. (1.177b). Let $\hat{\mathbf{e}}_b$ be the unit vector perpendicular to $\hat{\mathbf{e}}_t$ and $\hat{\mathbf{e}}_n$,

$$\hat{\mathbf{e}}_b \equiv \hat{\mathbf{e}}_t \times \hat{\mathbf{e}}_n,$$

so that

$$\boldsymbol{\omega} = \omega\hat{\mathbf{e}}_b, \qquad \frac{d\hat{\mathbf{e}}_n}{dt} = \boldsymbol{\omega}\times\hat{\mathbf{e}}_n = -\omega\hat{\mathbf{e}}_t, \qquad \frac{d\hat{\mathbf{e}}_t}{dt} = \boldsymbol{\omega}\times\hat{\mathbf{e}}_t = \omega\hat{\mathbf{e}}_n, \qquad \frac{d\hat{\mathbf{e}}_b}{dt} = 0.$$

We have

$$\frac{d^2\mathbf{R}}{dt^2}=(r_0-a)\omega^2\hat{\mathbf{e}}_n$$

$$\frac{d\boldsymbol{\omega}}{dt}\times\mathbf{r}=0 \qquad \left(\text{since}\frac{d\boldsymbol{\omega}}{dt}=0\right)$$

$$\boldsymbol{\omega}\times(\boldsymbol{\omega}\times\mathbf{r})=-a\omega^2(\sin\phi\hat{\mathbf{e}}_t+\cos\phi\hat{\mathbf{e}}_n)$$

$$\left(\frac{d^2\mathbf{r}}{dt^2}\right)_{\text{rot}}=-a(\dot{\phi})^2(\sin\phi\hat{\mathbf{e}}_t+\cos\phi\hat{\mathbf{e}}_n)=-\frac{r_0^2\omega^2}{a}(\sin\phi\hat{\mathbf{e}}_t+\cos\phi\hat{\mathbf{e}}_n)$$

$$2\boldsymbol{\omega}\times\left(\frac{d\mathbf{r}}{dt}\right)_{\text{rot}}=2r_0\omega^2(\sin\phi\hat{\mathbf{e}}_t+\cos\phi\hat{\mathbf{e}}_n).$$

Hence the acceleration of the point P is

$$\mathbf{a}=\left(2r_0-a-\frac{r_0^2}{a}\right)\omega^2\sin\phi\hat{\mathbf{e}}_t+\left[\left(2r_0-a-\frac{r_0^2}{a}\right)\omega^2\cos\phi+(r_0-a)\omega^2\right]\hat{\mathbf{e}}_n. \quad \blacksquare$$

1.4.15 Derivatives of Unitary Basis Vectors

In a general curvilinear coordinate system we are interested in derivatives of the basis vectors with respect to the coordinates. Consider the partial derivative of the ith unitary vector with respect to the jth coordinate. We can express this vector as follows:

$$\frac{\partial\mathbf{e}_i}{\partial q^j}=\left(\frac{\partial\mathbf{e}_i}{\partial q^j}\cdot\mathbf{e}^k\right)\mathbf{e}_k$$

$$=\left\{\begin{matrix}k\\ i\ j\end{matrix}\right\}\mathbf{e}_k, \tag{1.179}$$

where

$$\left\{\begin{matrix}k\\ i\ j\end{matrix}\right\}\equiv\frac{\partial\mathbf{e}_i}{\partial q^j}\cdot\mathbf{e}^k. \tag{1.180}$$

The symbol $\left\{\begin{matrix}k\\ i\ j\end{matrix}\right\}$ is called the *Christoffel symbol of the second kind*. Note that the index k refers to the summation index, whereas i and j refer to the ith unitary vector and the jth coordinate. Equation (1.179) is a generalization of Eqs. (1.169) which apply to orthonormal basis systems.

The Christoffel symbol can be evaluated in terms of the metric coefficients. To do this, we first write $\mathbf{e}^k = g^{kr}\mathbf{e}_r$ so that only unitary basis vectors are involved:

$$\left\{ \begin{matrix} k \\ i\ j \end{matrix} \right\} = g^{kr} \frac{\partial \mathbf{e}_i}{\partial q^j} \cdot \mathbf{e}_r. \tag{1.181}$$

We now divide this into two equal parts:

$$\left\{ \begin{matrix} k \\ i\ j \end{matrix} \right\} = \frac{1}{2} g^{kr} \left[\frac{\partial \mathbf{e}_i}{\partial q^j} \cdot \mathbf{e}_r + \frac{\partial \mathbf{e}_i}{\partial q^j} \cdot \mathbf{e}_r \right]. \tag{1.182}$$

We observe that

$$\frac{\partial \mathbf{e}_i}{\partial q^j} \equiv \frac{\partial}{\partial q^j}\left(\frac{\partial \mathbf{r}}{\partial q^i} \right) = \frac{\partial}{\partial q^i}\left(\frac{\partial \mathbf{r}}{\partial q^j} \right) = \frac{\partial \mathbf{e}_j}{\partial q^i}, \tag{1.183}$$

and thus the indices in this derivative can be interchanged. With this result we can rewrite Eq. (1.182) as

$$\begin{aligned} \left\{ \begin{matrix} k \\ i\ j \end{matrix} \right\} &= \frac{1}{2} g^{kr} \left[\frac{\partial \mathbf{e}_j}{\partial q^i} \cdot \mathbf{e}_r + \frac{\partial \mathbf{e}_i}{\partial q^j} \cdot \mathbf{e}_r \right] \\ &= \frac{1}{2} g^{kr} \left[\frac{\partial}{\partial q^i}(\mathbf{e}_j \cdot \mathbf{e}_r) + \frac{\partial}{\partial q^j}(\mathbf{e}_i \cdot \mathbf{e}_r) - \mathbf{e}_j \cdot \frac{\partial \mathbf{e}_r}{\partial q^i} - \mathbf{e}_i \cdot \frac{\partial \mathbf{e}_r}{\partial q^j} \right] \\ &= \frac{1}{2} g^{kr} \left[\frac{\partial}{\partial q^i}(g_{jr}) + \frac{\partial}{\partial q^j}(g_{ir}) - \frac{\partial}{\partial q^r}(g_{ij}) \right], \end{aligned} \tag{1.184}$$

where we accounted for the definition $g_{ij} \equiv \mathbf{e}_i \cdot \mathbf{e}_j$. Thus only metric coefficients are involved, and the summation is only over the index r. Expression (1.184) is symmetric in i and j.

It is useful to find alternative expressions for the Christoffel symbol. We note that

$$\mathbf{e}_i \cdot \mathbf{e}^k = \delta_i^k. \tag{1.185}$$

Thus

$$\frac{\partial}{\partial q^j}(\mathbf{e}_i \cdot \mathbf{e}^k) = 0$$

and

$$\mathbf{e}_i \cdot \frac{\partial \mathbf{e}^k}{\partial q^j} = -\frac{\partial \mathbf{e}_i}{\partial q^j} \cdot \mathbf{e}^k$$

$$= -\left\{ \begin{matrix} k \\ i\ j \end{matrix} \right\}. \tag{1.186}$$

It follows that the derivative of the kth dual unitary vector is given by

$$\frac{\partial \mathbf{e}^k}{\partial q^j} = -\left\{ \begin{matrix} k \\ i\ j \end{matrix} \right\} \mathbf{e}^i \tag{1.187a}$$

or

$$\frac{\partial \mathbf{e}^i}{\partial q^j} = -\left\{ \begin{matrix} i \\ k\ j \end{matrix} \right\} \mathbf{e}^k. \tag{1.187b}$$

It can be shown that

$$\left\{ \begin{matrix} k \\ i\ k \end{matrix} \right\} = \frac{\partial}{\partial q^i} \ln \sqrt{g}, \tag{1.188}$$

where g is the determinant of g_{ij}. This result is very useful in numerous manipulations. To prove this we proceed as follows:

We first observe that the evaluation of a determinant by Cramer's rule gives (see Exercise 1.3.18)

$$g = \sum_{k=1}^{3} g_{kr} \tilde{M}_{kr}, \tag{1.189}$$

where $\tilde{M}_{kr}$ is the cofactor of the matrix g_{kr}. We note that $\tilde{M}_{kr}$ does not depend on the (k, r) element g_{kr} itself. Thus we have

$$\frac{\partial g}{\partial g_{kr}} = \tilde{M}_{kr}. \tag{1.190}$$

The partial derivative of g with respect to a coordinate q^i is

$$\frac{\partial g}{\partial q^i} = \frac{\partial g}{\partial g_{kr}} \frac{\partial g_{kr}}{\partial q^i} = \tilde{M}_{kr} \frac{\partial g_{kr}}{\partial q^i} = g g^{kr} \frac{\partial g_{kr}}{\partial q^i}, \tag{1.191}$$

where we utilized Eq. (1.98) in the last step. Now from Eq. (1.181) we see that

$$\left\{ \begin{matrix} k \\ i\ k \end{matrix} \right\} = g^{kr} \frac{\partial \mathbf{e}_i}{\partial q^k} \cdot \mathbf{e}_r = g^{kr} \frac{\partial \mathbf{e}_k}{\partial q^i} \cdot \mathbf{e}_r = g^{rk} \frac{\partial \mathbf{e}_r}{\partial q^i} \cdot \mathbf{e}_k. \tag{1.192}$$

Because the metric tensor is symmetric, we also have

$$\left\{ {k \atop i\ k} \right\} = \frac{1}{2} g^{kr} \frac{\partial}{\partial q^i} (\mathbf{e}_k \cdot \mathbf{e}_r), \tag{1.193}$$

and substitution of this result into Eq. (1.191) leads directly to Eq. (1.188). In the general case where g might be negative, $\sqrt{g}$ should be replaced by $\sqrt{|g|}$ or $\sqrt{-g}$.

The so-called Christoffel symbol of the first kind is defined by

$$[ij,k] = \frac{\partial \mathbf{r}}{\partial q^k} \cdot \frac{\partial^2 \mathbf{r}}{\partial q^i \partial q^j} \tag{1.194}$$

and is related to the Christoffel symbol of the second kind by

$$[ij,m] = g_{km} \left\{ {k \atop i\ j} \right\} = \frac{1}{2} \left\{ \frac{\partial g_{ik}}{\partial q^j} + \frac{\partial g_{jk}}{\partial q^i} - \frac{\partial g_{ij}}{\partial q^k} \right\}. \tag{1.195}$$

Note that the Christoffel symbols depend only on the metric tensor and its derivatives.

In orthogonal curvilinear coordinates the Christoffel symbols may be identified as

$$\left\{ {1 \atop 1\ 2} \right\} = \frac{1}{h_1} \frac{\partial h_1}{\partial q^2}, \quad \left\{ {2 \atop 1\ 1} \right\} = -\frac{h_1}{h_2^2} \frac{\partial h_1}{\partial q^2}, \quad \left\{ {1 \atop 1\ 1} \right\} = \frac{1}{h_1} \frac{\partial h_1}{\partial q^1}, \quad \left\{ {3 \atop 1\ 2} \right\} = 0$$

$$[11,1] = h_1 \frac{\partial h_1}{\partial q^1}, \qquad [11,2] = -h_1 \frac{\partial h_1}{\partial q^2}, \qquad [12,1] = h_1 \frac{\partial h_1}{\partial q^2}, \qquad [12,3] = 0, \tag{1.196}$$

and so on. The derivatives of the orthogonal base vectors become

$$\begin{aligned} \frac{\partial \mathbf{e}_1}{\partial q^1} &= \frac{1}{h_1} \frac{\partial h_1}{\partial q^1} \mathbf{e}_1 - \frac{h_1}{h_2^2} \frac{\partial h_1}{\partial q^2} \mathbf{e}_2 - \frac{h_1}{h_3^2} \frac{\partial h_1}{\partial q^3} \mathbf{e}_3 \\ \frac{\partial \mathbf{e}_1}{\partial q^2} &= \frac{1}{h_1} \frac{\partial h_1}{\partial q^2} \mathbf{e}_1 + \frac{1}{h_2} \frac{\partial h_2}{\partial q^1} \mathbf{e}_2 \\ \frac{\partial \mathbf{e}_1}{\partial q^3} &= \frac{1}{h_1} \frac{\partial h_1}{\partial q^3} \mathbf{e}_1 + \frac{1}{h_3} \frac{\partial h_3}{\partial q^1} \mathbf{e}_3, \text{ etc.} \end{aligned} \tag{1.197}$$

These reduce to Eqs. (1.169) when

$$\mathbf{e}_1 = h_1\hat{\mathbf{e}}_1$$

$$\mathbf{e}_2 = h_2\hat{\mathbf{e}}_2$$

$$\mathbf{e}_3 = h_3\hat{\mathbf{e}}_3$$

are taken into account.

Exercises 1.5

1 The position vector in cylindrical coordinates is given by

$$\mathbf{r} = R\hat{\mathbf{e}}_R + z\hat{\mathbf{e}}_z.$$

Show that the velocity and Newton's second law of motion

$$\mathbf{V} = \frac{d\mathbf{r}}{dt}$$

$$\mathbf{F} = m\frac{d^2\mathbf{r}}{dt^2}$$

for a point particle of mass m can be written in component form in cylindrical coordinates as

$$V_R = \dot{R}$$

$$V_\phi = R\dot{\phi}$$

$$V_z = \dot{z}$$

$$F_R = m(\ddot{R} - R\dot{\phi}^2)$$

$$F_\phi = m(R\ddot{\phi} + 2\dot{R}\dot{\phi})$$

$$F_z = m\ddot{z}.$$

2 The position vector in spherical coordinates is given by

$$\mathbf{r} = r\hat{\mathbf{e}}_r.$$

Show that the velocity and Newton's second law of motion can be written in component form in spherical coordinates as

$$V_r = \dot{r}$$

$$V_\theta = r\dot{\theta}$$

$$V_\phi = r\dot{\phi}\sin\theta$$

$$F_r = m(\ddot{r} - r\dot{\theta}^2 - r\dot{\phi}^2\sin^2\theta)$$

$$F_\theta = m(r\ddot{\theta} + 2\dot{r}\dot{\theta} - r\dot{\phi}^2\cos\theta\sin\theta)$$

$$F_\phi = m(r\ddot{\phi}\sin\theta + 2\dot{r}\dot{\phi}\sin\theta + 2r\dot{\phi}\dot{\theta}\cos\theta).$$

3 In connection with Exercises 1 and 2 let a noninertial reference system be attached to the basis vectors. Then these coordinate systems rotate with the angular velocities

$$\boldsymbol{\omega} = \frac{d\boldsymbol{\Phi}}{dt} = \dot{\phi}\hat{\mathbf{e}}_z$$

$$\boldsymbol{\omega} = \frac{d\boldsymbol{\Phi}}{dt} = \dot{\phi}\cos\theta\,\hat{\mathbf{e}}_r - \dot{\phi}\sin\theta\,\hat{\mathbf{e}}_\theta + \dot{\theta}\hat{\mathbf{e}}_\phi$$

for the cylindrical and spherical systems, respectively. Show that when the equations for a noninertial reference system are utilized,

$$\mathbf{V} = \left(\frac{d\mathbf{r}}{dt}\right)_{\text{rot}} + \boldsymbol{\omega}\times\mathbf{r}$$

$$\mathbf{F} = m\left[\left(\frac{d^2\mathbf{r}}{dt^2}\right)_{\text{rot}} + 2\boldsymbol{\omega}\times\left(\frac{d\mathbf{r}}{dt}\right)_{\text{rot}} + \boldsymbol{\omega}\times(\boldsymbol{\omega}\times\mathbf{r}) + \frac{d\boldsymbol{\omega}}{dt}\times\mathbf{r}\right],$$

the same results as for Exercises 1 and 2 are obtained. Identify the Coriolis and centrifugal forces.

4 Consider a proton of mass m projected into a constant magnetic field given by $\mathbf{B} = B_z\hat{\mathbf{e}}_z$ with an initial velocity of $\mathbf{V}(0) = \dot{x}_0\hat{\mathbf{e}}_x + \dot{z}_0\hat{\mathbf{e}}_z$. From elementary physics the force on the proton is given by the Lorentz force

$$\mathbf{F} = \frac{e}{c}(\mathbf{V}\times\mathbf{B}) \qquad \text{(Gaussian cgs units)}$$

where e is the charge on the proton and c is the speed of light.

(a) By means of Newton's second law of motion, find the differential equations that the Cartesian components of the position vector $\mathbf{r} = x\hat{\mathbf{e}}_x + y\hat{\mathbf{e}}_y + z\hat{\mathbf{e}}_z$ must satisfy.

(b) Show that a solution to these equations that satisfies the velocity initial condition is

$$x = \frac{\dot{x}_0}{\omega}\sin\omega t, \qquad y = \frac{\dot{x}_0}{\omega}\cos\omega t, \qquad z = \dot{z}_0 t,$$

where $\omega^2 \equiv eB_z/mc$. Show that this solution describes a circular helix in space.

5 Consider two point particles interacting with each other by a force directed along their line of centers with a magnitude depending on the distance between them. Let $\mathbf{r}_1$ denote the position of the particle with mass m_1, and $\mathbf{r}_2$ the position of the particle with mass m_2 as shown in the sketch.

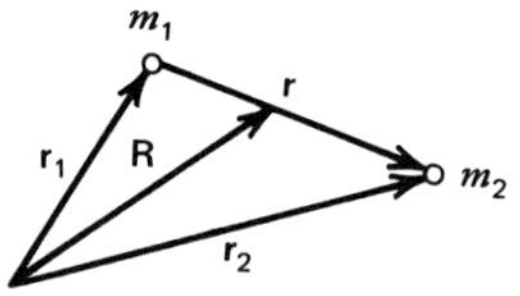

Define

$$\mathbf{r} \equiv \mathbf{r}_2 - \mathbf{r}_1$$

$$\mathbf{R} \equiv \frac{m_1\mathbf{r}_1 + m_2\mathbf{r}_2}{m_1 + m_2},$$

where $\mathbf{r}$ is the position of particle 2 relative to particle 1 and $\mathbf{R}$ is the center of mass of the two particles. The equations of motion are

$$m_1\frac{d^2\mathbf{r}_1}{dt^2} = F(r)\frac{\mathbf{r}}{r}$$

$$m_2\frac{d^2\mathbf{r}_2}{dt^2} = -F(r)\frac{\mathbf{r}}{r}.$$

(a) Deduce that in terms of $\mathbf{r}$ and $\mathbf{R}$ these equations reduce to

$$\ddot{\mathbf{R}} = 0$$

$$\mu\ddot{\mathbf{r}} = -F(r)\frac{\mathbf{r}}{r},$$

where $\mu \equiv m_1 m_2/(m_1 + m_2)$ is the *reduced mass*. Thus the center-of-mass of the two particles is unaccelerated, and the motion of particle 2 relative to particle 1 behaves as if particle 2 had a reduced mass μ with the same force exerted on it as before.

(b) Let $\mathbf{L} \equiv \mathbf{r} \times \dot{\mathbf{r}}$ denote the angular momentum of particle 2 relative to particle 1 per unit mass. Deduce by means of the equation of relative motion that

$$\frac{d\mathbf{L}}{dt} = 0$$

and hence that the angular momentum $\mathbf{L}$ is a constant and the motion takes place in a plane perpendicular to $\mathbf{L}$.

(c) The positions of $\mathbf{r}(t)$ and $\mathbf{r}(t+dt)$ at two neighboring instants of time are shown below:

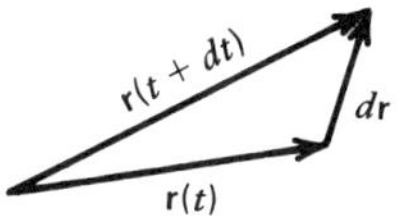

The area of this infinitesimal triangle is

$$d\mathbf{A} \equiv \tfrac{1}{2}(\mathbf{r} \times d\mathbf{r})$$

Deduce that

$$\frac{d\mathbf{A}}{dt} = \text{constant}$$

and thus that equal areas are swept out by the radius vector $\mathbf{r}$ in equal times. [For planetary motion this is one of Kepler's laws, but the result does not depend on a specific function for $F(r)$.]

(d) Scalar multiply the equation for relative motion in (a) by $\dot{\mathbf{r}}$ and integrate once to obtain

$$\frac{\mu}{2}(\dot{\mathbf{r}})^2 + U = E = \text{constant},$$

where

$$U(r) \equiv \int F(r)\, dr$$

is the potential energy and is taken here as an indefinite integral. This result states that the sum of the kinetic and potential energies is a constant.

(e) Take the cross product of the equation for relative motion in (a) by **L** and deduce that

$$\mu \frac{d}{dt}(\dot{\mathbf{r}} \times \mathbf{L}) = r^2 F(r) \frac{d}{dt}\left(\frac{\mathbf{r}}{r}\right)$$

since **L** is a constant.

(f) Newton's law of gravity states that

$$F(r) = \frac{G m_1 m_2}{r^2}$$

where G is the universal constant of gravitation. Integrate the equation in (e) and obtain

$$\mu(\dot{\mathbf{r}} \times \mathbf{L}) = G m_1 m_2 \left(\frac{\mathbf{r}}{r} + \mathbf{e}\right)$$

where the dimensionless vector **e** is the constant of integration called the *eccentricity* or the *apse vector*. Deduce that $\mathbf{e} \cdot \mathbf{L} = 0$.

(g) Solve for **e** in the preceding equation and square it, making use of the results of (d), to obtain

$$e^2 = 1 + \frac{2\mu L^2 E}{(G m_1 m_2)^2}.$$

Thus the magnitude of e is obtained in terms of the constants of integration L and E.

(h) Scalar multiply the second equation in (f) by **r** and solve for r to obtain the orbit equation

$$r = \frac{\mu L^2 / G m_1 m_2}{1 + e\cos\theta},$$

where $\cos\theta \equiv (\mathbf{r} \cdot \mathbf{e})/re$ is the cosine of the angle between **r** and **e**.

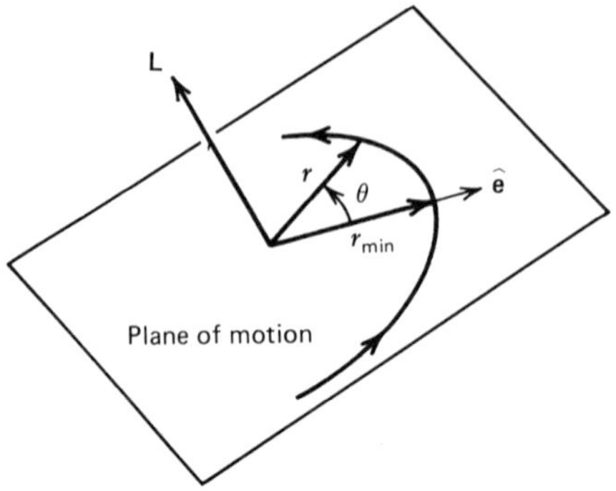

Deduce that **e** points in the direction where r is a minimum, as shown in the accompanying figure. When $e=0$, the orbit is a circle; when $0<e<1$, the orbit is an ellipse; when $e=1$, the orbit is a parabola; and when $e>1$, the orbit is a hyperbola.

6 It is useful to describe the rotational motion of rigid bodies by three angles ϕ, θ, and ψ, called the Euler angles, as shown in the figure. The coordinates X, Y, Z constitute an inertial rectangular Cartesian coordinate system. The x, y, z coordinates constitute a noninertial rectangular Cartesian coordinate system attached to a rotating rigid body. The two coordinate systems have a common origin. The noninertial x, y, z system rotates with the angular velocity $\boldsymbol{\omega}$. The Euler angles are delineated by three successive rotations. Let the two systems at first coincide, such that the x and X, y and Y, and z and Z axes are identical. First rotate the noninertial system about the common z and Z axes by the angle ϕ. Then rotate the noninertial system about its x axis, which now coincides with the line of nodes shown in the figure, by the angle θ. Finally, rotate the noninertial system about its z axis by the angle ψ. The two angles ϕ and θ thus describe the orientation of the z axis of the noninertial system with respect to the inertial system. The angle ψ describes how much the rigid body is rotated about its z axis.

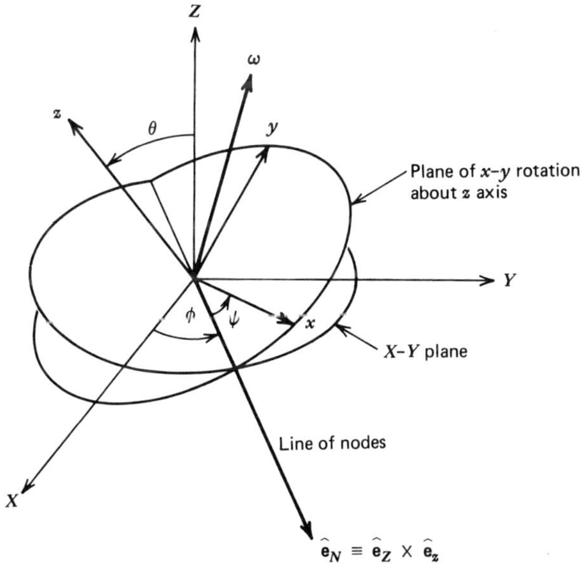

The rate of rotation vector $\boldsymbol{\omega}$ can be written in terms of the three rates of change of the angles ϕ, θ, and ψ as

$$\boldsymbol{\omega}=\dot{\phi}\hat{\mathbf{e}}_Z+\dot{\theta}\hat{\mathbf{e}}_N+\dot{\psi}\hat{\mathbf{e}}_z.$$

This description involves the unit vector $\hat{\mathbf{e}}_Z$ pointing along the inertial Z

axis, the unit vector $\hat{\mathbf{e}}_z$ pointing along the noninertial z axis, and the unit vector $\hat{\mathbf{e}}_N \equiv \hat{\mathbf{e}}_Z \times \hat{\mathbf{e}}_z$ pointing along the line of nodes, which is the intersection of the plane of $x-y$ rotation about the z axis with the $X-Y$ plane. This representation is simple physically, but is inconvenient because the mixed basis $\{\hat{\mathbf{e}}_Z, \hat{\mathbf{e}}_N, \hat{\mathbf{e}}_z\}$ is not orthogonal.

In applications of rigid-body dynamics it is useful to write $\boldsymbol{\omega}$ in various different basis systems, in particular, both the inertial and the body-fixed basis systems. We thus need to establish transformations between the X, Y, Z and the x, y, z systems. We first note from geometrical considerations that

$$\begin{aligned}\hat{\mathbf{e}}_N &= \cos\phi\,\hat{\mathbf{e}}_X + \sin\phi\,\hat{\mathbf{e}}_Y \\ &= \cos\psi\,\hat{\mathbf{e}}_x - \sin\psi\,\hat{\mathbf{e}}_y.\end{aligned}$$

(a) By similar considerations deduce that

$$\begin{aligned}\hat{\mathbf{e}}_z &= \cos\theta\,\hat{\mathbf{e}}_Z + \sin\theta(\hat{\mathbf{e}}_N \times \hat{\mathbf{e}}_Z) \\ \hat{\mathbf{e}}_x &= \cos\psi\,\hat{\mathbf{e}}_N + \sin\psi(\hat{\mathbf{e}}_z \times \hat{\mathbf{e}}_N) \\ \hat{\mathbf{e}}_y &= \hat{\mathbf{e}}_z \times \hat{\mathbf{e}}_x,\end{aligned}$$

and hence by the use of $\hat{\mathbf{e}}_N$ that

$$\begin{Bmatrix}\hat{\mathbf{e}}_x\\ \hat{\mathbf{e}}_y\\ \hat{\mathbf{e}}_z\end{Bmatrix} = \begin{bmatrix} \cos\phi\cos\psi - \cos\theta\sin\phi\sin\psi & \sin\phi\cos\psi + \cos\theta\cos\phi\sin\psi & \sin\theta\sin\psi \\ -\cos\phi\sin\psi - \cos\theta\sin\phi\cos\psi & -\sin\phi\sin\psi + \cos\theta\cos\phi\cos\psi & \sin\theta\cos\psi \\ \sin\theta\sin\phi & -\sin\theta\cos\phi & \cos\theta \end{bmatrix} \begin{Bmatrix}\hat{\mathbf{e}}_X\\ \hat{\mathbf{e}}_Y\\ \hat{\mathbf{e}}_Z\end{Bmatrix}.$$

This transformation is orthogonal.

(b) Deduce that the inverse transformation is

$$\begin{Bmatrix}\hat{\mathbf{e}}_X\\ \hat{\mathbf{e}}_Y\\ \hat{\mathbf{e}}_Z\end{Bmatrix} = \begin{bmatrix} \cos\phi\cos\psi - \cos\theta\sin\phi\sin\psi & -\cos\phi\sin\psi - \cos\theta\sin\phi\cos\psi & \sin\theta\sin\phi \\ \sin\phi\cos\psi + \cos\theta\cos\phi\sin\psi & -\sin\phi\sin\psi + \cos\theta\cos\phi\cos\psi & -\sin\theta\cos\phi \\ \sin\theta\sin\psi & \sin\theta\cos\psi & \cos\theta \end{bmatrix} \begin{Bmatrix}\hat{\mathbf{e}}_x\\ \hat{\mathbf{e}}_y\\ \hat{\mathbf{e}}_z\end{Bmatrix}.$$

(c) Deduce that the angular velocity can now be written in body-fixed components as

$$\omega = (\dot{\phi}\sin\theta\sin\psi + \dot{\theta}\cos\psi)\hat{\mathbf{e}}_x + (\dot{\phi}\sin\theta\cos\psi - \dot{\theta}\sin\psi)\hat{\mathbf{e}}_y$$
$$+ (\dot{\phi}\cos\theta + \dot{\psi})\hat{\mathbf{e}}_z$$

and in inertial components as

$$\omega = (\dot{\psi}\sin\theta\sin\phi + \dot{\theta}\cos\phi)\hat{\mathbf{e}}_X + (-\dot{\psi}\sin\theta\cos\phi + \dot{\theta}\sin\phi)\hat{\mathbf{e}}_Y$$
$$+ (\dot{\psi}\cos\theta + \dot{\phi})\hat{\mathbf{e}}_Z.$$

1.4.16 Derivative of a Scalar Function of a Vector

The basic notions of vector and scalar calculus, especially with regard to physical applications, are closely related to the rate of change of a scalar field with distance. Let us denote a scalar field by $\phi = \phi(\mathbf{r})$. In general coordinates we can write $\phi = \phi(q^1, q^2, q^3)$. The differential change is given by

$$d\phi = \frac{\partial\phi}{\partial q^1}dq^1 + \frac{\partial\phi}{\partial q^2}dq^2 + \frac{\partial\phi}{\partial q^3}dq^3. \tag{1.198}$$

The differentials dq^1, dq^2, dq^3 are components of $d\mathbf{r}$, that is,

$$d\mathbf{r} = dq^1\mathbf{e}_1 + dq^2\mathbf{e}_2 + dq^3\mathbf{e}_3.$$

We would now like to write $d\phi$ in such a way that we elucidate *the direction* as well as the magnitude of $d\mathbf{r}$. Since $\mathbf{e}^1\cdot\mathbf{e}_1 = 1, \mathbf{e}^2\cdot\mathbf{e}_2 = 1$, and $\mathbf{e}^3\cdot\mathbf{e}_3 = 1$, we can write

$$d\phi = \mathbf{e}^1\frac{\partial\phi}{\partial q^1}\cdot\mathbf{e}_1 dq^1 + \mathbf{e}^2\frac{\partial\phi}{\partial q^2}\cdot\mathbf{e}_2 dq^2 + \mathbf{e}^3\frac{\partial\phi}{\partial q^3}\cdot\mathbf{e}_3 dq^3$$

or

$$d\phi = \left(dq^1\mathbf{e}_1 + dq^2\mathbf{e}_2 + dq^3\mathbf{e}_3\right)\cdot\left(\mathbf{e}^1\frac{\partial\phi}{\partial q^1} + \mathbf{e}^2\frac{\partial\phi}{\partial q^2} + \mathbf{e}^3\frac{\partial\phi}{\partial q^3}\right) \tag{1.199}$$

or

$$d\phi = d\mathbf{r}\cdot\left(\mathbf{e}^1\frac{\partial\phi}{\partial q^1} + \mathbf{e}^2\frac{\partial\phi}{\partial q^2} + \mathbf{e}^3\frac{\partial\phi}{\partial q^3}\right).$$

Let us now denote the magnitude of $d\mathbf{r}$ by $ds \equiv |d\mathbf{r}|$. Then $\hat{\mathbf{e}} = d\mathbf{r}/ds$ is a unit vector in the direction of $d\mathbf{r}$, and we have

$$\left(\frac{d\phi}{ds}\right)_{\hat{\mathbf{e}}} = \hat{\mathbf{e}}\cdot\left(\mathbf{e}^1\frac{\partial\phi}{\partial q^1} + \mathbf{e}^2\frac{\partial\phi}{\partial q^2} + \mathbf{e}^3\frac{\partial\phi}{\partial q^3}\right). \tag{1.200}$$

The derivative $(d\phi/ds)_{\hat{\mathbf{e}}}$ is called the *directional derivative of* ϕ. We see that it is the *rate of change of* ϕ *with respect to distance and that it depends on the direction* $\hat{\mathbf{e}}$ *in which the distance is taken.*

The vector that is scalar multiplied by $\hat{\mathbf{e}}$ can be obtained immediately whenever the scalar field is given. Because the magnitude of this vector is equal to the maximum value of the directional derivative, it is called the *gradient vector* and is denoted by *grad* ϕ:

$$\text{grad}\,\phi \equiv \mathbf{e}^1\frac{\partial\phi}{\partial q^1} + \mathbf{e}^2\frac{\partial\phi}{\partial q^2} + \mathbf{e}^3\frac{\partial\phi}{\partial q^3}. \tag{1.201}$$

From this representation it can be seen that

$$\frac{\partial\phi}{\partial q^1}, \quad \frac{\partial\phi}{\partial q^2}, \quad \frac{\partial\phi}{\partial q^3}$$

are the *covariant components* of the gradient vector.

When the scalar function $\phi(\mathbf{r})$ is set equal to a constant, $\phi(\mathbf{r})=$constant, a family of surfaces is generated. A different surface is designated by different values of the constant, and each surface is called a *level surface* (see Fig. 1.33).

If the direction in which the directional derivative is taken lies within a level surface, then $d\phi/ds$ is zero, since ϕ is a constant on a level surface. In this case the unit vector $\hat{\mathbf{e}}$ is tangent to a level surface. It follows, therefore, that if $d\phi/ds$ is zero, then grad ϕ must be perpendicular to $\hat{\mathbf{e}}$ and thus *perpendicular to a level surface*. Thus if any surface is given by $\phi(\mathbf{r})=$constant, the unit normal to the surface is determined by

$$\hat{\mathbf{n}} = \pm\frac{\text{grad}\,\phi}{|\text{grad}\,\phi|}. \tag{1.202}$$

The plus or minus sign appears because the direction of $\hat{\mathbf{n}}$ may point in either direction away from the surface. *If the surface is closed, the usual convention is to take* $\hat{\mathbf{n}}$ *pointing outward.*

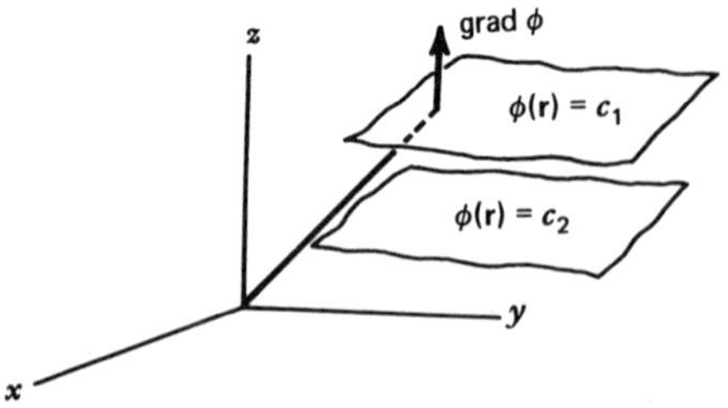

Figure 1.33 Level surfaces.

1.4.17 The del Operator

It is convenient to write the gradient vector as

$$\operatorname{grad}\phi \equiv \left(\mathbf{e}^1\frac{\partial}{\partial q^1}+\mathbf{e}^2\frac{\partial}{\partial q^2}+\mathbf{e}^3\frac{\partial}{\partial q^3}\right)\phi \tag{1.203}$$

and interpret grad ϕ as some operator operating on ϕ, that is, grad $\phi = \nabla\phi$. This operator is denoted by

$$\nabla \equiv \mathbf{e}^1\frac{\partial}{\partial q^1}+\mathbf{e}^2\frac{\partial}{\partial q^2}+\mathbf{e}^3\frac{\partial}{\partial q^3} \tag{1.204}$$

and is called the del operator. The del operator is a *vector differential* operator, and the "components" $\partial/\partial q^1, \partial/\partial q^2$, and $\partial/\partial q^3$ appear as covariant components.

It is important to note that whereas the del operator has some of the properties of a vector, it does not have them all, because it is an operator. For instance, $\nabla\cdot\mathbf{A}$ is a scalar (called the *divergence* of **A**) whereas $\mathbf{A}\cdot\nabla$ is a scalar *differential operator*. Thus the del operator does not commute in this sense.

In orthogonal curvilinear coordinates, del can be expressed in terms of the orthonormal basis $\hat{\mathbf{e}}_1, \hat{\mathbf{e}}_2, \hat{\mathbf{e}}_3$. Since $\mathbf{e}^1 = \hat{\mathbf{e}}_1/h_1, \mathbf{e}^2 = \hat{\mathbf{e}}_2/h_2$, and $\mathbf{e}^3 = \hat{\mathbf{e}}_3/h_3$ for an orthonormal basis, we have

$$\nabla \equiv \frac{\hat{\mathbf{e}}_1}{h_1}\frac{\partial}{\partial q^1}+\frac{\hat{\mathbf{e}}_2}{h_2}\frac{\partial}{\partial q^2}+\frac{\hat{\mathbf{e}}_3}{h_3}\frac{\partial}{\partial q^3}. \tag{1.205}$$

Since $ds_1 = h_1 dq^1$, $ds_2 = h_2 dq^2$, and $ds_3 = h_3 dq^3$, we also can write

$$\nabla \equiv \hat{\mathbf{e}}_1\frac{\partial}{\partial s_1}+\hat{\mathbf{e}}_2\frac{\partial}{\partial s_2}+\hat{\mathbf{e}}_3\frac{\partial}{\partial s_3}. \tag{1.206}$$

This form shows that the del operator involves derivatives with respect to distance in the respective coordinate directions.

In Cartesians we have the simple form

$$\nabla \equiv \hat{\mathbf{e}}_x\frac{\partial}{\partial x}+\hat{\mathbf{e}}_y\frac{\partial}{\partial y}+\hat{\mathbf{e}}_z\frac{\partial}{\partial z}. \tag{1.207}$$

In the summation convention we have

$$\nabla \equiv \mathbf{e}^i\frac{\partial}{\partial q^i}, \tag{1.208}$$

or in Cartesians,

$$\nabla \equiv \hat{\mathbf{i}}_i \frac{\partial}{\partial x^i}. \tag{1.209}$$

1.4.18 The Divergence of a Vector

The dot product of del with a vector is called the *divergence of a vector* and denoted by

$$\nabla \cdot \mathbf{A} \equiv \operatorname{div} \mathbf{A}. \tag{1.210}$$

Suppose that $\mathbf{A}$ is given in contravariant components. Then $\mathbf{A} = A^i \mathbf{e}_i$, and we have

$$\begin{aligned} \operatorname{div} \mathbf{A} &= \mathbf{e}^j \frac{\partial}{\partial q^j} \cdot A^i \mathbf{e}_i \\ &= \mathbf{e}^j \cdot \left(\frac{\partial A^i}{\partial q^j} \mathbf{e}_i + A^i \frac{\partial \mathbf{e}_i}{\partial q^j} \right) \\ &= \mathbf{e}^j \cdot \left(\frac{\partial A^i}{\partial q^j} \mathbf{e}_i + A^i \left\{ \begin{matrix} k \\ i\ j \end{matrix} \right\} \mathbf{e}_k \right). \end{aligned} \tag{1.211}$$

Recalling that $\mathbf{e}^j \cdot \mathbf{e}_i = \delta_i^j$, we have

$$\operatorname{div} \mathbf{A} = \frac{\partial A^i}{\partial q^i} + A^i \left\{ \begin{matrix} j \\ i\ j \end{matrix} \right\}. \tag{1.212}$$

Use of Eq. (1.188), however, gives

$$\begin{aligned} \operatorname{div} \mathbf{A} &= \frac{\partial A^i}{\partial q^i} + \frac{A^i}{\sqrt{g}} \frac{\partial \sqrt{g}}{\partial q^i} \\ &= \frac{1}{\sqrt{g}} \frac{\partial}{\partial q^i} \left(\sqrt{g} A^i \right). \end{aligned} \tag{1.213}$$

In orthogonal curvilinear coordinates we have, since $\mathbf{e}_i = h_i \hat{\mathbf{e}}_i$,

$$\begin{aligned} \mathbf{A} &= A^1 h_1 \hat{\mathbf{e}}_1 + A^2 h_2 \hat{\mathbf{e}}_2 + A^3 h_3 \hat{\mathbf{e}}_3 \\ &= \hat{A}_1 \hat{\mathbf{e}}_1 + \hat{A}_2 \hat{\mathbf{e}}_2 + \hat{A}_3 \hat{\mathbf{e}}_3, \end{aligned} \tag{1.214}$$

where we have set

$$\hat{A}_1 \equiv h_1 A^1, \qquad \hat{A}_2 \equiv h_2 A^2, \qquad \hat{A}_3 \equiv h_3 A^3. \tag{1.215}$$

Here $\hat{A}_1, \hat{A}_2, \hat{A}_3$ are the physical components corresponding to the contravariant components, and are not to be confused with the covariant components. *They are not the covariant components.* Further, since $g = h_1^2 h_2^2 h_3^2$, the divergence of **A** in orthogonal curvilinear coordinates becomes

$$\operatorname{div}\mathbf{A} = \frac{1}{h_1 h_2 h_3}\left[\frac{\partial}{\partial q^1}(h_2 h_3 \hat{A}_1) + \frac{\partial}{\partial q^2}(h_1 h_3 \hat{A}_2) + \frac{\partial}{\partial q^3}(h_1 h_2 \hat{A}_3)\right]. \tag{1.216}$$

In Cartesians we have simply

$$\begin{aligned}\operatorname{div}\mathbf{A} &= \frac{\partial A_1}{\partial x^1} + \frac{\partial A_2}{\partial x^2} + \frac{\partial A_3}{\partial x^3} \\ &= \frac{\partial A_i}{\partial x^i}.\end{aligned} \tag{1.217}$$

The physical interpretation of the divergence of a vector field is more easily perceived from the integral definition in Section 1.4.21.

1.4.19 The Laplacian of a Scalar

If we take the divergence of the gradient vector, we have

$$\begin{aligned}\operatorname{div}\operatorname{grad}\phi &\equiv \nabla\cdot\nabla\phi \\ &\equiv (\nabla\cdot\nabla)\phi \\ &\equiv \nabla^2\phi.\end{aligned} \tag{1.218}$$

The notation $\nabla^2 = \nabla\cdot\nabla$ is called the *Laplacian operator*. To make use of the previous expression to determine $\nabla^2\phi$ in general coordinates, we must express the covariant components $\partial\phi/\partial q^i$ as contravariant components. If A^i now denotes the contravariant components of $\nabla\phi$, then

$$A^i \equiv g^{ij}\frac{\partial\phi}{\partial q^j}. \tag{1.219}$$

Then we have

$$\nabla^2\phi = \frac{1}{\sqrt{g}}\frac{\partial}{\partial q^i}\left(\sqrt{g}\,g^{ij}\frac{\partial\phi}{\partial q^j}\right). \tag{1.220}$$

In orthogonal curvilinear components we have $g^{11}=h_1^{-2}, g^{22}=h_2^{-2}$, and $g^{33}=h_3^{-2}$, and thus

$$\nabla^2\phi=\frac{1}{h_1h_2h_3}\left[\frac{\partial}{\partial q^1}\left(\frac{h_2h_3}{h_1}\frac{\partial\phi}{\partial q^1}\right)+\frac{\partial}{\partial q^2}\left(\frac{h_1h_3}{h_2}\frac{\partial\phi}{\partial q^2}\right)+\frac{\partial}{\partial q^3}\left(\frac{h_1h_2}{h_3}\frac{\partial\phi}{\partial q^3}\right)\right]. \tag{1.221}$$

In Cartesians this reduces to the simple form

$$\nabla^2\phi=\frac{\partial^2\phi}{\partial x^2}+\frac{\partial^2\phi}{\partial y^2}+\frac{\partial^2\phi}{\partial z^2}$$

$$=\frac{\partial^2\phi}{\partial x^i\partial x^i}.$$

The Laplacian of a scalar appears frequently in the partial differential equations governing physical phenomena.

1.4.20 The Curl of a Vector

The curl of a vector is defined as the del operator operating on a vector by means of the cross product:

$$\operatorname{curl}\mathbf{A}\equiv\nabla\times\mathbf{A}. \tag{1.222}$$

Suppose that $\mathbf{A}$ is given in terms of its contravariant components so that

$$\operatorname{curl}\mathbf{A}=\mathbf{e}^j\frac{\partial}{\partial q^j}\times A^i\mathbf{e}_i$$

$$=\mathbf{e}^j\times\left[\frac{\partial A^i}{\partial q^j}\mathbf{e}_i+A^i\frac{\partial\mathbf{e}_i}{\partial q^j}\right]$$

$$=\mathbf{e}^j\times\left[\frac{\partial A^i}{\partial q^j}\mathbf{e}_i+A^i\left\{{k \atop i\ j}\right\}\mathbf{e}_k\right]$$

$$=\mathbf{e}^j\times\left[\frac{\partial A^k}{\partial q^j}+A_i\left\{{k \atop i\ j}\right\}\right]\mathbf{e}_k$$

$$=\left[\frac{\partial A^k}{\partial q^j}+A^i\left\{{k \atop i\ j}\right\}\right](\mathbf{e}^j\times\mathbf{e}_k). \tag{1.223}$$

The terms in brackets occur many times in operations such as these, and it is useful to give them a special notation:

$$A^k_{,j} \equiv \frac{\partial A^k}{\partial q^j} + A^i \begin{Bmatrix} k \\ i\ j \end{Bmatrix}. \tag{1.224}$$

The symbol $A^k_{,j}$ is called the *covariant derivative* of the contravariant component A^k. The comma before the index j means differentiation with respect to the jth coordinate. Thus we have

$$\text{curl}\,\mathbf{A} = A^k_{,j}(\mathbf{e}^j \times \mathbf{e}_k). \tag{1.225}$$

We must now evaluate the cross-product term. This is convenient to do in terms of the covariant unitary basis. Thus we write $\mathbf{e}^j = g^{ji}\mathbf{e}_i$, so that

$$\text{curl}\,\mathbf{A} = g^{ji}A^k_{,j}(\mathbf{e}_i \times \mathbf{e}_k). \tag{1.226}$$

It is easy to get an understanding of the notation associated with the cross-product term if we work in Cartesians. In this case the Christoffel symbol is zero (because the Cartesian basis is constant), and $g^{ji} = \delta_{ij}$. Thus we have

$$\begin{aligned} \text{curl}\,\mathbf{A} &= \hat{\mathbf{e}}_i \frac{\partial}{\partial x_i} \times A_j \hat{\mathbf{e}}_j \\ &= \frac{\partial A_j}{\partial x_i}(\hat{\mathbf{e}}_i \times \hat{\mathbf{e}}_j) \end{aligned} \tag{1.227}$$

which can be obtained easily from the general expression. We have previously defined [see Eq. (1.45)] the cross product $\hat{\mathbf{e}}_i \times \hat{\mathbf{e}}_j$ for a right-handed system as follows:

$$\hat{\mathbf{e}}_i \times \hat{\mathbf{e}}_j = \varepsilon_{ijk}\hat{\mathbf{e}}_k,$$

where ε_{ijk} is the permutation symbol defined in Eq. (1.46). We then have in Cartesians

$$\text{curl}\,\mathbf{A} = \frac{\partial A_j}{\partial x^i}\varepsilon_{ijk}\hat{\mathbf{e}}_k. \tag{1.228}$$

Consider the corresponding result for a general curvilinear coordinate system. From Exercise 1.4.5 we have (for a right-handed system)

$$g \equiv \det(g_{ij}) = [\mathbf{e}_1 \cdot \mathbf{e}_2 \times \mathbf{e}_3]^2, \tag{1.229}$$

and from Eqs. (1.100) and (1.101),

$$\mathbf{e}_i \times \mathbf{e}_j = \sqrt{g}\,\varepsilon_{ijk}\mathbf{e}^k \equiv \mathcal{E}_{ijk}\mathbf{e}^k. \tag{1.230}$$

With this result the general expression for the curl of **A** can be expressed as

$$\operatorname{curl}\mathbf{A} = \sqrt{g}\,g^{ji}A^k_{\ ,j}\varepsilon_{ikm}\mathbf{e}^m. \tag{1.231}$$

In orthogonal curvilinear coordinates, where $\hat{A}_1$, $\hat{A}_2$, and $\hat{A}_3$ denote the physical components of a vector, curl **A** can be expressed in terms of the easy-to-remember determinant

$$\operatorname{curl}\mathbf{A} = \frac{1}{h_1h_2h_3}\begin{vmatrix} h_1\hat{\mathbf{e}}_1 & h_2\hat{\mathbf{e}}_2 & h_3\hat{\mathbf{e}}_3 \\ \dfrac{\partial}{\partial q^1} & \dfrac{\partial}{\partial q^2} & \dfrac{\partial}{\partial q^3} \\ h_1\hat{A}_1 & h_2\hat{A}_2 & h_3\hat{A}_3 \end{vmatrix}. \tag{1.232}$$

The meaning and usefulness of the curl vector can be obtained from the notion of a rigid-body rotation. The curl of the velocity relation $\mathbf{V} = \boldsymbol{\omega} \times \mathbf{r}$ yields

$$\begin{aligned} \operatorname{curl}\mathbf{V} &= \operatorname{curl}(\boldsymbol{\omega} \times \mathbf{r}) = \nabla \times (\boldsymbol{\omega} \times \mathbf{r}) \\ &= (\nabla \cdot \mathbf{r} + \mathbf{r} \cdot \nabla)\boldsymbol{\omega} - (\nabla \cdot \boldsymbol{\omega} + \boldsymbol{\omega} \cdot \nabla)\mathbf{r} \\ &= 3\boldsymbol{\omega} - \boldsymbol{\omega} = 2\boldsymbol{\omega}. \end{aligned} \tag{1.233}$$

This result is obtained since $\boldsymbol{\omega}$ is a constant in a rigid-body rotation, and hence $(\mathbf{r} \cdot \nabla)\boldsymbol{\omega} = 0$ and $\nabla \cdot \boldsymbol{\omega} = 0$. Also $\nabla \cdot \mathbf{r} = 3$ and $(\boldsymbol{\omega} \cdot \nabla)\mathbf{r} = \boldsymbol{\omega}$. The value of curl $(\boldsymbol{\omega} \times \mathbf{r})$ may be obtained by any other straightforward expansion. Finally we have

$$\boldsymbol{\omega} = \tfrac{1}{2}\operatorname{curl}\mathbf{V}. \tag{1.234}$$

Thus the curl of a vector field is equal to *twice the local rotation of the field*. A summary of vector operations in both general vector notation and in Cartesian component form is given in Table 1.1. Some useful vector operations for cylindrical and spherical coordinate systems are shown in Table 1.2.

Example 1.14 Using the index-summation notation, we prove the following identity in a Cartesian system:

$$\nabla \times (\nabla \times \mathbf{v}) \equiv \nabla(\nabla \cdot \mathbf{v}) - \nabla^2\mathbf{v},$$

Table 1.1 Vector Terms and Operations in Cartesian Component Form

Term or Operation	Vector Form	Cartesian Component Form
1. Vector	$\mathbf{A}$	A_i
2. Scalar (or dot) product	$\mathbf{A}\cdot\mathbf{B}$	A_iB_i
3. Vector (or cross) product	$\mathbf{A}\times\mathbf{B}$	$\varepsilon_{ijk}A_jB_k$
4. Scalar triple product	$\mathbf{A}\cdot(\mathbf{B}\times\mathbf{C})$	$\varepsilon_{ijk}A_iB_jC_k$
5. Vector triple product	$\mathbf{A}\times(\mathbf{B}\times\mathbf{C})=\mathbf{B}(\mathbf{A}\cdot\mathbf{C})-\mathbf{C}(\mathbf{A}\cdot\mathbf{B})$	$\varepsilon_{ijk}\varepsilon_{klm}A_jB_lC_m$
6. Divergence of **A**	$\nabla\cdot\mathbf{A}$	$\frac{\partial A_i}{\partial x_i}$
7. Divergence of $U\mathbf{A}$ (U, scalar function)	$\nabla U\cdot\mathbf{A}+U\nabla\cdot\mathbf{A}$	$U\frac{\partial A_i}{\partial x_i}+A_i\frac{\partial U}{\partial x_i}$
8. Divergence of $(\mathbf{A}+\mathbf{B})$	$\nabla\cdot\mathbf{A}+\nabla\cdot\mathbf{B}$	$\frac{\partial A_i}{\partial x_i}+\frac{\partial B_i}{\partial x_i}$
9. Divergence of $(\mathbf{A}\times\mathbf{B})$	$\mathbf{B}\cdot(\nabla\times\mathbf{A})-\mathbf{A}\cdot(\nabla\times\mathbf{B})$	$\frac{\partial}{\partial x_i}(\varepsilon_{ijk}A_jB_k)$
10. Curl of **A**	$\nabla\times\mathbf{A}$	$\varepsilon_{ijk}\frac{\partial A_k}{\partial x_j}$
11. Curl of $U\mathbf{A}$	$\nabla U\times\mathbf{A}+U\nabla\times\mathbf{A}$	$\varepsilon_{ijk}U\frac{\partial A_k}{\partial x_j}+\varepsilon_{ijk}A_k\frac{\partial U}{\partial x_j}$
12. Curl of $(\mathbf{A}+\mathbf{B})$	$\nabla\times\mathbf{B}+\nabla\times\mathbf{A}$	$\varepsilon_{ijk}\frac{\partial A_k}{\partial x_j}+\varepsilon_{ijk}\frac{\partial B_k}{\partial x_j}$
13. Curl of $(\mathbf{A}\times\mathbf{B})$	$\mathbf{A}(\nabla\cdot\mathbf{B})-\mathbf{B}(\nabla\cdot\mathbf{A})$ $+(\mathbf{B}\cdot\nabla)\mathbf{A}-(\mathbf{A}\cdot\nabla)\mathbf{B}$	$\frac{\partial}{\partial x_i}(\varepsilon_{ijk}\varepsilon_{klm}A_lB_m)$
14. (Curl of **A**) $\times$ **B**	$(\nabla\times\mathbf{A})\times\mathbf{B}=\mathbf{B}\cdot[\nabla\mathbf{A}-(\nabla\mathbf{A})^{\mathrm{T}}]$	$B_j\frac{\partial A_i}{\partial x_j}-B_i\frac{\partial A_j}{\partial x_j}$
15. ∇^2U	$\nabla^2U=\nabla\cdot(\nabla U)$	$\frac{\partial}{\partial x_i}\frac{\partial}{\partial x_i}U$
16. $\nabla^2\mathbf{A}$	$\nabla^2\mathbf{A}=(\nabla\cdot\nabla)\mathbf{A}$	$\frac{\partial}{\partial x_i}\frac{\partial}{\partial x_i}A_j$
17. $\nabla\times(\nabla\times\mathbf{A})$	$\nabla(\nabla\cdot\mathbf{A})-(\nabla\cdot\nabla)\mathbf{A}$	$\frac{\partial}{\partial x_j}\frac{\partial}{\partial x_i}A_j-\frac{\partial}{\partial x_j}\frac{\partial}{\partial x_j}A_i$
18. $(\mathbf{A}\cdot\nabla)\mathbf{B}$	$(\mathbf{A}\cdot\nabla)\mathbf{B}$	$A_j\frac{\partial B_i}{\partial x_j}$
19. $\mathbf{A}(\nabla\cdot\mathbf{B})$	$\mathbf{A}(\nabla\cdot\mathbf{B})$	$A_i\frac{\partial B_j}{\partial x_j}$

where $\mathbf{v}$ is a vector function of the coordinates x_i. Observe that

$$\nabla\times(\nabla\times\mathbf{v})\equiv\hat{\mathbf{e}}_i\frac{\partial}{\partial x_i}\times\left(\hat{\mathbf{e}}_j\frac{\partial}{\partial x_j}\times v_k\hat{\mathbf{e}}_k\right)$$

$$\hat{\mathbf{e}}_i\frac{\partial}{\partial x_i}\times\left(\varepsilon_{jkl}\frac{\partial v_k}{\partial x_j}\hat{\mathbf{e}}_l\right)\equiv\varepsilon_{ilm}\varepsilon_{jkl}\frac{\partial^2 v_k}{\partial x_i\partial x_j}\hat{\mathbf{e}}_m.$$

Table 1.2 The del and Laplace Operators in Cylindrical and Spherical Coordinate Systems

---

Cylindrical (R, ϕ, z)

$$x = R\cos\phi, \qquad y = R\sin\phi, \qquad z = z$$
$$\hat{\mathbf{e}}_R = \cos\phi\,\hat{\mathbf{e}}_x + \sin\phi\,\hat{\mathbf{e}}_y$$
$$\hat{\mathbf{e}}_\phi = -\sin\phi\,\hat{\mathbf{e}}_x + \cos\phi\,\hat{\mathbf{e}}_y$$
$$\hat{\mathbf{e}}_z = \hat{\mathbf{e}}_z$$
$$\frac{\partial \hat{\mathbf{e}}_R}{\partial\phi} = -\sin\phi\,\hat{\mathbf{e}}_x + \cos\phi\,\hat{\mathbf{e}}_y = \hat{\mathbf{e}}_\phi$$
$$\frac{\partial \hat{\mathbf{e}}_\phi}{\partial\phi} = -\cos\phi\,\hat{\mathbf{e}}_x - \sin\phi\,\hat{\mathbf{e}}_y = -\hat{\mathbf{e}}_R \text{ (all other derivatives of the base vectors are zero)}$$
$$\nabla = \hat{\mathbf{e}}_R\frac{\partial}{\partial R} + \frac{\hat{\mathbf{e}}_\phi}{R}\frac{\partial}{\partial\phi} + \hat{\mathbf{e}}_z\frac{\partial}{\partial z}$$
$$\nabla^2 F = \frac{1}{R}\left[\frac{\partial}{\partial R}\left(R\frac{\partial F}{\partial R}\right) + \frac{1}{R}\frac{\partial^2 F}{\partial\phi^2} + R\frac{\partial^2 F}{\partial z^2}\right]$$

Spherical (r, θ, ϕ)

$$x = r\sin\theta\cos\phi$$
$$y = r\sin\theta\sin\phi$$
$$z = r\cos\theta$$
$$\hat{\mathbf{e}}_r = \sin\theta\cos\phi\,\hat{\mathbf{e}}_x + \sin\theta\sin\phi\,\hat{\mathbf{e}}_y + \cos\theta\,\hat{\mathbf{e}}_z$$
$$\hat{\mathbf{e}}_\theta = \cos\theta\cos\phi\,\hat{\mathbf{e}}_x + \cos\theta\sin\phi\,\hat{\mathbf{e}}_y - \sin\theta\,\hat{\mathbf{e}}_z$$
$$\hat{\mathbf{e}}_\phi = -\sin\phi\,\hat{\mathbf{e}}_x + \cos\phi\,\hat{\mathbf{e}}_y$$
$$\frac{\partial \hat{\mathbf{e}}_r}{\partial\theta} = \hat{\mathbf{e}}_\theta, \qquad \frac{\partial \hat{\mathbf{e}}_r}{\partial\phi} = \sin\theta\,\hat{\mathbf{e}}_\phi, \qquad \frac{\partial \hat{\mathbf{e}}_\theta}{\partial\theta} = -\hat{\mathbf{e}}_r, \qquad \frac{\partial \hat{\mathbf{e}}_\theta}{\partial\phi} = \cos\theta\,\hat{\mathbf{e}}_\phi$$
$$\frac{\partial \hat{\mathbf{e}}_\phi}{\partial\phi} = -\sin\theta\,\hat{\mathbf{e}}_r - \cos\theta\,\hat{\mathbf{e}}_\theta \text{ (all other derivatives of the base vectors are zero)}$$
$$\nabla = \hat{\mathbf{e}}_r\frac{\partial}{\partial\theta} + \frac{\hat{\mathbf{e}}_\theta}{r}\frac{\partial}{\partial\theta} + \frac{1}{r\sin\theta}\hat{\mathbf{e}}_\phi\frac{\partial}{\partial\phi}$$
$$\nabla^2 F = \frac{1}{r^2}\frac{\partial}{\partial r}\left(r^2\frac{\partial F}{\partial r}\right) + \frac{1}{r^2\sin\theta}\frac{\partial}{\partial\theta}\left(\sin\theta\frac{\partial F}{\partial\theta}\right) + \frac{1}{r^2\sin^2\theta}\frac{\partial^2 F}{\partial\phi^2}$$

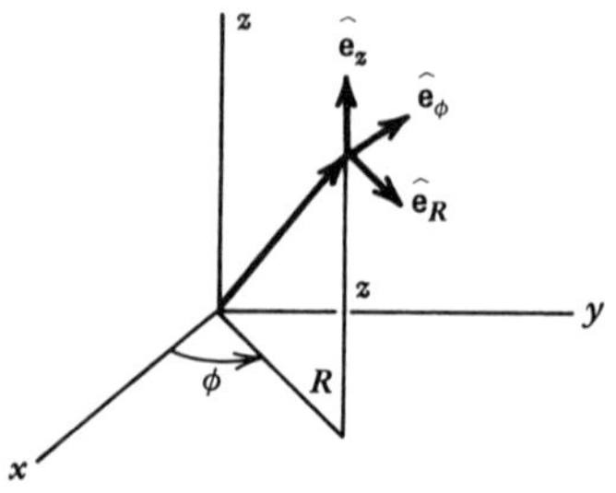

Cylindrical coordinate system.

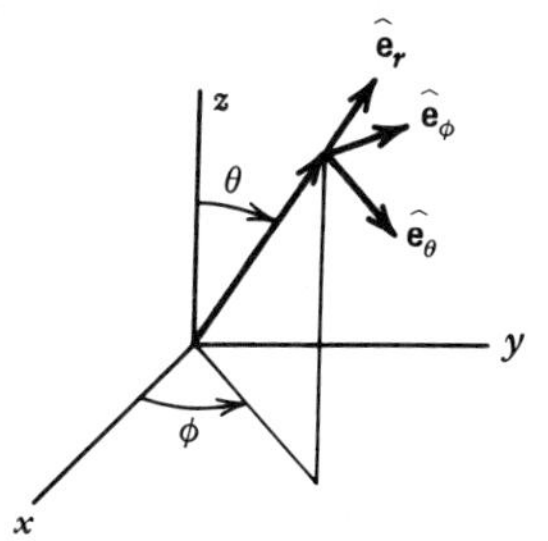

Spherical coordinate system.

Using the ε–δ identity, we get

$$\nabla\times(\nabla\times\mathbf{v})\equiv(\delta_{mj}\delta_{ik}-\delta_{mk}\delta_{ij})\frac{\partial^2 v_k}{\partial x_i\partial x_j}\hat{\mathbf{e}}_m$$

$$\equiv\frac{\partial^2 v_i}{\partial x_i\partial x_j}\hat{\mathbf{e}}_j-\frac{\partial^2 v_k}{\partial x_i\partial x_i}\hat{\mathbf{e}}_k=\hat{\mathbf{e}}_j\frac{\partial}{\partial x_j}\left(\frac{\partial v_i}{\partial x_i}\right)-\frac{\partial^2}{\partial x_i\partial x_i}(v_k\hat{\mathbf{e}}_k)$$

$$\equiv\nabla(\nabla\cdot\mathbf{v})-\nabla^2\mathbf{v}.$$

This result is sometimes used as the definition of the Laplacian of a vector, that is,

$$\nabla^2\mathbf{v}\equiv\text{grad}(\text{div}\,\mathbf{v})-\text{curl curl}\,\mathbf{v}. \quad \blacksquare$$

Exercises 1.6

1 Determine the directional derivative of the scalar function $\phi=2xy+z^2$ in the direction of the vector $\mathbf{s}=\hat{\mathbf{e}}_x+2\hat{\mathbf{e}}_y+2\hat{\mathbf{e}}_z$ at the point $(1,-1,3)$.

2 Consider the ellipsoid surface defined by

$$F(x,y,z)\equiv\frac{x^2}{a^2}+\frac{y^2}{b^2}+\frac{z^2}{c^2}-1=0.$$

Deduce that the unit outward normal at a point on this surface is given by

$$\hat{\mathbf{n}}=\frac{\dfrac{x}{a^2}\hat{\mathbf{e}}_x+\dfrac{y}{b^2}\hat{\mathbf{e}}_y+\dfrac{z}{c^2}\hat{\mathbf{e}}_z}{\left(\dfrac{x^2}{a^4}+\dfrac{y^2}{b^4}+\dfrac{z^2}{c^4}\right)^{1/2}}.$$

3 Let $\mathbf{r}$ denote a position vector. Show that:

(a) $\text{grad}\,(r^n) = nr^{n-2}\mathbf{r}$.

(b) $\nabla^2(r^n) = n(n+1)r^{n-2}$.

(c) $\text{div}\,(\mathbf{r}) = 3$.

(d) $\text{curl}\,(f(r)\mathbf{r}) = \mathbf{0}$, where $f(r)$ is an arbitrary continuous function of r with continuous first derivatives.

4 Let $\mathbf{A}$ and $\mathbf{B}$ be continuous vector functions of $\mathbf{r}$ with continuous first derivatives, and let F and G be continuous scalar functions of $\mathbf{r}$ with continuous first and second derivatives. Show that:

(a) $\text{div}\,(\mathbf{A} \times \mathbf{B}) \equiv \mathbf{B} \cdot \text{curl}\,\mathbf{A} - \mathbf{A} \cdot \text{curl}\,\mathbf{B}$.

(b) $\nabla^2(FG) \equiv F\nabla^2 G + 2\nabla F \cdot \nabla G + G\nabla^2 F$.

5 Let $\mathbf{r}$ be the position vector and $\mathbf{A}$ an arbitrary constant vector. Show that:

(a) $\text{grad}\,(\mathbf{r} \cdot \mathbf{A}) = \mathbf{A}$.

(b) $\text{div}\,(\mathbf{r} \times \mathbf{A}) = 0$.

(c) $\text{curl}\,(\mathbf{r} \times \mathbf{A}) = -2\mathbf{A}$.

(d) $\text{div}\,(r\mathbf{A}) = r^{-1}\mathbf{r} \cdot \mathbf{A}$.

(e) $\text{curl}\,(r\mathbf{A}) = r^{-1}\mathbf{r} \times \mathbf{A}$.

6 Verify the following identities, where $\phi = \phi(\mathbf{r})$, $\psi = \psi(\mathbf{r})$, and $\mathbf{A} = \mathbf{A}(\mathbf{r})$:

(a) $\text{curl}\,(\text{grad}\,\phi) \equiv 0$.

(b) $\text{div}\,(\text{curl}\,\mathbf{A}) \equiv 0$.

(c) $\text{div}\,(\text{grad}\,\phi \times \text{grad}\,\psi) \equiv 0$.

7 Verify the following identities:

(a) $\text{grad}\,(\phi\psi) \equiv \psi\,\text{grad}\,\phi + \phi\,\text{grad}\,\psi$.

(b) $\text{div}\,(\phi\mathbf{A}) \equiv \mathbf{A} \cdot \text{grad}\,\phi + \phi\,\text{div}\,\mathbf{A}$.

(c) $\text{curl}\,(\phi\mathbf{A}) \equiv \phi\,\text{curl}\,\mathbf{A} - \mathbf{A} \times \text{grad}\,\phi$.

(d) $\text{grad}\,(\mathbf{A} \cdot \mathbf{B}) \equiv \mathbf{A} \cdot \text{grad}\,\mathbf{B} + \mathbf{B} \cdot \text{grad}\,\mathbf{A} + \mathbf{A} \times \text{curl}\,\mathbf{B} + \mathbf{B} \times \text{curl}\,\mathbf{A}$.

(e) $\text{div}\,(\mathbf{A} \times \mathbf{B}) \equiv \mathbf{B} \cdot \text{curl}\,\mathbf{A} - \mathbf{A} \cdot \text{curl}\,\mathbf{B}$.

(f) $\text{curl}\,(\mathbf{A} \times \mathbf{B}) \equiv \mathbf{A}\,\text{div}\,\mathbf{B} - \mathbf{B}\,\text{div}\,\mathbf{A} + \mathbf{B} \cdot \text{grad}\,\mathbf{A} - \mathbf{A} \cdot \text{grad}\,\mathbf{B}$.

8 If g is the determinant of $\{g_{ij}\}$, whose elements are functions of q^i, prove that

$$\frac{\partial g}{\partial q^i} = g g^{jk} \frac{\partial g_{jk}}{\partial q^i}.$$

1.4.21 Integral Relations

Useful expressions for the gradient vector, divergence of a vector, and curl of a vector can be established from integral relations between volume integrals and surface integrals. Let R denote a region in space surrounded by the surface S.

Let dS be a differential element of surface and $\hat{\mathbf{n}}$ the unit outward normal, and let $d\tau$ be a differential volume element. The following relations are taken from advanced calculus (see [1,2,4,9]):

Gradient Theorem

$$\iiint_R \operatorname{grad}\phi\, d\tau = \oiint_S \hat{\mathbf{n}}\phi\, dS. \tag{1.235a}$$

Divergence Theorem

$$\iiint_R \operatorname{div}\mathbf{A}\, d\tau = \oiint_S \hat{\mathbf{n}}\cdot\mathbf{A}\, dS. \tag{1.235b}$$

Curl Theorem

$$\iiint_R \operatorname{curl}\mathbf{A}\, d\tau = \oiint_S \hat{\mathbf{n}}\times\mathbf{A}\, dS. \tag{1.235c}$$

Now let $\mathbf{A}=\operatorname{grad}\phi$ in Eq. (1.235b). Then the divergence theorem gives

$$\iiint_R \operatorname{div}(\operatorname{grad}\phi)\, d\tau \equiv \iiint_R \nabla^2\phi\, d\tau = \oiint_S \hat{\mathbf{n}}\cdot\operatorname{grad}\phi\, dS. \tag{1.236}$$

The quantity $\hat{\mathbf{n}}\cdot\operatorname{grad}\phi$ is called the *normal derivative* of ϕ on the surface S, and is denoted by

$$\frac{\partial\phi}{\partial n} \equiv \hat{\mathbf{n}}\cdot\operatorname{grad}\phi = \hat{\mathbf{n}}\cdot\nabla\phi. \tag{1.237}$$

In a Cartesian system this becomes

$$\frac{\partial\phi}{\partial n} = \frac{\partial\phi}{\partial x}n_x + \frac{\partial\phi}{\partial y}n_y + \frac{\partial\phi}{\partial z}n_z, \tag{1.238}$$

where n_x, n_y, and n_z are the direction cosines of the unit normal:

$$\hat{\mathbf{n}} = n_x\hat{\mathbf{e}}_x + n_y\hat{\mathbf{e}}_y + n_z\hat{\mathbf{e}}_z. \tag{1.239}$$

Now consider the situation when the region R is shrunk to a point. Let $\Delta\tau$ be the volume of R as the limit $\Delta\tau\to 0$ is approached and ΔS the small area enclosing $\Delta\tau$. Then as $\Delta\tau\to 0$, the above expressions become approximately

$$\Delta\tau\operatorname{grad}\phi \simeq \oiint_{\Delta S} \phi\hat{\mathbf{n}}\, dS \tag{1.240a}$$

$$\Delta\tau\operatorname{div}\mathbf{A} \simeq \oiint_{\Delta S} \mathbf{A}\cdot\hat{\mathbf{n}}\, dS \tag{1.240b}$$

$$\Delta\tau\operatorname{curl}\mathbf{A} \simeq \oiint_{\Delta S} \hat{\mathbf{n}}\times\mathbf{A}\, dS. \tag{1.240c}$$

These forms of the integral relations are also equivalent to use of the mean-value theorem. In the limit $\Delta\tau \to 0$, we thus have the following definitions for grad ϕ, div $\mathbf{A}$, and curl $\mathbf{A}$:

$$\operatorname{grad}\phi = \lim_{\Delta\tau\to 0} \frac{1}{\Delta\tau} \oiint_{\Delta S} \phi\hat{\mathbf{n}}\, dS \tag{1.241a}$$

$$\operatorname{div}\mathbf{A} = \lim_{\Delta\tau\to 0} \frac{1}{\Delta\tau} \oiint_{\Delta S} \mathbf{A}\cdot\hat{\mathbf{n}}\, dS \tag{1.241b}$$

$$\operatorname{curl}\mathbf{A} = \lim_{\Delta\tau\to 0} \frac{1}{\Delta\tau} \oiint_{\Delta S} \hat{\mathbf{n}}\times\mathbf{A}\, dS. \tag{1.241c}$$

These forms are known as the *invariant forms* since they do not depend in any way upon defined coordinate systems.

The first two relations above allow the gradient and divergence operations to be interpreted physically, and as such, the names *gradient* and *divergence* are appropriate. We note at first that the above definitions are akin to the definition of any ordinary derivative:

$$\frac{du}{dx} = \lim_{\Delta x\to 0} \frac{u(x+\Delta x)-u(x)}{\Delta x}. \tag{1.242}$$

The gradient operation is seen to be the weighing of the surface area $\hat{\mathbf{n}}\, dS$ by the weight function ϕ all around a closed surface, which is then divided by the volume of the enclosed region, in the limit as the region is shrunk to a point. It is sort of a three-dimensional version of the derivative du/dx, defined above, and accounts for direction as well as magnitude.

The combination $\mathbf{A}\cdot\hat{\mathbf{n}}\, dS$ is called the *outflow of* **A** *through the differential surface dS*. The integral is called the total or net outflow through the surrounding surface ΔS. This is easiest to see if one imagines that **A** is a velocity vector and the outflow is an amount of fluid flow. In the limit as the region shrinks to a point, the net outflow per unit volume is associated therefore with the *divergence of the vector field.*

The curl vector is also closely related to the circulation of a vector field by *Stokes' theorem*:

$$\iint_{CS} (\operatorname{curl}\mathbf{A})\cdot\hat{\mathbf{n}}\, dS = \oint_C \mathbf{A}\cdot d\mathbf{s}. \tag{1.243}$$

The line integral is called the *circulation of* **A** *about the curve C*. The surface involved "caps" the bounding curve and is called the *capping surface* (see Fig. 1.34).

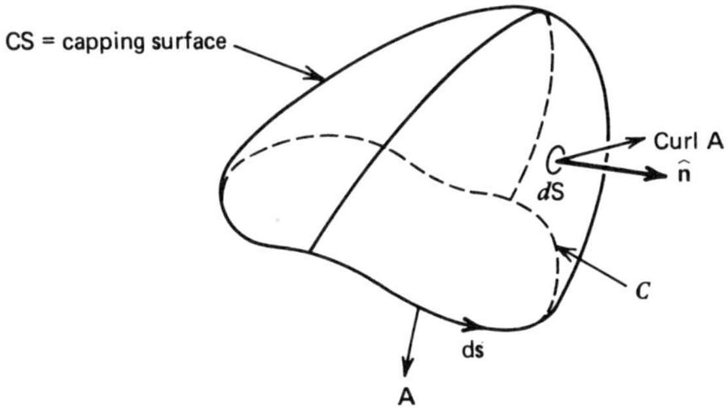

Figure 1.34 The Capping surface bounded by curve C.

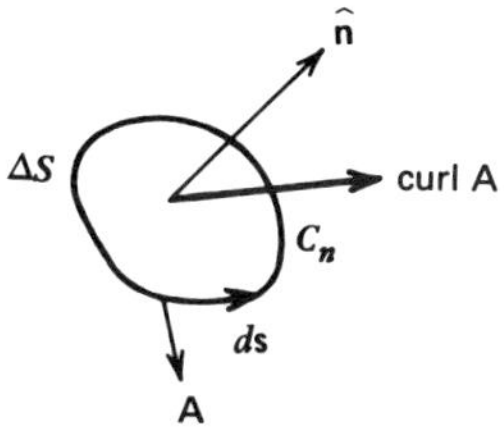

Figure 1.35 Infinitesimal surface element.

If we now let curve C shrink so that it surrounds an infinitesimal plane area, which is taken to be the capping surface (Fig. 1.35), then we have approximately

$$\Delta S \hat{\mathbf{n}} \cdot \operatorname{curl} \mathbf{A} \simeq \oint_C \mathbf{A} \cdot d\mathbf{s}. \tag{1.244}$$

In the limit $\Delta S \to 0$ we then obtain a definition for the component of curl **A** in the direction of $\hat{\mathbf{n}}$:

$$\hat{\mathbf{n}} \cdot \operatorname{curl} \mathbf{A} = \lim_{\Delta S \to 0} \frac{1}{\Delta S} \oint_{C_n} \mathbf{A} \cdot d\mathbf{s}, \tag{1.245}$$

where C_n is the curve surrounding the small area ΔS associated with the direction $\hat{\mathbf{n}}$.

Example 1.15 This example illustrates the relation between the integral relations (1.235) and the so-called integration by parts. Consider a rectangular region $R = \{(x, y): \ 0<x<a, 0<y<b\}$ with boundary C, which is the union of disjoint sets C_1, C_2, C_3, and C_4 (see Fig. 1.36).

Suppose that we wish to evaluate the integral $\int_R \nabla^2 \phi \, dx \, dy$. From Eq. (1.236) we have

$$\int_R \nabla^2 \phi \, dx \, dy = \oint_C \frac{\partial \phi}{\partial n} dS$$

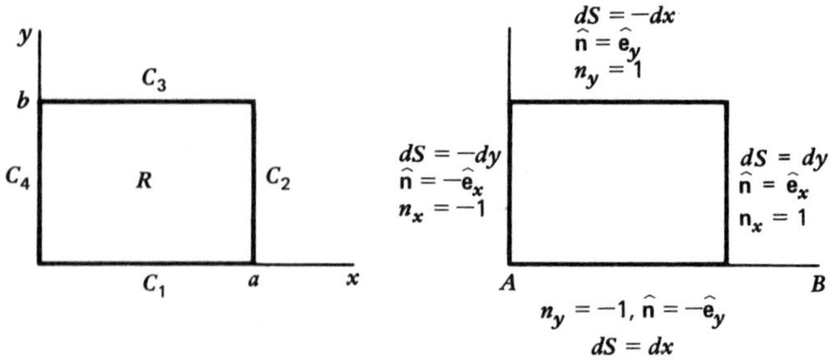

Figure 1.36 Integration over rectangular regions.

The line integral can be simplified for the region under consideration as follows (note that in two dimensions, the volume integral becomes a surface integral):

$$\oint_C \frac{\partial\phi}{\partial n}dS = \int_{C_1}\frac{\partial\phi}{\partial n}dS + \int_{C_2}\frac{\partial\phi}{\partial n}dS + \int_{C_3}\frac{\partial\phi}{\partial n}dS + \int_{C_4}\frac{\partial\phi}{\partial n}dS$$

$$= \int_0^a \left(-\frac{\partial\phi}{\partial y}\right)\bigg|_{y=0} dx + \int_0^b \left(\frac{\partial\phi}{\partial x}\right)\bigg|_{x=a} dy$$

$$+ \int_a^0 \left(\frac{\partial\phi}{\partial y}\right)\bigg|_{y=b} dx + \int_b^0 \left(-\frac{\partial\phi}{\partial x}\right)\bigg|_{x=0} dy$$

$$= \int_0^a \left[\left(\frac{\partial\phi}{\partial y}\right)_{y=b} - \left(\frac{\partial\phi}{\partial y}\right)_{y=0}\right] dx + \int_0^b \left[\left(\frac{\partial\phi}{\partial x}\right)_{x=a} - \left(\frac{\partial\phi}{\partial x}\right)_{x=0}\right] dy.$$

The same result can be obtained by means of integration by parts:

$$\int_R \nabla^2\phi\, dx\, dy = \int_0^b\int_0^a \left(\frac{\partial^2\phi}{\partial x^2} + \frac{\partial^2\phi}{\partial y^2}\right) dx\, dy$$

$$= \int_0^b\int_0^a \frac{\partial}{\partial x}\left(\frac{\partial\phi}{\partial x}\right) dx\, dy + \int_0^a\int_0^b \frac{\partial}{\partial y}\left(\frac{\partial\phi}{\partial y}\right) dy\, dx$$

$$= \int_0^b \left(\frac{\partial\phi}{\partial x}\right)\bigg|_{x=0}^{x=a} dy + \int_0^a \left(\frac{\partial\phi}{\partial y}\right)\bigg|_{y=0}^{y=b} dx$$

$$= \int_0^b \left[\left(\frac{\partial\phi}{\partial x}\right)_{x=a} - \left(\frac{\partial\phi}{\partial x}\right)_{x=0}\right] dy + \int_0^a \left[\left(\frac{\partial\phi}{\partial y}\right)_{y=b} - \left(\frac{\partial\phi}{\partial y}\right)_{y=0}\right] dx.$$

Thus integration by parts is a special case of the gradient or the divergence theorem. However, for arbitrary domains it is not feasible to use integration by parts explicitly. ■

Exercises 1.7

1 Show that the vector area of a closed surface is zero, that is,

$$\oiint_S \hat{\mathbf{n}}\, dS \equiv 0.$$

2 Show that the volume enclosed by a surface S is

$$\text{volume} = \tfrac{1}{6}\oiint_S \operatorname{grad}(r^2)\cdot \hat{\mathbf{n}}\, dS$$

or

$$\text{volume} = \tfrac{1}{3}\oiint_S \mathbf{r}\cdot \hat{\mathbf{n}}\, dS.$$

3 Let $\phi(\mathbf{r})$ be a scalar field. Show that

$$\iiint_R \nabla^2\phi\, d\tau = \oiint_S \frac{\partial\phi}{\partial n}\, dS$$

where $\partial\phi/\partial n \equiv \hat{\mathbf{n}}\cdot \operatorname{grad}\phi$ is the derivative of ϕ in the outward direction normal to the surface.

4 Deduce that the Laplacian of a scalar function of position can be expressed independently of all coordinate systems by the invariant form

$$\nabla^2\phi \equiv \lim_{\Delta\tau\to 0} \frac{1}{\Delta\tau}\oiint_{\Delta S} \frac{\partial\phi}{\partial n}\, dS.$$

5 In the divergence theorem, set $\mathbf{A} = \phi \operatorname{grad}\psi$ and $\mathbf{A} = \psi \operatorname{grad}\phi$ successively and obtain the integral forms

$$\iiint_R [\phi\nabla^2\psi + \nabla\phi\cdot\nabla\psi]\, d\tau = \oiint_S \phi\frac{\partial\psi}{\partial n}\, dS$$

$$\iiint_R [\phi\nabla^2\psi - \psi\nabla^2\phi]\, d\tau = \oiint_S \left[\phi\frac{\partial\psi}{\partial n} - \psi\frac{\partial\phi}{\partial n}\right] dS$$

$$\iiint_R [\phi\nabla^4\psi - \nabla^2\phi\nabla^2\psi]\, d\tau = \oiint_S \left[\phi\frac{\partial}{\partial n}(\nabla^2\psi) - \nabla^2\psi\frac{\partial\phi}{\partial n}\right] dS.$$

The first two identities are sometimes called Green's first and second theorems.

6 Define the Dirac delta function $\delta(\mathbf{r}-\boldsymbol{\xi})=\delta(\boldsymbol{\xi}-\mathbf{r})$ as a function that is singular at $\mathbf{r}=\boldsymbol{\xi}$, zero for all $\mathbf{r}\neq\boldsymbol{\xi}$, and having the integral property

$$\iiint_R \phi(\mathbf{r})\delta(\mathbf{r}-\boldsymbol{\xi})\,d\tau=\phi(\boldsymbol{\xi}), \qquad \boldsymbol{\xi} \text{ inside } R$$

$$=0, \qquad \boldsymbol{\xi} \text{ not inside } R.$$

By means of the divergence theorem, with the region R equal to a sphere of radius r centered at the point $\boldsymbol{\xi}$, verify that $\psi(\mathbf{r})\equiv|\mathbf{r}-\boldsymbol{\xi}|^{-1}$ satisfies the singular Poisson equation

$$\nabla^2\psi=-4\pi\delta(\mathbf{r}-\boldsymbol{\xi}).$$

7 Consider a function $\phi(\mathbf{r})$ governed by the general Poisson equation

$$\nabla^2\phi=Q(\mathbf{r}),$$

where $Q(\mathbf{r})$ is a known function and where $\phi=\phi(\mathbf{r})$ vanishes and $\nabla\phi$ vanishes faster than r^{-1} at infinity. Deduce that the solution for the Poisson equation, obtained by means of Green's second theorem and the function $\psi(\mathbf{r})=-|\mathbf{r}-\boldsymbol{\xi}|^{-1}$, is

$$\phi(\boldsymbol{\xi})=-\frac{1}{4\pi}\iiint_R \frac{Q(\mathbf{r})}{|\mathbf{r}-\boldsymbol{\xi}|}\,d\tau,$$

where the integration is over all of space. Here there are no boundary conditions imposed on any finite surfaces, and hence the above solution can be regarded as the particular integral of the Poisson equation evaluated over all of space.

8 The static pressure field $p(\mathbf{r})$ in a fluid medium with mass density $\rho(\mathbf{r})$ under the influence of a gravitational acceleration $\mathbf{g}$ is governed by the equation

$$\operatorname{grad} p=\rho\mathbf{g}.$$

The buoyancy force exerted by the fluid medium on an object bounded by the surface S is given by

$$\mathbf{F}=-\oiint_S p\,\hat{\mathbf{n}}\,dS.$$

By means of the gradient theorem, show that the buoyancy force $\mathbf{F}$ is opposite to $\mathbf{g}$ and equal to the weight of the displaced fluid, that is,

$$\mathbf{F}=-\mathbf{g}\iiint_R \rho\,d\tau.$$

9 Let an arbitrary region in a continuous medium be denoted by R and the bounding, closed surface of this region be continuous and denoted by S. Let each point on the bounding surface move with the velocity $\mathbf{V}_s$. It can be shown that the time derivative of the volume integral over some continuous function $Q(\mathbf{r}, t)$ is given by

$$\frac{d}{dt}\iiint_R Q(\mathbf{r}, t)\, d\tau \equiv \iiint_R \frac{\partial Q}{\partial t} d\tau + \oiint_S Q\mathbf{V}_s \cdot \hat{\mathbf{n}}\, dS.$$

This expression for the differentiation of a volume integral with variable limits is sometimes known as the three-dimensional *Leibnitz rule*.

Let each element of mass in the medium move with the velocity $\mathbf{V}(\mathbf{r}, t)$ and consider a special region R such that the bounding surface S is attached to a fixed set of material elements. Then each point of this surface moves itself with the material velocity, that is, $\mathbf{V}_s = \mathbf{V}$, and the region R thus contains a fixed total amount of mass since no mass crosses the boundary surface S. To distinguish the time rate of change of an integral over this material region, we replace d/dt by D/Dt and write

$$\frac{D}{Dt}\iiint_R Q(\mathbf{r}, t)\, d\tau \equiv \iint_S \frac{\partial Q}{\partial t} d\tau + \oiint_S Q\mathbf{V} \cdot \hat{\mathbf{n}}\, dS,$$

which holds for a material region, that is, a region of fixed total mass.

Show that the relation between the time derivative following an arbitrary region and the time derivative following a material region (fixed total mass) is

$$\frac{d}{dt}\iiint_R Q(\mathbf{r}, t)\, d\tau \equiv \frac{D}{Dt}\iiint_R Q(\mathbf{r}, t)\, d\tau + \oiint_S Q(\mathbf{V}_s - \mathbf{V}) \cdot \hat{\mathbf{n}}\, dS.$$

The velocity difference $\mathbf{V} - \mathbf{V}_s$ is the velocity of the material measured relative to the velocity of the surface. The surface integral

$$\oiint_S Q(\mathbf{V} - \mathbf{V}_s) \cdot \hat{\mathbf{n}}\, dS$$

thus measures the total *outflow* of the property Q from the region R.

10 Let $Q = \rho(\mathbf{r}, t)$ denote the mass density of a continuous region. Then conservation of mass for a *material* region requires that

$$\frac{D}{Dt}\iiint_R \rho\, d\tau = 0.$$

Show that for a *fixed region* ($\mathbf{V}_s = \mathbf{0}$) conservation of mass can also be stated as

$$\frac{d}{dt}\iiint_R \rho\, d\tau = -\oiint_S \rho \mathbf{V}\cdot \hat{\mathbf{n}}\, dS$$

or

$$\iiint_R \frac{\partial \rho}{\partial t}\, d\tau = -\oiint_S \rho \mathbf{V}\cdot \hat{\mathbf{n}}\, dS.$$

Interpret these equations physically.

11 In the last equation of Exercise 10 convert the surface integral to a volume integral by means of the divergence theorem, and convert the equation to a single volume integral that vanishes. Since this integral vanishes, for a continuous medium, for any arbitrary region R, deduce that this can be true only if the integrand itself vanishes identically, and thus that

$$\frac{\partial \rho}{\partial t} + \operatorname{div}(\rho \mathbf{V}) = 0.$$

This equation, called the *continuity equation*, expresses local conservation of mass at any point in a continuous medium. Express this equation in appropriate forms for contravariant and covariant components of $\mathbf{V}$. Write the equation in appropriate forms for orthogonal curvilinear coordinates and for Cartesian coordinates.

12 In the field description of a continuous variable $Q = Q(\mathbf{r}, t)$ let the field position be a function of time such that $\mathbf{r} = \mathbf{r}(t)$. Deduce that the total time derivative of Q can be written

$$\frac{dQ}{dt} = \frac{\partial Q}{\partial t} + \frac{d\mathbf{r}}{dt}\cdot \operatorname{grad} Q.$$

This arbitrary total time derivative corresponds to a change in Q following a change in $\mathbf{r}$ with time. If we let $\mathbf{r}$ correspond to the position of a fixed material element, then $d\mathbf{r}/dt = \mathbf{V}$ corresponds to the velocity of the material element. To distinguish the time rates of change following a material element from other arbitrary changes, we write

$$\frac{DQ}{Dt} \equiv \frac{\partial Q}{\partial t} + \mathbf{V}\cdot \operatorname{grad} Q.$$

This is the differential time rate of change of a field variable $Q(\mathbf{r}, t)$ following a material element. It is referred to as the *material derivative*, the *substantial derivative*, or the *Eulerian derivative*. The corresponding material derivative for integrals was defined in Exercise 9.

By means of vector identities show that the continuity equation for mass conservation can be written as

$$\frac{D\rho}{Dt}+\rho\,\mathrm{div}\,\mathbf{V}=0.$$

13 The material derivative operator D/Dt corresponds to changes with respect to a fixed mass, that is, $\rho\,d\tau$ is constant with respect to this operator. Show formally by means of Leibnitz's rule, the divergence theorem, and conservation of mass that

$$\frac{D}{Dt}\iiint_R \rho Q\,d\tau \equiv \iiint_R \rho\frac{DQ}{Dt}\,d\tau.$$

14 Letting a finite volume R shrink to an infinitesimal volume $\Delta\tau$, show by setting $Q=1$ in Leibnitz's rule and by use of the divergence theorem that div $\mathbf{V}$ can be interpreted as

$$\mathrm{div}\,\mathbf{V}\equiv \lim_{\Delta\tau\to 0}\frac{1}{\Delta\tau}\frac{D}{Dt}(\Delta\tau).$$

The right-hand side can be interpreted as the rate of volumetric strain following a material particle, called the dilatation rate.

15 The acceleration of a material element in a continuum is described by

$$\frac{D\mathbf{V}}{Dt}\equiv\frac{\partial\mathbf{V}}{\partial t}+\mathbf{V}\cdot\mathrm{grad}\,\mathbf{V}.$$

Show by means of vector identities that the acceleration can also be written as

$$\frac{D\mathbf{V}}{Dt}\equiv\frac{\partial\mathbf{V}}{\partial t}+\mathrm{grad}\left(\frac{\mathbf{V}^2}{2}\right)-\mathbf{V}\times\mathrm{curl}\,\mathbf{V}.$$

This form displays the role of the *vorticity vector* curl $\mathbf{V}$.

16 Deduce that

$$\mathbf{V}\cdot\frac{D\mathbf{V}}{Dt}\equiv\frac{D}{Dt}\left(\frac{\mathbf{V}^2}{2}\right).$$

Note that $V^2/2$ is the kinetic energy per unit mass of a material particle.

17 Deduce that

$$\mathrm{curl}\left(\frac{D\mathbf{V}}{Dt}\right)\equiv\frac{D\boldsymbol{\Omega}}{Dt}+\boldsymbol{\Omega}\,\mathrm{div}\,\mathbf{V}-\boldsymbol{\Omega}\cdot\nabla\mathbf{V}$$

where $\boldsymbol{\Omega}\equiv\mathrm{curl}\,\mathbf{V}$ as the vorticity vector.

18 Newton's second law of motion in its elementary form $m\mathbf{a}=\mathbf{F}$ holds strictly for a point particle of fixed mass m. For a material region of continuously distributed mass, Newton's second law reads

$$\frac{D}{Dt}\iiint_R \rho\mathbf{V}\,d\tau=\mathbf{F},$$

where $\mathbf{F}$ is the sum of all the forces acting on the material in R. Deduce by means of the results of Exercise 9 that the corresponding law of motion for a region of *variable* mass is

$$\frac{d}{dt}\iiint_R \rho\mathbf{V}\,d\tau=\mathbf{F}+\oiint_S \rho\mathbf{V}\,(\mathbf{V}_s-\mathbf{V})\cdot\hat{\mathbf{n}}\,dS,$$

where $\mathbf{V}_s$ is the velocity of points on the surface S bounding the variable mass region R.

19 Define the mass in an arbitrary region by

$$m\equiv\iiint_R \rho\,d\tau.$$

From conservation of mass it follows that

$$\dot{m}\equiv\frac{d}{dt}\iiint_R \rho\,d\tau=\oiint_S \rho(\mathbf{V}_s-\mathbf{V})\cdot\hat{\mathbf{n}}\,dS.$$

Define the mass average velocity inside the arbitrary region R by

$$\mathbf{V}_a\equiv\frac{\iiint_R \rho\mathbf{V}\,d\tau}{\iiint_R \rho\,d\tau}$$

and the average efflux velocity by

$$\mathbf{V}_e\equiv\frac{\oiint_S \rho\mathbf{V}(\mathbf{V}_s-\mathbf{V})\cdot\hat{\mathbf{n}}\,dS}{\oiint_S \rho(\mathbf{V}_s-\mathbf{V})\cdot\hat{\mathbf{n}}\,dS}.$$

Show that the equation of motion, corresponding to Newton's second law, for a body of variable mass $m(t)$ can be written

$$\frac{d}{dt}(m\mathbf{V}_a)=\mathbf{F}+\dot{m}\mathbf{V}_e.$$

This can also be written as

$$m\frac{d\mathbf{V}_a}{dt}=\mathbf{F}+\dot{m}\mathbf{V}_{ea},$$

where $\mathbf{V}_{ea}\equiv\mathbf{V}_e-\mathbf{V}_a$ is the average efflux velocity measured relative to the mass average velocity of the body. This last equation is sometimes referred to as the *rocket equation*. The relative momentum flux $\dot{m}\mathbf{V}_{ea}$ plays the role of an apparent thrust force.

1.5 DYADICS AND TENSORS

1.5.1 Dyadics in Physical Applications

We shall consider two physical examples in which dyadics appear and are defined because of their usefulness.

Angular Momentum of a Rotating Rigid Body

If ρ is the density of a continuous medium and $d\tau$ a small volume element, such that $\rho\, d\tau$ is the mass contained in the volume element, then the angular momentum about some point O is denoted by

$$d\mathbf{H}_O=\mathbf{r}\times(\rho\, d\tau)\mathbf{V}, \tag{1.246}$$

where $\mathbf{r}$ is the distance from the point O to the volume element $d\tau$. If R is the entire region of a rotating rigid body, then the angular momentum of the entire body is given by

$$\mathbf{H}_O=\iiint_R(\mathbf{r}\times\mathbf{V})\rho\, d\tau. \tag{1.247}$$

The vector $\mathbf{H}_O$ is called the *angular momentum* about the point O (see Fig. 1.37).

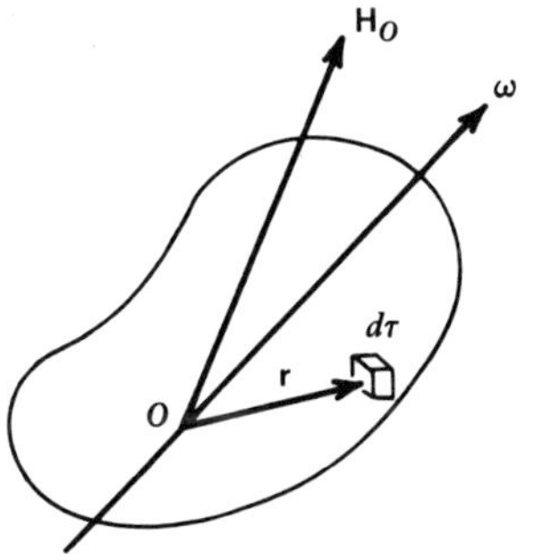

Figure 1.37 Angular momentum for a rotating rigid body.

The velocity of the element at $\mathbf{r}$ is given by $\mathbf{V}(\mathbf{r})=\boldsymbol{\omega}\times\mathbf{r}$, where $\boldsymbol{\omega}$ does not depend on $\mathbf{r}$ for a rigid body but may depend upon the time t. We wish to write $\mathbf{H}_O$ in such a way as to display explicitly the dependence on the angular rotation $\boldsymbol{\omega}$, which in general is in a different direction from $\mathbf{H}_O$. We first note that

$$\mathbf{H}_O=\iiint_R[\mathbf{r}\times(\boldsymbol{\omega}\times\mathbf{r})]\rho\,d\tau. \qquad (1.248)$$

Use of Eq. (1.31) for the vector triple product gives

$$\mathbf{H}_O=\iiint_R[r^2\boldsymbol{\omega}-(\mathbf{r}\cdot\boldsymbol{\omega})\mathbf{r}]\rho\,d\tau. \qquad (1.249)$$

Because $\boldsymbol{\omega}$ does not depend on $\mathbf{r}$, we can factor it out of the integral. This is easily done for the first term, and it can be done for the second term if we agree to write $(\mathbf{r}\cdot\boldsymbol{\omega})\mathbf{r}$ as $\boldsymbol{\omega}\cdot(\mathbf{rr})$ or $(\mathbf{rr})\cdot\boldsymbol{\omega}$. To do this, however, we must introduce the notion of two vectors standing side by side and acting as a single entity, such as $(\mathbf{rr})$. This combination of two vectors, with no operation such as a dot or cross between them, is called a *dyad*. A linear combination of dyads is called a *dyadic*. Using this new notion, we can interpret $(\mathbf{rr})\cdot\boldsymbol{\omega}$ as meaning $\mathbf{r}(\mathbf{r}\cdot\boldsymbol{\omega})$. Thus when a dyad operates on a vector with a dot product, the dot operation is associated only with the adjacent vectors.

It is convenient to treat the other term $r^2\boldsymbol{\omega}$ in an analogous fashion. To do this we introduce the notion of a unit dyadic $\overleftrightarrow{\mathbf{I}}$ defined as follows:

$$\overleftrightarrow{\mathbf{I}}=\mathbf{e}_1\mathbf{e}^1+\mathbf{e}_2\mathbf{e}^2+\mathbf{e}_3\mathbf{e}^3. \qquad (1.250)$$

It follows that

$$\begin{aligned}\overleftrightarrow{\mathbf{I}}\cdot\boldsymbol{\omega}&=\mathbf{e}_1(\mathbf{e}^1\cdot\boldsymbol{\omega})+\mathbf{e}_2(\mathbf{e}^2\cdot\boldsymbol{\omega})+\mathbf{e}_3(\mathbf{e}^3\cdot\boldsymbol{\omega})\\&=\omega^1\mathbf{e}_1+\omega^2\mathbf{e}_2+\omega^3\mathbf{e}_3\\&=\boldsymbol{\omega}\end{aligned} \qquad (1.251)$$

Thus the unit dyadic when it operates on a vector with a dot product changes the vector into itself. This is consistent with the notion of unity in the scalar number system. We can also define $\overleftrightarrow{\mathbf{I}}$ with the basis vectors reversed and obtain the same results, that is,

$$\overleftrightarrow{\mathbf{I}}=\mathbf{e}^1\mathbf{e}_1+\mathbf{e}^2\mathbf{e}_2+\mathbf{e}^3\mathbf{e}_3. \qquad (1.252)$$

When the vectors in the dyads can be interchanged without affecting the dyadic, the dyadic is said to be *symmetric*. Thus in summation notation, we have

$$\overleftrightarrow{\mathbf{I}}=\mathbf{e}^i\mathbf{e}_i=\mathbf{e}_i\mathbf{e}^i.$$

Moreover, since $\overleftrightarrow{\mathbf{I}}$ is symmetric, we also have

$$\boldsymbol{\omega}\cdot\overleftrightarrow{\mathbf{I}}=\overleftrightarrow{\mathbf{I}}\cdot\boldsymbol{\omega}=\boldsymbol{\omega}. \tag{1.253}$$

With the above definitions we can now write the angular-momentum vector as

$$\mathbf{H}_0=\overleftrightarrow{\mathfrak{I}}\cdot\boldsymbol{\omega}, \tag{1.254}$$

where

$$\overleftrightarrow{\mathfrak{I}}\equiv\iiint_R[r^2\overleftrightarrow{\mathbf{I}}-\mathbf{rr}]\rho\,d\tau. \tag{1.255}$$

The entity $\overleftrightarrow{\mathfrak{I}}$ is called the *moment-of-inertia dyadic*, or *moment-of-inertia tensor*. If the density is uniform, it depends entirely on the geometric shape of the rigid body and not on how it is rotating. Because $\mathbf{H}_0$ is a vector that in general has a different length and orientation from $\boldsymbol{\omega}$, we can interpret $\overleftrightarrow{\mathfrak{I}}$ as an *operator that changes* $\boldsymbol{\omega}$ *into* $\mathbf{H}_0$.

The dynamical equation for angular momentum states that

$$\frac{d\mathbf{H}_0}{dt}=\mathbf{M}_0, \tag{1.256}$$

where $\mathbf{M}_0$ is the resultant moment about the point O acting on the body. This can also be written as

$$\frac{d}{dt}(\overleftrightarrow{\mathfrak{I}}\cdot\boldsymbol{\omega})=\mathbf{M}_0. \tag{1.257}$$

In this equation we must take note that both $\overleftrightarrow{\mathfrak{I}}$ and $\boldsymbol{\omega}$ depend on time t. If we take a coordinate system fixed in the rigid body, however, $\overleftrightarrow{\mathfrak{I}}$ is a constant and can be computed once and for all. Thus for a reference system fixed in the rotating rigid body we can recall Eq. (1.176) for the time derivative of vectors in rotating coordinate systems,

$$\frac{d}{dt}\to\frac{d}{dt}+\boldsymbol{\omega}\times, \tag{1.258}$$

and thus write the dynamical equation as

$$\overleftrightarrow{\mathfrak{I}}\cdot\frac{d\boldsymbol{\omega}}{dt}+\boldsymbol{\omega}\times(\overleftrightarrow{\mathfrak{I}}\cdot\boldsymbol{\omega})=\mathbf{M}_0. \tag{1.259}$$

This equation is known as *Euler's equation*, which is a vector differential equation for $\boldsymbol{\omega}$ when $\mathbf{M}_0$ is regarded as known. In body-fixed coordinates, the moment-of-inertia tensor is a constant.

It is useful in these problems to determine the kinetic energy of the rotation. We obtain

$$\begin{aligned} E &= \tfrac{1}{2}\iiint_R \mathbf{V}^2 \rho\, d\tau \\ &= \tfrac{1}{2}\iiint_R (\boldsymbol{\omega}\times\mathbf{r})\cdot \mathbf{V}\rho\, d\tau \\ &= \tfrac{1}{2}\iiint_R \boldsymbol{\omega}\cdot(\mathbf{r}\times\mathbf{V})\rho\, d\tau \\ &= \tfrac{1}{2}\boldsymbol{\omega}\cdot \iiint_R (\mathbf{r}\times\mathbf{V})\rho\, d\tau. \end{aligned} \tag{1.260}$$

Thus we finally get

$$E = \tfrac{1}{2}\boldsymbol{\omega}\cdot \overleftrightarrow{\mathfrak{I}}\cdot \boldsymbol{\omega}. \tag{1.261}$$

Even the most pedestrian among us will have to admit there is a certain amount of beauty in this representation!

Stress Dyadic for a Continuous Medium

The surface force acting on a small element of area in a continuous medium depends not only on the magnitude of the area but also upon the orientation of the area. If we denote by $\mathbf{F}(\hat{\mathbf{n}})$ the force on a small area $\hat{\mathbf{n}}\Delta S$ located at the position $\mathbf{r}$ (Fig. 1.38), the *stress vector* can be defined as follows:

$$\boldsymbol{\pi}(\hat{\mathbf{n}}) = \lim_{\Delta S\to 0} \frac{\mathbf{F}(\hat{\mathbf{n}})}{\Delta S}. \tag{1.262}$$

We see that the stress vector is a point function of the unit normal $\hat{\mathbf{n}}$ which denotes the orientation of the surface ΔS. The component of $\boldsymbol{\pi}$ that is in the direction of $\hat{\mathbf{n}}$ is called the *normal* stress. The component of $\boldsymbol{\pi}$ that is normal to

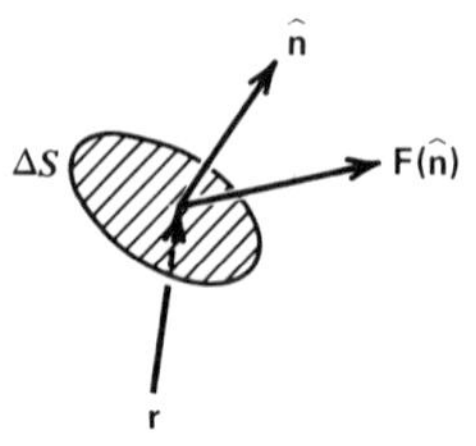

Figure 1.38 Force on an area element.

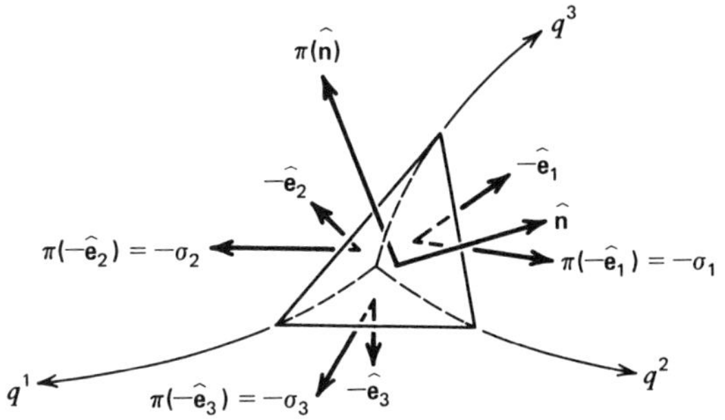

Figure 1.39 Tetrahedral element in orthogonal curvilinear coordinates.

$\hat{\mathbf{n}}$ is called a *shear* stress. Because of Newton's third law for action and reaction, we see that $\boldsymbol{\pi}(-\hat{\mathbf{n}}) = -\boldsymbol{\pi}(\hat{\mathbf{n}})$.

The stress at a point $\mathbf{r}$ is defined in terms of two vectors: at a fixed point $\mathbf{r}$ for each given unit vector $\hat{\mathbf{n}}$ there is a stress vector $\boldsymbol{\pi}(\hat{\mathbf{n}})$ acting on the plane normal to $\hat{\mathbf{n}}$. It is fruitful to establish a relationship between $\boldsymbol{\pi}$ and $\hat{\mathbf{n}}$. To do this we now set up an infinitesimal tetrahedron in orthogonal curvilinear coordinates as shown in Fig. 1.39.

If $-\boldsymbol{\sigma}_1, -\boldsymbol{\sigma}_2, -\boldsymbol{\sigma}_3$, and $\boldsymbol{\pi}$ denote the stress vectors in the outward directions on the faces of the infinitesimal tetrahedron whose areas are $\Delta S_1, \Delta S_2, \Delta S_3$, and ΔS, we have by Newton's second law for the mass inside the tetrahedron,

$$\boldsymbol{\pi}\Delta S - \boldsymbol{\sigma}_1\Delta S_1 - \boldsymbol{\sigma}_2\Delta S_2 - \boldsymbol{\sigma}_3\Delta S_3 + \rho\Delta\tau\mathbf{f} = \rho\Delta\tau\mathbf{a}, \tag{1.263}$$

where $\Delta\tau$ is the volume of the tetrahedron, ρ the density, $\mathbf{f}$ the body force per unit mass, and $\mathbf{a}$ the acceleration.

Setting $\phi = 1$ in the gradient theorem (1.235a) shows that the total vector area of a closed surface is zero. Thus for the tetrahedron we have

$$\Delta S\hat{\mathbf{n}} - \Delta S_1\hat{\mathbf{e}}_1 - \Delta S_2\hat{\mathbf{e}}_2 - \Delta S_3\hat{\mathbf{e}}_3 = 0. \tag{1.264}$$

It follows that

$$\Delta S_1 = (\hat{\mathbf{n}}\cdot\hat{\mathbf{e}}_1)\Delta S, \qquad \Delta S_2 = (\hat{\mathbf{n}}\cdot\hat{\mathbf{e}}_2)\Delta S, \qquad \Delta S_3 = (\hat{\mathbf{n}}\cdot\hat{\mathbf{e}}_3)\Delta S. \tag{1.265}$$

The volume of the element can be expressed as

$$\Delta\tau = \frac{\Delta h}{3}\Delta S, \tag{1.266}$$

where Δh is the normal distance from the origin to the slant face. [This result can be obtained by setting $\mathbf{A} = \mathbf{r}$ in the divergence theorem (1.235b).] These expressions substituted into Newton's second law reduce it to the form

$$\boldsymbol{\pi} = (\hat{\mathbf{n}}\cdot\hat{\mathbf{e}}_1)\boldsymbol{\sigma}_1 + (\hat{\mathbf{n}}\cdot\hat{\mathbf{e}}_2)\boldsymbol{\sigma}_2 + (\hat{\mathbf{n}}\cdot\hat{\mathbf{e}}_3)\boldsymbol{\sigma}_3 + \rho\frac{\Delta h}{3}(\mathbf{a} - \mathbf{f}). \tag{1.267}$$

In the limit when the tetrahedron shrinks to a point, $\Delta h \to 0$, we are left with

$$\boldsymbol{\pi} = (\hat{\mathbf{n}} \cdot \hat{\mathbf{e}}_1)\boldsymbol{\sigma}_1 + (\hat{\mathbf{n}} \cdot \hat{\mathbf{e}}_2)\boldsymbol{\sigma}_2 + (\hat{\mathbf{n}} \cdot \hat{\mathbf{e}}_3)\boldsymbol{\sigma}_3.$$

It is now convenient to display the unit normal $\hat{\mathbf{n}}$ by writing the last equation as

$$\boldsymbol{\pi} = \hat{\mathbf{n}} \cdot [\hat{\mathbf{e}}_1\boldsymbol{\sigma}_1 + \hat{\mathbf{e}}_2\boldsymbol{\sigma}_2 + \hat{\mathbf{e}}_3\boldsymbol{\sigma}_3]. \tag{1.268}$$

The terms in the brackets are to be treated as a dyadic, called a *stress dyadic*, or *stress tensor*:

$$\overleftrightarrow{\boldsymbol{\sigma}} \equiv \hat{\mathbf{e}}_1\boldsymbol{\sigma}_1 + \hat{\mathbf{e}}_2\boldsymbol{\sigma}_2 + \hat{\mathbf{e}}_3\boldsymbol{\sigma}_3. \tag{1.269}$$

The stress tensor is a property of the medium that is independent of $\hat{\mathbf{n}}$. Thus we have

$$\boldsymbol{\pi}(\hat{\mathbf{n}}) = \hat{\mathbf{n}} \cdot \overleftrightarrow{\boldsymbol{\sigma}}, \tag{1.270}$$

and the dependence on $\hat{\mathbf{n}}$ has been explicitly displayed. The dot operation of the stress dyadic on $\hat{\mathbf{n}}$ can be viewed as transforming the vector $\hat{\mathbf{n}}$ into the vector $\boldsymbol{\pi}$.

It is useful to resolve the stress vectors $\boldsymbol{\sigma}_1, \boldsymbol{\sigma}_2$, and $\boldsymbol{\sigma}_3$ into their respective orthonormal components. We have

$$\begin{aligned} \boldsymbol{\sigma}_i &= \sigma_{i1}\hat{\mathbf{e}}_1 + \sigma_{i2}\hat{\mathbf{e}}_2 + \sigma_{i3}\hat{\mathbf{e}}_3 \\ &= \sigma_{ij}\hat{\mathbf{e}}_j. \end{aligned} \tag{1.271}$$

Hence the stress dyadic can be written in summation notation as

$$\begin{aligned} \overleftrightarrow{\boldsymbol{\sigma}} &= \hat{\mathbf{e}}_i\hat{\boldsymbol{\sigma}}_i \\ &= \sigma_{ij}\hat{\mathbf{e}}_i\hat{\mathbf{e}}_j. \end{aligned} \tag{1.272}$$

The component σ_{ij} represents the stress on an area perpendicular to the ith coordinate and in the jth direction. The stress vector $\boldsymbol{\pi}$ represents the *vectorial* stress on an area perpendicular to the direction $\hat{\mathbf{n}}$. The character of the stress dyadic, or stress tensor, $\overleftrightarrow{\boldsymbol{\sigma}}$ depends on the properties of a particular medium.

1.5.2 General Properties of Dyadics

Because of its utilization in physical applications, as we have seen, a dyad is defined as two vectors standing side by side and acting as a unit. A linear combination of dyads is called a dyadic. Let $\mathbf{A}_1, \mathbf{A}_2, \ldots, \mathbf{A}_n$ and $\mathbf{B}_1, \mathbf{B}_2, \ldots, \mathbf{B}_n$ be arbitrary vectors. Then we can represent a dyadic as

$$\overleftrightarrow{\boldsymbol{\Phi}} = \mathbf{A}_1\mathbf{B}_1 + \mathbf{A}_2\mathbf{B}_2 + \cdots + \mathbf{A}_n\mathbf{B}_n. \tag{1.273}$$

One of the properties of a dyadic is defined by the dot product with a vector, say **V**:

$$\overleftrightarrow{\mathbf{\Phi}}\cdot\mathbf{V}=\mathbf{A}_1(\mathbf{B}_1\cdot\mathbf{V})+\mathbf{A}_2(\mathbf{B}_2\cdot\mathbf{V})+\cdots+\mathbf{A}_n(\mathbf{B}_n\cdot\mathbf{V})$$

$$\mathbf{V}\cdot\overleftrightarrow{\mathbf{\Phi}}=(\mathbf{V}\cdot\mathbf{A}_1)\mathbf{B}_1+(\mathbf{V}\cdot\mathbf{A}_2)\mathbf{B}_2+\cdots+(\mathbf{V}\cdot\mathbf{A}_n)\mathbf{B}_n. \qquad (1.274)$$

The dot operation with a vector produces another vector. In the first case the dyadic acts as a *prefactor* and in the second case as a *postfactor*. The two operations in general produce different vectors.

The conjugate, or transpose, of a dyadic is defined as the result obtained by the interchange of the two vectors in each of the dyads:

$$\overleftrightarrow{\mathbf{\Phi}}^t=\mathbf{B}_1\mathbf{A}_1+\mathbf{B}_2\mathbf{A}_2+\cdots+\mathbf{B}_n\mathbf{A}_n. \qquad (1.275)$$

It is clear that we have

$$\mathbf{V}\cdot\overleftrightarrow{\mathbf{\Phi}}=\overleftrightarrow{\mathbf{\Phi}}^t\cdot\mathbf{V}$$

$$\overleftrightarrow{\mathbf{\Phi}}\cdot\mathbf{V}=\mathbf{V}\cdot\overleftrightarrow{\mathbf{\Phi}}^t. \qquad (1.276)$$

1.5.3 Nonion Form of a Dyadic

Let each of the vectors in the dyadic be represented in a given basis system. Then we have

$$\mathbf{A}_i=A_i^j\mathbf{e}_j$$

$$\mathbf{B}_i=B_i^k\mathbf{e}_k. \qquad (1.277)$$

Here we have selected a cogredient basis or unitary basis in each case. We could have mixed the basis systems. For this case we have

$$\overleftrightarrow{\mathbf{\Phi}}=\mathbf{A}_i\mathbf{B}_i$$

$$=A_i^jB_i^k\mathbf{e}_j\mathbf{e}_k$$

$$=\phi^{jk}\mathbf{e}_j\mathbf{e}_k, \qquad (1.278)$$

where

$$\phi^{jk}\equiv A_i^jB_i^k. \qquad (1.279)$$

The summations on i, j, and k are implied by the repeated indices. We can display all of the components by letting the k index run to the right and the j

index run downward:

$$\overleftrightarrow{\mathbf{\Phi}} = \phi^{11}\mathbf{e}_1\mathbf{e}_1 + \phi^{12}\mathbf{e}_1\mathbf{e}_2 + \phi^{13}\mathbf{e}_1\mathbf{e}_3$$
$$+ \phi^{21}\mathbf{e}_2\mathbf{e}_1 + \phi^{22}\mathbf{e}_2\mathbf{e}_2 + \phi^{23}\mathbf{e}_2\mathbf{e}_3$$
$$+ \phi^{31}\mathbf{e}_3\mathbf{e}_1 + \phi^{32}\mathbf{e}_3\mathbf{e}_2 + \phi^{33}\mathbf{e}_3\mathbf{e}_3. \tag{1.280}$$

This form is called the *nonion* form.

Equation (1.280) illustrates that a dyadic in three-space, or what we shall call a second-order tensor, has nine independent components in general, each component associated with a certain dyad pair. The components are thus said to be ordered. When the ordering is understood, such as suggested by the nonion form (1.280), the explicit writing of the dyads can be suppressed and the dyadic written as an array:

$$[\Phi] = \begin{bmatrix} \phi^{11} & \phi^{12} & \phi^{13} \\ \phi^{21} & \phi^{22} & \phi^{33} \\ \phi^{31} & \phi^{32} & \phi^{33} \end{bmatrix}, \qquad \overleftrightarrow{\mathbf{\Phi}} = \begin{Bmatrix} \mathbf{e}_1 \\ \mathbf{e}_2 \\ \mathbf{e}_3 \end{Bmatrix}^T [\Phi] \begin{Bmatrix} \mathbf{e}_1 \\ \mathbf{e}_2 \\ \mathbf{e}_3 \end{Bmatrix}. \tag{1.281}$$

This representation is simpler than Eq. (1.280), but it is taken to mean the same.

Example 1.16 With reference to a Cartesian system (x_1, x_2, x_3), the components of the stress dyadic at a certain point of a continuous medium are given by

$$[\sigma] = \begin{bmatrix} 200 & 400 & 300 \\ 400 & 0 & 0 \\ 300 & 0 & -100 \end{bmatrix} \quad \text{psi.}$$

Determine the stress vector $\boldsymbol{\pi}$ at the point acting on a plane passing through the point and parallel to the plane, $x_1 + 2x_2 + 2x_3 - 6 = 0$. What are the normal and tangential components of the stress vector at the point?

First we should find the unit normal to the plane on which we are required to find the stress vector. The unit normal is given by [see Eq. (1.202)]

$$\hat{\mathbf{n}} = \frac{\nabla P}{|\nabla P|}, \qquad P(x_1, x_2, x_3) = x_1 + 2x_2 + 2x_3 - 6$$

$$\hat{\mathbf{n}} = \frac{\hat{\mathbf{e}}_1 + 2\hat{\mathbf{e}}_2 + 2\hat{\mathbf{e}}_3}{3}.$$

The components of the stress vector are given by the matrix form of Eq. (1.270):

$$\begin{Bmatrix} \pi_1 \\ \pi_2 \\ \pi_3 \end{Bmatrix} = \begin{bmatrix} 200 & 400 & 300 \\ 400 & 0 & 0 \\ 300 & 0 & -100 \end{bmatrix} \begin{Bmatrix} \frac{1}{3} \\ \frac{2}{3} \\ \frac{2}{3} \end{Bmatrix} = \begin{Bmatrix} \frac{1600}{3} \\ \frac{400}{3} \\ \frac{100}{3} \end{Bmatrix} \text{psi},$$

or

$$\boldsymbol{\pi}(\hat{\mathbf{n}}) = \frac{1600\hat{\mathbf{e}}_1 + 400\hat{\mathbf{e}}_2 + 100\hat{\mathbf{e}}_3}{3} \text{ psi}.$$

The normal component σ_n of the stress vector on the plane is given by

$$\sigma_n = \boldsymbol{\pi}(\hat{\mathbf{n}}) \cdot \hat{\mathbf{n}} = \tfrac{2600}{9} \text{ psi}$$

and the tangential component is given by (the Pythagorean theorem)

$$\sigma_t = \sqrt{|\boldsymbol{\pi}|^2 - \sigma_n^2} = \frac{10^2}{9}\sqrt{(256+16+1)9 - 26 \times 26} \text{ psi}$$

$$= 100\frac{\sqrt{1781}}{81} \text{ psi.} \quad \blacksquare$$

1.5.4 Components of a Dyadic

Just as a vector can be represented in terms of different basis vectors, so can a dyadic. Analogously to vectors, we have

$$\begin{aligned} \overleftrightarrow{\mathbf{\Phi}} &= \phi^{jk}\mathbf{e}_j\mathbf{e}_k, && \text{contragredient components} \\ &= \phi_{jk}\mathbf{e}^j\mathbf{e}^k, && \text{cogredient components} \\ &= \phi^j{}_k\mathbf{e}_j\mathbf{e}^k, && \text{mixed components} \\ &= \phi_j{}^k\mathbf{e}^j\mathbf{e}_k, && \text{mixed components.} \end{aligned} \tag{1.282}$$

The terminology is similar to that of vectors, but because the cogredient and contragredient basis vectors can be mixed, the corresponding components are called *mixed components*.

All of the above forms represent the same dyadic. When a *unitary basis* is utilized and rules are set up to guarantee *invariance of the dyadic*, the notions of *covariance* and *contravariance* come into play, just as for vectors. In the general scheme that is thus developed, vectors are called *first-order tensors* and dyadics are called *second-order tensors*. Scalars are called *zeroth-order tensors*. The generalization to *third-order tensors* thus leads, or is derived from, *triadics*, or three vectors standing side by side. It follows that higher order tensors are developed from *polyadics*.

1.5.5 Symmetric and Antisymmetric Dyadics

A dyadic (second-order tensor) is *symmetric* if it is equal to its transpose:

$$\overleftrightarrow{\boldsymbol{\Phi}} = \overleftrightarrow{\boldsymbol{\Phi}}^{\mathrm{t}}$$

$$\phi^{ij}\mathbf{e}_i\mathbf{e}_j = \phi^{ij}\mathbf{e}_j\mathbf{e}_i$$

$$= \phi^{ji}\mathbf{e}_i\mathbf{e}_j.$$

Hence if a dyadic is symmetric, then

$$\phi^{ij} = \phi^{ji}. \tag{1.283}$$

Interchanging the indices on the components is equivalent to interchanging basis vectors in the dyads. A symmetric dyadic has only six independent components.

A dyadic is *antisymmetric* if it is equal to the negative of its transpose:

$$\overleftrightarrow{\boldsymbol{\Phi}} = -\overleftrightarrow{\boldsymbol{\Phi}}^{\mathrm{t}}$$

$$\phi^{ij}\mathbf{e}_i\mathbf{e}_j = -\phi^{ij}\mathbf{e}_j\mathbf{e}_i = -\phi^{ji}\mathbf{e}_i\mathbf{e}_j$$

or

$$\phi^{ij} = -\phi^{ji}. \tag{1.284}$$

Clearly when $i = j$, we must have

$$\begin{aligned} \phi^{11} &= -\phi^{11} = 0 \\ \phi^{22} &= -\phi^{22} = 0 \\ \phi^{33} &= -\phi^{33} = 0. \end{aligned} \tag{1.285}$$

Thus an antisymmetric dyadic has only *three* independent components.

1.5.6 Separation of a Dyadic into Its Symmetric and Antisymmetric Parts

Any dyadic can be separated into its symmetric and antisymmetric parts. To show this, we add and subtract $\frac{1}{2}$ of the transpose dyadic in the following way:

$$\overleftrightarrow{\boldsymbol{\Phi}} = \tfrac{1}{2}\underbrace{[\overleftrightarrow{\boldsymbol{\Phi}} + \overleftrightarrow{\boldsymbol{\Phi}}^{\mathrm{t}}]}_{\text{symmetric}} + \tfrac{1}{2}\underbrace{[\overleftrightarrow{\boldsymbol{\Phi}} - \overleftrightarrow{\boldsymbol{\Phi}}^{\mathrm{t}}]}_{\text{antisymmetric}}. \tag{1.286a}$$

From the basic definitions the two parts shown are easily seen to be symmetric and antisymmetric. In component form we have

$$\phi^{ij} = \underbrace{\tfrac{1}{2}\left[\phi^{ij} + \phi^{ji}\right]}_{\text{symmetric}} + \underbrace{\tfrac{1}{2}\left[\phi^{ij} - \phi^{ji}\right]}_{\text{antisymmetric}}. \tag{1.286b}$$

1.5.7 Transformations of Second-Order Tensors (Dyadics)

Depending on the bases used, a dyadic can be written in any of the following forms:

$$\begin{aligned}\overleftrightarrow{\mathbf{\Phi}} &= \phi^{ij}\mathbf{e}_i\mathbf{e}_j \\ &= \phi_{ij}\mathbf{e}^i\mathbf{e}^j \\ &= \phi^i{}_j\mathbf{e}_i\mathbf{e}^j \\ &= \phi_i{}^j\mathbf{e}^i\mathbf{e}_j.\end{aligned} \tag{1.287}$$

In a barred coordinate system we would have

$$\begin{aligned}\overleftrightarrow{\mathbf{\Phi}} &= \bar{\phi}^{mn}\bar{\mathbf{e}}_m\bar{\mathbf{e}}_n \\ &= \bar{\phi}_{mn}\bar{\mathbf{e}}^m\bar{\mathbf{e}}^n \\ &= \bar{\phi}^m{}_n\bar{\mathbf{e}}_m\bar{\mathbf{e}}^n \\ &= \bar{\phi}_m{}^n\bar{\mathbf{e}}^m\bar{\mathbf{e}}_n.\end{aligned} \tag{1.288}$$

All of these forms represent the same tensor. To guarantee that the dyadic (or tensor) is invariant, the values of the components must be unchanged when transformed to the new coordinate system. To establish the rules of the transformation, we note from Section 1.4.11 that

$$\begin{aligned}\bar{\mathbf{e}}_s &= \frac{\partial q^j}{\partial \bar{q}^s}\mathbf{e}_j \qquad \text{or} \qquad \mathbf{e}_s = \frac{\partial \bar{q}^j}{\partial q^s}\bar{\mathbf{e}}_j \\ \bar{\mathbf{e}}^s &= \frac{\partial \bar{q}^s}{\partial q^i}\mathbf{e}^i \qquad \text{or} \qquad \mathbf{e}^s = \frac{\partial q^s}{\partial \bar{q}^i}\bar{\mathbf{e}}^i.\end{aligned} \tag{1.289}$$

Substituting the expressions for $\mathbf{e}_i$ and $\mathbf{e}^i$ into the unbarred forms, we get for Eq. (1.287)

$$\overleftrightarrow{\mathbf{\Phi}} = \phi^{ij}\frac{\partial \bar{q}^m}{\partial q^i}\frac{\partial \bar{q}^n}{\partial q^j}\bar{\mathbf{e}}_m\bar{\mathbf{e}}_n. \tag{1.290}$$

In order for this form to be equal to the barred form

$$\overset{\leftrightarrow}{\mathbf{\Phi}} = \bar{\phi}^{mn}\bar{\mathbf{e}}_m\bar{\mathbf{e}}_n, \tag{1.288a}$$

we must have

$$\bar{\phi}^{mn} = \phi^{ij}\frac{\partial \bar{q}^m}{\partial q^i}\frac{\partial \bar{q}^n}{\partial q^j}, \qquad \text{contravariant law.} \tag{1.291a}$$

In similar fashion we deduce that

$$\bar{\phi}_{mn} = \phi_{ij}\frac{\partial q^i}{\partial \bar{q}^m}\frac{\partial q^j}{\partial \bar{q}^n}, \qquad \text{covariant law} \tag{1.291b}$$

$$\bar{\phi}^m{}_n = \phi^i{}_j\frac{\partial \bar{q}^m}{\partial q^i}\frac{\partial q^j}{\partial \bar{q}^n}, \qquad \text{mixed components} \tag{1.291c}$$

$$\bar{\phi}_m{}^n = \phi_i{}^j\frac{\partial q^i}{\partial \bar{q}^m}\frac{\partial \bar{q}^n}{\partial q^j}, \qquad \text{mixed components.} \tag{1.291d}$$

The above rules now can be taken as the *definitions* of *contravariant*, *covariant*, and *mixed* second-order tensors. When only *rectangular Cartesian coordinates* are involved, the above four forms are all equivalent, and there is no distinction between the four kinds of components. In this case a second-order tensor is called a *Cartesian tensor.*

A (second-order) Cartesian tensor $\overset{\leftrightarrow}{\mathbf{\Phi}}$ transforms according to

$$\overset{\leftrightarrow}{\mathbf{\Phi}} = \phi_{ij}\hat{\mathbf{e}}_i\hat{\mathbf{e}}_j$$

$$= \bar{\phi}_{kl}\hat{\bar{\mathbf{e}}}_k\hat{\bar{\mathbf{e}}}_l.$$

The unit base vectors in the barred and unbarred systems are related by

$$\hat{\mathbf{e}}_i = \frac{\partial \bar{x}_j}{\partial x_i}\hat{\bar{\mathbf{e}}}_j \equiv \beta_{ji}\hat{\bar{\mathbf{e}}}_j, \tag{1.292}$$

where the β_{ij}'s denote the direction cosines between barred and unbarred systems [see Eq. (1.57)]. Thus we have

$$\bar{\phi}_{kl} = \phi_{ij}\beta_{ki}\beta_{lj} \qquad \text{or} \qquad [\bar{\phi}] = [\beta][\phi][\beta]^{\mathrm{T}}. \tag{1.293}$$

In right-handed orthogonal systems the determinant of the transformation matrix is unity, and

$$[\beta]^{-1} = [\beta]^{\mathrm{T}}. \tag{1.294}$$

1.5.8 Unit Tensor

The unit tensor has already been defined as

$$\overleftrightarrow{\mathbf{I}}=\mathbf{e}^i\mathbf{e}_i$$

$$=\mathbf{e}_i\mathbf{e}^i. \tag{1.295}$$

With the use of the Kronecker delta symbol, this can be written alternatively as

$$\overleftrightarrow{\mathbf{I}}=\delta^i{}_j\mathbf{e}_i\mathbf{e}^j$$

$$=\delta_i{}^j\mathbf{e}^i\mathbf{e}_j.$$

The unit tensor has the property that it changes a vector into itself, that is,

$$\mathbf{A}\cdot\overleftrightarrow{\mathbf{I}}=\overleftrightarrow{\mathbf{I}}\cdot\mathbf{A}=\mathbf{A}. \tag{1.296}$$

Clearly the unit tensor is symmetric.

In the above form the components are the *mixed components* of the tensor. Because the mixed components are the Kronecker delta symbols, and as such in any coordinate system, they are *invariant* with coordinate transformations. Thus the mixed components of the unit tensor are also called the *invariant* components.

It is also useful to find the covariant and contravariant components of the unit tensor. To do this, we note

$$\mathbf{e}_i=g_{im}\mathbf{e}^m \qquad \text{and} \qquad \mathbf{e}^i=g^{im}\mathbf{e}_m. \tag{1.297}$$

Substituting these expressions into the above formulas for the unit tensor, we deduce finally the following four forms:

$$\begin{aligned}
\overleftrightarrow{\mathbf{I}}&=g^{ij}\mathbf{e}_i\mathbf{e}_j, &&\text{contravariant components}\\
&=g_{ij}\mathbf{e}^i\mathbf{e}^j, &&\text{covariant components}\\
&=g_i{}^j\mathbf{e}^i\mathbf{e}_j, &&\text{invariant components}\\
&=g^i{}_j\mathbf{e}_i\mathbf{e}^j, &&\text{invariant components.}
\end{aligned} \tag{1.298}$$

Here we have $g^i{}_j=\delta^i{}_j$ and $g_i{}^j=\delta_i{}^j$. The components are the same as the fundamental metric tensor. Hence the *fundamental metric tensor is a unit tensor*.

The unit tensor is also called the *unit dyadic* or *idemfactor*.

1.5.9 Contraction of a Second-Order Tensor

The contraction of a second-order tensor or dyadic is the result obtained by placing a dot between the vectors in each of the dyads:

$$\begin{aligned}\phi_s &\equiv \phi^{ij}\mathbf{e}_i \cdot \mathbf{e}_j = \phi^{ij}g_{ij}\\ &\equiv \phi_{ij}\mathbf{e}^i \cdot \mathbf{e}^j = \phi_{ij}g^{ij}\\ &\left.\begin{aligned}&\equiv \phi_i{}^j\mathbf{e}^i \cdot \mathbf{e}_j = \phi_i{}^i\\ &\equiv \phi^i{}_j\mathbf{e}_i \cdot \mathbf{e}^j = \phi^i{}_i,\end{aligned}\right\}\quad \text{sum of the diagonal terms.}\end{aligned} \tag{1.299}$$

The contraction operation lowers the order of the tensor by 2. Thus the contraction of a dyadic is sometimes called the *scalar of the dyadic.*

1.5.10 Vector of a Second-Order Tensor

The vector of a second-order tensor is obtained by placing a cross between the vectors in the basis dyads:

$$\begin{aligned}\mathbf{\Phi}_V &\equiv \phi^{ij}\mathbf{e}_i \times \mathbf{e}_j = \sqrt{g}\,\phi^{ij}\varepsilon_{ijk}\mathbf{e}^k\\ &\equiv \phi_{ij}\mathbf{e}^i \times \mathbf{e}^j = \sqrt{g}\,\phi_{ij}g^{im}g^{jn}\varepsilon_{mnk}\mathbf{e}^k\\ &\equiv \phi^i{}_j\mathbf{e}_i \times \mathbf{e}^j = \sqrt{g}\,\phi^i{}_j g^{jm}\varepsilon_{imk}\mathbf{e}^k\\ &\equiv \phi_i{}^j\mathbf{e}^i \times \mathbf{e}_j = \sqrt{g}\,\phi_i{}^j g^{im}\varepsilon_{mjk}\mathbf{e}^k.\end{aligned} \tag{1.300}$$

This aspect of a dyadic has a number of important applications. It can be shown, for instance, that

$$\mathbf{A}\cdot[\overleftrightarrow{\mathbf{\Phi}} - \overleftrightarrow{\mathbf{\Phi}}^{\mathrm{t}}] \equiv \mathbf{\Phi}_V \times \mathbf{A}. \tag{1.301}$$

Since $(\overleftrightarrow{\mathbf{\Phi}} - \overleftrightarrow{\mathbf{\Phi}}^{\mathrm{t}})$ is an antisymmetric tensor and has only three independent components, it is related to some vector, and that vector is $\mathbf{\Phi}_V$. The relation is that shown above. In particular, if $\overleftrightarrow{\mathbf{\Phi}} = \operatorname{grad}\mathbf{B}$ then $\mathbf{\Phi}_V = \operatorname{curl}\mathbf{B}$, and we have

$$\mathbf{A}\cdot\left[\nabla\mathbf{B} - (\nabla\mathbf{B})^{\mathrm{t}}\right] \equiv (\nabla\times\mathbf{B})\times\mathbf{A}.$$

1.5.11 Invariants of a Second-Order Tensor

The sum of the diagonal terms of the mixed form of a second-order tensor is called the *trace of the tensor* (see Section 1.3.2):

$$\text{trace of } \overleftrightarrow{\mathbf{\Phi}} = \phi^i{}_i = \phi_i{}^i. \tag{1.302a}$$

The trace of a tensor is an *invariant*, called the first invariant of $\overleftrightarrow{\mathbf{\Phi}}$, and is denoted by I_1; that is, it is invariant with coordinate transformations. The second and third invariants of a second-order tensor are given by

$$I_2 = -\tfrac{1}{2}\left[\phi^j{}_i\phi^i{}_j - \phi^i{}_i\phi^j{}_j\right] \tag{1.302b}$$

$$I_3 = \det \phi^j{}_i. \tag{1.302c}$$

1.5.12 Dyadics with Orthonormal Bases

In an orthonormal basis system it is not necessary to distinguish between the different kinds of components. In a unit orthogonal basis system we can write

$$\overleftrightarrow{\mathbf{\Phi}} = \phi_{ij}\hat{\mathbf{e}}_i\hat{\mathbf{e}}_j. \tag{1.303}$$

The components ϕ_{ij} here are the physical components and should, probably, be designated by $\hat{\phi}_{ij}$, but it is conventional not to do so. They should not be confused with the *covariant* components.

The same representation holds for a *Cartesian tensor*. For a Cartesian tensor the basis vectors are *constants* and thus do not take roles as variables in differentiation and integration.

The invariants and the scalar for this case become

$$I_1 = \phi_{ii}, \qquad I_2 = \tfrac{1}{2}(\phi_{ij}\phi_{ij} - \phi_{ii}\phi_{jj}), \qquad I_3 = \det \phi_{ij}. \tag{1.304}$$

The vector of the tensor becomes

$$\begin{aligned}\mathbf{\Phi}_V &= \phi_{ij}\hat{\mathbf{e}}_i \times \hat{\mathbf{e}}_j \\ &= \phi_{ij}\varepsilon_{ijk}\hat{\mathbf{e}}_k.\end{aligned} \tag{1.305}$$

1.5.13 Double-Dot Product

The double-dot product between two dyadics is very useful in many problems. The double-dot product between a dyad (**AB**) and another dyad (**CD**) is defined as

$$(\mathbf{AB}):(\mathbf{CD}) \equiv (\mathbf{B}\cdot\mathbf{C})(\mathbf{A}\cdot\mathbf{D}). \tag{1.306}$$

The double-dot product, by this definition, is commutative. The original definition by Gibbs was not commutative. This product is seen to be a scalar.

The double-dot product between two dyadics $\overleftrightarrow{\mathbf{\Phi}}$ and $\overleftrightarrow{\mathbf{\Psi}}$ is given by

$$\begin{aligned}\overleftrightarrow{\mathbf{\Phi}}:\overleftrightarrow{\mathbf{\Psi}} &= (\phi^{ij}\mathbf{e}_i\mathbf{e}_j):(\psi^{mn}\mathbf{e}_m\mathbf{e}_n) \\ &= \phi^{ij}\psi^{mn}(\mathbf{e}_j\cdot\mathbf{e}_m)(\mathbf{e}_i\cdot\mathbf{e}_n) \\ &= \phi^{ij}\psi^{mn}g_{jm}g_{in}.\end{aligned} \tag{1.307}$$

A simpler form is given by

$$\overleftrightarrow{\mathbf{\Phi}}:\overleftrightarrow{\mathbf{\Psi}}=\left(\phi^{ij}\mathbf{e}_i\mathbf{e}_j\right):\psi_{mn}\mathbf{e}^m\mathbf{e}^n$$

$$=\phi^{ij}\psi_{mn}\,\delta_j^m\,\delta_i^n$$

$$=\phi^{ij}\psi_{ji}. \tag{1.308}$$

Different forms appear depending on which kind of components are used for the dyadics $\overleftrightarrow{\mathbf{\Phi}}$ and $\overleftrightarrow{\mathbf{\Psi}}$.

If one of the dyadics, say $\overleftrightarrow{\mathbf{\Psi}}$, is the unit dyadic, we have

$$\overleftrightarrow{\mathbf{\Phi}}:\overleftrightarrow{\mathbf{I}}=\phi^{ij}\mathbf{e}_i\mathbf{e}_j:g^{mn}\mathbf{e}_m\mathbf{e}_n$$

$$=\phi^{ij}g^{mn}g_{jm}g_{in}.$$

From Eq. (1.94), however, we have

$$g^{mn}g_{jm}=\delta_j{}^n.$$

Thus, it follows that

$$\overleftrightarrow{\mathbf{\Phi}}:\overleftrightarrow{\mathbf{I}}=\phi^{ij}\,\delta_j{}^n g_{in}$$

$$=\phi^{ij}g_{ij}. \tag{1.309}$$

This result is simply the contraction of $\overleftrightarrow{\mathbf{\Phi}}$, and is one useful way of representing it.

1.5.14 The Tensor Gradient

Let a vector field be represented by its contravariant components $\mathbf{A}=A^i\mathbf{e}_i$. The gradient of the vector field is defined as

$$\operatorname{grad}\mathbf{A}=\nabla\mathbf{A}=\mathbf{e}^j\frac{\partial}{\partial q^j}\left(A^i\mathbf{e}_i\right)$$

$$=\mathbf{e}^j\left[\frac{\partial A^i}{\partial q^j}\mathbf{e}_i+A^i\left\{\begin{matrix}k\\ i\;j\end{matrix}\right\}\mathbf{e}_k\right]$$

$$=\left[\frac{\partial A^k}{\partial q^j}+A^i\left\{\begin{matrix}k\\ i\;j\end{matrix}\right\}\right]\mathbf{e}^j\mathbf{e}_k$$

$$=A^k{}_{,j}\mathbf{e}^j\mathbf{e}_k. \tag{1.310}$$

The term $A^k_{,j}$, which appeared earlier in Eq. (1.224), is called the *covariant derivative of the contravariant component* A^k. We see that these terms are the mixed components of a second-order tensor called the *tensor gradient*.

Suppose that the vector **A** is represented by its covariant components, $\mathbf{A}=A_i\mathbf{e}^i$. Then we have

$$\begin{aligned}\operatorname{grad}\mathbf{A}=\nabla\mathbf{A}&=\mathbf{e}^j\left[\frac{\partial A_i}{\partial q^j}\mathbf{e}^i+A_i\frac{\partial \mathbf{e}^i}{\partial q^j}\right]\\ &=\mathbf{e}^j\left[\frac{\partial A_i}{\partial q^j}\mathbf{e}^i-A_i\left\{\begin{matrix} i \\ k\ j\end{matrix}\right\}\mathbf{e}^k\right]\\ &=\left[\frac{\partial A_k}{\partial q^j}-A_i\left\{\begin{matrix} i \\ k\ j\end{matrix}\right\}\right]\mathbf{e}^j\mathbf{e}^k\\ &=A_{k,j}\mathbf{e}^j\mathbf{e}^k,\end{aligned} \tag{1.311}$$

where

$$A_{k,j}\equiv\frac{\partial A_k}{\partial q^j}-A_i\left\{\begin{matrix} i \\ k\ j\end{matrix}\right\}. \tag{1.312}$$

The components $A_{k,j}$ are called the *covariant derivative of the covariant component* A_i. We see that they are the covariant components of the tensor gradient $\nabla\mathbf{A}$. The components $A_{k,j}$, taken as an ordered group, are thus referred to as the *covariant tensor gradient*.

The contraction of the tensor gradient is given by

$$\begin{aligned}(\nabla\mathbf{A})_s&=A^k_{,j}\mathbf{e}^j\cdot\mathbf{e}_k\\ &=A^k_{,k}.\end{aligned} \tag{1.313a}$$

This expression, however, is simply the divergence of the vector **A**. Thus

$$(\nabla\mathbf{A})_s=A^k_{,k}=\operatorname{div}\mathbf{A}. \tag{1.313b}$$

It is easy to deduce also that the *trace of* ∇**A** *is equal to div* **A**.

The vector of the tensor gradient is given by

$$\begin{aligned}(\nabla\mathbf{A})_V&=A^k_{,j}\mathbf{e}^j\times\mathbf{e}_k\\ &=A^k_{,j}g^{ji}\mathbf{e}_i\times\mathbf{e}_k\\ &=\sqrt{g}\,g^{ji}A^k_{,j}\varepsilon_{ikm}\mathbf{e}^m.\end{aligned} \tag{1.314}$$

This expression is equal to the curl of **A**. Thus

$$(\nabla \mathbf{A})_V = \operatorname{curl} \mathbf{A}. \tag{1.315}$$

Consequently, the vector of $\nabla \mathbf{A}$, as well as the antisymmetric part of grad **A**, is associated with the *local rotational characteristics of the vector field* **A**. Thus we have, for instance,

$$\mathbf{B} \cdot \left[\tfrac{1}{2} \{ \nabla \mathbf{A} - (\nabla \mathbf{A})^t \} \right] \equiv \tfrac{1}{2} (\operatorname{curl} \mathbf{A}) \times \mathbf{B}. \tag{1.316}$$

By comparison with the velocity in a rigid-body rotation,

$$\mathbf{V} = \tfrac{1}{2} (\operatorname{curl} \mathbf{V}) \times \mathbf{r}, \tag{1.317}$$

deduced by means of Eq. (1.234), we interpret expression (1.316) as a change in **A** over the distance **B** brought about by the rotation of **A**.

In Cartesians the expression for $\nabla \mathbf{A}$ is simple. It can be deduced from the above results or from the following straightforward analysis:

$$\begin{aligned} \nabla \mathbf{A} &= \hat{\mathbf{i}}_i \frac{\partial}{\partial x^i} A_j \hat{\mathbf{i}}_j \\ &= \frac{\partial A_j}{\partial x^i} \hat{\mathbf{i}}_i \hat{\mathbf{i}}_j. \end{aligned} \tag{1.318}$$

The symmetric and antisymmetric parts are

$$\nabla \mathbf{A} = \frac{1}{2} \underbrace{\left[\frac{\partial A_j}{\partial x^i} + \frac{\partial A_i}{\partial x^j} \right]}_{\text{symmetric}} \hat{\mathbf{i}}_i \hat{\mathbf{i}}_j + \frac{1}{2} \underbrace{\left[\frac{\partial A_j}{\partial x^i} - \frac{\partial A_i}{\partial x^j} \right]}_{\text{antisymmetric}} \hat{\mathbf{i}}_i \hat{\mathbf{i}}_j. \tag{1.319}$$

In orthogonal curvilinear coordinates with an orthonormal basis system, we have to account for changes in the basis vectors. We have

$$\begin{aligned} \operatorname{grad} \mathbf{A} = \nabla \mathbf{A} &= \frac{\hat{\mathbf{e}}_i}{h_i} \frac{\partial}{\partial q^i} \hat{A}_j \hat{\mathbf{e}}_j \\ &= \frac{1}{h_i} \frac{\partial \hat{A}_j}{\partial q^i} \hat{\mathbf{e}}_i \hat{\mathbf{e}}_j + \frac{\hat{A}_j}{h_i} \hat{\mathbf{e}}_i \frac{\partial \hat{\mathbf{e}}_j}{\partial q^i}. \end{aligned} \tag{1.320}$$

The partial derivatives $\partial \hat{\mathbf{e}}_j / \partial q^i$ can be obtained from the differential formulas given by Eqs. (1.169), or they can be obtained from the Christoffel symbol in the following way:

$$\frac{\partial \hat{\mathbf{e}}_j}{\partial q^i} = \frac{\partial}{\partial q^i}(h_j \mathbf{e}_j)$$

$$= h_j \frac{\partial \hat{\mathbf{e}}_j}{\partial q^i} + \hat{\mathbf{e}}_j \frac{\partial h_j}{\partial q^i} \qquad \text{(no summation on } j) \tag{1.321a}$$

or

$$\frac{\partial \hat{\mathbf{e}}_j}{\partial q^i} = \frac{1}{h_j}\left[\frac{\partial \mathbf{e}_j}{\partial q^i} - \hat{\mathbf{e}}_j \frac{\partial h_j}{\partial q^i}\right]$$

$$= \frac{1}{h_j}\left[h_k \left\{ {k \atop i\ j} \right\} - \delta_{jk} \frac{\partial h_j}{\partial q^i}\right] \hat{\mathbf{e}}_k. \tag{1.321b}$$

To evaluate the Christoffel symbol, we observe for orthogonal coordinates that

$$g_{ij} = h_i^2 \delta_{ij} \qquad \text{and} \qquad g^{ij} = \frac{\delta_{ij}}{h_i^2}, \tag{1.322}$$

where the *indices on the scale factors are not included in the summation convention*. Thus we have

$$\left\{ {k \atop i\ j} \right\} \equiv \frac{1}{2} g^{kr}\left[\frac{\partial(g_{jr})}{\partial q^i} + \frac{\partial(g_{ir})}{\partial q^j} - \frac{\partial(g_{ij})}{\partial q^r}\right]$$

$$= \frac{1}{2}\frac{\delta_{kr}}{h_k^2}\left[\delta_{jr}\frac{\partial h_j^2}{\partial q^i} + \delta_{ir}\frac{\partial h_i^2}{\partial q^j} - \delta_{ij}\frac{\partial h_i^2}{\partial q^r}\right]$$

$$= \frac{1}{2}\left[\frac{\delta_{jk}}{h_j^2}\frac{\partial h_j^2}{\partial q^i} + \frac{\delta_{ik}}{h_i^2}\frac{\partial h_i^2}{\partial q^j} - \frac{\delta_{ij}}{h_k^2}\frac{\partial h_j^2}{\partial q^k}\right]. \tag{1.323}$$

We can obtain the final simple form

$$\frac{\partial \hat{\mathbf{e}}_j}{\partial q^i} = \left[\frac{\delta_{ik}}{h_j}\frac{\partial h_i}{\partial q^j} - \frac{\delta_{ij}}{h_k}\frac{\partial h_j}{\partial q^k}\right]\mathbf{e}_k, \tag{1.324}$$

where the *indices on the scale factors do not take part in the summation convention*. This expression is a compact form for Eqs. (1.169).

With this result we can now write

$$\operatorname{grad}\mathbf{A}=\nabla\mathbf{A}=\frac{1}{h_i}\left[\frac{\partial\hat{A}_k}{\partial q^i}+\hat{A}_j\left\{\frac{\delta_{ik}}{h_j}\frac{\partial h_i}{\partial q^j}-\frac{\delta_{ij}}{h_k}\frac{\partial h_j}{\partial q^k}\right\}\right]\hat{\mathbf{e}}_i\hat{\mathbf{e}}_k$$

$$=\frac{1}{h_i}\left[\frac{\partial\hat{A}_k}{\partial q^i}+\frac{\hat{A}_j}{h_j}\frac{\partial h_i}{\partial q^j}\delta_{ik}-\frac{\hat{A}_i}{h_k}\frac{\partial h_i}{\partial q^k}\right]\hat{\mathbf{e}}_i\hat{\mathbf{e}}_k. \tag{1.325}$$

Here again the scale factors do not take part in the summation convention.

For the special case of cylindrical coordinates we have $q^1=R$, $q^2=\phi$, $q^3=z$, $h_1=1$, $h_2=R$, $h_3=1$, and hence

$$\begin{aligned}\nabla\mathbf{A}= &\frac{\partial\hat{A}_R}{\partial R}\hat{\mathbf{e}}_R\hat{\mathbf{e}}_R+\frac{\partial\hat{A}_\phi}{\partial R}\hat{\mathbf{e}}_R\hat{\mathbf{e}}_\phi+\frac{\partial\hat{A}_z}{\partial R}\hat{\mathbf{e}}_R\hat{\mathbf{e}}_z\\ &+\frac{1}{R}\left(\frac{\partial\hat{A}_R}{\partial\phi}-\hat{A}_\phi\right)\hat{\mathbf{e}}_\phi\hat{\mathbf{e}}_R+\frac{1}{R}\left(\frac{\partial\hat{A}_\phi}{\partial\phi}+\hat{A}_R\right)\hat{\mathbf{e}}_\phi\hat{\mathbf{e}}_\phi+\frac{1}{R}\frac{\partial\hat{A}_z}{\partial\phi}\hat{\mathbf{e}}_\phi\hat{\mathbf{e}}_z\\ &+\frac{\partial\hat{A}_R}{\partial z}\hat{\mathbf{e}}_z\hat{\mathbf{e}}_R+\frac{\partial\hat{A}_\phi}{\partial z}\hat{\mathbf{e}}_z\hat{\mathbf{e}}_\phi+\frac{\partial\hat{A}_z}{\partial z}\hat{\mathbf{e}}_z\hat{\mathbf{e}}_z.\end{aligned} \tag{1.326}$$

For the special case of spherical coordinates we have $q^1=r$, $q^2=\Theta$, $q^3=\phi$, $h_1=1$, $h_2=r$, $h_3=r\sin\Theta$, and thus

$$\begin{aligned}\nabla\mathbf{A}= &\frac{\partial\hat{A}_r}{\partial r}\hat{\mathbf{e}}_r\hat{\mathbf{e}}_r+\frac{\partial\hat{A}_\Theta}{\partial r}\hat{\mathbf{e}}_r\hat{\mathbf{e}}_\Theta+\frac{\partial\hat{A}_\phi}{\partial r}\hat{\mathbf{e}}_r\hat{\mathbf{e}}_\phi\\ &+\frac{1}{r}\left(\frac{\partial\hat{A}_r}{\partial\Theta}-\hat{A}_\Theta\right)\hat{\mathbf{e}}_\Theta\hat{\mathbf{e}}_r+\frac{1}{r}\left(\frac{\partial\hat{A}_\Theta}{\partial\Theta}+\hat{A}_r\right)\hat{\mathbf{e}}_\Theta\hat{\mathbf{e}}_\Theta+\frac{1}{r}\frac{\partial\hat{A}_\phi}{\partial\Theta}\hat{\mathbf{e}}_\Theta\hat{\mathbf{e}}_\phi\\ &+\frac{\dfrac{\partial\hat{A}_r}{\partial\phi}-\hat{A}_\phi\sin\Theta}{r\sin\Theta}\hat{\mathbf{e}}_\phi\hat{\mathbf{e}}_r+\frac{\dfrac{\partial\hat{A}_\Theta}{\partial\phi}-\hat{\mathbf{A}}_\phi\cos\Theta}{r\sin\Theta}\hat{\mathbf{e}}_\phi\hat{\mathbf{e}}_\Theta+\frac{\dfrac{\partial\hat{A}_\phi}{\partial\phi}+\hat{A}_r\sin\Theta+\hat{A}_\Theta\cos\Theta}{r\sin\Theta}\hat{\mathbf{e}}_\phi\hat{\mathbf{e}}_\phi.\end{aligned} \tag{1.327}$$

1.5.15 Divergence of a Second-Order Tensor

Analogously to the divergence of a vector, the divergence of a dyadic (or second-order tensor) is defined as

$$\operatorname{div}\overleftrightarrow{\boldsymbol{\Phi}}=\nabla\cdot\overleftrightarrow{\boldsymbol{\Phi}}. \tag{1.328}$$

If the tensor is given in contravariant components, we have

$$\operatorname{div}\overleftrightarrow{\mathbf{\Phi}} = \nabla \cdot \overleftrightarrow{\mathbf{\Phi}} = \mathbf{e}^i \frac{\partial}{\partial q^i} \cdot \phi^{jk}\mathbf{e}_j\mathbf{e}_k$$

$$= \mathbf{e}^i \cdot \left[\frac{\partial \phi^{jk}}{\partial q^i}\mathbf{e}_j\mathbf{e}_k + \phi^{jk}\frac{\partial \mathbf{e}_j}{\partial q^i}\mathbf{e}_k + \phi^{jk}\mathbf{e}_j\frac{\partial \mathbf{e}_k}{\partial q^i} \right]$$

$$= \mathbf{e}^i \cdot \left[\frac{\partial \phi^{jk}}{\partial q^i}\mathbf{e}_j\mathbf{e}_k + \phi^{jk}\begin{Bmatrix} m \\ i\ j \end{Bmatrix}\mathbf{e}_m\mathbf{e}_k + \phi^{jk}\begin{Bmatrix} n \\ i\ k \end{Bmatrix}\mathbf{e}_j\mathbf{e}_n \right]$$

$$= \mathbf{e}^i \cdot \left[\frac{\partial \phi^{rs}}{\partial q^i} + \phi^{js}\begin{Bmatrix} r \\ i\ j \end{Bmatrix} + \phi^{rk}\begin{Bmatrix} s \\ i\ k \end{Bmatrix} \right]\mathbf{e}_r\mathbf{e}_s$$

$$= \left[\frac{\partial \phi^{is}}{\partial q^i} + \phi^{js}\begin{Bmatrix} i \\ i\ j \end{Bmatrix} + \phi^{ik}\begin{Bmatrix} s \\ i\ k \end{Bmatrix} \right]\mathbf{e}_s. \tag{1.329}$$

The divergence of a dyadic is thus seen to be a vector. The divergence operation reduces the order (or rank) of the tensor by 1.

In Cartesians, if $\overleftrightarrow{\mathbf{\Phi}}$ is given by $\overleftrightarrow{\mathbf{\Phi}} = \phi_{ij}\hat{\mathbf{i}}_i\hat{\mathbf{i}}_j$, we have

$$\operatorname{div}\overleftrightarrow{\mathbf{\Phi}} = \nabla \cdot \overleftrightarrow{\mathbf{\Phi}} = \frac{\partial \phi_{ij}}{\partial x^i}\hat{\mathbf{i}}_j. \tag{1.330}$$

This simple form arises because the basis vectors are constant, and the Christoffel symbols thus vanish.

It is interesting to see the implications of this operation in terms of the stress tensor. If $\hat{\mathbf{n}} \cdot \overleftrightarrow{\boldsymbol{\sigma}}\, dS$ represents the surface force acting on the surface element dS of a small volume $\Delta\tau$, then the total force on the volume element $\Delta\tau$ is

$$\oint\!\!\!\oint_{\Delta S} \hat{\mathbf{n}} \cdot \overleftrightarrow{\boldsymbol{\sigma}}\, dS, \tag{1.331}$$

where ΔS is the surface that encloses $\Delta\tau$. In the limit as $\Delta\tau \to 0$, we have analogously to the case for vectors,

$$\operatorname{div}\overleftrightarrow{\boldsymbol{\sigma}} = \lim_{\Delta\tau \to 0} \frac{1}{\Delta\tau} \oint\!\!\!\oint_{\Delta S} \hat{\mathbf{n}} \cdot \overleftrightarrow{\boldsymbol{\sigma}}\, dS. \tag{1.332}$$

Thus the physical interpretation of $\operatorname{div}\overleftrightarrow{\boldsymbol{\sigma}}$ is the *resultant force per unit volume in a continuous medium arising because of surface stresses.*

Example 1.17 Express the vector equation

$$\nabla \cdot \overleftrightarrow{\boldsymbol{\sigma}} + \mathbf{f} = \mathbf{0}, \qquad \overleftrightarrow{\boldsymbol{\sigma}} = \overleftrightarrow{\boldsymbol{\sigma}}^{\mathrm{t}},$$

which represents the stress equilibrium equation for a continuous medium, in cylindrical coordinates.

First we express $\overleftrightarrow{\boldsymbol{\sigma}}$ in its nonion form in cylindrical coordinates, and ∇ in cylindrical coordinates:

$$\boldsymbol{\sigma} = \sigma_{rr}\hat{\mathbf{e}}_r\hat{\mathbf{e}}_r + \sigma_{r\phi}(\hat{\mathbf{e}}_r\hat{\mathbf{e}}_\phi + \hat{\mathbf{e}}_\phi\hat{\mathbf{e}}_r) + \sigma_{rz}(\hat{\mathbf{e}}_r\hat{\mathbf{e}}_z + \hat{\mathbf{e}}_z\hat{\mathbf{e}}_r) + \sigma_{\phi\phi}\hat{\mathbf{e}}_\phi\hat{\mathbf{e}}_\phi$$
$$+ \sigma_{\phi z}(\hat{\mathbf{e}}_\phi\hat{\mathbf{e}}_z + \hat{\mathbf{e}}_z\hat{\mathbf{e}}_\phi) + \sigma_{zz}\hat{\mathbf{e}}_z\hat{\mathbf{e}}_z$$

$$\nabla = \hat{\mathbf{e}}_r\frac{\partial}{\partial r} + \hat{\mathbf{e}}_\phi\frac{1}{r}\frac{\partial}{\partial \phi} + \hat{\mathbf{e}}_z\frac{\partial}{\partial z}, \qquad \mathbf{f} = f_r\hat{\mathbf{e}}_r + f_\phi\hat{\mathbf{e}}_\phi + f_z\hat{\mathbf{e}}_z.$$

Substituting into the vector equation, we have

$$\nabla \cdot \overleftrightarrow{\boldsymbol{\sigma}} = \hat{\mathbf{e}}_r \cdot \left[\frac{\partial \sigma_{rr}}{\partial r}\hat{\mathbf{e}}_r\hat{\mathbf{e}}_r + \frac{\partial \sigma_{r\phi}}{\partial r}\hat{\mathbf{e}}_r\hat{\mathbf{e}}_\phi + \frac{\partial \sigma_{rz}}{\partial r}\hat{\mathbf{e}}_r\hat{\mathbf{e}}_z\right]$$
$$+ \frac{\hat{\mathbf{e}}_\phi}{r} \cdot \left[\sigma_{rr}\frac{\partial \hat{\mathbf{e}}_r}{\partial \phi}\hat{\mathbf{e}}_r + \frac{\partial \sigma_{r\phi}}{\partial \phi}\hat{\mathbf{e}}_\phi\hat{\mathbf{e}}_r + \phi_{r\phi}\left(\frac{\partial \hat{\mathbf{e}}_r}{\partial \phi}\hat{\mathbf{e}}_\phi + \hat{\mathbf{e}}_\phi\frac{\partial \hat{\mathbf{e}}_r}{\partial \phi}\right) + \frac{\partial \sigma_{\phi z}}{\partial \phi}\hat{\mathbf{e}}_\phi\hat{\mathbf{e}}_z\right.$$
$$\left. + \sigma_{rz}\frac{\partial \hat{\mathbf{e}}_r}{\partial \phi}\hat{\mathbf{e}}_z + \frac{\partial \sigma_{\phi\phi}}{\partial \phi}\hat{\mathbf{e}}_\phi\hat{\mathbf{e}}_\phi + \sigma_{\phi\phi}\hat{\mathbf{e}}_\phi\frac{\partial \hat{\mathbf{e}}_\phi}{\partial \phi}\right]$$
$$+ \hat{\mathbf{e}}_z \cdot \left[\frac{\partial \sigma_{rz}}{\partial z}\hat{\mathbf{e}}_z\hat{\mathbf{e}}_r + \frac{\partial \sigma_{\phi z}}{\partial z}\hat{\mathbf{e}}_z\hat{\mathbf{e}}_\phi + \frac{\partial \sigma_{zz}}{\partial z}\hat{\mathbf{e}}_z\hat{\mathbf{e}}_z\right],$$

where all other terms do not contribute to the result. We have

$$\left[\frac{\partial \sigma_{rr}}{\partial r} + \frac{\sigma_{rr} - \sigma_{\phi\phi}}{r} + \frac{1}{r}\frac{\partial \sigma_{r\phi}}{\partial \phi} + \frac{\partial \sigma_{rz}}{\partial z} + f_r\right]\hat{\mathbf{e}}_r$$
$$+ \left[\frac{\partial \sigma_{r\phi}}{\partial r} + \frac{2\sigma_{r\phi}}{r} + \frac{1}{r}\frac{\partial \sigma_{\phi\phi}}{\partial \phi} + \frac{\partial \sigma_{\phi z}}{\partial z} + f_\phi\right]\hat{\mathbf{e}}_\phi$$
$$+ \left[\frac{\partial \sigma_{rz}}{\partial r} + \frac{\sigma_{rz}}{r} + \frac{1}{r}\frac{\partial \sigma_{\phi z}}{\partial \phi} + \frac{\partial \sigma_{zz}}{\partial z} + f_z\right]\hat{\mathbf{e}}_z = \mathbf{0}.$$

This implies that the coefficient of each base vector (i.e., the expression in square brackets) is zero separately, giving three scalar equations. ■

1.5.16 Integral Theorems for Dyadics

Analogously for the vector integral theorems [see Eqs. (1.235)], we have the following dyadic integral theorems [22]:

$$\iiint_R \operatorname{grad} \mathbf{A}\, d\tau = \oiint_S \hat{\mathbf{n}}\mathbf{A}\, dS \tag{1.333a}$$

$$\iiint_R \operatorname{div} \overleftrightarrow{\mathbf{\Phi}}\, d\tau = \oiint_S \hat{\mathbf{n}}\cdot\overleftrightarrow{\mathbf{\Phi}}\, dS \tag{1.333b}$$

$$\iiint_R \operatorname{curl} \overleftrightarrow{\mathbf{\Phi}}\, d\tau = \oiint_S \hat{\mathbf{n}}\times\overleftrightarrow{\mathbf{\Phi}}\, dS \tag{1.333c}$$

$$\iint_{CS} \hat{\mathbf{n}}\cdot\operatorname{curl} \overleftrightarrow{\mathbf{\Phi}}\, dS = \oint_C \mathbf{ds}\cdot\overleftrightarrow{\mathbf{\Phi}}. \tag{1.333d}$$

It is important that the order of the operations be observed in the above expressions. These expressions give rise to invariant definitions in the limit as the volume shrinks to zero (i.e., a point).

1.5.17 Taylor-Series Expansions

If a scalar function $u(x)$ of a single scalar x is analytic at the point x, it can be evaluated at a neighboring point $x+\delta$ by means of the Taylor-series expansion:

$$u(x+\delta)=u(x)+\left(\frac{du}{dx}\right)\bigg|_{\delta=0}\delta+\left(\frac{d^2u}{dx^2}\right)\bigg|_{\delta=0}\frac{\delta^2}{2!}+\cdots+\left(\frac{d^nu}{dx^n}\right)\bigg|_{\delta=0}\frac{\delta^n}{n!}+\cdots. \tag{1.334}$$

An analogous situation exists for a scalar function of a vector. We have in this case

$$\phi(\mathbf{r}+\boldsymbol{\delta})=\phi(\mathbf{r})+\boldsymbol{\delta}\cdot\nabla\phi+\frac{\boldsymbol{\delta}\cdot\nabla}{2!}(\boldsymbol{\delta}\cdot\nabla\phi)+\cdots. \tag{1.335}$$

In similar fashion it can be shown that the Taylor-series expansion for a vector function of a vector is

$$\mathbf{A}(\mathbf{r}+\boldsymbol{\delta})=\mathbf{A}(\mathbf{r})+\boldsymbol{\delta}\cdot\nabla\mathbf{A}+\frac{\boldsymbol{\delta}\cdot\nabla}{2!}(\boldsymbol{\delta}\cdot\nabla\mathbf{A})+\cdots, \tag{1.336}$$

where $\boldsymbol{\delta}$ is a constant vector displacement. Usually in practice only the first two terms of these expansions are required, the remaining terms being referred to as higher order terms.

These expansions can be written conveniently in operator form. Recall that the Taylor-series expansion for the exponential function is, about $x=0$,

$$e^x=1+x+\frac{x^2}{2!}+\cdots+\frac{x^n}{n!}+\cdots. \tag{1.337}$$

The above Taylor-series expansions can thus be written

$$u(x+\delta)=(e^{\delta(d/dx)})u(x) \tag{1.338a}$$

$$\phi(\mathbf{r}+\boldsymbol{\delta})=(e^{\boldsymbol{\delta}\cdot\nabla})\phi(\mathbf{r}) \tag{1.338b}$$

$$\mathbf{A}(\mathbf{r}+\boldsymbol{\delta})=(e^{\boldsymbol{\delta}\cdot\nabla})\mathbf{A}(\mathbf{r}), \tag{1.338c}$$

where $e^{\boldsymbol{\delta}\cdot\nabla}$ is to be regarded as an operator and expanded as follows:

$$e^{\boldsymbol{\delta}\cdot\nabla}=1+\boldsymbol{\delta}\cdot\nabla+\frac{(\boldsymbol{\delta}\cdot\nabla)^2}{2!}+\cdots+\frac{(\boldsymbol{\delta}\cdot\nabla)^n}{n!}+\cdots. \tag{1.339}$$

1.5.18 Relative Motion Between Two Neighboring Points in a Continuum Velocity Field

Consider a continuum velocity field given by $\mathbf{V}=\mathbf{V}(\mathbf{r})$. If the behavior of the velocity field is known in the vicinity of $\mathbf{r}$, then the velocity at a neighboring point $\mathbf{r}+\Delta\mathbf{r}$ is given by the first two terms of the Taylor series by

$$\mathbf{V}(\mathbf{r}+\Delta\mathbf{r})=\mathbf{V}(\mathbf{r})+\Delta\mathbf{r}\cdot\nabla\mathbf{V}+\text{higher order terms}. \tag{1.340}$$

The higher order terms become negligible as $\Delta\mathbf{r}\to 0$.

The tensor gradient $\nabla\mathbf{V}$ can be divided into its symmetric and antisymmetric parts:

$$\nabla\mathbf{V}=\tfrac{1}{2}\left[\nabla\mathbf{V}+(\nabla\mathbf{V})^t\right]+\tfrac{1}{2}\left[\nabla\mathbf{V}-(\nabla\mathbf{V})^t\right]. \tag{1.341}$$

The symmetric part is denoted by $\overleftrightarrow{\varepsilon}$ and called the rate of strain tensor. Thus we have

$$\mathbf{V}(\mathbf{r}+\Delta\mathbf{r})=\mathbf{V}(\mathbf{r})+\Delta\mathbf{r}\cdot\overleftrightarrow{\varepsilon}+\tfrac{1}{2}\Delta\mathbf{r}\cdot\left[(\nabla\mathbf{V})-(\nabla\mathbf{V})^t\right]. \tag{1.342}$$

The third term on the right can be written as

$$\tfrac{1}{2}\Delta\mathbf{r}\cdot\left[(\nabla\mathbf{V})-(\nabla\mathbf{V})^t\right]=\tfrac{1}{2}(\operatorname{curl}\mathbf{V})\times\Delta\mathbf{r}. \tag{1.343}$$

This term can thus be identified with a *rigid-body rotation*. We can thus write

$$\mathbf{V}(\mathbf{r}+\Delta\mathbf{r}) = \underbrace{\mathbf{V}(\mathbf{r})}_{\substack{\text{rate of}\\ \text{translation}}} + \underbrace{\Delta\mathbf{r}\cdot\overleftrightarrow{\varepsilon}}_{\substack{\text{rate of}\\ \text{strain}}} + \underbrace{\tfrac{1}{2}(\operatorname{curl}\mathbf{V})\times\Delta\mathbf{r}}_{\substack{\text{rate of}\\ \text{rotation}}}. \qquad (1.344)$$

Thus for small displacements the general rate of displacement of a small element of a continuum is made up of three parts: *rate of translation*, *rate of strain*, and *rate of rotation*.

Exercises 1.8

1 Consider a strain tensor $\overleftrightarrow{\mathbf{E}}$ pertaining to plane radially symmetric deformations. Write $\overleftrightarrow{\mathbf{E}}$ in cylindrical coordinates as

$$\overleftrightarrow{\mathbf{E}} = E_{RR}\hat{\mathbf{e}}_R\hat{\mathbf{e}}_R + E_{\phi\phi}\hat{\mathbf{e}}_\phi\hat{\mathbf{e}}_\phi,$$

where E_{RR} and $E_{\phi\phi}$ denote the components of the symmetric tensor associated with the gradient of the displacement field. Obtain the strain compatibility condition existing between E_{RR} and $E_{\phi\phi}$ by requiring that

$$\nabla\times(\nabla\times\overleftrightarrow{\mathbf{E}})^{\mathrm{t}} = \mathbf{0}$$

be satisfied.

2 In an incompressible velocity field, $\operatorname{div}\mathbf{V}=0$, the viscous stress tensor is given by

$$\overleftrightarrow{\tau} = \mu\left[\nabla\mathbf{V} + (\nabla\mathbf{V})^{\mathrm{t}}\right],$$

where μ is the coefficient of viscosity. Show that when μ is a constant, the resultant force per unit volume on a fluid particle, $\operatorname{div}\overleftrightarrow{\tau}$, is given by

$$\operatorname{div}\overleftrightarrow{\tau} = \mu\nabla^2\mathbf{V} = -\mu\operatorname{curl}\operatorname{curl}\mathbf{V}.$$

3 In a particular problem the electric field vector $\mathbf{E}$ is governed by the Maxwell equations

$$\operatorname{div}\mathbf{E} = 4\pi\rho(\mathbf{r})$$

$$\operatorname{curl}\mathbf{E} = \mathbf{0},$$

where $\rho(\mathbf{r})$ is the charge density. The electric field stress tensor is given by

$$\overleftrightarrow{\tau} = \frac{1}{4\pi}\left(\mathbf{E}\mathbf{E} - \frac{1}{2}E^2\overleftrightarrow{\mathbf{I}}\right).$$

Show that the resultant force per unit volume in the electric field, $\operatorname{div}\overleftrightarrow{\tau}$, is given by

$$\operatorname{div}\overleftrightarrow{\tau}=\rho\mathbf{E}.$$

4 Newton's second law of motion applied to a continuum states that the rate of change of momentum following a material region of fixed mass is equal to the sum of all the forces acting on the region. When the forces are divided into surface forces and body forces, Newton's second law reads:

$$\frac{D}{Dt}\iiint_R \rho\mathbf{V}\,d\tau=\oiint_S \hat{\mathbf{n}}\cdot\overleftrightarrow{\sigma}\,dS+\iiint_R \rho\mathbf{f}\,d\tau,$$

where $\overleftrightarrow{\sigma}$ is the surface stress tensor, $\mathbf{f}$ is the body force per unit mass, ρ is the mass density, and $\mathbf{V}$ is the material velocity. Since the material particle mass $\rho\,d\tau$ is constant with respect to the material time derivative D/Dt, make use of the divergence theorem and obtain the differential form of Newton's second law of motion for a continuum:

$$\rho\frac{D\mathbf{V}}{Dt}=\operatorname{div}\overleftrightarrow{\sigma}+\rho\mathbf{f}.$$

5 Multiply the continuity equation

$$\frac{\partial\rho}{\partial t}+\operatorname{div}(\rho\mathbf{V})=0$$

by the velocity $\mathbf{V}$ and add the result to the left-hand side of the momentum equation (Newton's second law) in Exercise 4. After use of vector identities, obtain the result

$$\frac{\partial}{\partial t}(\rho\mathbf{V})+\operatorname{div}(\rho\mathbf{V}\mathbf{V}-\overleftrightarrow{\sigma})=\rho\mathbf{f},$$

which is called the conservation form of the momentum equation. The combination $\rho\mathbf{V}\mathbf{V}$ is called the momentum-flux tensor.

6 Take the scalar product with $\mathbf{V}$ of the equation for Newton's second law in Exercise 4 and obtain with the use of Exercise 1.7.16 the equation of change for the kinetic energy of a material particle in a continuum:

$$\rho\frac{D}{Dt}\left(\frac{V^2}{2}\right)=\mathbf{V}\cdot\operatorname{div}\overleftrightarrow{\sigma}+\rho\mathbf{V}\cdot\mathbf{f}.$$

How do you interpret the rate-of-work terms on the right-hand side?

7 Let e denote the thermodynamic internal energy per unit mass of a material. Then the equation of change for total energy of a material region can be written:

$$\frac{D}{Dt}\iiint_R \rho\left(e+\frac{V^2}{2}\right)d\tau=\oiint_S \hat{\mathbf{n}}\cdot\overleftrightarrow{\sigma}\cdot\mathbf{V}\,d\tau+\iiint_R \rho\mathbf{f}\cdot\mathbf{V}\,d\tau-\oiint_S \mathbf{q}\cdot\hat{\mathbf{n}}\,dS.$$

The first two terms on the right-hand side describe the rate of work done on the material region by the surface stresses and the body forces. The third integral describes the net *outflow* of heat from the region, causing a decrease of energy inside the region. The heat-flux vector $\mathbf{q}$ describes the magnitude and direction of the flow of heat energy per unit time and per unit area.

By suitable operations obtain the differential form of the energy equation:

$$\rho\frac{D}{Dt}\left(e+\frac{V^2}{2}\right)=\operatorname{div}(\overleftrightarrow{\sigma}\cdot\mathbf{V})+\rho\mathbf{f}\cdot\mathbf{V}-\operatorname{div}\mathbf{q}.$$

Subtract the contribution from kinetic energy and obtain

$$\rho\frac{De}{Dt}=\operatorname{div}(\overleftrightarrow{\sigma}\cdot\mathbf{V})-\mathbf{V}\cdot\operatorname{div}\overleftrightarrow{\sigma}-\operatorname{div}\mathbf{q}.$$

This is called the *thermodynamic form* of the energy equation for a continuum.

8 The total rate of work done by the surface stresses per unit volume is given by $\operatorname{div}(\overleftrightarrow{\sigma}\cdot\mathbf{V})$. The rate of work done by the resultant of the surface stresses per unit volume is given by $\mathbf{V}\cdot\operatorname{div}\overleftrightarrow{\sigma}$. The difference between these two terms yields the rate of work done by the surface stresses in deformation of the material particle, per unit volume. Show that this can be written as

$$\begin{aligned}\operatorname{div}(\overleftrightarrow{\sigma}\cdot\mathbf{V})-\mathbf{V}\cdot\operatorname{div}\overleftrightarrow{\sigma}&=\overleftrightarrow{\sigma}:(\nabla\mathbf{V})^{t}\\&=\overleftrightarrow{\sigma}:\nabla\mathbf{V}\qquad(\overleftrightarrow{\sigma}\text{ symmetric})\\&=\tfrac{1}{2}\overleftrightarrow{\sigma}:\left[\nabla\mathbf{V}+(\nabla\mathbf{V})^{t}\right]\qquad(\overleftrightarrow{\sigma}\text{ symmetric}).\end{aligned}$$

9 Given the dyadic $\overleftrightarrow{\mathbf{D}}=\hat{\mathbf{e}}_R\hat{\mathbf{e}}_\phi+\hat{\mathbf{e}}_\phi\hat{\mathbf{e}}_z+\hat{\mathbf{e}}_z\hat{\mathbf{e}}_R$ in cylindrical coordinates determine $\operatorname{div}\overleftrightarrow{\mathbf{D}}$ and $\operatorname{curl}\overleftrightarrow{\mathbf{D}}$.

10 Euler's vector equation of motion for a rigid body is

$$\overleftrightarrow{\mathcal{I}}\cdot\dot{\omega}+\omega\times(\overleftrightarrow{\mathcal{I}}\cdot\omega)=\mathbf{M}.$$

Introduce Cartesian components such that

$$\overleftrightarrow{\mathfrak{I}} = I_{ij}\hat{\mathbf{e}}_i\hat{\mathbf{e}}_j, \qquad \omega = \omega_i\hat{\mathbf{e}}_i, \qquad \mathbf{M} = M_i\hat{\mathbf{e}}_i.$$

Show that Euler's equation can be written in Cartesian tensor form as

$$I_{ij}\dot{\omega}_j + \omega_l\omega_n I_{mn}\varepsilon_{lmi} = M_i, \qquad i = 1,2,3.$$

11 When the body-fixed Cartesian coordinates are aligned with the *principal axes*, the moment-of-inertia tensor takes the simpler form

$$\overleftrightarrow{\mathfrak{I}} = I_{11}\hat{\mathbf{e}}_1\hat{\mathbf{e}}_1 + I_{22}\hat{\mathbf{e}}_2\hat{\mathbf{e}}_2 + I_{33}\hat{\mathbf{e}}_3\hat{\mathbf{e}}_3.$$

Show that the Euler equations now can be written as

$$I_{11}\dot{\omega}_1 + \omega_2\omega_3(I_{33} - I_{22}) = M_1$$

$$I_{22}\dot{\omega}_2 + \omega_1\omega_3(I_{11} - I_{33}) = M_2$$

$$I_{33}\dot{\omega}_3 + \omega_1\omega_2(I_{22} - I_{11}) = M_3.$$

12 Show that the invariants in Eq. (1.302) are indeed invariant under coordinate transformations.

13 Consider the scalar function

$$\phi(\mathbf{r} - \boldsymbol{\delta}) = \frac{1}{|\mathbf{r} - \boldsymbol{\delta}|} = \frac{1}{\sqrt{r^2 - 2\mathbf{r}\cdot\boldsymbol{\delta} + \delta^2}}.$$

Show that the first three terms of a Taylor-series expansion for small $\boldsymbol{\delta}$ can be written as

$$\frac{1}{|\mathbf{r} - \boldsymbol{\delta}|} = \frac{1}{r} - \frac{\mathbf{r}\cdot\boldsymbol{\delta}}{r^3} + \frac{3(\mathbf{r}\cdot\boldsymbol{\delta})^2 - r^2\delta^2}{2r^5} + O(\delta^3).$$

14 The governing equation for the Newtonian potential is the Poisson equation

$$\nabla^2\phi = -4\pi\rho(\mathbf{r}),$$

where $\rho(\mathbf{r})$ is the mass-density distribution (or charge density for electrostatic problems). The solution for $\phi(\mathbf{r})$ where $\rho(\mathbf{r})$ is compactly distributed in a finite region R is (see Exercise 1.7.7).

$$\phi(\mathbf{r}) = \iiint_R \frac{\rho(\boldsymbol{\xi})\,d^3\xi}{|\mathbf{r} - \boldsymbol{\xi}|}.$$

When $\mathbf{r}$ is far away from any point inside the compactly distributed mass in region R, the dummy position $\boldsymbol{\xi}$ can be treated as small compared to $\mathbf{r}$, and the function $|\mathbf{r}-\boldsymbol{\xi}|$ can be expanded in a Taylor series for small $\boldsymbol{\xi}$. Show when this is done that the first three terms of the Taylor-series expansion for the Newtonian potential take the form

$$\phi(\mathbf{r})=\frac{M}{r}-\frac{\mathbf{r}\cdot\mathbf{M}_1}{r^3}+\frac{\mathbf{rr}}{r^5}:\left(\overleftrightarrow{\mathbf{I}}M_2-\frac{3}{2}\overleftrightarrow{\mathcal{I}}\right)+O\left(\frac{1}{r^4}\right),$$

where

$$M\equiv\iiint_R\rho(\boldsymbol{\xi})\,d^3\xi$$

$$\mathbf{M}_1\equiv\iiint_R\boldsymbol{\xi}\rho(\boldsymbol{\xi})\,d^3\xi$$

$$M_2\equiv\iiint_R\xi^2\rho(\boldsymbol{\xi})\,d^3\xi$$

$$\overleftrightarrow{\mathcal{I}}=\iiint_R(\xi^2\overleftrightarrow{\mathbf{I}}-\boldsymbol{\xi\xi})\rho(\boldsymbol{\xi})\,d^3\xi.$$

Here M is the total mass, $\mathbf{M}_1$ the moment of the mass distribution about the origin of coordinates, $\overleftrightarrow{\mathcal{I}}$ the moment of inertia tensor, and M_2 a second moment. In effect, the spatial dependence of $\mathbf{r}$ on the potential has been factored out, each succeeding term of the expansion becoming less significant as $\mathbf{r}$ becomes larger. When the center of mass is chosen as the origin of coordinates, the moment $\mathbf{M}_1$ and thus the second term vanish.

1.5.19 Eigenvectors Associated with Dyadics

It is conceptually useful to regard a dyadic as an operator that changes a vector into another vector (by means of the dot product). In this regard it is of interest to inquire whether there are certain vectors that have only their lengths, and not their orientations, changed when operated upon by a given dyadic or tensor. If such vectors exist, they must satisfy the equation

$$\overleftrightarrow{\boldsymbol{\Phi}}\cdot\mathbf{A}=\lambda\mathbf{A}. \tag{1.345}$$

The vectors $\mathbf{A}$ are called *characteristic vectors, or eigenvectors*, associated with $\overleftrightarrow{\boldsymbol{\Phi}}$. The parameter λ is called an *eigenvalue*, and it characterizes the change in length (and possibly sense) of the eigenvector $\mathbf{A}$ after it has been operated upon by $\overleftrightarrow{\boldsymbol{\Phi}}$ (see Exercise 1.3.21).

Since $\mathbf{A}$ can be expressed as $\mathbf{A}=\overleftrightarrow{\mathbf{I}}\cdot\mathbf{A}$, Eq. (1.345) can also be written as

$$(\overleftrightarrow{\boldsymbol{\Phi}}-\lambda\overleftrightarrow{\mathbf{I}})\cdot\mathbf{A}=0. \tag{1.346}$$

When written in matrix form for Cartesian components, this equation becomes

$$\begin{bmatrix} \phi_{11}-\lambda & \phi_{12} & \phi_{13} \\ \phi_{21} & \phi_{22}-\lambda & \phi_{23} \\ \phi_{31} & \phi_{32} & \phi_{33}-\lambda \end{bmatrix} \begin{Bmatrix} A_1 \\ A_2 \\ A_3 \end{Bmatrix} = 0. \tag{1.347}$$

Because this is a homogeneous set of equations for A_1, A_2, A_3, a nontrivial solution will not exist unless the determinant of the matrix for $\overleftrightarrow{\mathbf{\Phi}}-\lambda\mathbf{I}$ vanishes. The vanishing of this determinant yields a cubic equation for λ, called the *characteristic equation*, the solution of which yields three values of λ, that is, three eigenvalues λ_1, λ_2, and λ_3. The character of these eigenvalues depends on the character of the dyadic $\overleftrightarrow{\mathbf{\Phi}}$. At least one of the eigenvalues must be real. The other two may be real and distinct, real and repeated, or complex conjugates.

In the preponderance of practical problems, the dyadic $\overleftrightarrow{\mathbf{\Phi}}$ is symmetric, that is, $\overleftrightarrow{\mathbf{\Phi}}=\overleftrightarrow{\mathbf{\Phi}}^t$. Of course, $\overleftrightarrow{\mathbf{\Phi}}$ is always real in our considerations. For example, the moment-of-inertia dyadic $\overleftrightarrow{\mathcal{I}}$ is symmetric, and the stress tensor $\overleftrightarrow{\boldsymbol{\sigma}}$ is usually (but not always) symmetric. We thus limit our discussion to symmetric dyadics.

The vanishing of the determinant assures that three eigenvectors exist, $\mathbf{A}_1$, $\mathbf{A}_2$, and $\mathbf{A}_3$, each corresponding to an eigenvalue. The eigenvectors are not unique to within a multiplicative constant, however, and an infinite number of solutions exist having at least three different orientations. Since only orientation is important, it is thus useful to represent the three eigenvectors by three unit eigenvectors $\hat{\mathbf{e}}_1^*, \hat{\mathbf{e}}_2^*, \hat{\mathbf{e}}_3^*$, denoting three different orientations, each associated with a particular eigenvalue.

Suppose now that λ_1 and λ_2 are two distinct eigenvalues and $\mathbf{A}_1$ and $\mathbf{A}_2$ are their corresponding eigenvectors:

$$\overleftrightarrow{\mathbf{\Phi}}\cdot\mathbf{A}_1=\lambda_1\mathbf{A}_1 \tag{1.348a}$$

$$\overleftrightarrow{\mathbf{\Phi}}\cdot\mathbf{A}_2=\lambda_2\mathbf{A}_2. \tag{1.348b}$$

Scalar product of the first equation by $\mathbf{A}_2$ and the second by $\mathbf{A}_1$, and then subtraction, yields

$$\mathbf{A}_2\cdot\overleftrightarrow{\mathbf{\Phi}}\cdot\mathbf{A}_1-\mathbf{A}_1\cdot\overleftrightarrow{\mathbf{\Phi}}\cdot\mathbf{A}_2=(\lambda_1-\lambda_2)\mathbf{A}_1\cdot\mathbf{A}_2. \tag{1.349}$$

Since $\overleftrightarrow{\mathbf{\Phi}}$ is symmetric, one can establish that the left-hand side of this equation vanishes. Thus

$$0=(\lambda_1-\lambda_2)\mathbf{A}_1\cdot\mathbf{A}_2. \tag{1.350}$$

Now suppose that λ_1 and λ_2 are complex conjugates such that $\lambda_1-\lambda_2=2i\lambda_{1i}$, where $i=\sqrt{-1}$ and λ_{1i} is the imaginary part of λ_1. Then $\mathbf{A}_1\cdot\mathbf{A}_2$ is always positive since $\mathbf{A}_1$ and $\mathbf{A}_2$ are complex conjugate vectors associated with λ_1 and

λ_2. It then follows from Eq. (1.350) that $\lambda_{1_i}=0$, and hence *that the three eigenvalues associated with a symmetric dyadic are all real.*

Now assume that λ_1 and λ_2 are real and distinct such that $\lambda_1-\lambda_2$ is not zero. It then follows from Eq. (1.350) that $\mathbf{A}_1\cdot\mathbf{A}_2=0$. Thus the *eigenvectors associated with distinct eigenvalues of a symmetric dyadic are orthogonal.* If the three eigenvalues are all distinct, then the three eigenvectors are mutually orthogonal.

If λ_1 and λ_2 are distinct, but λ_3 is repeated, say $\lambda_3=\lambda_2$, then $\mathbf{A}_3$ must also be perpendicular to $\mathbf{A}_1$ as deduced by an argument similar to that for $\mathbf{A}_2$ stemming from Eq. (1.350). Neither $\mathbf{A}_2$ nor $\mathbf{A}_3$ is preferred, and they are both arbitrary, except insofar as they are both perpendicular to $\mathbf{A}_1$. It is useful, however, to select $\mathbf{A}_3$ such that it is perpendicular to both $\mathbf{A}_1$ and $\mathbf{A}_2$. We do this by choosing $\mathbf{A}_3=\mathbf{A}_1\times\mathbf{A}_2$ and thus establishing a mutually orthogonal set of eigenvectors. This sort of behavior arises when there is an axis of symmetry present in a problem.

For a symmetric real dyadic, three eigenvectors thus can be determined corresponding to the three eigenvalues λ_1, λ_2, and λ_3. Since these eigenvectors are nonunique because of an arbitrary multiplicative factor, it is useful to represent them as unit vectors:

$$\hat{\mathbf{e}}_1^*=\frac{\mathbf{A}_1}{|\mathbf{A}_1|},\qquad \hat{\mathbf{e}}_2^*=\frac{\mathbf{A}_2}{|\mathbf{A}_2|},\qquad \hat{\mathbf{e}}_3^*=\frac{\mathbf{A}_3}{|\mathbf{A}_3|}. \tag{1.351}$$

The unit eigenvectors now represent characteristic orthogonal directions. These directions are referred to as *principal directions* of the dyadic $\overleftrightarrow{\mathbf{\Phi}}$. When the axes of a coordinate system are aligned along the principal directions, these axes are called the *principal axes* associated with $\overleftrightarrow{\mathbf{\Phi}}$.

The set of orthonormal eigenvectors can be treated as a basis system, and the tensor $\overleftrightarrow{\mathbf{\Phi}}$, as well as any vector or tensor, can be expressed in this basis system. Thus we can write

$$\overleftrightarrow{\mathbf{\Phi}}=\phi^{ij}\mathbf{e}_i\mathbf{e}_j \tag{1.352a}$$

$$=\phi_{mn}^*\hat{\mathbf{e}}_m^*\hat{\mathbf{e}}_n^* \tag{1.352b}$$

where

$$\phi_{mn}^*=\phi^{ij}(\mathbf{e}_i\cdot\hat{\mathbf{e}}_m^*)(\mathbf{e}_j\cdot\hat{\mathbf{e}}_n^*) \tag{1.352c}$$

are the components of $\overleftrightarrow{\mathbf{\Phi}}$ in the orthonormal eigenvector basis or, in other words, in the principal-axis system. In this system the eigenvalue Eq. (1.345) can be written

$$(\phi_{mn}^*\hat{\mathbf{e}}_m^*\hat{\mathbf{e}}_n^*)\cdot\hat{\mathbf{e}}_j^*=\lambda_j\hat{\mathbf{e}}_j^*, \tag{1.353}$$

where there is no summation on the index of λ_j. Equation (1.353) reduces to

$$\phi^*_{mj}\hat{\mathbf{e}}^*_m=\lambda_j\hat{\mathbf{e}}^*_j \qquad \text{(no summation on } j\text{)}. \tag{1.354}$$

The scalar product with $\hat{\mathbf{e}}^*_i$ yields

$$\phi^*_{ij}=\lambda_j\delta_{ij} \qquad \text{(no summation on } j\text{)}. \tag{1.355}$$

We thus have

$$\begin{aligned}
\phi^*_{11}&=\lambda_1\\
\phi^*_{22}&=\lambda_2\\
\phi^*_{33}&=\lambda_3\\
\phi^*_{ij}&=0, \qquad i\neq j.
\end{aligned} \tag{1.356}$$

In the principal-axis system only the components of $\overleftrightarrow{\mathbf{\Phi}}$ along the diagonal are nonzero, and the diagonal terms are equal to the eigenvalues themselves. Use of the principal axes thus provides a simplification of the governing equations involving a given dyadic $\overleftrightarrow{\mathbf{\Phi}}$.

In a Cartesian system the characteristic equation associated with a dyadic can be expressed in the form (see Exercise 1.9.3)

$$\lambda^3-I_1\lambda^2-I_2\lambda-I_3=0, \tag{1.357}$$

where I_1, I_2, and I_3 are the invariants associated with the matrix of $\overleftrightarrow{\mathbf{\Phi}}$. The invariants can also be expressed in terms of the eigenvalues,

$$I_1=\lambda_1+\lambda_2+\lambda_3, \qquad I_2=-(\lambda_1\lambda_2+\lambda_2\lambda_3+\lambda_3\lambda_1), \qquad I_3=\lambda_1\lambda_2\lambda_3. \tag{1.358}$$

Finding the roots of the cubic Eq. (1.357) is not always easy. However, when the matrix under consideration is of the form

$$\begin{bmatrix}\phi_{11} & 0 & 0\\ 0 & \phi_{22} & \phi_{23}\\ 0 & \phi_{32} & \phi_{33}\end{bmatrix},$$

one of the roots is $\lambda_1=\phi_{11}$, and the remaining two roots can be found from the quadratic equation

$$(\phi_{22}-\lambda)(\phi_{33}-\lambda)-\phi_{23}\phi_{32}=0.$$

That is,

$$\lambda_{2,3}=\frac{\phi_{22}+\phi_{33}}{2}\pm\frac{1}{2}\sqrt{(\phi_{22}+\phi_{33})^2-4(\phi_{22}\phi_{33}-\phi_{23}\phi_{32})}\,. \tag{1.359}$$

In cases where one of the roots is not obvious, an alternate procedure given below proves to be useful.

In the alternate method we seek the eigenvalues of the so-called *deviatoric tensor* associated with $\overleftrightarrow{\mathbf{\Phi}}$:

$$\phi'_{ij}\equiv\phi_{ij}-\tfrac{1}{3}\phi_{kk}\delta_{ij}. \tag{1.360}$$

Note that

$$\phi'_{ii}=\phi_{ii}-\phi_{kk}=0. \tag{1.361}$$

That is, the first invariant I'_1 of the deviatoric tensor is zero. As a result the characteristic equation associated with the deviatoric tensor is of the form

$$(\lambda')^3-I'_2\lambda'-I'_3=0, \tag{1.362}$$

where λ' is the eigenvalue of the deviatoric tensor. The eigenvalues associated with ϕ_{ij} itself can be computed from

$$\lambda=\lambda'+\tfrac{1}{3}\phi_{kk}. \tag{1.363}$$

The cubic Eq. (1.362) is of a special form that allows a direct computation of its roots. Equation (1.362) can be solved explicitly by introducing the transformation

$$\lambda'=2\left(\tfrac{1}{3}I'_2\right)^{1/2}\cos\alpha, \tag{1.364}$$

which transforms it into

$$2\left(\tfrac{1}{3}I'_2\right)^{3/2}[4\cos^3\alpha-3\cos\alpha]=I'_3. \tag{1.365}$$

The expression in square brackets is equal to $\cos 3\alpha$. Hence

$$\cos 3\alpha=\frac{I'_3}{2}\left(\frac{3}{I'_2}\right)^{3/2}. \tag{1.366}$$

If α_1 is the angle satisfying $0\leq 3\alpha_1\leq\pi$ whose cosine is given by Eq. (1.366), then $3\alpha_1$, $3\alpha_1+2\pi$, and $3\alpha_1-2\pi$ all have the same cosine, and furnish three independent roots of Eq. (1.362),

$$\lambda'_i=2\cos\alpha_i\left(\tfrac{1}{3}I'_2\right)^{1/2},\qquad i=1,2,3, \tag{1.367}$$

where

$$\alpha_1=\frac{1}{3}\left\{\cos^{-1}\left[\frac{I_3'}{2}\left(\frac{3}{I_2'}\right)^{3/2}\right]\right\},\qquad \alpha_2=\alpha_1+\frac{2\pi}{3},\qquad \alpha_3=\alpha_1-\frac{2\pi}{3}. \tag{1.368}$$

Finally we can compute λ_i from Eq. (1.363).

Example 1.18 Suppose that we are required to determine the eigenvalues and eigenvectors of the matrix

$$[\phi]=\begin{bmatrix}2 & 1 & 0\\ 1 & 4 & 1\\ 0 & 1 & 2\end{bmatrix}.$$

The characteristic equation is obtained by setting $\det(\phi-\lambda\delta_{ij})$ to zero:

$$\begin{vmatrix}2-\lambda & 1 & 0\\ 1 & 4-\lambda & 1\\ 0 & 1 & 2-\lambda\end{vmatrix}=(2-\lambda)[(4-\lambda)(2-\lambda)-1]-1\cdot(2-\lambda)=0$$

or

$$(2-\lambda)[(4-\lambda)(2-\lambda)-2]=0.$$

Hence

$$\lambda_1=2,\qquad \lambda_{2,3}=3\pm\sqrt{9-6}\,.$$

Alternately,

$$[\phi']=\begin{bmatrix}2-\frac{8}{3} & 1 & 0\\ 1 & 4-\frac{8}{3} & 1\\ 0 & 1 & 2-\frac{8}{3}\end{bmatrix}$$

$$I_2'=\tfrac{1}{2}(\phi_{ij}'\phi_{ij}'-\phi_{ii}'\phi_{jj}')$$

$$=\tfrac{1}{2}\left[\left(-\tfrac{2}{3}\right)^2+\left(-\tfrac{2}{3}\right)^2+\left(\tfrac{4}{3}\right)^2+2+2\right]=\tfrac{10}{3}$$

$$I_3'=\det\phi_{ij}'=\tfrac{52}{27}.$$

From Eq. (1.368),

$$\alpha_1 = \frac{1}{3}\left\{\cos^{-1}\left[\frac{52}{54}\left(\frac{9}{10}\right)^{3/2}\right]\right\}$$

$$= 11.565°$$

$$\alpha_2 = 131.565°, \qquad \alpha_3 = -108.435°,$$

and from Eq. (1.367),

$$\lambda_1' = 2.065384, \qquad \lambda_2' = -1.3987, \qquad \lambda_3' = -0.66667.$$

Finally, using Eq. (1.363), we get

$$\lambda_1 = 4.7321, \qquad \lambda_2 = 1.2679, \qquad \lambda_3 = 2.00.$$

The eigenvector corresponding to $\lambda = 2$ is calculated as follows. From $(\phi_{ij} - \lambda\delta_{ij})e_j = 0$ we have

$$\begin{bmatrix} 2-2 & 1 & 0 \\ 1 & 4-2 & 1 \\ 0 & 1 & 2-2 \end{bmatrix}\begin{Bmatrix} e_1 \\ e_2 \\ e_3 \end{Bmatrix} = \begin{Bmatrix} 0 \\ 0 \\ 0 \end{Bmatrix}.$$

This gives

$$e_2 = 0, \qquad e_1 = -e_2.$$

Using $e_1^2 + e_2^2 + e_3^2 = 1$, we get

$$\hat{\mathbf{e}}_1 = \pm\frac{1}{\sqrt{2}}(1, 0, -1), \qquad \text{for } \lambda_1 = 2.$$

Similarly, the eigenvectors corresponding to $\lambda = 3 \pm \sqrt{3}$ are given by

$$\hat{\mathbf{e}}_2 = \pm\frac{\left(3-\sqrt{3}\right)}{12}\left(1, \left(1+\sqrt{3}\right), 1\right), \qquad \text{for } \lambda_2 = 3+\sqrt{3}$$

$$\hat{\mathbf{e}}_3 = \pm\frac{\left(3+\sqrt{3}\right)}{12}\left(1, \left(1-\sqrt{3}\right), 1\right), \qquad \text{for } \lambda_3 = 3-\sqrt{3}. \quad ■$$

Exercises 1.9

1 Determine the rotation transformation matrix such that the new base vector $\hat{\bar{\mathbf{e}}}_1$ is along $\hat{\mathbf{e}}_1 - \hat{\mathbf{e}}_2 + \hat{\mathbf{e}}_3$, and $\hat{\bar{\mathbf{e}}}_2$ is along the normal to the plane $2x_1 + 3x_2 + x_3 - 5 = 0$.

2 If $\overleftrightarrow{\mathbf{T}}$ is the dyadic whose components in the unbarred system are given by $T_{11}=1$, $T_{12}=0$, $T_{13}=-1$, $T_{22}=3$, $T_{23}=-2$, and $T_{33}=0$, find the components in the barred coordinates of Exercise 1.

3 Show that the characteristic equation for a second-order tensor σ_{ij} can be expressed as

$$\lambda^3 - I_1\lambda^2 - I_2\lambda - I_3 = 0,$$

where

$$I_1 = \sigma_{kk}$$

$$I_2 = -\tfrac{1}{2}(\sigma_{ii}\sigma_{jj} - \sigma_{ij}\sigma_{ji})$$

$$I_3 = \tfrac{1}{6}(2\sigma_{ij}\sigma_{jk}\sigma_{ki} - 3\sigma_{ij}\sigma_{ji}\sigma_{kk} + \sigma_{ii}\sigma_{jj}\sigma_{kk}) = \det\{\sigma_{ij}\}$$

are three invariants of the tensor.

4 Find the eigenvalues and eigenvectors of the following matrices:

(a) $\begin{bmatrix} 4 & -4 & 0 \\ -4 & 0 & 0 \\ 0 & 0 & 3 \end{bmatrix}$. **(b)** $\begin{bmatrix} 2 & -\sqrt{3} & 0 \\ -\sqrt{3} & 4 & 0 \\ 0 & 0 & 4 \end{bmatrix}$.

(c) $\begin{bmatrix} 1 & 0 & 0 \\ 0 & 3 & -1 \\ 0 & -1 & 3 \end{bmatrix}$. **(d)** $\begin{bmatrix} 2 & -1 & 1 \\ -1 & 0 & 1 \\ 1 & 1 & 2 \end{bmatrix}$.

(e) $\begin{bmatrix} 3 & 5 & 8 \\ 5 & 1 & 0 \\ 8 & 0 & 2 \end{bmatrix}$. **(f)** $\begin{bmatrix} 1 & -1 & 0 \\ -1 & 2 & -1 \\ 0 & -1 & 2 \end{bmatrix}$.

5 Evaluate the three invariants of the matrices in Exercise 4 and check them against the invariants obtained by using the eigenvalues [i.e., use Eq. (1.358)].

6 The components of a stress dyadic at a point, referred to as (X_1, X_2, X_3) system, are (in kips≡1000 psi):

$$\begin{bmatrix} 12 & 9 & 0 \\ 9 & -12 & 0 \\ 0 & 0 & 6 \end{bmatrix}, \quad \begin{bmatrix} 9 & 0 & 12 \\ 0 & -25 & 0 \\ 12 & 0 & 16 \end{bmatrix}, \quad \begin{bmatrix} 1 & -3 & \sqrt{2} \\ -3 & 1 & -\sqrt{2} \\ \sqrt{2} & -\sqrt{2} & 4 \end{bmatrix}.$$

Find the following:

(a) The stress vector acting on a plane perpendicular to the vector $2\hat{\mathbf{e}}_1 - 2\hat{\mathbf{e}}_2 + \hat{\mathbf{e}}_3$ passing through the point.

(b) The magnitude of the stress vector and the angle between the stress vector and the normal to the plane.

(c) The magnitudes of the normal and tangential components of the stress vector.

7 Let $\hat{\mathbf{e}}$ denote the direction of some axis passing through a point 0. Define a dyadic

$$\overleftrightarrow{\mathbf{\Phi}}=\hat{\mathbf{e}}\hat{\mathbf{e}}+\cos\theta(\overleftrightarrow{\mathbf{I}}-\hat{\mathbf{e}}\hat{\mathbf{e}})+\sin\theta\,\overleftrightarrow{\mathbf{I}}\times\hat{\mathbf{e}}.$$

This dyadic, when operating on some vector $\mathbf{A}$ by $\overleftrightarrow{\mathbf{\Phi}}\cdot\mathbf{A}$, rotates the vector $\mathbf{A}$ about the axis $\hat{\mathbf{e}}$ by the angle θ. Verify this by letting $\overleftrightarrow{\mathbf{\Phi}}$ operate on $\hat{\mathbf{e}}$ and a unit vector $\hat{\mathbf{b}}$ which is perpendicular to $\hat{\mathbf{e}}$, that is, by evaluating $\overleftrightarrow{\mathbf{\Phi}}\cdot\hat{\mathbf{e}}$ and $\overleftrightarrow{\mathbf{\Phi}}\cdot\hat{\mathbf{b}}$. Finally let $\mathbf{A}$ be some arbitrary vector and evaluate $\overleftrightarrow{\mathbf{\Phi}}\cdot\mathbf{A}$ and interpret the results.

8 Verify that

$$\overleftrightarrow{\mathbf{I}}\cdot\overleftrightarrow{\mathbf{\Phi}}=\overleftrightarrow{\mathbf{\Phi}}\cdot\overleftrightarrow{\mathbf{I}}=\overleftrightarrow{\mathbf{\Phi}}.$$

9 If $\mathbf{A}$ is an arbitrary vector and $\overleftrightarrow{\mathbf{\Phi}}$ is an arbitrary dyadic, verify that:

(a) $(\overleftrightarrow{\mathbf{I}}\times\mathbf{A})\cdot\overleftrightarrow{\mathbf{\Phi}}=\mathbf{A}\times\overleftrightarrow{\mathbf{\Phi}}$.

(b) $(\mathbf{A}\times\overleftrightarrow{\mathbf{I}})\cdot\overleftrightarrow{\mathbf{\Phi}}=\mathbf{A}\times\overleftrightarrow{\mathbf{\Phi}}$.

(c) $(\mathbf{\Phi}\times\mathbf{A})^{t}=-\mathbf{A}\times\overleftrightarrow{\mathbf{\Phi}}^{t}$.

10 What are the eigenvalues of a triangular matrix (see Section 1.3)?

11 Show that $[A]$ is singular if and only if zero is an eigenvalue of $[A]$.

12 Determine how the eigenvalues of $[A]^2$ are related to the eigenvalues of $[A]$.

13 Let $[A]$ be an $n\times n$ matrix with eigenvalues $\lambda_1, \lambda_2, \ldots, \lambda_n$. Show that if $[\Lambda]$ is the diagonal matrix,

$$[\Lambda]=\begin{bmatrix}\lambda_1 & 0 & \cdots & 0\\ 0 & \lambda_2 & & \\ \vdots & & \ddots & \vdots\\ 0 & & & \lambda_n\end{bmatrix},$$

and $[P]=[p_{ij}]$ is the matrix in which p_{ij} is the ith component of the jth eigenvector, then $[A][P]=[P][\Lambda]$. Further, show that if $[A]$ has n linearly independent eigenvalues, $[P]^{-1}[A][P]=[\Lambda]$.

14 Two matrices $[A]$ and $[B]$ are said to be *similar* if there exists a nonsingular matrix $[P]$ such that $[B]=[P]^{-1}[A][P]$. Show that similar matrices have the same characteristic polynomial (hence, the same eigenvalues and eigenvectors).

1.5.20 Higher Order Tensors

So far we have considered scalars, vectors, and dyadics, which we can regard in general as tensor quantities of order (or rank) 0, 1, and 2. These entities can be generalized even further. For instance, three vectors taken side by side as a single entity can be regarded as a *triad*, and a linear combination of triads as a *triadic*. When the notion of invariance with regard to coordinate transformations is introduced, it is customary to call a triadic a tensor of rank 3. In general a linear combination of n vectors taken side by side as an entity is called a polyadic and generalized to a tensor of rank n.

A triadic can be written in terms of a covariant basis as

$$\mathbf{\Phi}^{(3)} = \phi^{ijk}\mathbf{e}_i\mathbf{e}_j\mathbf{e}_k, \tag{1.369}$$

and ϕ^{ijk} are called the contravariant components of the triadic or third-order tensor. If we transform the coordinates to a new barred system, then the covariant base vectors transform according to the rule given by Eq. (1.145), that is,

$$\mathbf{e}_i = \frac{\partial \bar{q}^l}{\partial q^i}\bar{\mathbf{e}}_l. \tag{1.370}$$

In the new coordinates Eq. (1.369) takes the form

$$\mathbf{\Phi}^{(3)} = \bar{\phi}^{lmn}\,\bar{\mathbf{e}}_l\bar{\mathbf{e}}_m\bar{\mathbf{e}}_n, \tag{1.371}$$

where

$$\bar{\phi}^{lmn} = \phi^{ijk}\frac{\partial \bar{q}^l}{\partial q^i}\frac{\partial \bar{q}^m}{\partial q^j}\frac{\partial \bar{q}^n}{\partial q^k}. \tag{1.372}$$

The enforcement of the transformation law (1.372) guarantees the invariance of $\mathbf{\Phi}^{(3)}$ under the transformation law and thus can be used as the definition of the covariant components of a third-order tensor.

The corresponding results can also be established by means of the dual or contravariant basis system. Thus we can also write

$$\mathbf{\Phi}^{(3)} = \phi_{ijk}\mathbf{e}^i\mathbf{e}^j\mathbf{e}^k, \tag{1.373}$$

where ϕ_{ijk} are the covariant components of the triadic. If a transformation to a new barred system is made, then the contravariant basis vectors transform according to the law (1.145):

$$\mathbf{e}^i = \frac{\partial q^i}{\partial \bar{q}^l}\bar{\mathbf{e}}^l. \tag{1.374}$$

Equation (1.373) then takes the form in the new system

$$\mathbf{\Phi}^{(3)} = \bar{\phi}_{lmn}\bar{\mathbf{e}}^{l}\bar{\mathbf{e}}^{m}\bar{\mathbf{e}}^{n}, \tag{1.375}$$

where

$$\bar{\phi}_{lmn} = \phi_{ijk}\frac{\partial q^{i}}{\partial \bar{q}^{l}}\frac{\partial q^{j}}{\partial \bar{q}^{m}}\frac{\partial q^{k}}{\partial \bar{q}^{n}}. \tag{1.376}$$

Equation (1.376) constitutes the transformation law for the covariant components of a third-order tensor. There are also alternative representations of a triadic involving a mixed use of the covariant and contravariant basis vectors. The components are then called mixed components of the third-order tensor, and there are six different ways in which this can be done. Whichever the representation, a third-rank tensor has 27 independent components in general.

We have already encountered a third-order tensor when introducing the alternating symbol ε_{ijk} defined by Eq. (1.45). Let us use an orthonormal basis system associated with an orthogonal system of coordinates and introduce the triadic

$$\mathbf{E}^{(3)} = \varepsilon_{ijk}\hat{\mathbf{e}}_{i}\hat{\mathbf{e}}_{j}\hat{\mathbf{e}}_{k}. \tag{1.377}$$

This representation is also appropriate for rectangular Cartesian coordinates. We wish to find a representation for $\mathbf{E}^{(3)}$ in terms of the covariant basis system and in terms of the contravariant basis system. To proceed we first introduce notation appropriate for covariant and contravariant components. We define new alternating symbols e_{ijk} and e^{ijk} such that

$$e^{ijk} = e_{ijk} = \varepsilon_{ijk} \tag{1.378}$$

and thus reserve ε_{ijk} for the Cartesian tensor representation. For orthogonal coordinate systems we have

$$\begin{aligned}
\hat{\mathbf{e}}_{1} &= \frac{\mathbf{e}_{1}}{\sqrt{g_{11}}} = \sqrt{g_{11}}\,\mathbf{e}^{1}\\
\hat{\mathbf{e}}_{2} &= \frac{\mathbf{e}_{2}}{\sqrt{g_{22}}} = \sqrt{g_{22}}\,\mathbf{e}^{2}\\
\hat{\mathbf{e}}_{3} &= \frac{\mathbf{e}_{3}}{\sqrt{g_{33}}} = \sqrt{g_{33}}\,\mathbf{e}^{3}
\end{aligned} \tag{1.379}$$

and also

$$g = g_{11}g_{22}g_{33}. \tag{1.380}$$

Consequently $\mathbf{E}^{(3)}$ can be written in either contravariant or covariant components as

$$\mathbf{E}^{(3)} = \mathcal{E}^{ijk}\mathbf{e}_i\mathbf{e}_j\mathbf{e}_k \tag{1.381a}$$

$$= \mathcal{E}_{ijk}\mathbf{e}^i\mathbf{e}^j\mathbf{e}^k, \tag{1.381b}$$

where

$$\mathcal{E}^{ijk} = \frac{e^{ijk}}{\sqrt{g}} \tag{1.382a}$$

$$\mathcal{E}_{ijk} = \sqrt{g}\, e_{ijk}. \tag{1.382b}$$

These results hold also for the contravariant and covariant components of the alternating tensor $\mathbf{E}^{(3)}$ in general curvilinear coordinates.

The generalization of dyadics and triadics to polyadics is straightforward. For instance, the contravariant and covariant component representation of a fourth-rank tensor can be written as

$$\mathbf{\Phi}^{(4)} = \phi^{ijkl}\mathbf{e}_i\mathbf{e}_j\mathbf{e}_k\mathbf{e}_l$$

$$= \phi_{ijkl}\mathbf{e}^i\mathbf{e}^j\mathbf{e}^k\mathbf{e}^l. \tag{1.383}$$

There are 14 different ways the mixed components can be represented, and in general there are 81 independent components.

1.5.21 Isotropic Tensors

In physical situations there are variables that may have different properties in different directions. For instance, the elastic properties of a piece of wood may be different along the grain than they are transverse to the grain. The heat-transfer properties may also be different. In other situations the variables are independent of the directions considered. When a variable is independent of direction, it is described as *isotropic*. Here we consider tensor variables having properties that are independent of direction, that is, isotropic tensors.

A tensor is defined as isotropic at a point if its *components* are invariant with respect to rotation of the axes of the coordinate systems used to describe the tensor. Because at a given point only different directions are considered, it is useful to deal with rectangular Cartesian coordinates and thus Cartesian tensors. If an orthonormal Cartesian basis system is rotated to another barred Cartesian coordinate system, the transformation of the basis vectors is given by (see Eq. (1.62), $a_{ij} = \beta_{ji}$)

$$\hat{\mathbf{e}}_i = a_{ij}\hat{\bar{\mathbf{e}}}_j \quad \text{and} \quad \hat{\bar{\mathbf{e}}}_i = a_{ji}\hat{\mathbf{e}}_j, \qquad i=1,2,3, \tag{1.384}$$

where

$$a_{ij} \equiv \hat{\mathbf{e}}_i \cdot \hat{\bar{\mathbf{e}}}_j \tag{1.385}$$

is the direction cosine of the angle between the x_i axis and the $\bar{x}_j$ axis.

The lowest rank tensor is a tensor of rank 0, which is a scalar. A scalar by its nature must be isotropic because at a point its value cannot depend on any particular direction. Thus at a point, for instance, it would make no sense to say that temperature depended on a direction. A tensor of rank 1 is a vector, which of course is intimately associated with a direction. In a given Cartesian system a vector **A** is described by

$$\mathbf{A} = A_i \hat{\mathbf{e}}_i, \tag{1.386}$$

and in the rotated system it is described by

$$\mathbf{A} = \bar{A}_j \bar{\mathbf{e}}_j, \tag{1.387}$$

where

$$\bar{A}_j = A_i a_{ij}, \qquad j = 1,2,3. \tag{1.388}$$

If the components were invariant, then we must have $\bar{A}_j = A_j$, and the components A_j would satisfy the equation

$$A_j = A_i a_{ij}, \qquad j = 1,2,3. \tag{1.389a}$$

or

$$A_i(\delta_{ij} - a_{ij}) = 0, \qquad j = 1,2,3. \tag{1.389b}$$

This is a homogeneous set of equations for A_j, and the only solution is $A_1 = A_2 = A_3 = 0$. Thus the only isotropic vector is the null vector $\mathbf{A} = \mathbf{0}$.

A second-rank tensor can be represented by the dyadic

$$\overleftrightarrow{\boldsymbol{\Phi}} = \phi_{ij} \hat{\mathbf{e}}_i \hat{\mathbf{e}}_j \tag{1.390}$$

and in the rotated Cartesian system by

$$\overleftrightarrow{\boldsymbol{\Phi}} = \bar{\phi}_{mn} \hat{\bar{\mathbf{e}}}_m \hat{\bar{\mathbf{e}}}_n. \tag{1.391}$$

If this dyadic is to be isotropic, then we must have $\bar{\phi}_{mn} = \phi_{mn}$ and thus

$$\phi_{mn} = \phi_{ij} a_{im} a_{jn}, \qquad m, n = 1,2,3. \tag{1.392}$$

A typical term on the diagonal, say ϕ_{11}, can be written out as

$$\begin{aligned}\phi_{11} &= \phi_{ij}a_{i1}a_{j1} \\ &= \sum_{r=1}^{3} \phi_{rr}a_{r1}a_{r1} + a_{11}a_{21}(\phi_{12}+\phi_{21}) \\ &\quad + a_{11}a_{31}(\phi_{13}+\phi_{31}) + a_{21}a_{31}(\phi_{23}+\phi_{32}).\end{aligned} \tag{1.393a}$$

A typical term off the diagonal, say ϕ_{12}, can be written out as

$$\begin{aligned}\phi_{12} &= \phi_{ij}a_{i1}a_{j2} \\ &= \sum_{r=1}^{3} \phi_{rr}a_{r1}a_{r2} + a_{22}(\phi_{12}a_{11}+\phi_{32}a_{31}) \\ &\quad + a_{12}(\phi_{21}a_{21}+\phi_{31}a_{31}) + a_{32}(\phi_{13}a_{11}+\phi_{23}a_{21}).\end{aligned} \tag{1.393b}$$

We now recall that for rotation of coordinates we can establish from Eqs. (1.384) that

$$\delta_{ij} = a_{ir}a_{jr} = a_{ri}a_{rj}. \tag{1.394}$$

With Eqs. (1.392) and (1.394) as examples, it can now be established that the general solution of Eqs. (1.392) occurs when the off-diagonal terms are zero ($\phi_{ij}=0, i \neq j$) and the diagonal terms are all equal [$\phi_{11}=\phi_{22}=\phi_{22}\equiv\alpha(\mathbf{r})$]. Thus the most general isotropic second-rank tensor occurs when

$$\phi_{ij} = \alpha(\mathbf{r})\delta_{ij}, \tag{1.395}$$

where $\alpha(\mathbf{r})$ is an arbitrary scalar. The same result can be arrived at by writing Eq. (1.392) in the form,

$$\phi_{ij}(\delta_{im}\delta_{jn} - a_{im}a_{jn}) = 0.$$

For arbitrary (rotation) transformations (i.e., for arbitrary a_{ij}) the above identity holds only if ϕ_{ij} is of the form in Eq. (1.395). Thus in general form we have

$$\overleftrightarrow{\mathbf{\Phi}} = \alpha(\mathbf{r})\overleftrightarrow{\mathbf{I}}, \tag{1.396}$$

where $\overleftrightarrow{\mathbf{I}}$ is the unit dyadic. This is to be expected on general grounds, since $\overleftrightarrow{\mathbf{I}}$ converts a vector into itself. If $\overleftrightarrow{\mathbf{\Phi}}$ converted a vector into a vector with a different direction, then somehow $\overleftrightarrow{\mathbf{I}}$ would be dependent on direction and would therefore not be isotropic.

Consider a third-rank tensor given by its Cartesian form:

$$\mathbf{\Phi}^{(3)} = \phi_{ijk}\hat{\mathbf{e}}_i\hat{\mathbf{e}}_j\hat{\mathbf{e}}_k$$

$$= \bar{\phi}_{lmn}\hat{\bar{\mathbf{e}}}_l\hat{\bar{\mathbf{e}}}_m\hat{\bar{\mathbf{e}}}_n, \tag{1.397}$$

where

$$\bar{\phi}_{lmn} = \phi_{ijk}a_{il}a_{jm}a_{kn}. \tag{1.398}$$

If this tensor is to be isotropic, then $\bar{\phi}_{lmn} = \phi_{lmn}$. Let us note that when the scalar triple product $[\hat{\bar{\mathbf{e}}}_l\hat{\bar{\mathbf{e}}}_m\hat{\bar{\mathbf{e}}}_n]$ is evaluated in the barred and unbarred basis systems, we get the result

$$\varepsilon_{lmn} = \varepsilon_{ijk}a_{il}a_{jm}a_{kn}. \tag{1.399}$$

It is now possible to show that the only solution of Eq. (1.398) with $\bar{\phi}_{lmn} = \phi_{lmn}$ is $\phi_{ijk} = \beta(\mathbf{r})\varepsilon_{ijk}$, where $\beta(\mathbf{r})$ is an arbitrary scalar.

Alternately, rewrite Eq. (1.398) in the form (with $\bar{\phi}_{lmn} = \phi_{lmn}$)

$$\phi_{ijk}\left(\delta_{il}\delta_{jm}\delta_{kn} - a_{il}a_{jm}a_{kn}\right) = 0.$$

Since the determinant of a_{ij} is unity, from Exercise 1.3.13 we have

$$\tfrac{1}{6}\varepsilon_{ijk}\varepsilon_{lmn}a_{il}a_{jm}a_{kn} = 1.$$

Therefore for arbitrary transformations ϕ_{ijk} must be [note that, from Exercise 1.2.1(d), $\varepsilon_{lmn}\varepsilon_{lmn} = 6$] of the form

$$\phi_{ijk} = \beta(\mathbf{r})\varepsilon_{ijk}.$$

Thus the only isotropic third-rank tensor is a scalar multiple of the alternating tensor in Eq. (1.377).

Finally, consider a fourth-rank tensor given by its Cartesian forms:

$$\mathbf{\Phi}^{(4)} = \phi_{ijkl}\hat{\mathbf{e}}_i\hat{\mathbf{e}}_j\hat{\mathbf{e}}_k\hat{\mathbf{e}}_l$$

$$= \bar{\phi}_{mnpq}\hat{\bar{\mathbf{e}}}_m\hat{\bar{\mathbf{e}}}_n\hat{\bar{\mathbf{e}}}_p\hat{\bar{\mathbf{e}}}_q, \tag{1.400}$$

where

$$\bar{\phi}_{mnpq} = \phi_{ijkl}a_{im}a_{jn}a_{kp}a_{lq}. \tag{1.401}$$

For isotropy we must have $\bar{\phi}_{mnpq} = \phi_{mnpq}$. In the summation on the right-hand side of Eq. (1.401), only three different values of i, j, k, and l are possible, and hence at least two of them must be equal. There are three classes of solutions, depending on which pairs of i, j, k, and l are set equal. Consider first the class out of which we set $k=l$ or $i=j$. This suggests a product solution of the form $\phi_{ijkl}=\tilde{\phi}_{ij}\tilde{\tilde{\phi}}_{kl}$, and Eq. (1.401) can be written

$$\tilde{\phi}_{mn}\tilde{\tilde{\phi}}_{pq}=\left(\tilde{\phi}_{ij}a_{im}a_{jm}\right)\left(\tilde{\tilde{\phi}}_{kl}a_{kp}a_{lq}\right). \tag{1.402}$$

This reduces to a product of solutions that are the same as for Eq. (1.392), and thus for this class of solutions we would have

$$\phi_{ijkl}=\lambda_1(\mathbf{r})\delta_{ij}\delta_{kl}, \tag{1.403}$$

where λ_1 is an arbitrary scalar multiplier. A similar line of reasoning can be applied to the class of solutions out of which we set $i=k$ or $j=l$. Then we would obtain for the second class

$$\phi_{ijkl}=\lambda_2(\mathbf{r})\delta_{ik}\delta_{jl}, \tag{1.404}$$

where λ_2 is an arbitrary scalar multiplier. The third class of solutions is obtained when we set $i=l$ or $j=k$. Then we get

$$\phi_{ijkl}=\lambda_3(\mathbf{r})\delta_{il}\delta_{jk}, \tag{1.405}$$

where λ_3 is an arbitrary scalar multiplier. It can be shown that the general fourth-rank isotropic tensor is a linear combination of these three special cases:

$$\phi_{ijkl}=\lambda_1\delta_{ij}\delta_{kl}+\lambda_2\delta_{ik}\delta_{jl}+\lambda_3\delta_{il}\delta_{jk}. \tag{1.406}$$

Thus the general fourth-rank isotropic tensor has been reduced to only three independent components.

It is useful to rewrite the isotropic tensor of rank 4 in alternative forms in order to separate the symmetric and antisymmetric parts. We set $\lambda_1=\lambda$, $\lambda_2=\mu+\nu$, and $\lambda_3=\mu-\nu$, where λ, μ, and ν are three new scalars. We then have

$$\begin{aligned}\phi_{ijkl}&=\lambda\delta_{ij}\delta_{kl}+\mu\left(\delta_{ik}\delta_{jl}+\delta_{il}\delta_{jk}\right)+\nu\left(\delta_{ik}\delta_{jl}-\delta_{il}\delta_{jk}\right)\\&=\lambda\delta_{ij}\delta_{kl}+\mu\left(\delta_{ik}\delta_{jl}+\delta_{il}\delta_{jk}\right)+\nu\varepsilon_{ijs}\varepsilon_{kls}.\end{aligned} \tag{1.407}$$

The coefficients of λ and μ are symmetric in $i-j$ and $k-l$, whereas the coefficient of ν is antisymmetric.

Physical Examples Consider two examples where considerations of isotropic tensors enter physical problems. The first example concerns conductive heat transfer. It is assumed that the flow of heat is proportional to the gradient of the temperature ∇T. If it is assumed that the components of the heat-flux vector $\mathbf{q}$ are linear combinations of the temperature gradient, then we write

$$\mathbf{q}=-\overleftrightarrow{\mathbf{K}}\cdot\nabla T, \tag{1.408}$$

where $\overleftrightarrow{\mathbf{K}}$ is the heat conductivity dyadic or tensor. The negative sign implies that $\mathbf{q}$ is approximately opposite in direction to ∇T, and the heat conductivity tensor can be regarded as changing the vector ∇T into the vector $\mathbf{q}$, the two of which in general may not be parallel. If the conducting medium is isotropic, then it has no preferential directions, and in this case we would have $\overleftrightarrow{\mathbf{K}}=k\overleftrightarrow{\mathbf{I}}$, where k is now a positive scalar conductivity. Then for an isotropic medium Eq. (1.408) becomes

$$\mathbf{q}=-k\nabla T, \tag{1.409}$$

and $\mathbf{q}$ is exactly opposite in direction to ∇T. Equation (1.409) is Fourier's law for an isotropic medium.

As a second example, consider the constitutive relation for the relation between stress and strain for solid mechanics, and viscous stress and rate of strain for fluid mechanics. Let τ_{ij} denote the Cartesian component of stress for solid mechanics or viscous stress for fluid mechanics. Let ε_{ij} denote the Cartesian component of infinitesimal strain for solid mechanics or the rate of strain for fluid mechanics, which we write

$$\varepsilon_{ij}=\frac{1}{2}\left[\frac{\partial V_j}{\partial x_i}+\frac{\partial V_i}{\partial x_j}\right]. \tag{1.410}$$

Here V_j denotes a Cartesian displacement for a solid element or a Cartesian velocity component for a fluid element. In linear elasticity and Newtonian fluid mechanics we assume that τ_{ij} is a linear combination of ε_{ij}, that is,

$$\tau_{ij}=C_{ijmn}\varepsilon_{mn}, \tag{1.411}$$

where C_{ijmn} is the modulus of elasticity tensor or the viscosity tensor. A fluid medium is usually assumed to be isotropic, and a wide class of solid materials can also be assumed to be isotropic. Moreover, the stresses τ_{ij} are assumed to be symmetric along with the strains. Thus for these cases we assume that C_{ijmn} is isotropic and symmetric in $i-j$ and $m-n$, and thus takes the form of (1.407):

$$C_{ijmn}=\lambda\delta_{ij}\delta_{mn}+\mu(\delta_{im}\delta_{jn}+\delta_{in}\delta_{jm}), \tag{1.412}$$

where $\nu=0$ owing to symmetry. In solid mechanics μ and λ are called the Lamé coefficients of elasticity, and in fluid mechanics μ and λ are called the first and second coefficients of viscosity. Thus τ_{ij} can be written for an isotropic medium as

$$\tau_{ij}=2\mu\varepsilon_{ij}+\lambda\varepsilon_{mm}\delta_{ij}. \tag{1.413}$$

REFERENCES AND ADDITIONAL READING

1 Aris, R., *Vectors, Tensors, and the Basic Equations in Fluid Mechanics*, Prentice-Hall, Englewood Cliffs, NJ, 1962.

2 Bedford, F. W., and T. D. Dwivedi, *Vector Calculus*, McGraw-Hill, New York, 1970.

3 Bellman, R. E., *Introduction to Matrix Analysis*, 2nd ed., McGraw-Hill, New York, 1970.

4 Borisenko, A. I., and I. E. Tarapov, *Vector and Tensor Analysis with Applications*, R. A. Silverman, Transl. and Ed., Prentice-Hall, Englewood Cliffs, NJ, 1968.

5 Bourne, D. E., and P. C. Kendall, *Vector Analysis and Cartesian Tensors*, 2nd ed., Academic Press, New York, 1977.

6 Bowen, R. M., and C. C. Wang, *Introduction to Vectors and Tensors, Plenum, New York*, 1976.

7 Campbell, H. G., *An Introduction to Matrices, Vectors and Linear Programming*, Prentice-Hall, Englewood Cliffs, NJ, 1977.

8 Chambers, L. G., *A Course in Vector Analysis*, Chapman and Hall, London, 1969.

9 Chisholm, J. S. R., *Vectors in Three-Dimensional Space*, Cambridge University Press, Cambridge, 1978.

10 Chorlton, F., *Vector and Tensor Methods*, Halsted Press, New York, 1976.

11 Coburn, N., *Vector and Tensor Analysis*, Macmillan, New York, 1964.

12 Eisele, J. A., *Applied Matrix and Tensor Analysis*, Wiley-Interscience, New York, 1970.

13 Gantmacher, F. R., *The Theory of Matrices*, Chelsea, New York, 1959.

14 Graham, A., *Matrix Theory and Applications for Engineers and Mathematicians*, Halsted Press, New York 1979.

15 Haskell, R. E., *Introduction to Vectors and Cartesian Tensors*, Prentice-Hall, Englewood Cliffs, NJ, 1972.

16 Hauge, B., *An Introduction to Vector Analysis for Physicists and Engineers*, 6th ed., revised by D. Martin, Methuen, London, 1970.

17 Jeffreys, H., *Cartesian Tensors*, Cambridge University Press, London, 1965.

18 Karamcheti, K., *Vector Analysis and Cartesian Tensors*, Holden Day, San Francisco, 1967.

19 Lanczos, C., "Tensor Calculus," in *Handbook of Physic*, London and Odishaw, Eds., McGraw-Hill, New York, 1958.

20 Lichnerowicz, A., *Linear Algebra and Analysis*, Holden-Day, San Francisco, 1967.

21 Moon, P. H., and D. E. Spencer, *Vectors*, Van Nostrand, Princeton, NJ, 1965.

22 Phillips, H. B., *Vector Analysis*, Wiley, New York, 1933.

23 Sokolnikoff, I. S., *Tensor Analysis*, Wiley, New York, 1964.

24 Wills, A. P., *Vector Analysis with an Introduction to Tensor Analysis*, Dover, New York, 1953.

25 Wilson, E. B., *Vector Analysis* (founded on the lecture notes of J. W. Gibbs), Dover, New York, 1969.

26 Wrede, R. C., *Introduction to Vector and Tensor Analysis*, Dover, New York, 1972.

27 Young, E. C., *Vector and Tensor Analysis*, Marcel Dekker, New York, 1978.

Chapter Two

ELEMENTS OF FUNCTIONAL ANALYSIS

2.1 INTRODUCTORY COMMENTS

In the study of solutions to various physical problems that are unrelated, we discover that the problems share certain common mathematical properties. For example, in the approximate solution of problems by numerical methods we are lead ultimately to the solution of an algebraic set of equations. Then it is necessary to know under what conditions a given set of equations, without regard to a specific problem, possesses a unique solution. That is, instead of studying a particular kind of equation or problem, we study the particular in the context of the general. A systematic study of the mathematical (and computational) properties of a given *class* of problems leads to a formal treatment that is more abstract than the treatment of a specific problem. This in turn leads to deeper understanding and, more importantly, insight into the essential features of the problem. On the other hand, if we have the necessary tools at our disposal to study an abstract problem, it is a simple matter to specialize the results to any specific problem that falls into the general category. A good example of this kind is provided by the notion of an (abstract) vector, which includes as a special case the physical vector we studied in the previous chapter. The notion of "direction" is generalized to include a "property" that the vector should possess, among other things, in order to qualify as a vector. The collection of all vectors which share a common property, such as satisfying a given differential equation, and satisfying certain rules of addition and scalar multiplication, is termed a *vector space*. An *abstract* (vector) *space* contains points that could be numbers, points in ordinary Euclidean space, functions, and so on. The study of those properties of a completely arbitrary collection of points in a *space* which hold independently of their particular nature, whether they be numbers, points, or functions, is the subject of functional analysis.

Functional analysis is a generalization of concepts and methods from elementary analysis, algebra, and geometry. The subject of functional analysis has its origins in the calculus of variations, the theory of differential equations, and the theory of approximations of functions. In functional analysis the fundamental concepts and methods of analysis are studied from a unified point of view and for more general objects. For example, the definitions of a functional, of an extremum of a functional, and the conditions for the existence of an extremum in the calculus of variations are entirely analogous to the definitions of a function, of an extremum of a function, and the conditions for the existence of such an extremum in the differential calculus. Another example of analogy between functional analysis and geometry is given by the development of functions with respect to elements of an orthogonal system, which resemble orthogonal systems of vectors in the Euclidean space. Also, the decomposition of a function into a Fourier series corresponds to the decomposition of a vector into its components. Thus functional analysis is an indispensable tool in approximation theory and numerical analysis.

To keep the scope of the book within reasonable limits, only the most fundamental concepts from linear algebra and analysis are included here. For further study the reader can consult any one of many books on functional analysis (see the references at the end of this chapter).

2.2 ELEMENTS OF LINEAR ALGEBRA

2.2.1 Introduction and Notation

The set theoretic concepts pervade all of mathematics in a most natural way. The idea of a collection of numbers is indispensible to the organization and establishment of the theory of the real and natural numbers. To understand the theory of Fourier-series approximation, a topic rich in the applications of mathematics to electrical engineering and heat conduction problems, one must know the theory of sets, as well as other theories which would not exist without the theory of sets. As in any subject of mathematics, the use of a formal language (i.e., terminology), logic, and notation results in abstraction. For example, to restate the trivial fact "all bears are animals" in terms of the set theory, we write

$$x \in B \subset A \quad (\text{meaning:} \quad x \text{ an element in } B \text{ which is a subset of } A) \tag{2.1}$$

where A is the set of all animals and B is the set of all bears. However, the same abstraction makes it simple to express complicated thoughts. We use the following notation, very standard in mathematics, which proves to be convenient in our study:

$\in$ means "a member of"
$\notin$ means "not a member of"

2.2.2 Sets and Set Operations

A collection of objects sharing a common property is grouped together and is called *a set*. We use the notation

$$A = \{a: \ a \text{ has property } P\} \tag{2.2}$$

to define the set. Here A is a set, a a typical element of set A, and P is the property shared by the elements of set A. Obviously it is implied that there is a larger collection of elements and A is only a *subset* of the elements which share property P. For example, the set of all students enrolled in the course "Advanced Engineering Analysis" (at a university) can be expressed as

$$A = \{a: \quad a \text{ is a student in the "Advanced Engineering Analysis" course}\}. \tag{2.3}$$

Obviously A is a subset of the students enrolled in all courses. Further, the set

$$B = \{b: \ b \text{ is a female student in the "Advanced Engineering Analysis" course}\} \tag{2.4}$$

is a subset of A. This fact is expressed as

$$B \subset A \text{ (means "}B\text{ is a subset of }A\text{")}$$

or

$$A \supset B \text{ (means "}A\text{ contains }B\text{")}.$$

That is, every element of B is also an element of A.

Since every set is a subset of itself, we use the notation $B \subset A$ to indicate that B is a subset of A. A set B is said to be a *proper subset* of another set A if and only if $B \subset A$ and $A \neq B$. Two sets are said to be equal, $A = B$, if and only if

$$A \subset B \qquad \text{and} \qquad B \subset A.$$

A set that contains no elements is called the *null* or *empty set*, and is denoted by $\varnothing$.

The *union* of two sets A and B is the set of all elements that belong to A or B, and is denoted by

$$A \cup B = \{x: \quad x \in A \quad \text{or} \quad x \in B\}. \tag{2.5}$$

Clearly, the possibility that x belongs to both A and B is not ruled out in the above definition; it implies that x belongs *at least* to one of the two sets. If B is a subset of A, then $A \cup B = A$.

The *intersection* of two sets A and B is the set of elements that belong to A and B and is denoted by

$$A \cap B = \{y: \quad y \in A \qquad \text{and} \qquad y \in B\}. \tag{2.6}$$

If B is a subset of A, then $A \cap B = B$. Two sets A and B are said to be *disjoint* if they have no element in common; that is, their intersection is the empty set,

$$A \cap B = \varnothing .$$

The *difference* of two sets A and B (note that the order is important) is the set of all elements that belong to A but not to B:

$$A - B = \{x: \quad x \in A, x \notin B\}. \tag{2.7}$$

If $A \cap B$ is empty, then $A - B = A$. Note also that $A - B \neq B - A$.

The *complement* of a set A (subset of a larger set S) *with respect to* S is the set of elements in S that do not belong to A. That is, the complement of A with respect to S, denoted by A/S or A', is the difference of S and A. Obviously, $S/S = \varnothing$, $\varnothing/S = S$.

Figure 2.1 shows geometrical representations (called *Venn diagrams*) of various set operations.

The following associative and distributive laws hold:

$$\begin{aligned} &1 \quad A \cup (B \cup C) = (A \cup B) \cup C. \\ &2 \quad A \cap (B \cap C) = (A \cap B) \cap C. \\ &3 \quad A \cup (B \cap C) = (A \cup B) \cap (A \cup C). \\ &4 \quad A \cap (B \cup C) = (A \cap B) \cup (A \cap C). \end{aligned} \tag{2.8}$$

Example 2.1 We prove the first of the two de Morgan's laws:

1 $A - (B \cup C) = (A - B) \cap (A - C)$.

2 $A - (B \cap C) = (A - B) \cup (A - C)$.

Let $x \in A - (B \cup C)$. This implies that $x \in A$ and $x \notin B \cup C$. That is, $x \notin B$ and $x \notin C$. Since $x \in A$ and $x \notin B$, it follows that $x \in A - B$. Since $x \in A$ and $x \notin C$, we have $x \in A - C$. Thus $x \in (A - B) \cap (A - C)$, and we have shown that an arbitrary element in $A - (B \cup C)$ also belongs to $(A - B) \cap (A - C)$. Hence $A - (B \cup C) \subset (A - B) \cap (A - C)$.

Now let $y \in (A - B) \cap (A - C)$. This implies that $y \in A - B$ and $y \in A - C$. That is, $y \in A$ and $y \notin B$, $y \notin C$. This means that $y \notin B \cup C$. Therefore $y \in A - (B \cup C)$. Thus we have $(A - B) \cap (A - C) \subset A - (B \cup C)$. In view of the definition of the equality of sets, it follows that $A - (B \cup C) = (A - B) \cap (A - C)$. The second of de Morgan's laws can be proved along similar lines. ■

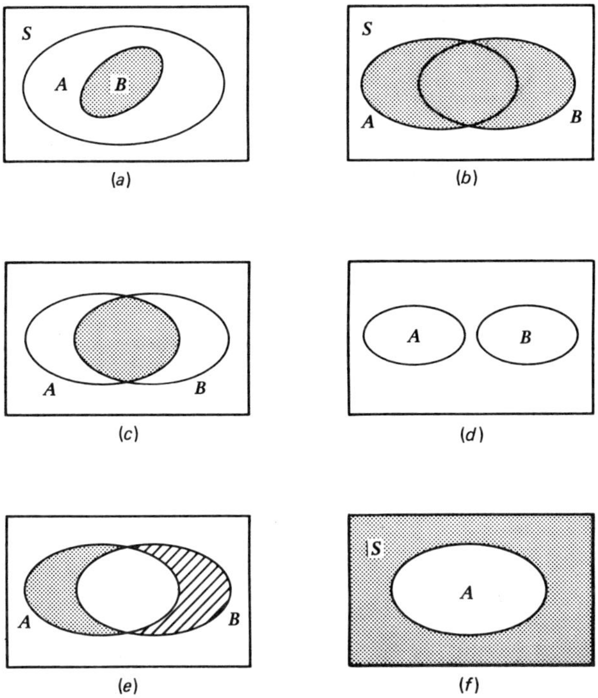

Figure 2.1 Geometrical representation of various set operations. (a) Subset B of A, $B \subset A$. (b) Union of A and B, $A \cup B$. (c) Nonempty intersection, $A \cap B$. (d) Empty intersection, $A \cap B = \varnothing$. (e) Difference of A and B, ▩ $A - B$, ▨ $B - A$. (f) Complement of A with respect to S, A/S.

Sets of Numbers Some of the frequently used sets of numbers are listed below. The same notation is employed throughout the text:

$$\mathsf{R} = \{x: \quad x \text{ is a real number, } -\infty < x < \infty\}$$

$$\mathsf{C} = \left\{z: \quad z \text{ is a complex number, } z = x + iy,\ i = \sqrt{-1}, -\infty < x, y < \infty\right\} \tag{2.9}$$

$$Z = \{\ldots, -3, -2, -1, 0, 1, 2, \ldots\} = \text{set of } \textit{integers}$$

$$J = \{1, 2, 3, 4, \ldots\} = \text{set of } \textit{natural numbers} \text{ (i.e. positive integers)}$$

$$Q = \{r: \quad r = p/q, \text{ where } p, q \in Z, q \neq 0\} = \text{set of } \textit{rational numbers}.$$

Clearly, $J \subset Z \subset \mathsf{R}$ and $Q \subset \mathsf{R}$. The complement of Q with respect to R is called the set of *irrational numbers*.

In addition to the above sets, frequent use is made of the following notation for intervals of real numbers:

$$[a,b]=\{x:\ x \text{ is real}, a\leq x\leq b\}$$

$$[a,b)=\{x:\ x \text{ is real}, a\leq x< b\}$$

$$(a,b]=\{x:\ x \text{ is real}, a< x\leq b\}$$

$$(a,b)=\{x:\ x \text{ is real}, a< x< b\}.$$

The set $[a, b]$ is called a *closed interval* and (a, b) is called an *open interval.* In each case $b-a$ is called the *length* of the interval.

The following relations among the real numbers prove to be useful in subsequent sections. Let a, b, and c be any real numbers. Then

1 $|a+b|\leq|a|+|b|$.
2 $|ab|\leq|a|\cdot|b|$.
3 $||a|-|b||\leq|a-b|$.
4 $\dfrac{a}{1+a}\leq\dfrac{b}{1+b}$ for all $b\geq a\geq 0$.
5 $\dfrac{|a+b|}{1+|a+b|}\leq\dfrac{|a|}{1+|a|}+\dfrac{|b|}{1+|b|}$. (2.10)
6 $ab\leq\dfrac{1}{4\varepsilon}a^2+\varepsilon b^2,\quad \varepsilon>0$.
7 $a\leq a^p\leq 1$ for all $0\leq a\leq 1$, $0<p<1$.
8 $(|a|+|b|)^p\leq|a|^p+|b|^p,\quad 0<p<1$.

A set $A\subset \mathsf{R}$ is said to be *bounded from above* if there exists a real number μ_0 such that $a\leq\mu_0$ for all $a\in A$. The real number μ_0 is said to be an *upper bound* of the set A. Similarly, a set A is said to be *bounded from below* if there exists a real number γ_0 such that $a\geq\gamma_0$ for all $a\in A$. The real number γ_0 is said to be *lower bound* of the set A. If a set A is bounded from above and from below, we say that A is bounded. An upper (lower) bound $M(m)$ for A is said to be the maximum (minimum) of $A\subset\mathsf{R}$ if $M\in A(m\in A)$. It should be noted that even a bounded set need not have a maximum or a minimum.

Every nonempty set of real numbers bounded from above has a least upper bound, and every nonempty set of real numbers bounded from below has a greatest lower bound. Proof of these statements can be found in [10]. It should be noted that the above statements do not hold for the set of rational numbers Q.

Let A be a set that is bounded from above, and let X be the set of all upper bounds of A. The "least upper bound" of A is the minimum of X. This minimum of X is called the *supremum* of A and is denoted by $\sup A$. Similarly, for a nonempty set A that is bounded below, let Y denote the set of all lower bounds of A. The "greatest lower bound" of A is the maximum of Y. The maximum of Y is called the *infimum* of A and is denoted by $\inf A$.

Example 2.2

1 Let A and B be the subsets of the real-number field R:

$$A=\{a:\quad a\in\mathsf{R}, 0\le a\le 1\}$$

$$B=\{b:\quad b\in\mathsf{R}, 0<b<1\}.$$

The set X_A of all upper bounds and the set of Y_A of all lower bounds of A are given by

$$X_A=\{x:\quad 1\le x<\infty\},\qquad Y_A=\{y:\quad -\infty<y\le 0\}.$$

Note that $X_A=X_B$ and $Y_A=Y_B$, where X_B and Y_B are sets of upper bounds and lower bounds for B. The maximum of A is 1 and the minimum of A is 0. However, B has no maximum or minimum. The least upper bound of A is 1 and the greatest lower bound of A is 0, and $\sup B=\sup A$ and $\inf B=\inf A$.

2 Let Q be the set of rational numbers,

$$Q=\{q:\quad q=m/n, \text{ where } m \text{ and } n \text{ are integers}, n\ne 0\}.$$

Let $A\subset Q$ be the set

$$A=\left\{a:\quad a\in Q, 0<a<\sqrt{2}\right\}.$$

The set A has neither a maximum nor a supremum. The infimum of A is 0.

3 Let X be the set of functions $f_n(t)=\sin n\pi t$, $n=0,1,2,\ldots$, and $-\infty<t<\infty$. Since $-1\le f_n(t)\le 1$ for all values of n and t, the set $\{f_n(t)\}$ has a maximum and a minimum. ■

Partitions of Sets In engineering analyses we often encounter situations in which a given set X must be divided into a family of disjoint subsets of X. In such cases X is said to be *partitioned*. A family $\{S_n\}$ of nonempty subsets of a set X is said to be a *partition* of X if

1 $S_n\cap S_m=\varnothing$, for $n\ne m$. (2.11)

2 $\cup_n S_n\equiv S_1\cup S_2\cup\cdots=X$ (i.e., the union of all S_n's is equal to X).

Example 2.3

1 Consider the set $\Omega=\{(x,y):\quad x^2+y^2\le r^2\}\subset\mathsf{R}^2$, which represents a circle of radius r. The set Ω can be partitioned into a family of two subsets:

$$S_1=\left\{(x,y):\quad (x,y)\in\Omega, x^2+y^2<r^2\right\}$$

$$S_2=\left\{(x,y):\quad (x,y)\in\Omega, x^2+y^2=r^2\right\}.$$

Clearly, $S_1 \cap S_2 = \varnothing$, $S_1 \cup S_2 = \Omega$. Note that S_1 represents the interior of Ω, and S_2 represents the boundary of Ω.

2 Let X be the set of all continuous functions $x(t)$ defined on the interval $0 \le t \le 1$, and define $S_n, 0 \le n < \infty$, by

$$S_n = \left\{ x: \quad x \in X, \int_0^1 |x(t)|^2 \, dt = n^2 \right\}.$$

Clearly, $S_n \cap S_m = \varnothing$ for any two distinct real numbers n and m, and since

$$\int_0^1 |x(t)|^2 \, dt < \infty,$$

it follows that $X = \cup_{n=0}^{\infty} S_n$. ■

2.2.3 Cartesian Products, Relations, and Equivalence Classes

The *Cartesian product* of two sets A and B, denoted by $A \times B$, is the set of all ordered pairs (a, b), where $a \in A$ and $b \in B$:

$$A \times B = \{(a, b): \quad a \in A, b \in B\}. \tag{2.12}$$

Note that the Cartesian product is not commutative.

A *relation* from A to B can be defined to be any subset of the Cartesian product $A \times B$. If $a \in A$ and $b \in B$, then "a has relation R to b" is denoted by aRb. Thus R is a set of pairs of elements from A and B which satisfy the relation. Note that here we are using R to denote the set as well as the relation.

Let R be a relation from A into A (i.e., $R \subset A \times A$). Then R is said to be

1 *Reflexive* if and only if for every $a \in A$, $(a, a) \in R$ (i.e., a is related to a, aRa, for every $a \in A$).
2 *Symmetric* if and only if $(a, b) \in R$ implies $(b, a) \in R$ for every $a, b \in A$ (i.e., aRb implies bRa).
3 *Transitive* if and only if $(a, b) \in R$ and $(b, c) \in R$ implies $(a, c) \in R$ (i.e., aRb and bRc implies aRc).

A relation R is called an *equivalence relation* if and only if R is reflexive, symmetric, and transitive.

Example 2.4 Let A be the set of three people: John, aged 79, Karen, aged 45, and Laura, aged 75. John and Laura are the parents of Karen. Let B be the set of five people: Robert, aged 55, Olive, aged 50, Paul, aged 25, Mack, aged 22, and Neal, aged 20. Robert and Olive are the parents of Paul, Mack, and Neal. For simplicity we shall use only the first letter of each name.

The Cartesian product of A and B is the set $\{(J, R), (J, O), (J, P), (J, M), (J, N), (K, R), (K, O), (K, P), (K, M), (K, N), (L, R), (L, O), (L, P),$

(L, M), $(L, N)\}$. Note that there is no relation between the components of each pair.

Define the relation R from A to B by aRb = "a is older than b." Then the set R is given by $R = \{(J, R), (J, O), (J, P), (J, M), (J, N), (K, P), (K, M), (K, N), (L, R), (L, O), (L, P), (L, M), (L, N)\}$. Note that R is a subset of $A \times B$.

Now let R be the same relation from A into itself. Then the subset R of $A \times A$ is given by $\{(J, K), (L, K), (J, L)\}$. Note that this relation is *not* symmetric nor reflexive.

Finally, let R be the relation on B (i.e., $R \subseteq B \times B$) defined by

$$aRb \Leftrightarrow \text{“}a \text{ is a brother of } b\text{”} \text{ for all } a, b \in B.$$

Then the set R is given by (assuming that one is a brother of himself)

$$R = \{(P, P), (P, M), (P, N), (M, P), (M, M), (M, N),$$

$$(N, P), (N, M), (N, N)\}.$$

Clearly R is reflexive (by assumption), symmetric, and transitive. Hence R is an equivalence relation. ■

Example 2.5 Let $A = (-5, 2, 5, 8)$. Let R be defined by $aRb = a$ is smaller than b. Then the relation R from A into A is the set

$$R = \{(-5,2), (-5,5), (-5,8), (2,5), (2,8), (5,8)\}.$$

Note that R is not reflexive or symmetric, since $a < b$ does not imply $b < a$. However, it is transitive since $a < b$, and $b < c$ implies $a < c$. ■

Equivalence Classes Let A be a set and R be an equivalence relation defined on A (i.e., A into A). If $a \in A$, the set of elements x satisfying the relation xRa constitutes a subset of A, called an *equivalence class*. It is denoted by $R[a]$:

$$R[a] = \{x: \quad x \in A, xRa\}. \tag{2.13}$$

Example 2.6 Let $A = \{-5, -2, 0, 1, 4\}$, and let R be the relation such that the pair (a, b) is in R if and only if $a - b$ is an even number. Assuming that zero is an even number, we write the set R

$$R = \{(-5,5), (-5,1), (-2,-2), (-2,0), (-2,4), (0,4), (4,4), (1,1),$$

$$(1,-5), (0,-2), (4,-2), (0,0), (4,0)\}.$$

Clearly, R is an equivalence relation. The equivalence classes are given by

$$R[-5]=\{-5,1\}, \quad R[-2]=\{-2,0,4\},$$

$$R[0]=\{-2,0,4\}, \quad R[1]=\{-5,1\}, \quad R[4]=\{-2,0,4\}. \blacksquare$$

2.2.4 Functions and Inverses

Most basic mathematical concepts are but abstractions of the ordinary expression in a day-to-day life. For example, take the notion of correspondence, which is the root of a function that we will shortly define. Suppose that we are concerned with the correspondence between the automobile accidents and the number of automobiles now on the earth. This correspondence is not so simple, being a function of the driver, speed limits, and so on. Next, take a rubber strip X of unit length and stretch it "uniformly" to twice its length and call the resulting strip Y. We can write the correspondence between points y in Y and points x in X as: $x \in X$ corresponds to the number $y \in Y$ such that $y=2x$. The correspondence can also be described, more concisely, as the set $R \subset X \times Y$ defined by

$$R=\{(x,y):\ \ y=2x\}=\{(x,2x):\ \ x\in X\}. \tag{2.14}$$

Yet another way of looking at the correspondence is to think of it as being a relation between the points of X and Y.

A function is a relation (or correspondence) that is "single valued." That is, if $y_1=2x$ and $y_2=2x$ for $x \in X$, then $y_1=y_2$. We use the notation $f(x)$ to denote the value of the function at x, and the pair $(x, f(x))$ to denote the element of the relation R. We now give a formal definition of a function.

Let A and B be two sets, and suppose that f is a rule that assigns to each element in A *exactly one* element of B. Then f is called a *function* defined on A with values in B. The terms *mapping*, *transformation*, and *operator* are also used in place of function.

We denote a function from A into B by $f\colon\ A\to B$ to mean that for $a\in A$, $f(a)\in B$. The set A is called the *domain* of f. If $b=f(a)$, we say that b is the image of a. The *range* of a function $f\colon\ A\to B$, denoted by $\mathfrak{R}(f)$, is the set of all images of f. If $\mathfrak{R}(f)\subset B$, f is said to map A *into* B. If $\mathfrak{R}(f)=B$, then f is said to map A *onto* B.

If the domain of a function f does not contain two elements with the same image, then f is called a *one-to-one* function or mapping. In other words, f is a one-to-one mapping if and only if every point in $\mathfrak{R}(f)$ has only one pre-image in A. When a function is one-to-one and onto its range, a mapping of the range into the domain can be defined. We will elaborate on this shortly. When the range of a function contains only one element, the function is called a *constant function* (because every element in the domain is mapped into the same element).

Suppose that f: $A \to B$ and g: $B \to C$ are two functions. Then f and g define a *product* or *composite* function, denoted by $g \cdot f$ (or simply gf), from A into C. The composition of f and g is defined by

$$g \cdot f(a) = g(f(a)). \tag{2.15}$$

Obviously this makes sense only if $\mathcal{R}(f) \subset \mathcal{D}(g)$. Otherwise we say that gf does not exist. In general we have $g \cdot f \neq f \cdot g$ (in fact, $g \cdot f$ may exist even if $f \cdot g$ does not). The composition of functions is associative:

$$(h \cdot g) \cdot f = h \cdot (g \cdot f).$$

If $S_A \subset A$, then its *direct image* under the mapping f is given by

$$f(S_A) = \{ f(a): \quad a \in S_A \subset A \}. \tag{2.16}$$

Here it is understood that $f(S_A)$ is a set rather than an element. Clearly, $f(S_A)$ is a subset of $\mathcal{R}(f)$. Figure 2.2 shows various classifications of a function.

Now suppose that f: $A \to B$ and S_B is a subset of B. Then the set

$$g(S_B) = \{ a: \quad a \in A, f(a) \in S_B \} \subset A$$

is called the *inverse image* of S_B under the mapping f. That is, $g(S_B)$ consists of all pre-image points in S_B. The mapping g is referred to as the *inverse* mapping and denoted by $f^{-1}(S_B) \equiv g(S_B)$. It should be noted that if $f(S_A) = S_B$, then it does *not* necessarily follow that $f^{-1}(S_B) = S_A$, since S_B may contain images of other elements in A which are not in S_A. Thus in order for $f(S_A) = S_B$ it is necessary and sufficient that f is one-to-one and onto S_B.

Recall that a function f associates with each element of its domain $\mathcal{D}(f)$ exactly one element in the range $\mathcal{R}(f)$. Both $\mathcal{R}(f)$ and $\mathcal{D}(f)$ are sets. Let f: $U \to V$, and $A \subseteq U$, $B \subseteq U$, $C \subseteq V$, and $D \subseteq V$. Then the following rules hold for a function:

$$\begin{aligned}
&1 \quad f(A \cup B) = f(A) \cup f(B) \\
&2 \quad f(A \cap B) \subseteq f(A) \cap f(B) \\
&3 \quad f^{-1}(C \cup D) = f^{-1}(C) \cup f^{-1}(D) \\
&4 \quad f^{-1}(C \cap D) = f^{-1}(C) \cap f^{-1}(D).
\end{aligned} \tag{2.17}$$

Example 2.7 Let f be a function such that it assigns to each real number its square, and let $A = \{-6, -3, -2, 0, 1, 2, 4\}$ be its domain. That is, $f(a) = a^2$ for every a in A. The direct image of A is

$$\begin{aligned}
f(A) &= \{ f(-6), f(-3), f(-2), f(0), f(1), f(2), f(4) \} \\
&= \{0, 1, 4, 9, 16, 36\} \equiv B.
\end{aligned}$$

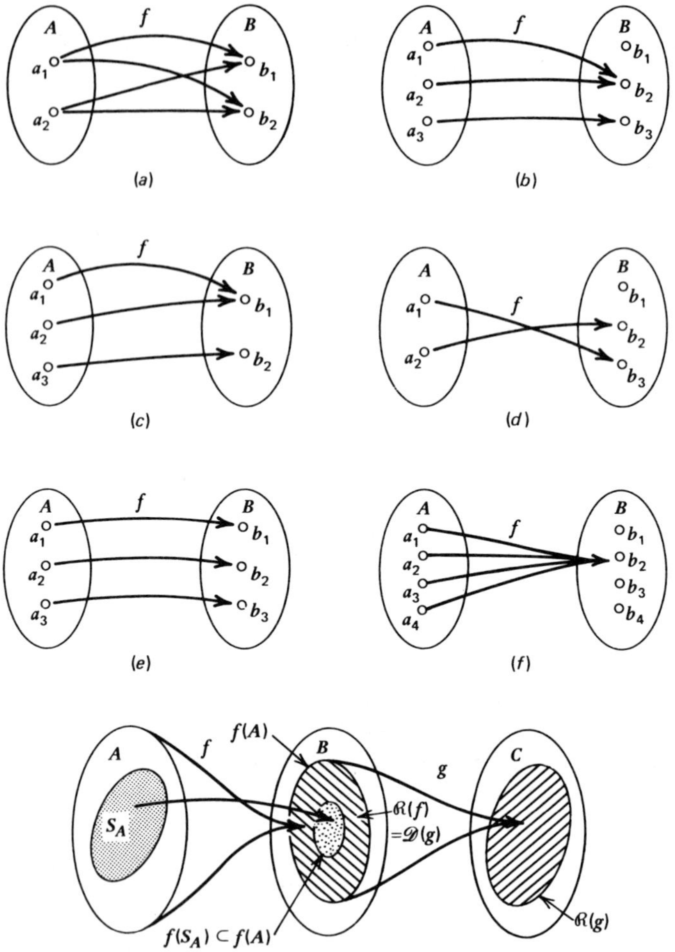

Figure 2.2 Relations, classification of functions, and composition of functions. (a) A relation; *not* a function. (b) Function (into); *not* onto B. (c) Onto function; *not* one-to-one. (d) One-to-one function; *not* onto B. (e) One-to-one and onto function. (f) Constant function.

Thus f: $A \rightarrow B$. Note that f is not one-to-one since it maps $f(a)$ and $f(-a)$ into the same image. However, f maps A onto B since every element in B has a pre-image in A. Let $S_B = \{0, 1, 9, 16\}$ be the subset of B and S_A the subset $\{0, 1, -3, 4\}$ of A. Then $f(S_A) = S_B$.

Now let us define another function g from B into $D = \{3, 6, 11, 18\}$ by $g(b) = b + 2$. Then the domain of g is $\mathcal{D}(g) = \{1, 4, 9, 16\} \subset B$. The composition $g \cdot f$ is given by

$$(g \cdot f)(a) = g(f(a)) = a^2 + 2.$$

Hence $g \cdot f$ maps the subset $\{-3, -2, 1, 2, 4\}$ of A into D. On the other hand $(f \cdot g)$ is given by

$$(f \cdot g)(d) = f(g(d)) = (d+2)^2 = d^2 + 4d + 4.$$

Hence $(f \cdot g)$ maps $\mathfrak{D}(g)$ into $\{9, 36, 117, 326\}$. ■

It is useful to define the *identity function*. The function u_A: $A \to A$ is called the *identity function* for A if for every $a \in A$, $u_A(a) = a$. We now use the identity function to define various *inverse* functions.

A function f: $A \to B$ is said to be *left invertible* if there exists a function g: $B \to A$ such that

$$g \cdot f = u_A \tag{2.18}$$

and g is called a *left inverse* of f. The function f is said to be *right invertible* if

$$f \cdot g = u_B \tag{2.19}$$

and g is called a *right inverse* of f. A function f: $A \to B$ is invertible if and only if there exists a function g: $A \to B$ such that Eqs. (2.18) and (2.19) hold. Note that a function is right invertible if and only if it is onto, and it is left invertible if and only if it is one-to-one mapping.

Example 2.8 Let $X = \mathsf{R}^4$ and $Y = \mathsf{R}^3$ be the sets of all ordered quadruples and all ordered triples of real numbers, respectively. Define the mapping f: $X \to Y$ by the following matrix equation:

$$\begin{bmatrix} 1 & 0 & 0 & 1 \\ 2 & 1 & 0 & -1 \\ 3 & 2 & 1 & 2 \end{bmatrix} \begin{Bmatrix} x_1 \\ x_2 \\ x_3 \\ x_4 \end{Bmatrix} = \begin{Bmatrix} y_1 \\ y_2 \\ y_3 \end{Bmatrix}.$$

The range of the mapping is $Y = \mathsf{R}^3$, and therefore a right inverse exists. Two right inverses are given below:

$$\begin{bmatrix} 0 & -1 & -1 \\ 1 & 4 & 3 \\ -4 & -7 & -4 \\ 1 & 1 & 1 \end{bmatrix}, \quad \begin{bmatrix} -1 & 1 & -1 \\ 4 & -2 & 4 \\ -9 & 3 & -4 \\ -2 & -1 & 1 \end{bmatrix}.$$

Of course there are many other right inverses. ■

The following result is concerned with the conditions for the existence of the inverse of a function.

Theorem 2.1 A function f: $A \to B$ is invertible if and only if f is one-to-one and onto.

Proof First note that the statement "f is invertible *if* f is one-to-one and onto" implies that one-to-one and onto are sufficient conditions. On the other hand the statement "f is invertible *only if* f is one-to-one and onto" implies that one-to-one and onto are necessary conditions for f to have the inverse. We prove sufficiency first.

Sufficiency (*if*) To prove sufficiency we assume that f is one-to-one and onto and then show that f has an inverse. Let $b \in B$. Since f is one-to-one and onto, there exists a unique element $a \in A$ such that $f(a)=b$. Since for every b there is a unique element $a \in A$, the correspondence can be expressed as $g(b)=a$, where $g\colon\ B \to A$. Hence, $f(a)=f\cdot(g(b))=b$, which implies that $f\cdot g=u_B$. Similarly, $g(b)=g(f(a))=a$ implies that $g\cdot f=u_A$. Hence f is invertible and $g=f^{-1}$ is the inverse of f.

Necessity (*only if*) To prove the necessity, we assume that f is invertible, and show that it is one-to-one and onto. Let g be the inverse of f, $g\colon\ B \to A$. Suppose that there is at least one element $b_0 \in B$ such that $b_0 \notin \mathcal{R}(f)$. Since g is a function, with B as its domain, $b_0 \in \mathcal{D}(g)$ and $g(b_0)=a_0 \in A$. Hence $f(a_0)=f(g(b_0))=(f\cdot g)(b_0)=b_0$, since $f\cdot g$ is the identity function. Thus $b_0 \in \mathcal{R}(f)$, which is a contradiction. Hence $B=\mathcal{R}(f)$, or f is onto. We prove that f is one-to-one by contradiction. Let $a_1, a_2 \in A$ be such that $f(a_1)=f(a_2)=b$. That is, f maps a_1 and $a_2 (a_1 \neq a_2)$ into the same image b. Since f is invertible,

$$a_1=u_B(a_1)=(g\cdot f)(a_1)=(f^{-1}\cdot f)(a_1)$$

$$=f^{-1}(f(a_1))=f^{-1}(f(a_2))=a_2,$$

which is a contradiction. Hence f is one-to-one. ■

Example 2.9

1 Let $f\colon\ A \to A$ be defined by $f(x)=x^2$, where $A=[-1,1]$, the set of real numbers, $-1 \le x \le 1$. Since f maps $-a$ and $+a$ into the same image a^2, it is not one-to-one. It is *not* onto, since there exist elements in A whose pre-images do not belong to A. For example, a negative number does not have a pre-image. Suppose that $g\colon\ A \to B$, where $B=[0,1]$. Then g is onto since every element in B has a pre-image in A. However, g is not one-to-one.

2 Let $f\colon\ A \to A$ be defined by $f(x)=\sin x$, where $A=[-1,1]$. It is one-to-one but *not* onto.

3 Let $f\colon\ \mathsf{R} \to \mathsf{R}$ (R being the set of real numbers) be defined by $f(x)=x^3+a$, where a is a fixed number. Note that f is one-to-one and onto. Hence the inverse f^{-1} of f exists. To find f^{-1}, we set $y=f(x)=x^3+a$ and solve for $x=\sqrt[3]{y-a}$. Then $f^{-1}(x)=\sqrt[3]{x-a}$, so that

$$(f^{-1}\cdot f)(x)=f^{-1}(f(x))=f^{-1}(x^3+a)=\sqrt[3]{(x^3+a)-a}=x.$$

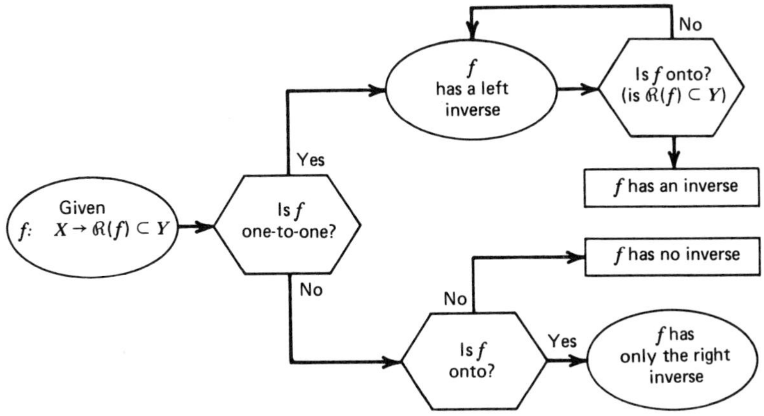

Figure 2.3 Existence of right and left inverses, and the inverse.

4 The transformation T: $\mathsf{R}^2 \to \mathsf{R}^2$, where $\mathsf{R}^2 = \mathsf{R} \times \mathsf{R}$, which maps the Cartesian rectangular coordinates into other Cartesian rectangular coordinates oriented at the angle θ (counterclockwise), is given by

$$\begin{Bmatrix} \bar{x} \\ \bar{y} \end{Bmatrix} = \begin{bmatrix} \cos\theta & \sin\theta \\ -\sin\theta & \cos\theta \end{bmatrix} \begin{Bmatrix} x \\ y \end{Bmatrix}.$$

It can be easily verified that the mapping is one-to-one and onto. Indeed, the inverse of the matrix representing the mapping is equal to its transpose. In Section 2.6 we shall see that linear transformations on finite-dimensional spaces can be represented by matrices. ■

We conclude this section with a summary of the results in a flow-chart form (see Fig. 2.3).

Exercises 2.1

1 Let $A = \{a, b, d, f, h\}$, $B = \{b, c, e, f, g\}$, and $C = \{a, b, d, f, k\}$. Find:

(a) $A \cup B$.

(b) $A \cap B$.

(c) $(A \cup B) \cap C$.

(d) $B - A$.

2 State whether the following sets have upper and lower bounds, maximum and minimum, supremum and infimum:

(a) $A = \{a\colon\ a < 1\}$.

(b) $B = \{2, 4, 8, 16, 32, \ldots\}$.

(c) $C = \{10^6, 10^3, 0, 10^{-3}, 10^{-6}\}$.

(d) $D=\{1,2,1,2,\cdot,\cdot\}$.

(e) $E=\{1,\frac{1}{2},\frac{1}{3},\frac{1}{4},\cdots\}$.

3 Prove the second of de Morgan's laws in Example 2.1.

4 Let $A=\{1,2,4,5\}$ and R be the relation on $A\times A$ defined by $aRb=$"a is divisible by b." Find the set R. Is it symmetric, reflexive, and transitive?

5 Let $A=\{2,4,6,\ldots\}$ and let R be the relation on A such that aRb implies $a-b$ is divisible by 2. Find the equivalence classes $R[2]$, $R[6]$, and $R[8]$.

6 Check whether the following relations are symmetric, reflexive, or transitive on set X:

(a) $aRb \Leftrightarrow a\le b$, $X=\mathsf{R}$.

(b) $aRb \Leftrightarrow |a|=|b|$, $X=\mathsf{R}$.

(c) $xRy \Leftrightarrow \int_0^1 |x(t)|^2\,dt = \int_0^1 |y(t)|^2\,dt$, $X=$ set of continuous functions in $[0,1]$.

7 Let X be the set of continuous functions defined in the interval $[0,T]$, and define f: $X\to X$ by

$$f(x(t))\equiv y(t)=\begin{cases} x_0, & \text{if } x(t)\ge x_0 \\ x(t), & \text{if } -x_0<x(t)<x_0 \\ -x_0, & \text{if } x(t)\le -x_0 \end{cases}$$

where $x_0>0$ is a real number. Show that the statement defines a mapping of X into itself. Describe the range of the mapping. Is the mapping one-to-one and/or onto X? Plot the function for a sinusoidal input $x(t)$.

8 Prove the relations in Eqs. (2.17).

9 Let $A=\{-2,-1,0,2,4,5\}$, and f is defined by $f(a)=a^2-2a+1$. Find the direct image of the set A under the mapping f. Is the mapping one-to-one and onto its range?

10 Let f: $\mathsf{R}\to\mathsf{R}$, and g: $\mathsf{R}\to\mathsf{R}$ be defined by

$$f(x)=x^3+1, \qquad g(x)=x-1.$$

Find:

(a) $f\cdot g$.

(b) $g\cdot f$.

(c) The inverse of f (if it exists).

11 Let X be the set of all real-valued functions $x(t)$ with continuous first derivatives defined on the closed interval $[0,T]$ such that $x(0)=0$. Let Y

be the set of all real-valued continuous functions $y(t)$ defined on $[0, T]$. Prove that f: $X \to Y$ defined by

$$y = f(x) \equiv \frac{dx}{dt}$$

is a function. Is f one-to-one and onto? If so, determine the inverse.

12 Prove that the composition gf of two invertible mappings f: $X \to Y$ and g: $Y \to Z$ is invertible and that $(g \cdot f)^{-1} = f^{-1} \cdot g^{-1}$.

13 Determine whether the following functions (f: $X \to Y$) are one-to-one or/and onto (if f is one-to-one and onto, find its inverse):

(a) $X = \{-1, 0, 1, 2, 3\}$, $Y = \mathsf{R}$, $f(x) = (1+x)^2 + 1$.

(b) $X = Y = [-1, 1]$, $f(x) = \sin x$.

(c) $X = \mathsf{R} - \{1\}$, $Y = \mathsf{R} - \{2\}$, $f(x) = (2x-1)/(x-1)$.

(d) $X = Y = \mathsf{R}$, $f(x) = 5x - 3$.

(e) $X = Y = \mathsf{R}$, $f(x) = x^2 - 2x + 1$.

Here R denotes the set of real numbers.

14 Prove the statement that a function is left (right) invertible if and only if it is one-to-one (onto).

15 (*Equivalent Sets*) If X and Y are sets, we say that X is *equivalent* to Y, written $X \sim Y$, if there is a one-to-one function f from X onto Y. Show for any sets X, Y and Z that the following properties hold:

(a) $X \sim X$.

(b) If $X \sim Y$, then $Y \sim X$.

(c) If $X \sim Y$ and $Y \sim X$, then $X \sim Z$.

(*Hint*: show the existence of a one-to-one function in each case.)

16 Let f: $X \to Y$ and g: $Y \to Z$.

(a) If f and g are one-to-one, show that the composition $g \cdot f$ is also one-to-one.

(b) If f and g are onto (i.e. $\mathcal{R}(f) = Y$ and $\mathcal{R}(g) = Z$), show that the composition $g \cdot f$ is also onto.

17 (*Countable Sets*) A set X is *countable* if X is equivalent to the set J of positive integers. A set X is *infinite* if X is either empty or there is an $n \in J$ such that X is equivalent to the set $\{1, 2, \ldots, n\}$, that is, there is a one-to-one correspondance between J and X. Show that:

(a) Any infinite subset of a countable set is countable.

(b) If X and Y are countable, then $X \times Y$ is countable.

2.3 METRIC AND METRIC SPACES

2.3.1 Metric and Pseudometric

Thus far in our discussion of sets the concept of "closeness" of one element of a set to another element of the same set was not considered. The measure of closeness is called a *metric*, or *distance*, function. Analogous to the length of a vector connecting two points in space, a metric must satisfy certain conditions, called *axioms*. Let $d(x, y)$ denote the distance (function) between two elements x, y of a set X. Then $d(x, y)$ must satisfy:

1. $d(x, y) \geq 0$, and $d(x, y)=0$ if and only if $x=y$, (nonnegative).
2. $d(x, y)=d(y, x)$, (symmetry).
3. $d(x, y)+d(y, z) \geq d(x, z)$ for all $x, y, z \in X$, (triangle inequality).

All of these three conditions are consistent with the concept of distance in the Euclidean space. Indeed, if $\mathbf{x}$ and $\mathbf{y}$ are the position vectors of points (x_1, x_2, x_3) and (y_1, y_2, y_3) in three-dimensional Euclidean space, the distance between the two points can be defined by

$$d(\mathbf{x},\mathbf{y})=\sqrt{(\mathbf{x}-\mathbf{y}) \cdot (\mathbf{x}-\mathbf{y})}$$

$$=\left[(x_1-y_1)^2+(x_2-y_2)^2+(x_3-y_3)^2\right]^{1/2}. \tag{2.20}$$

Clearly, $d(\cdot, \cdot)$ defined above satisfies the axioms of a metric.

It is possible to define more than one metric on a nonempty set. A real-valued function $p(x, y)$ is said to be a *pseudometric* if it satisfies all of the conditions of a metric, except $p(x, y)=0$ does not necessarily imply $x=y$.

In order to prove that a given quantity is a metric or pseudometric, it is often difficult to establish that the triangle inequality holds. In most cases use is made of certain standard inequalities associated with sums and of integrals of real-valued functions.

The Holder and Minkowski inequalities are useful in establishing the triangle inequality. We state and prove them in the next section.

2.3.2 Hölder and Minkowski Inequalities

Hölder Inequalities

We first state the inequalities for finite sums, infinite sums, and integrals and then prove a typical one in detail.

Finite Sums

$$\sum_{i=1}^{n} |x_i y_i| \leq \left(\sum_{i=1}^{n} |x_i|^p \right)^{1/p} \left(\sum_{i=1}^{n} |y_i|^q \right)^{1/q}. \tag{2.21}$$

Infinite Sums For $\Sigma_{i=1}^{\infty}|x_i|^p<\infty$ and $\Sigma_{i=1}^{\infty}|y_i|^q<\infty$, we have

$$\sum_{i=1}^{\infty}|x_i y_i| \leq \left(\sum_{i=1}^{\infty}|x_i|^p\right)^{1/p}\left(\sum_{i=1}^{\infty}|y_i|^q\right)^{1/q}. \tag{2.22}$$

Integrals For $\int_\Omega|x|^p\,dt<\infty$ and $\int_\Omega|y|^q\,dt<\infty$, we have

$$\int_\Omega|xy|\,dt \leq \left(\int_\Omega|x|^p\,dt\right)^{1/p}\left(\int_\Omega|y|^q\,dt\right)^{1/q}. \tag{2.23}$$

For $p=q=2$, the above inequalities yield the *Schwarz inequality*. The following equality is needed to establish the Hölder inequalities. Let a and b be nonnegative real numbers. Then

$$ab \leq \frac{a^p}{p}+\frac{b^q}{q} \tag{2.24}$$

for all p, q such that

$$1<p<\infty, \qquad \frac{1}{p}+\frac{1}{q}=1. \tag{2.25}$$

The inequality in Eq. (2.24) can be interpreted geometrically as follows. Consider the curves $y=x^{p-1}$ and $x=y^{q-1}$ in the x, y plane. Let

$$A_1=\int_0^a x^{p-1}\,dx=\frac{a^p}{p}, \qquad A_2=\int_0^b y^{q-1}\,dy=\frac{b^q}{q}.$$

Then the sum of the areas A_1 and A_2 (see Fig. 2.4) is greater than or equal to the product of a and b.

We now prove the Hölder inequality for finite sums. The proofs of the Hölder inequalities for infinite sums and integrals follow in a similar way. Since these inequalities are trivial for $x=0$ or $y=0$, we assume that x and y are nonzero. Let

$$\|x\|_p \equiv \left(\sum_{i=1}^{n}|x_i|^p\right)^{1/p}.$$

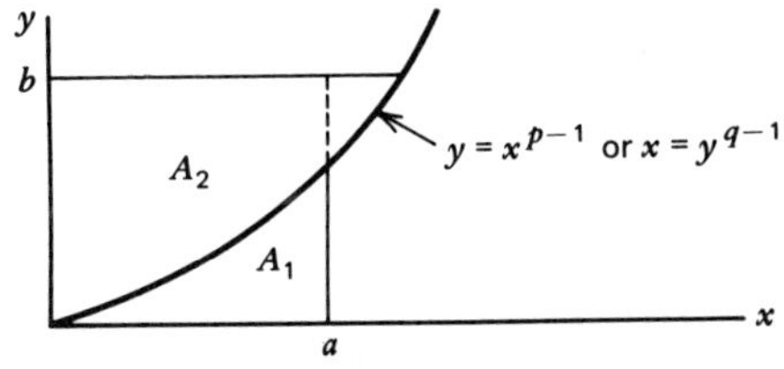

Figure 2.4 Geometric interpretation of Eq. (2.24).

Since

$$\frac{|x_i||y_i|}{\|x\|_p\|y\|_q} \le \frac{|x_i|^p}{p\|x\|_p^p} + \frac{|y_i|^q}{q\|y\|_q^q},$$

we have, on summing,

$$\sum_{i=1}^{n} \frac{|x_i||y_i|}{\|x\|_p\|y\|_q} \le \frac{1}{p\|x\|_p^p}\sum_{i=1}^{n}|x_i|^p + \frac{1}{q\|y\|_q^q}\sum_{i=1}^{n}|y_i|^q$$

$$= \frac{1}{p} + \frac{1}{q} = 1.$$

from which inequality (2.21) follows.

Minkowski Inequalities

Again we state the inequalities for finite sums, infinite sums, and integrals. Let $1 \le p < \infty$.

Finite Sums

$$\left(\sum_{i=1}^{n} |x_i \pm y_i|^p\right)^{1/p} \le \left(\sum_{i=1}^{n} |x_i|^p\right)^{1/p} + \left(\sum_{i=1}^{n} |y_i|^p\right)^{1/p}. \tag{2.26}$$

Infinite Sums For $\sum_{i=1}^{\infty}|x_i|^p < \infty$ and $\sum_{i=1}^{\infty}|y_i|^p < \infty$, we have

$$\left(\sum_{i=1}^{\infty} |x_i \pm y_i|^p\right)^{1/p} \le \left(\sum_{i=1}^{n} |x_i|^p\right)^{1/p} + \left(\sum_{i=1}^{n} |y_i|^p\right)^{1/p}. \tag{2.27}$$

Integrals For $\int_\Omega |x|^p\,dt < \infty$ and $\int_\Omega |y|^p\,dt < \infty$, we have

$$\left(\int_\Omega |x \pm y|^p\,dt\right)^{1/p} \le \left(\int_\Omega |x|^p\,dt\right)^{1/p} + \left(\int_\Omega |y|^p\,dt\right)^{1/p}. \tag{2.28}$$

Minkowski inequalities follow from the Hölder inequalities. We establish this for finite sums. First note that

$$(|a|+|b|)^p = |a|(|a|+|b|)^{p-1} + |b|(|a|+|b|)^{p-1}.$$

Setting $a = x_i$ and $b = y_i$, and summing over i, we obtain

$$\sum_{i=1}^{n} |x_i \pm y_i|^p \le \sum_{i=1}^{n} (|x_i|+|y_i|)^p$$

$$= \sum_{i=1}^{n} (|x_i|+|y_i|)^{p-1}|x_i| + \sum_{i=1}^{n} (|x_i|+|y_i|)^{p-1}|y_i|.$$

Now applying the Hölder inequality to each of the sums on the right-hand side of the above equation, we get, since $(p-1)q=p$,

$$\sum_{i=1}^{n}(|x_i|+|y_i|)^p \le \left(\sum_{i=1}^{n}(|x_i|+|y_i|)^p\right)^{1/q}\left(\sum_{i=1}^{n}|x_i|^p\right)^{1/p}$$
$$+\left(\sum_{i=1}^{n}(|x_i|+|y_i|)^p\right)^{1/q}\left(\sum_{i=1}^{n}|y_i|^p\right)^{1/p}.$$

Dividing both sides of the above equation by $(\Sigma_{i=1}^{n}(|x_i|+|y_i|)^p)^{1/q}$, we obtain the Minkowski inequality (2.26):

$$\left(\sum_{i=1}^{n}|x_i \pm y_i|^p\right)^{1/p} \le \left(\sum_{i=1}^{n}(|x_i|+|y_i|)^p\right)^{1/p}$$
$$\le \left(\sum_{i=1}^{n}|x_i|^p\right)^{1/p}+\left(\sum_{i=1}^{n}|y_i|^p\right)^{1/p}.$$

2.3.3 Metric Space, Subspace, and Product Spaces

A *metric space* is a nonempty set on which a metric can be defined. We shall denote a metric space X by the pair (X, d), d being the metric on X.

Subspace A subset A of a metric space (X, d) is called a *subspace* of X, with metric d; that is, the same metric d is used on A also. Although a different metric can be defined on A, when we say A is a subspace of X, it is understood that the metric on A is the same as that on X.

Product Space Let (X, d_x) and (Y, d_y) be two metric spaces. The product space $Z=X\times Y$ is defined to be the collection of ordered pairs (x, y), $x\in X$ and $y\in Y$. The space Z can be endowed with any one of the following metrics:

$$\begin{aligned}
&1\quad d_z((x_1, y_1),(x_2, y_2))=d_x(x_1, x_2)+d_y(y_1, y_2)\\
&2\quad d_z((x_1, y_1),(x_2, y_2))=[d_x^2(x_1, x_2)+d_y^2(y_1, y_2)]^{1/2}\\
&3\quad d_z((x_1, y_1),(x_2, y_2))=\max\{d_x(x_1, x_2), d_y(y_1, y_2)\}.
\end{aligned} \tag{2.29}$$

Example 2.10 (*Space* $C[0, t_0]$) Let $X=C[0, t_0]$ be the set of all real-valued continuous functions of t, $0\le t\le t_0$. Then X is a metric space with respect to the metric, called the *sup-metric*,

$$d_\infty(x, y)=\sup\{|x(t)-y(t)|:\ 0\le t\le t_0\}. \tag{2.30a}$$

Conditions 1 and 2 of a metric are obviously satisfied by $d_\infty(\cdot,\cdot)$. To prove the triangle inequality, consider

$$\begin{aligned}
d_\infty(x,y) &= \sup_{0\le t\le t_0} |x(t)-z(t)+z(t)-y(t)| \\
&\le \sup_{0\le t\le t_0} \{|x(t)-z(t)|+|z(t)-y(t)|\} \\
&\le \sup_{0\le t\le t_0} |x(t)-z(t)| + \sup_{0\le t\le t_0} |z(t)-y(t)| \\
&= d_\infty(x,z)+d_\infty(z,y).
\end{aligned}$$

Thus $d_\infty(\cdot,\cdot)$ satisfies the triangle inequality, and hence $d_\infty(\cdot,\cdot)$ is a metric on $C[0,t_0]$.

Another metric can be defined on $C[0,t_0]$, called the *L_p-metric*,

$$d_p(x,y)=\left(\int_0^{t_0}|x(t)-y(t)|^p\,dt\right)^{1/p}, \qquad p\ge 1. \tag{2.30b}$$

By means of the Minkowski inequality for integrals (2.28), it can be readily verified that the triangle inequality is satisfied by $d_p(x,y)$.

Since $C[0,t_0]$ is a space of continuous functions $x(t)$, $t\in[0,t_0]$, the sup-metric d_∞ gives the largest pointwise difference between two functions over the interval, while the L_2-metric gives the net area between the two functions. In Fig. 2.5 the sup-metric and the L_p-metric for $p=1$ are shown. ■

Example 2.11 Consider the transverse motion of a string of length L, fixed at its ends (see Fig. 2.6). Let $C[0,L]$ denote the set of all real-valued continuous functions $y(x,t)$ defined on the closed interval $0\le x\le L$ for any time t, $0\le t\le t_0$, and let

$$d(y_1,y_2)=\sup\{|y_1(x,t)-y_2(x,t)|:\ \ 0\le x\le L\}.$$

The transverse deflection $y(\cdot,t)$ of the string at any given time t can be viewed as a point in the metric space $(C[0,L],d)$. Hence the motion of the string can be viewed as a curve $y\in C[0,L]$ parameterized by t. Note that not every point in $C[0,L]$ can lie on the curve y because all possible motions y should vanish at $x=0$, and $x=L$. Thus all possible deflection curves of the string form only a subspace of $C[0,L]$. Therefore the subspace

$$A(t)=\{y(\cdot,t):\ \ y\in C[0,L],\ y(0,t)=y(L,t)=0\}$$

contains all possible deflections of the string. Of these deflections, we are interested in only the one that satisfies the governing differential equation of the string.

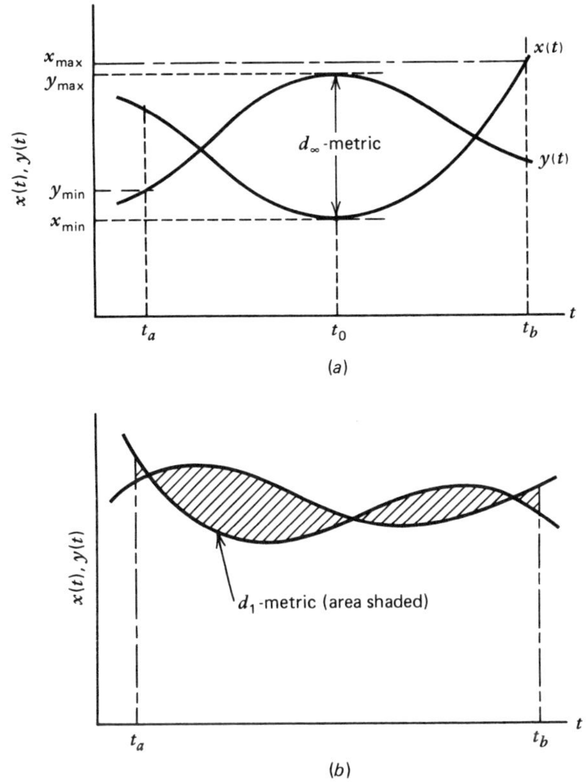

Figure 2.5 Various measures of the difference of two functions. (a) $d_\infty(x, y)$-metric. (b) $d_1(x, y)$-metric.

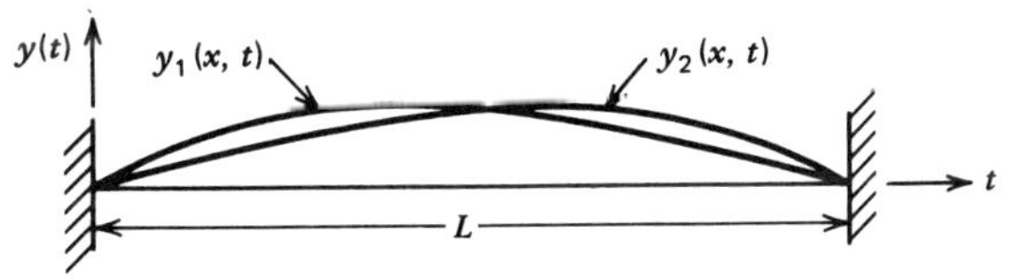

Figure 2.6 Motion of a vibrating string.

Suppose that $v \equiv dy/dt$ is the rate of change of deflection with time. Then $v(\cdot, t)$ (assumed to exist) can be viewed as a point in the metric space $(C[0, L], d)$ at any time t. The ordered pair $(y, dy/dt) = (y, v)$ can be thought of as a point in the product space $C[0, L] \times C[0, L]$. ■

2.3.4 Continuity, Convergence, and Completeness in Metric Spaces

Continuity Let f be a mapping from the metric space (X, d_x) into the metric space (Y, d_y). The mapping f is said to be *continuous at the point* x_0 in X if for every real number $\varepsilon>0$ there exists a real number $\delta=\delta(x_0, \varepsilon)>0$ such that

$$d_y(f(x), f(x_0))<\varepsilon \qquad \text{whenever } d_x(x, x_0)<\delta. \tag{2.31}$$

The mapping f is said to be continuous if it is continuous at each point in its domain X. The mapping f: $(X, d_x)\to(Y, d_y)$ is said to be *uniformly continuous* if for each $\varepsilon>0$ there exists a $\delta=\delta(\varepsilon)>0$ (i.e., independent of x_0) such that for any x_0 Eq. (2.31) holds.

Note that the definition of continuity is the same as that in elementary analysis, with the exception that the absolute value $|f(x)-f(x_0)|$ is replaced by the metric $d(f(x), f(x_0))$.

Convergence First let us recall the definition of a convergent sequence of real numbers. We say that a sequence of real numbers $\{x_k\}$ converges to x if for each real number $\varepsilon>0$ there is an integer N such that $|x_k-x|<\varepsilon$ whenever $k>N=N(\varepsilon)$. Intuitively we observe that N gets larger as ε gets smaller. The notion of convergence in a metric space is analogous to that of a sequence of real numbers. We say that a sequence $\{x_k\}$ of elements of a metric space (X, d_x) converges to $x\in X$ if to each $\varepsilon>0$ there exists an N such that

$$d_x(x_k, x)<\varepsilon \qquad \text{whenever } k>N.$$

This is sometimes written as $\lim_{k\to\infty}x_k=x$, and x is said to be the limit of the sequence $\{x_k\}$.

A sequence of elements $\{x_k\}$ in a metric space (X, d) is said to be a *Cauchy sequence* in X if for each $\varepsilon>0$ there exists an N such that

$$d(x_m, x_n)\le\varepsilon \qquad \text{for all } m, n\ge N.$$

One should note that the definition of Cauchy sequence involves a double limit, and that every convergent sequence is a Cauchy sequence, but not every Cauchy sequence is convergent. If $\{x_k\}$ is a convergent sequence with the limit x, then there exists an N such that

$$d(x_n, x)\le\frac{\varepsilon}{2} \qquad \text{and} \qquad d(x_m, x)\le\frac{\varepsilon}{2}, \qquad \text{for } n, m\ge N.$$

Then by the triangle inequality we have

$$d(x_n, x_m)\le d(x_n, x)+d(x_m, x)\le\varepsilon,$$

which implies that $\{x_k\}$ is a Cauchy sequence.

Example 2.12 Consider the sequence $\{1/n\}_{n=1}^{\infty}$. First note that the notation implies that

$$\left\{\frac{1}{n}\right\}_{n=1}^{\infty}=\left\{1, \frac{1}{2}, \frac{1}{3}, \dots\right\}.$$

Intuitively the sequence should converge to zero (and we shall prove it shortly). As pointed out earlier, the choice of N may depend on the choice of ε. For example, given $\varepsilon>0$, there is an N such that for $n\geq N$

$$\left|\frac{1}{n}-0\right|=\frac{1}{n}<\varepsilon.$$

For example, if $\varepsilon=0.01$, $N=101$ satisfies the conditions of the definition of convergence, since $n\geq 101$ gives $1/n<0.01$. If $\varepsilon=0.001$, $N=1001$ satisfies the conditions of the definition.

To prove that our intuition is correct concerning the limit of the sequence, choose $\varepsilon>0$. There is a positive integer N such that $N>1/\varepsilon$ or $\varepsilon>1/N$. Thus for $n\geq N$ we have $|1/n-0|=1/n\leq 1/N<\varepsilon$. ■

Note again that the definition of a Cauchy sequence involves the consideration of elements from the sequence only, whereas the definition of convergence involves the consideration of elements outside the sequence, namely, the limit of the sequence. In view of this, in many situations it is easier to show that a given sequence is a Cauchy sequence than to show that it is convergent.

If a function f: $(X,d_x)\to(Y,d_y)$ is continuous at x_0, then for every sequence $\{x_n\}$ converging to x_0 (i.e., $\lim_{n\to\infty}x_n=x_0$), we have

$$\lim_{n\to\infty} f(x_n)=f(x_0).$$

The converse is also true. That is, the function f is continuous if and only if it preserves convergent sequences.

Completeness A metric space (X,d_x) is said to be *complete* if every Cauchy sequence is a convergent sequence in (X,d_x).

A closed subspace of a complete space is also a complete subspace.

We will see some examples of complete spaces shortly, but let us first note that the sets of real numbers R and complex numbers C with the usual metric (i.e., absolute value) form complete metric spaces. However, the space of rational numbers with the absolute-value metric is not a complete metric space. For additional details the reader is referred to Refs. 13 and 14.

The property of completeness is very important in the study of the existence of solutions to differential equations. Therefore it is of fundamental importance to construct spaces that are complete. In a certain sense an incomplete metric space can be completed, called *completion* of the original metric space. In order to state the meaning of being completed, we need the concepts of a dense set and isometry.

Countable and Dense Sets A set is said to be *countable* if its elements can be placed in one-to-one correspondence with the set J of positive integers. Every set containing a finite number of elements is countable. (See Exercise 2.1.17.)

Let (X,d) be a metric space, and let S and D be two sets in X with $D\subset X$. The set D is said to be *dense in S* if, to every element x in D and to every $\varepsilon>0$, we can find an element $x_0 \in S$ such that $d(x, x_0) < \varepsilon$. In other words, to any element of S we can find an arbitrarily close element in D. Clearly, the concept of denseness depends on the metric. As will be seen in later chapters, the notion of denseness plays an important role in approximation theory.

The set of rational numbers is dense and countable.

Separable Metric Space A metric space (X, d) is said to be separable if it contains a dense set D that is countable.

An example of separable metric space is provided by the set of real numbers with the usual metric. The set contains the set of rational numbers, which is dense and countable.

The notion of isometry is needed to define the completion of a metric space.

Isometry Two metric spaces (X, d_x) and (Y, d_y) are said to be *isometric* (to one another) if there exists a mapping f of X onto Y such that $d_x(x, y) = d_y(f(x), f(y))$. The mapping f is called an *isometry*.

Completion of Metric Spaces Let (X, d_x) be a metric space. A metric space (Y, d_y) is said to be a completion of (X, d_x) if (1) (Y, d_y) is complete, and (2) (X, d_x) is isometric with a dense subspace (Z, d_z) of (Y, d_y). Note that a completion of (X, d_x) does not necessarily contain (X, d_x), but it contains a dense subspace that is isometric to (X, d_x).

The requirement that Z be dense in Y implies that every point of Y is either a point or a limit point of Z, which means that given any point y in Y, there exists a sequence of points in Z that converge to y.

Now we consider some examples of complete metric spaces.

Example 2.13

1 Consider the metric space consisting of real numbers between (not including) 0 and (including) 1. We use the notation $X = (0,1] = \{x: \ x \in \mathsf{R}, 0 < x \leq 1\}$. Let d_x be the metric, $d_x(x, y) = |x - y|$. The sequence $\{1/k\} = \{1, 1/2, 1/3, \ldots\}$ lies in X. The sequence $\{1/k\}$ is a Cauchy sequence, as was shown in Example 2.12. However, the sequence $\{x_k\}$ is not convergent in X since the limit 0 does belong to X. If we "add" this limit point to the set X, then $\{1/k\}$ would converge in $\bar{X} = (0,1] \cup \{0\}$. The set $\bar{X}$ is the completion of the set $(0,1]$.

2 Let $X = C[0,1]$ with the *sup-metric* d_∞ (see Example 2.10). We shall show that $(C[0,1], d_\infty)$ is complete. Let $\{x_n(t)\}$ be an arbitrary Cauchy sequence in $(C[0,1], d_\infty)$. That is, given any $\varepsilon > 0$, there is an $N(\varepsilon)$ such that $m, n > N$ implies that

$$|x_n(t) - x_m(t)| \leq d_\infty(x_n, x_m) \leq \varepsilon, \qquad \text{for all } t.$$

Since t is arbitrary, the sequence $\{x_n(\cdot)\}$ converges pointwise to a function $x(\cdot)$. But $N = N(\varepsilon)$, being independent of t, implies that $\{x_n(\cdot)\}$ converges uniformly to $x(\cdot)$. Since $x(t)$ is the limit of a uniformly convergent sequence of continuous functions, it is also continuous on [0, 1]. Therefore $x(t) \in C[0,1]$ and $\lim_{n\to\infty} d_\infty(x_n, x) \to 0$.

3 Consider the space $C[-1,1]$ again. Now we use a different metric than the one used above, namely, the L_2-metric [see Eq. (2.30b)]:

$$d_2(x, y) = \left(\int_{-1}^{1} [x(t) - y(t)]^2 \, dt \right)^{1/2}.$$

We can show that $(C[-1,1], d_2)$ is a metric space that is *not* complete. We show this by a counterexample (note that, in any case, proving that something holds requires a general proof, whereas proving that something does not hold requires an example of contradiction). Consider the sequence of continuous functions from $C[-1,1]$:

$$x_k(t) = \frac{1}{2} + \frac{1}{\pi} \arctan kt, \qquad -1 \le t \le 1.$$

This sequence is a Cauchy sequence, $\lim_{m,n\to\infty} \int_{-1}^{1} \{x_m(t) - x_n(t)\}^2 \, dt = 0$. However, the limit of this sequence is the *discontinuous* function

$$x(t) = \begin{cases} 1, & 0 < t \le 1 \\ \frac{1}{2}, & t = 0 \\ 0, & -1 \le t < 0. \end{cases}$$ ■

The above example illustrates the fact that the completeness of a space depends on the metric used.

Example 2.14 (*Space* $L_2[0,1]$) Let $X = C[0,1]$ be the space defined in Example 2.10 with L_2-metric, $d_2(x, y)$. Then the completion of X is $L_2[0,1]$, which is the space of all square-integrable (in Lebesque sense; see Ref. 14) functions $x(t)$ defined on [0, 1] with the property

$$\int_0^1 |x(t)|^2 \, dt < \infty.$$

Then the completion of $C[0,1]$ is $L_2[0,1]$. The space $C[0,1]$ of continuous functions is not extensive enough to accommodate the L_2-metric. The space is completed (not described here) by adding the limits of all convergent Cauchy sequences to $C[0,1]$. The resulting space is called $L_2[0,1]$-space. ■

Every metric space has a completion, and all completions of a metric space are isometric with one another (i.e., completion is essentially unique). In Example 2.13 the completion of the set $X = (0,1]$ is shown to be $\overline{X} = [0,1]$. In this case $\overline{X}$ contains X as a subset.

From the previous discussion it follows that if two metric spaces (X, d_x) and (Y, d_y) are isometric to one another, then they are, except for the names of the points in the sets, the same metric space. In this connection the following result is important. For a proof the reader is referred to references at the end of this chapter.

Theorem 2.2 Let (X, d_x) and (Y, d_y) be two metric spaces that are isometric to one another. Then (X, d_x) is complete if and only if (Y, d_y) is complete. ■

In Section 2.6 we deal primarily with linear mappings defined on linear vector spaces. We close this section with the concept of contractive mapping and the fixed-point theorem. Since the theorem is not used in the present text, the proof of the theorem is not presented here.

Let (X, d) be a metric space. A mapping f of X into itself is called *contractive* if there exists a number $\alpha, 0<\alpha<1$, such that for every pair of elements x, $y \in X$ we have

$$d(f(x), f(y)) \leq \alpha d(x, y).$$

Thus a contractive mapping "contracts distances": the distance of the images $f(x)$, $f(y)$ is smaller, by a scale factor of α, than the distance between the elements x, y. The mapping brings points uniformly closer together.

When α in the above equation takes any positive value (i.e., not necessarily between 0 and 1), the mapping is called *Lipschitz continuous*. It is clear that a Lipschitz continuous (and hence, a contractive) mapping is continuous.

Theorem 2.3 (*The Banach fixed-point theorem.*) Let f be a contractive mapping on a complete metric space X. Then

(i) The equation $x = f(x)$ has one and only one solution in X.

(ii) The unique solution x of the equation $x = f(x)$ can be obtained as the limit of a sequence of elements $x_n \in X$,

$$x = \lim_{n \to \infty} x_n \qquad \text{where } x_n = f(x_{n-1}), \qquad n = 1, 2, \ldots$$

and $x_0 \in X$ can be chosen arbitrarily. ■

The theorem not only presents an existence and uniqueness result but also gives an algorithm to obtain the solution by an iterative procedure. For an application of the theorem, the reader is referred to Exercise 2.7.3.

2.3.5 Some Additional Concepts from Metric Spaces

In this section we present definitions of certain concepts that are needed in the sequel. This section can be omitted in the first reading of the book, and the material can be filled in as needed in the later portions.

Open and Closed Balls Let (X, d) be a metric space with metric d, and let x_0 be an arbitrary point in X. The set

$$B_r(x_0)=\{x\in X:\quad d(x_0,x)<r\},\qquad 0<r<\infty$$

is called the *open ball* of radius r centered at x_0. The set

$$B_r[x_0]=\{x\in X:\quad d(x_0,x)\le r\},\qquad 0\le r<\infty$$

is called the *closed ball* of radius r centered at x_0. The set

$$S_r[x_0]=\{x\in X:\quad d(x_0,x)=r\},\qquad 0\le r<\infty$$

is referred to as the *sphere* of radius r centered at x_0.

In three-dimensional Euclidean space ($X=\mathsf{R}^3$), with the usual Euclidean norm (2.20), the above definitions correspond, respectively, to the "interior," "interior and boundary," and "boundary" of the sphere centered at x_0. It is clear that $B_r[x_0]$ is the union of $B_r(x_0)$ and $S_r[x_0]$; it is sometimes referred to as the *closure* of $B_r(x_0)$.

Local Neighborhoods Let (X, d) be a metric space. A subset N of X is said to be a *local neighborhood* of a point x_0 in X if N is either $B_r(x_0)$ or $B_r[x_0]$ for $r\neq 0$. The positive number r is said to be the *radius* of the local neighborhood N. The open balls $B_r(x_0)$ are called *open local neighborhoods*, and the closed balls $B_r[x_0]$ are called *closed local neighborhoods*.

Example 2.15 Here we give two examples of local neighborhoods.

1 Let $X=C[0,T]$ with the sup-metric d_∞ (see Example 2.10). Let $x_0(t)$ be an arbitrary point (i.e., function) in (X, d_∞). The set of all continuous functions x such that $|x(t)-x_0(t)|<r_0\neq 0$ for all $t\in[0,T]$ is an open local neighborhood of x_0. This set contains all continuous functions between $x_0(t)-r_0$ and $x_0(t)+r_0$.

2 Let $X=C[0,T]$ with the L_2-metric (see Example 2.10). For an arbitrary point x_0 in X, the set of all x in X such that $d_2(x,x_0)<r_0$ is an open local neighborhood of x_0. The set contains all continuous functions $x(t)$ such that the area $d_2(x,x_0)$ is less than r_0. ■

The concepts of continuity and convergence can be characterized in terms of local neighborhoods. However, we shall not attempt to get any deeper than necessary into the subject here. An interested reader can consult Refs. 5 and 18.

Open and Closed Sets A set S in a metric space (X, d) is said to be *open* if S contains a local neighborhood of each one of its points, that is, S is open if for every x in S there is a local neighborhood N of x with $N\subset S$. A subset $S\subset X$ is said to be *closed* if its complement $S'=X-S$ is an open set.

All open local neighborhoods are open sets, and all closed local neighborhoods are closed sets. It should be pointed out that there exist sets that are both open and closed; it is also possible for a set to be neither open nor closed. It is important to state which space X is being considered when stating that a set is open or closed.

In passing we state, without proof, a fact concerning closed sets. A set S in a metric space (X, d) is a closed set if and only if every convergent sequence $\{x_n\}$ with $\{x_n\} \subset S$ has its limit in S.

Exercises 2.2

1 If $d(x, y)$ is a metric on a metric space X, show that the following are also metrics on X:

(a) $d_1(x, y) \equiv \dfrac{d(x, y)}{1 + d(x, y)}$.

(b) $d_2(x, y) \equiv \min(1, d(x, y))$.

2 Let $X = \mathsf{R}^2$ and let S be the subset

$$S = \left\{\mathbf{x} = (x_1, x_2):\ \sqrt{x_1^2 + x_2^2} \leq 1\right\}$$

with

$$d_2(\mathbf{x}, \mathbf{y}) = \sqrt{(x_1 - y_1)^2 + (x_2 - y_2)^2}.$$

Is (S, d_2) a metric space?

3 Show that $d_0(\mathbf{x}, \mathbf{y}) = |x_1 - y_1|$, where $\mathbf{x} = (x_1, x_2)$ and $\mathbf{y} = (y_1, y_2)$ is a pseudometric on $X = \mathsf{R} \times \mathsf{R} = \mathsf{R}^2$.

4 Let $X = C[0, T]$ be the space of all real-valued functions $x(t)$ with continuous derivatives defined on $[0, T]$ such that $x(0) = 0$. Define a metric on $C[0, T]$ by

$$d(x, y) = \left\{\int_0^T \left|\frac{dx}{dt} - \frac{dy}{dt}\right|^2 dt\right\}^{1/2}.$$

Is $d(x, y)$ a metric or a pseudometric? Answer the same question when the functions are not required to vanish at $t = 0$.

5 Show that the metrics defined in Eq. (2.29) are indeed satisfying the axioms of a metric.

6 Let X be the set of real numerical sequences of the form $\mathbf{x} = \{x_1, x_2, \ldots, x_n, \ldots\}$. Show that

$$d(\mathbf{x}, \mathbf{y}) = \sum_{n=1}^{\infty} \frac{1}{2^n} \frac{|x_n - y_n|}{1 + |x_n - y_n|}$$

is a metric on X.

7 (*The l_p space*) Let l_p denote the set of all bounded sequences of real numbers of the form $\mathbf{x}=\{x_1, x_2,\ldots,x_n,\ldots\}$. Show that

$$d_p(\mathbf{x},\mathbf{y})=\left[\sum_{i=1}^{\infty}|x_i-y_i|^p\right]^{1/p}, \qquad 1\le p<\infty$$

is a metric on l_p.

8 (*The L_p space*) Let X be the set of all functions $x(t)$ defined and measurable (the concept of the *measure* of a function can be found in Ref. 14 at the end of this chapter) in the closed interval [0, 1] with the property

$$\int_0^1 |x(t)|^p\,dt<\infty.$$

The space is generally denoted by $L_p(0,1)$. Show that $L_p(0,1)$ is a metric space with

$$d(x,y)=\left(\int_0^1 |x(t)-y(t)|^p\,dt\right)^{1/p}.$$

9 Let l_∞ be the set of all bounded sequences $\{x_n\}$ (that is, there exists a real number M such that $|x_n|\le M$ for all n) of real numbers. Let

$$d_\infty(x,y)=\sup|x_n-y_n|.$$

Show that (l_∞, d_∞) is a metric space. If we replace sup with max in the definition of $d_\infty(x,y)$, does it satisfy the axioms of a metric? (*Hint*: In Example 2.10 one can replace sup with max, but here one cannot.)

10 Show that $d(x,y)$ defined by

$$d(x,y)=\begin{cases}1, & x\neq y\\ 0, & x=y\end{cases}$$

is a metric on any nonempty set X. (This metric can be used in a "hit-or-miss" type mathematical model.)

11 Discuss and define appropriate metrics in the following cases:
- **(a)** X is a flower shop carrying five different kinds of flowers.
- **(b)** X is the set containing the cities Chicago, Dallas, Oklahoma City, and New York; define a metric appropriate for:
  - **(i)** Air travel
  - **(ii)** Automobile travel
  - **(iii)** Train travel.
- **(c)** X is a set of targets.

12 Show that the displacement vector $\mathbf{u}=(u, v, w)$ of a particle in a material body can be modeled as a point in the product of metric spaces.

13 Describe the solution (u, v) to the following differential equations as a point in the product of appropriate metric spaces:

$$\nabla^2 u + c_0 \frac{\partial}{\partial y}\left(\frac{\partial u}{\partial x} + \frac{\partial v}{\partial y}\right) = 0 \qquad \text{in } \Omega$$

$$\nabla^2 v + c_0 \frac{\partial}{\partial x}\left(\frac{\partial u}{\partial x} + \frac{\partial v}{\partial y}\right) = 0 \qquad \text{in } \Omega,$$

where $\Omega = \{(x, y):\ 0 < x, y < 1\}$, c_0 is a constant, and $u = v = 0$ at $x = 0, 1$; $v = 0$ at $y = 0, 1$; $u = 0$ at $y = 0$, and $u = 1$ at $y = 1$.

14 Two Cauchy sequences $\{x_n\}$ and $\{y_n\}$ in a metric space (X, d) are said to be *equivalent* if $d(x_n, y_n) \to 0$ as $n \to \infty$. Then we write $\{x_n\} \sim \{y_n\}$. Show that the set of all Cauchy sequences in X can be partitioned into equivalence classes. Here $x_n R y_n \Leftrightarrow \{x_n\} \sim \{y_n\}$.

15 Consider the sequence $\{x_k(t)\}, -1 \le t \le 1$,

$$x_k(t) = \begin{cases} 0, & -1 \le t \le 0 \\ kt, & 0 < t \le 1/k \\ 1, & 1/k < t \le 1. \end{cases}$$

Prove that $\{x_k\}$ is a Cauchy sequence in the L_2-metric.

16 Prove or disprove that the following sequence is a Cauchy sequence in the L_2-metric and/or the sup-metric on $[0, 1]$:

$$x_k(t) = \begin{cases} 1 - kt, & 0 \le t \le 1/k \\ 0, & 1/k < t \le 1. \end{cases}$$

17 A sequence $\{x_n\}_{n=1}^{\infty}$ is *bounded from above* if there is a real number M such that $x_n \le M$ for all n, and the sequence is *bounded from below* if there is a real number m such that $m \le x_n$ for all n. A sequence is *bounded* if it is both bounded from above and bounded from below. Prove that every convergent sequence is bounded.

18 Prove whether the following infinite sequences of real numbers are convergent. If so, find the interval in which these sequences are convergent:

(a) $\left\{1 + \frac{1}{n}\right\}$.

(b) $\left\{\frac{1}{n^2}\right\}$.

(c) $\left\{\frac{1}{n^3}\right\}$.

(d) $\left\{\frac{1}{\sqrt{n}}\right\}$.

(e) $\left\{\frac{n^2}{1+n^2}\right\}$.

(f) $\left\{\frac{n^3-50n^2+n-10}{3n^3+2n+1}\right\}$.

(g) $\left\{\frac{n^3-25n^2+2n-60}{3n^3+65n+70}\right\}$.

(h) $\left\{\frac{\sin(n\pi/2)}{n}\right\}$.

(i) $\left\{\frac{t}{n}e^{-t/n}\right\}$, $\quad 0 \le t < \infty$.

19 Show that the l_p space of Exercise 7 is complete.

20 Show that the L_p space of Exercise 8 is complete.

21 Prove the assertion that for every metric space X there exists, up to isometry, one complete metric space Y with a subspace X_0 which is dense in Y and is isometric to X.

22 Prove that the l_p space of Exercise 7 is separable.

23 Show that the $L_p(0,1)$ space is separable.

24 (*Closed set theorem.*) Prove the statement that a set S in a metric space (X,d) is a closed set if and only if every convergent sequence $\{x_n\} \subset S$ has its limit in S.

2.4 LINEAR VECTOR SPACES

A generalization of the familiar concept of the ordinary physical vector (to study more general objects) forms the subject of linear vector spaces, or simply linear spaces. Recall that in previous discussions $f\colon\ X \to Y$ denoted a transformation from set X into set Y. The sets X and Y in practical applications are equipped with algebraic (as opposed to set-theoretic) as well as metric structures. That is, X and Y are function spaces with a measure of the difference or distance between any two functions defined over X and Y. As an example consider an electrical network. A network receives an input and produces an output. The network can be viewed as a transformation which transforms a given input (unambiguously) into an output. One can combine two inputs by *adding* them and putting their sum through the system. This will produce an output. If the output is the sum of the outputs of the individual inputs, the system is said to be *additive*. One can also change the input by multiplying its magnitude by a *constant* (scalar) factor. If the resulting output is also multiplied by the same factor, the system is said to be *homogeneous*. If the system is both additive and homogeneous, it is said to be *linear*. Further, one can also measure the difference, called *distance*, between two inputs. Thus it seems essential to equip the sets with certain algebraic structures so that the sum of two elements and the scalar multiplication of an element from a set are defined. These ideas are formally expressed in the definition of a linear vector space.

The rules of vector addition and scalar multiplication, to be given shortly, are analogous to those given in Sections 1.2.2 and 1.2.4 for the ordinary physical vector. The notion "direction" takes on a new interpretation in going from ordinary vectors to vectors from a linear vector space. The norm and inner products are generalizations of the magnitude and scalar products defined in Section 1.2.8.

2.4.1 Definition of a Linear Vector Space

A collection of elements, called *vectors*, $u, v, w, \ldots$ is called a *real linear vector space* (or simply *vector space* or *linear space*) V over the real number field R if the following rules are satisfied by the elements of the vector space:

Vector Addition To every pair of vectors u and v there corresponds a unique vector $u+v$ in V, called the *sum* of u and v, with the properties:

1a $u+v=v+u$, (commutative).

1b $(u+v)+w=u+(v+w)$, (associative).

1c There exists a unique vector θ independent of u such that $u+\theta=u$ for $u \in V$, (existence of an identity element).

1d To every u there exists a unique vector (that depends on u), denoted by I_u (or $-u$) such that $u+I_u=\theta$, (existence of the additive inverse element).

Scalar Multiplication To every vector u and every real number $\alpha \in \mathsf{R}$, there corresponds a unique vector αu in V, called the *product* of u and α, such that

2a $\alpha(\beta u)=(\alpha\beta)u$, associative.

2b $(\alpha+\beta)u=\alpha u+\beta u$, distributive with respect to scalar addition.

2c $\alpha(u+v)=\alpha u+\alpha v$, distributive with respect to vector addition.

2d $1 \cdot u=u$, $1 \in \mathsf{R}$.

A set of elements satisfying axioms (1b)–(1d) is called a *group*. A set satisfying axioms (1a)–(1d) is called a *commutative* group or *Abelian* group. Note that the above axioms are identical to those that are satisfied by the vectors considered in Chapter 1. However, there is one exception: magnitude is not a basic property of the abstract vector. Indeed, the set of the physical vectors considered in Chapter 1 forms a special linear vector space. The vector θ is called the *identity element*, and the vector I_u associated with each element $u \in V$ is called the *inverse* element of u. In order to check whether a given set V qualifies as a linear vector space, one must first define the rules of vector addition and scalar multiplication of a vector over the set. Then the *closure property* must be verified: If $u, v \in V$, then $\alpha u + \beta v \in V$ for all $\alpha, \beta \in \mathsf{R}$.

Cartesian Space The set of all ordered n-tuples $(a_1, a_2, \ldots, a_n)$ of real numbers $a_1, a_2, \ldots, a_n$ in the real number field R is called the Cartesian space, and is denoted by R^n. The Cartesian space is a linear vector space with respect to the usual rules of addition and scalar multiplication. An example of R^3 follows.

Example 2.16

1 (*Ordinary physical vector.*) Consider the set V of ordinary vectors in R^3. An element $\mathbf{a} \in V$ represents an ordered triple (a_1, a_2, a_3). The set of ordinary vectors is a linear vector space (in fact, it is more than just a linear vector space, as will be seen later, it is an *inner product space*) if we define the vector addition and scalar multiplication by

$$\mathbf{a}+\mathbf{b}=(a_1+b_1, a_2+b_2, a_3+b_3), \qquad \text{addition}$$

$$\alpha\mathbf{a}=(\alpha a_1, \alpha a_2, \alpha a_3), \qquad \text{scalar multiplication}$$

where $\mathbf{a}, \mathbf{b} \in V$ and $\alpha \in \mathsf{R}$. Clearly, both operations are *closed* in V (i.e., the vector sum and scalar multiple of a vector are also in V). It is a simple matter to verify the axioms of a linear vector space. Note that the vector addition defined above is, geometrically, equivalent to the *parallelogram law of vector addition* (see Section 1.2). The identity element θ is the *zero* vector $\mathbf{0}=(0,0,0)$, and the inverse of an element is its negative.

2 (*Second-order tensors.*) Consider the set V of all dyadics in R^3 studied in Section 1.5. An element a in V represents an ordered 3×3 array of real numbers (see Eq. (1.281)):

$$a=\begin{bmatrix} a_{11} & a_{12} & a_{13} \\ a_{21} & a_{22} & a_{23} \\ a_{31} & a_{32} & a_{33} \end{bmatrix}.$$

We define (a) the vector addition and (b) the scalar multiplication of a vector by the following rules:

$$\textbf{(a)} \qquad a+b=\begin{bmatrix} a_{11}+b_{11} & a_{12}+b_{12} & a_{13}+b_{13} \\ a_{21}+b_{21} & a_{22}+b_{22} & a_{23}+b_{23} \\ a_{31}+b_{31} & a_{32}+b_{32} & a_{33}+b_{33} \end{bmatrix}.$$

(b) $$\alpha a = \begin{bmatrix} \alpha a_{11} & \alpha a_{12} & \alpha a_{13} \\ \alpha a_{21} & \alpha a_{22} & \alpha a_{23} \\ \alpha a_{31} & \alpha a_{32} & \alpha a_{33} \end{bmatrix}.$$

The vector addition and scalar multiplication are closed in V by virtue of the closure property of real numbers. The identity element is the zero dyadic represented by the zero matrix, and the inverse of an element is represented by the negative matrix. Thus V is a linear vector space. As will be seen later, a dyadic represents a linear transformation from R^n into itself (i.e., the set of all linear transformations is a linear vector space). ■

Example 2.17

1 Let P be the set of polynomials in an independent variable x over the real number field. Define the vector addition to be the ordinary addition of polynomials, and the multiplication as the ordinary multiplication of a polynomial by an element of R. Then P is a linear vector space.

2 Let V be the set of all $m \times n$ matrices defined over the real number field. Then V is a linear vector space with respect to the addition and scalar multiplication of matrices defined in Section 1.3.5. Indeed, the properties of matrix addition described there are analogous to the rules of a linear vector space. The identity element is the $m \times n$ matrix of zeros, and the inverse element is the negative of a matrix.

3 The set of all real-valued functions on R, with the vector addition and scalar multiplication defined by the rules

$$(f+g)(x) = f(x) + g(x)$$

$$(\alpha f)(x) = \alpha f(x), \tag{2.32}$$

is a linear vector space.

4 Let P be the set of polynomials with real coefficients and *positive* constant term, and let the vector addition and scalar multiplication be as defined above. Then P is *not* a space because if $p(x)$ is in P, then $-p(x)$ is not in P.

5 Consider the set V of elements $\{\mathbf{a}, \mathbf{b}, \mathbf{c}, \mathbf{d}, \dots\}$. Each element in V is an ordered pair of real numbers, $\mathbf{a} = (a_1, a_2)$. Suppose that the vector addition and scalar multiplication are defined by

$$\mathbf{a} + \mathbf{b} = (a_1, a_2) + (b_1, b_2) = (a_1 b_1, a_2 b_2)$$

$$\alpha \mathbf{a} = (\alpha a_1^2, \alpha a_2^2).$$

Then the addition is commutative and associative. The identity element can be determined from

$$\mathbf{a} + \boldsymbol{\theta} \equiv (a_1 H_1, a_2 H_2) = (a_1, a_2),$$

which implies that $\boldsymbol{\theta}=(1,1)$. Similarly, the inverse element can be determined from

$$\mathbf{a}+\mathbf{I}_a \equiv (a_1 I_{a1}, a_2 I_{a2}) = \boldsymbol{\theta} = (1,1),$$

which gives $\mathbf{I}_a = (1/a_1, 1/a_2)$. Thus V is an Abelian group.

Now check the associativity of scalar multiplication of a vector:

$$\alpha(\beta\mathbf{a}) = \alpha(\beta a_1^2, \beta a_2^2) = \left(\alpha(\beta a_1^2)^2, \alpha(\beta a_2^2)^2\right)$$

$$= \left(\alpha\beta^2 a_1^4, \alpha\beta^2 a_2^4\right) \neq (\alpha\beta a_1^2, \alpha\beta a_2^2) = (\alpha\beta)\mathbf{a}.$$

Thus V is *not* a linear vector space with respect to rules of vector addition and scalar multiplication defined above.

6 Let V be the set of all functions with continuous second derivatives defined over the interval $[0, L]$ and satisfying the differential equation

$$\frac{d^2u}{dx^2} - k^2 u = 0, \qquad k > 0.$$

The set V is a linear vector space with respect to the usual rules of addition and scalar multiplication, and a typical element in V has the form $v = \alpha e^{kx} + \beta e^{-kx}$, where α and β are scalars. ■

Linear Subspaces A linear subspace S of a given linear vector space V over R is a nonempty subset of V which is itself a linear vector space with respect to the operations of addition and scalar multiplication defined over V.

To determine whether a given subset of a linear vector space is a subspace, one must check the existence of the sum and scalar product of vectors in the subspace. In other words, a subset S of a linear space V is a subspace if and only if $0 \in S$ and $\alpha u + \beta v \in S$ for all $u, v \in S$ and scalars $\alpha, \beta \in \mathsf{R}$.

Example 2.18

1 Let $V = \mathsf{R}^3$, and consider the set $S_1 = \{\mathbf{x} \in \mathsf{R}^3\colon\ x_1 + x_2 + x_3 = c\}$, where c is a real number. Clearly, the zero element $\mathbf{0} = (0,0,0)$ is not in S_1 unless $c = 0$. Also $\alpha\mathbf{x} + \beta\mathbf{y} \neq \alpha c + \beta c \neq c$ is not in S_1 for any nonzero values of c. Thus S_1 is a subspace of V only if $c = 0$. Next consider the set $S_2 = \{\mathbf{x} \in \mathsf{R}^3\colon x_1^2 + x_2^2 + x_3^2 = r^2\}$, where r is a real number. Again, the zero element is not in S_2 unless $r = 0$. However, for *any* r the sum $\alpha\mathbf{x} + \beta\mathbf{y}$ is not in S_2. In other words, the set of the (Euclidean) metrics of all elements in R^3 is not a subspace.

2 Let $V = C^2[0, L]$ be the set of all real-valued twice differentiable functions $u(x)$ defined on $0 \le x \le L$. The set $C^2[0, L]$ is a linear vector space with the usual definitions of addition and scalar multiplication (see part 3 of

Example 2.17). Let S_f be the subset

$$S_f = \left\{ u \in C^2[0, L]: \quad \frac{d^2u}{dx^2} - k^2u = f, f \text{ is a real number} \right\}.$$

Clearly, the zero element is not in S_f unless f is equal to zero. Further, if u_1 and u_2 are in S_f, then $\alpha u_1 + \beta u_2 \notin S_f$. Of course, S_0 (i.e., for $f=0$) is a subspace of $C^2[0, L]$. This means that the set of all solutions to the homogeneous differential equation is a linear vector space. Also note that S_0 consists of infinitely many elements of the type $u = Ae^{-kx} + Be^{kx}$, where A and B are real numbers.

3 Let $V = C^2[0, L]$ be the linear vector space defined above. Define the set

$$S = \{ p(x) \in C^2[0, L]: \quad p(0) = \alpha, p(L) = \beta \text{ for any } \alpha, \beta \in \mathsf{R} \}.$$

The set is not a linear vector space for any nonzero α or β, and it is a subspace for $\alpha = \beta = 0$. The subspace contains infinitely many elements, which are divisible by $x(x-L)$.

4 Let $V = \mathsf{R}^n$ and consider any homogeneous system of linear equations in n unknowns with real coefficients

$$\sum_{j=1}^{n} a_{ij}x_j = 0, \qquad i = 1, 2, \ldots, n$$

for $a_{ij} \in \mathsf{R}$ and $\mathbf{x} = (x_1, x_2, \ldots, x_n) \in \mathsf{R}^n$. The set S of all solutions of the homogeneous system is a subspace of R^n, called the *solution space*. ■

Analogous to the notions of union and intersection of sets, we can define the sum and intersection of subspaces. If U and W are subspaces of a linear space V, then the set of all vectors of the form $u + w$, with $u \in U$ and $w \in W$, is also a linear space, called the *sum*, $U + W$, of U and W. The set of all vectors common to both U and W also forms a linear space, called the *intersection* of U and W, and denoted by $U \cap W$. A linear space V is said to be the *direct sum* of its subspaces U and W if and only if

$$V = U + W \quad \text{and} \quad U \cap W = \{0\} \tag{2.33}$$

where $\{0\}$ denotes the set (or space) with only the zero vector. We shall denote the direct sum of U and V by $U \oplus V$.

Example 2.19

1 The intersection of set S_0 defined in part 2 and S defined in part 3 (with $\alpha = \beta = 0$) of Example 2.18 is $\{0\}$. This means that the solution to the differential equation $d^2u/dx^2 - k^2u = 0$ that satisfies the conditions $u(0) = u(L) = 0$ is the trivial solution.

2 Let V be the set of all polynomials with real coefficients. It was already shown that V is a vector space. Let the subspaces S_1 and S_2 be defined by

$$S_1=\{p(x)\in V:\ p(1)=0\},\qquad S_2=\{p(x)\in V:\ p(2)=0\}.$$

The intersection $S_1\cap S_2$ contains all polynomials divisible by $(x-1)(x-2)$. The sum S_1+S_2 is equal to whole space V. ■

2.4.2 Linear Dependence and Independence of Vectors

Let V be a linear space. If an element u in V can be expressed as $u=\Sigma\alpha_i u_i$, $\alpha_i\in\mathrm{R}$ and $u_i\in V$, we say that u is a linear combination of u_i's. An expression of the form $\Sigma\alpha_i u_i=0$ is called a *linear relation* among the u_i's. A relation with all $\alpha_i=0$ is called a *trivial relation*, and a relation with at least one coefficient nonzero is called a *nontrivial linear relation*.

Analogous to the definition of linear dependence of the ordinary vector (see Section 1.2.7), a set of vectors is said to be *linearly dependent* if there exists a nontrivial linear relation among them. Otherwise the set is said to be *linearly independent*.

The following observations are simple consequences of the definitions:

1 If $(u_1,u_2,\dots,u_n)$ are dependent, at least one vector is a linear combination of the others.
2 If the zero vector belongs to a set of vectors, the set is linearly dependent (since for $\alpha\in\mathrm{R}$ and $0\in V$ we have $\alpha 0=0$).
3 A set consisting of exactly one nonzero vector is linearly independent (since for $\alpha\in\mathrm{R}$ and $u\in V$, $\alpha u=0$ implies $\alpha=0$).
4 If $(u_1,u_2,\dots,u_n)$ is a linearly dependent set, so is any set that includes $(u_1,u_2,\dots,u_n)$.
5 If $(u_1,u_2,\dots,u_n)$ is a linearly independent set, any subset of it is also linearly independent.

The reader is asked to prove these to himself.

Example 2.20

1 Let P be the (vector) space of polynomials in x, and let $p_1(x)=1+x+x^2$, $p_2(x)=x^2-x-2$, $p_3(x)=x^2+x-1$, and $p_4(x)=x-1$. Then the linear combination of the vector p_i, $i=1,2,3,4$ is given by

$$\sum_{i=1}^{4} p_i\alpha_i=\alpha_1(1+x+x^2)+\alpha_2(x^2-x-2)+\alpha_3(x^2+x-1)+\alpha_4(x-1),$$

$$=(\alpha_1-2\alpha_2-\alpha_3-\alpha_4)+(\alpha_1-\alpha_2+\alpha_3+\alpha_4)x+(\alpha_1+\alpha_2+\alpha_3)x^2.$$

Now suppose that $\Sigma_{i=1}^{4} p_i\alpha_i = 0$. This implies that

$$\alpha_1 - 2\alpha_2 - \alpha_3 - \alpha_4 = 0$$

$$\alpha_1 - \alpha_2 + \alpha_3 + \alpha_4 = 0$$

$$\alpha_1 + \alpha_2 + \alpha_3 = 0,$$

which has infinitely many solutions. For example, $\alpha_1 = 3$, $\alpha_2 = 2$, $\alpha_3 = -5$, and $\alpha_4 = 4$ is a solution. Thus the linear combination of p_i is nontrivial and therefore is a dependent set. Indeed, we can express $p_3(x)$ in terms of p_1, p_2, and p_4:

$$p_3(x) = \frac{(3p_1(x) + 2p_2(x) + 4p_4(x))}{5}.$$

2 The set $p_1(x) = 1$, $p_2(x) = x - 1$, $p_3 = 1 + x + x^2$ is linearly independent, because

$$\alpha p_1(x) + \beta p_2(x) + \gamma p_3(x) = 0$$

implies that $\alpha = \beta = \gamma = 0$. ■

2.4.3 Span, Basis, and Dimension

Span A set of vectors $\{u_1, u_2, \dots\}$ of a vector space V over R is said to span V if every vector $u \in V$ can be expressed as a linear combination of the set $\{u_1, u_2, \dots\}$.

A vector space can have more than one span. For example, the vector $u = (\alpha, \beta, \gamma) \in \mathsf{R}^3$ can be expressed as a linear combination of vectors $\{u_1, u_2, u_3\}$, $\{u_4, u_5, u_3\}$, or $\{u_1, u_2, u_3, u_6\}$:

$$\begin{aligned}(\alpha, \beta, \gamma) &= \alpha u_1 + (\beta - 2\alpha)u_2 + (\gamma - 2\beta + \alpha)u_3 \\ &= \alpha u_4 + \beta u_5 + \gamma u_3 \\ &= 2\alpha u_1 + (\beta - 3\alpha)u_2 + (\gamma + 2\alpha - 2\beta)u_3 - \alpha u_6\end{aligned}$$

where $u_1 = (1,2,3)$, $u_2 = (0,1,2)$, $u_3 = (0,0,1)$, $u_4 = (1,0,0)$, $u_5 = (0,1,0)$, and $u_6 = (1,-1,2)$. Note that the sets $\{u_1, u_2, u_3\}$ and $\{u_4, u_5, u_3\}$ are linearly independent, whereas the set $\{u_1, u_2, u_3, u_6\}$ is linearly dependent.

Basis A linearly independent set spanning a linear space V is called *basis* or *base* of V.

A basis is essentially a coordinate system (not necessarily rectangular). Note that there can be more than one basis for a vector space. For example, in three-dimensional Euclidean space R^3 the Cartesian base vectors, cylindrical base vectors, and spherical base vectors each form a basis (that is, an arbitrary vector can be represented by any one of these bases). In fact, any three noncoplanar vectors form a basis in three-dimensional Euclidean space. For example, the set $\{(1,0,0),(0,1,0),(0,0,1)\}$ is a basis for $\mathbf{R}^3$. First, the set is linearly independent since

$$\alpha(1,0,0)+\beta(0,1,0)+\gamma(0,0,1)=(\alpha,\beta,\gamma),$$

and the linear combination is zero only if $\alpha=\beta=\gamma=0$. Second, if (α,β,γ) is any vector in $\mathbf{R}^3$, it can be expressed in terms of the basis as shown above. Another basis of $\mathbf{R}^3$ is given by the set $\{(1,0,0),(1,1,0),(1,1,1)\}$.

Dimension A linear space V is called *n-dimensional* if it possesses a set of n linearly independent vectors, but every set of $n+1$ vectors is a dependent set. A linear space with a finite basis is called a *finite-dimensional* vector space, and the number of elements in a basis is called the *dimension* of the space. If there is no upper bound to the number of linearly independent vectors in a vector space, then the vector space is said to be an *infinite-dimensional* vector space.

The following statements can be proved and are left as exercises to the reader.

1 If a vector space has one basis with a finite number of elements, then all other bases are finite and have the same number of elements.
2 In a finite-dimensional vector space, every spanning set contains a basis.
3 In a finite-dimensional vector space any linearly dependent set of vectors can be extended to a basis.

Thus any vector $\mathbf{u}$ in an n-dimensional vector space V can be expressed as

$$\mathbf{u}=\alpha_1\mathbf{u}_1+\alpha_2\mathbf{u}_2+\cdots+\alpha_n\mathbf{u}_n$$

where $\{\mathbf{u}_1,\mathbf{u}_2,\ldots,\mathbf{u}_n\}$ is a set of linearly independent vectors of V. We refer to the vectors $\mathbf{u}_1,\mathbf{u}_2,\ldots,\mathbf{u}_n$ that form a basis of V as *base vectors*, to the scalars $\alpha_1,\alpha_2,\ldots,\alpha_n$ as the *coordinates* of $\mathbf{u}$ relative to the basis $\{\mathbf{u}_1,\mathbf{u}_2,\ldots,\mathbf{u}_n\}$, and to the vectors $\alpha_i u_i$ as the ith *component* of $\mathbf{u}$. The base vectors are said to *generate* the space V.

There are many interesting infinite-dimensional vector spaces; for example, the space P of all polynomials is infinite dimensional.

The set of all vectors in an n-dimensional vector space V_n which can be expressed as linear combinations of a given set of k vectors forms a subspace V_r of the vector space V_n. The integer r, the dimension of V_r, is called the *rank* of the given set of k vectors. Obviously r is less than or equal to both k and n. If the given set of vectors happens to be linearly independent, then its rank equals its order.

In Section 2.6 it will be shown that linear transformations on finite-dimensional vector spaces can be represented by matrices. In particular, the transformation between any two bases of a vector space can be represented by a nonsingular (why?) matrix. The order (or rank) of this matrix is the dimension of the vector space.

Example 2.21

1 Let V be the linear vector space of all real-valued twice differentiable functions $u(x)$ defined on $[0,1]$ such that

$$\frac{d^2u}{dx^2}-k^2u=0, \qquad k>0.$$

We know that e^{-kx} and e^{kx} are in V. It can be easily verified that the pair is linearly independent. Consider the linear combination

$$c_1e^{-kx}+c_2e^{kx}=0.$$

Since we need another equation to determine c_1 and c_2, we differentiate the linear combination,

$$-c_1e^{-kx}+c_2e^{kx}=0.$$

Solving for c_1 and c_2, we get $c_1=c_2=0$. Hence $\{e^{-kx}, e^{kx}\}$ is a linearly independent set.

Next consider a set $\{u_1, u_2, u_3\}$ of three elements from V. Again we set

$$c_1u_1+c_2u_2+c_3u_3=0$$

$$c_1\frac{du_1}{dx}+c_2\frac{du_2}{dx}+c_3\frac{du_3}{dx}=0$$

$$c_1\frac{d^2u_1}{dx^2}+c_2\frac{d^2u_2}{dx^2}+c_3\frac{d^2u_3}{dx^2}=0.$$

In matrix form ($u_1'\equiv du_1/dx$, etc.) we have

$$\begin{bmatrix} u_1 & u_2 & u_3 \\ u_1' & u_2' & u_3' \\ u_1'' & u_2'' & u_3'' \end{bmatrix}\begin{Bmatrix} c_1 \\ c_2 \\ c_3 \end{Bmatrix}=\begin{Bmatrix} 0 \\ 0 \\ 0 \end{Bmatrix}.$$

These equations have a zero solution only if the determinant is nonzero:

$$D\equiv\begin{vmatrix} u_1 & u_2 & u_3 \\ u_1' & u_2' & u_3' \\ u_1'' & u_2'' & u_3'' \end{vmatrix}$$

$$=u_1(u_2'u_3''-u_2''u_3')+u_2(u_1''u_3'-u_3''u_1')+u_3(u_2''u_1'-u_2'u_1'').$$

Since each u_i satisfies the differential equation $u_i'' - k^2 u_i = 0$, we have

$$D = k^2 u_1(u_2' u_3 - u_2 u_3') + k^2 u_2(u_1 u_3' - u_3 u_1') + k^2 u_3(u_2 u_1' - u_2' u_1) = 0.$$

Thus there are many solutions to the set, and therefore *any* three elements in V are linearly dependent. In other words, the differential equation has only two linearly independent solutions. Hence the dimension of V is 2, $\{e^{kx}, e^{-kx}\}$ is the basis of V, and any other element in V is a linear combination of elements e^{kx} and e^{-kx}.

2 Let S_1 be the subspace of R^4 spanned by $\{(1,1,0,0), (1,0,1,1)\}$ and let S_2 be the subspace of R^4 spanned by $\{(2,-1,3,3), (0,1,-1,-1)\}$. We first characterize the subspaces S_1 and S_2. Let $(a_1, a_2, a_3, a_4) \in S_1$. Then

$$(a_1, a_2, a_3, a_4) = \alpha_1(1,1,0,0) + \alpha_2(1,0,1,1).$$

This gives

$$\alpha_1 + \alpha_2 = a_1, \qquad \alpha_1 = a_2, \qquad \alpha_2 = a_3, \qquad \alpha_2 = a_4.$$

Thus S_1 is defined by

$$S_1 = \{\mathbf{x} \in \mathsf{R}^4: \ x_1 = x_2 + x_3, x_4 = x_3\}.$$

A typical element of S_1 is given by $\mathbf{x} = (x_1, x_2, x_1 - x_2, x_1 - x_2)$. Next let $(b_1, b_2, b_3, b_4) \in S_2$. Then

$$(b_1, b_2, b_3, b_4) = \beta_1(2,-1,3,3) + \beta_2(0,1,-1,-1),$$

which gives

$$2\beta_1 = b_1, \qquad \beta_2 - \beta_1 = b_2, \qquad 3\beta_1 - \beta_2 = b_3, \qquad 3\beta_1 - \beta_2 = b_4.$$

Solving we get $b_3 = b_4$, $b_1 = b_2 + b_3$. Thus S_2 is the same subspace as S_1. In other words, both sets span the same subspace of R^4.

Next let us check whether the set is linearly independent. Consider the linear relation

$$\alpha_1(1,1,0,0) + \alpha_2(1,0,1,1) = 0$$

$$\alpha_1 + \alpha_2 = 0, \qquad \alpha_1 = 0, \qquad \alpha_2 = 0.$$

Thus the set is linearly independent. Hence $S_1 = S_2$ is a two-dimensional subspace of R^4, and the set $\{(1,1,0,0), (1,0,1,1)\}$ is a basis for S_1. ■

Example 2.22 Let S_1 be the vector space spanned by the set $\{(1,0,2), (1,2,2)\}$, and S_2 be the vector space spanned by $\{(1,1,0), (0,1,1)\}$. Let us first check whether these sets are linearly independent. Consider the first set of vectors:

$$\alpha_1(1,0,2) + \alpha_2(1,2,2) = 0.$$

This gives $\alpha_1=\alpha_2=0$. Hence the set is linearly independent. That is, S_1 is a two-dimensional space (i.e., subspace of R^3). Similarly we have

$$\alpha_1(1,1,0)+\alpha_2(0,1,1)=0 \Rightarrow \alpha_1=0, \qquad \alpha_2=0.$$

Hence S_2 is also a two-dimensional vector space (a subspace of R^3). Both are subspaces of dimension 2. Indeed, S_1 and S_2 represent planes in a three-dimensional space. Their intersection (cannot be empty since the zero vector is in both) is either a line or, if they coincide, a plane. Let us find a basis for $S_1 \cap S_2$. Any $u=(\alpha,\beta,\gamma)\in S_1\cap S_2$ can be expressed as

$$u=(\alpha,\beta,\gamma)=\alpha_1(1,0,2)+\alpha_2(1,2,2) \qquad \text{since } u\in S_1$$

$$u=(\alpha,\beta,\gamma)=\alpha_3(1,1,0)+\alpha_4(0,1,1) \qquad \text{since } u\in S_2.$$

The components α_1 and α_2 in S_1 are related to the components α_3 and α_4 in S_2 by

$$\alpha_1(1,0,2)+\alpha_2(1,2,2)=\alpha_3(1,1,0)+\alpha_4(0,1,1).$$

This leads to

$$\alpha_1+\alpha_2=\alpha_3, \qquad 2\alpha_2=\alpha_3+\alpha_4, \qquad 2(\alpha_1+\alpha_2)=\alpha_4$$

whose solution is

$$\alpha_2=-3\alpha_1, \qquad \alpha_3=-2\alpha_1, \qquad \alpha_4=-4\alpha_1.$$

Thus

$$(\alpha,\beta,\gamma)=\alpha_1[(1,0,2)-3(1,2,2)]=\alpha_1(-2,-6,-4) \quad \text{in } S_1$$

$$(\alpha,\beta,\gamma)=-2\alpha_1(1,1,0)-4\alpha_1(0,1,1)=\alpha_1(-2,-6,-4) \quad \text{in } S_2.$$

Hence $(-2,-6,-4)$ or $(1,3,2)$ is a basis for $S_1\cap S_2$. Note also that $\{(1,3,2),(1,0,2)\}$ is a basis of S_1, and $\{(1,1,0),(1,3,2)\}$ is a basis of S_2. ■

Example 2.23 Let S_1 be the space spanned by $\{(1,2,3,6),(4,-1,3,6),(5,1,6,12)\}$, and let S_2 be the space spanned by $\{(1,-1,1,1),(2,-1,4,5)\}$, which are subspaces of R^4. We wish to find a basis for $S_1\cap S_2$.

From the preceding example it is clear that the transpose of the augmented matrix associated with the linear equations (obtained from a linear combination of the vectors from the spanning set) must be reduced to upper echelon form. We have

$$[M_1]=\begin{bmatrix} 1 & 2 & 3 & 6\\ 4 & -1 & 3 & 6\\ 5 & 1 & 6 & 12\\ \hline \alpha & \beta & \gamma & \lambda \end{bmatrix}, \qquad [M_2]=\begin{bmatrix} 1 & -1 & 1 & 1\\ 2 & -1 & 4 & 5\\ \hline \alpha & \beta & \gamma & \lambda \end{bmatrix}.$$

Then we reduce each of these matrices to the so-called *upper echelon* form, which contains only zero elements below the diagonal.

Starting with $[M_1]$, by elementary row operations (i.e., adding a scalar multiple of one row to another row), we can reduce it to the upper echelon form

$$\begin{bmatrix} 1 & 2 & 3 & 6 \\ 4 & -1 & 3 & 6 \\ 5 & 1 & 6 & 12 \\ \hdashline \alpha & \beta & \gamma & \lambda \end{bmatrix} \Rightarrow \begin{bmatrix} 1 & 2 & 3 & 6 \\ 0 & 1 & 1 & 2 \\ 0 & 1 & 1 & 2 \\ \hdashline 0 & \beta-2\alpha & \gamma-3\alpha & \lambda-6\alpha \end{bmatrix}$$

$$\Rightarrow \begin{bmatrix} 1 & 2 & 3 & 6 \\ 0 & 1 & 1 & 2 \\ 0 & \beta-2\alpha & \gamma-3\alpha & \lambda-6\alpha \\ \hdashline 0 & 0 & 0 & 0 \end{bmatrix}$$

$$\Rightarrow \begin{bmatrix} 1 & 2 & 3 & 6 \\ 0 & 1 & 1 & 2 \\ 0 & 0 & \gamma-\beta-\alpha & \lambda-2\beta-2\alpha \\ \hdashline 0 & 0 & 0 & 0 \end{bmatrix}.$$

From this it is clear that the dimension of S_1 is 2 (since there are only two linearly independent rows—except for the row corresponding to the arbitrary vector). Now set the elements of the third row to zero:

$$\gamma-\beta-\alpha=0$$

$$\lambda-2\beta-2\alpha=0.$$

Similarly, from $[M_2]$ we get

$$\begin{bmatrix} 1 & -1 & 1 & 1 \\ 2 & -1 & 4 & 5 \\ \alpha & \beta & \gamma & \lambda \end{bmatrix} \Rightarrow \begin{bmatrix} 1 & -1 & 1 & 1 \\ 0 & 1 & 2 & 3 \\ 0 & \alpha+\beta & \gamma-\alpha & \lambda-\alpha \end{bmatrix}$$

$$\Rightarrow \begin{bmatrix} 1 & -1 & 1 & 1 \\ 0 & 1 & 2 & 3 \\ 0 & 0 & \lambda-3\alpha-2\beta & \lambda-4\alpha-3\beta \end{bmatrix}.$$

Thus S_2 is also two-dimensional. We have

$$\lambda-3\alpha-2\beta=0$$

$$\lambda-4\alpha-3\beta=0.$$

Solving the four equations, we get

$$\alpha=-\frac{\lambda}{2}, \qquad \beta=\lambda, \qquad \gamma=\frac{\lambda}{2},$$

so that

$$(\alpha,\beta,\gamma,\lambda)=\frac{\lambda}{2}(-1,2,1,2)$$

is the basis of the intersection $S_1 \cap S_2$. ■

Change of Basis Let $(e_1, e_2, \dots, e_n)$ and $(\bar{e}_1, \bar{e}_2, \dots, \bar{e}_n)$ be two arbitrary bases of an n-dimensional vector space V_n. Then by definition of a basis, each vector of the second basis can be expressed as a linear combination of the vectors in the first one (and vice versa). Let A^i_j be the ith component of the vector $\bar{e}_j$. Then

$$\bar{e}_j = \sum_{i=1}^{n} A^i_j e_i, \qquad j=1,2,\dots,n. \tag{2.34a}$$

Similarly,

$$e_j = \sum_{i=1}^{n} B^i_j \bar{e}_i, \qquad j=1,2,\dots,n. \tag{2.34b}$$

Now let u be an arbitrary vector in V_n whose components with respect to the first basis are denoted by u^i and whose components with respect to the second basis are denoted by $\bar{u}^i$. Then

$$u = \sum_{i=1}^{n} u^i e_i = \sum_{j=1}^{n} \bar{u}^j \bar{e}_j$$

$$= \sum_{j=1}^{n} \bar{u}^j A^i_j e_i$$

or

$$u^i = \sum_{j=1}^{n} A^i_j \bar{u}^j. \tag{2.35}$$

Similarly,

$$\bar{u}^i = \sum_{j=1}^{n} B^i_j u^j. \tag{2.36}$$

Equations (2.35) and (2.36) are the fundamental formulas for the transformation of the components of a vector under the given change of basis. Note that Eq. (2.36) is identical to Eq. (1.55d).

Exercises 2.3

1 Let V be the set of elements $\mathbf{a}$ of the form $\mathbf{a}=(a_1, a_2)$. Define the vector addition and the scalar multiplication by

$$\mathbf{a}+\mathbf{b}=(a_1b_1, a_2b_2)$$

$$\alpha\mathbf{a}=(\alpha a_1, \alpha a_2), \qquad \alpha\in\mathsf{R}.$$

Is V:

(a) A group.

(b) An Abelian group?

(c) A vector space on R? Determine the identity and inverse elements.

2 Consider the set V of elements $\{\mathbf{a},\mathbf{b},\mathbf{c},\ldots\}$ with each element of the form $\mathbf{a}=(a_1, a_2, a_3)$. Define the vector addition and scalar multiplication by

$$\mathbf{a}+\mathbf{b}=(a_2b_3-b_2a_3, b_1a_3-a_1b_3, a_1b_2-b_1a_2)$$

$$\alpha\mathbf{a}=(\alpha a_1, \alpha a_2, \alpha a_3), \qquad \alpha\in\mathsf{R}.$$

What axioms of the vector space are violated by the vector addition and scalar multiplication?

3 Consider the set S_f defined in part 2 of Example 2.18. Define the set

$$\hat{S}=\{w: \quad w=u-f, u\in S_f, f \text{ a fixed real number}\}.$$

Is $\hat{S}$ a linear vector space?

4 (*Quotient Space*) Consider the vector space $V=C^2[0,1]$. Define a relation R on V such that uRv implies that $d^2u/dx^2=d^2v/dx^2$ for u, v in V. Is the set R a linear vector space? Show that R is an equivalence relation on V, and that R partitions V into equivalence classes of the type

$$[u]=\left\{v\in V: \quad uRv\Leftrightarrow\frac{d^2u}{dx^2}-\frac{d^2v}{dx^2}=0\right\}.$$

Show that the set of all such equivalence classes is a linear vector space with respect to the addition and scalar multiplication defined by

$$[\alpha u_1+\beta u_2]=\alpha[u_1]+\beta[u_2]$$

for all $\alpha, \beta\in\mathsf{R}$ and $u_1, u_2\in V$. This space is called the *quotient space*.

5 Let U be a vector space and S be a subspace of U. For each $u\in U$ let $[u]$ denote the set of all $v\in U$ such that $u-v\in S$. Prove that the set of all $[u]$

is a vector space with respect to vector addition ($[u]=[v]$ if uRv):

$$[u]+[v]=[u+v]$$

$$\alpha[u]=[\alpha u].$$

6 Let U be a vector space and S be a subspace of U. Let R be the relation on U defined according to

$$uRv \Leftrightarrow u-v \in S \subset U.$$

Show that R is an equivalence relation.

7 Determine which of the following subsets of $\mathbf{R}^n$ are subspaces:

(a) $S_1=\{\mathbf{x}\in\mathbf{R}^n\colon\ x_1\geq 0\}$.
(b) $S_2=\{\mathbf{x}\in\mathbf{R}^n\colon\ 2x_1+3x_2=0\}$.
(c) $S_3=\{\mathbf{x}\in\mathbf{R}^n\colon\ 2x_1+3x_2=1\}$.
(d) $S_4=\{\mathbf{x}\in\mathbf{R}^n\colon\ 4x_1+3x_2-2x_3-x_4=0,\ x_1+2x_2-3x_3-4x_4=0\}$.

8 Let P be the space of all polynomials with real coefficients. Determine which of the following subsets of P are subspaces:

(a) $S_1=\{p(x)\colon\ p(1)=0\}$.
(b) $S_2=\{p(x)\colon$ degree of $p(x)=3\}$.
(c) $S_3=\{p(x)\colon$ degree of $p(x)\leq 3\}$.
(d) $S_4=\{p(x)\colon$ constant term is zero$\}$.

9 Let W be a linear vector space and let U and V be subspaces of W. Prove that the following assertions are equivalent:

(a) If $W=U\oplus V$, then the subspaces U and V have only the zero (null) element of the space in common.
(b) If every element $w\in W$ can be represented in the form

$$w=u+v, \qquad u\in U, v\in V,$$

and if $U\cap V=\{0\}$, then $W=U\oplus V$.

10 Determine the sum and intersection of the following subspaces of $\mathbf{R}^3$. Also state whether the sum is a direct sum.

(a) $U=\{\mathbf{x}\in\mathbf{R}^3\colon\ x_1=x_2=x_3\}$, $V=\{\mathbf{x}\in\mathbf{R}^3\colon\ x_1=0\}$.
(b) $U=\{\mathbf{x}\in\mathbf{R}^3\colon\ x_3=0\}$, $V=\{\mathbf{x}\in\mathbf{R}^3\colon\ x_1=0\}$.
(c) $U=\{\mathbf{x}\in\mathbf{R}^3\colon\ x_1+x_2+x_3=0\}$, $V=\{\mathbf{x}\in\mathbf{R}^3\colon\ x_1=x_3\}$.
(d) $U=\{\mathbf{x}\in\mathbf{R}^3\colon\ x_1=x_2=0\}$, $V=\{\mathbf{x}\in\mathbf{R}^3\colon\ x_1=x_3\}$.

11 Determine which of the following sets in $\mathbf{R}^3$ are linearly independent over $\mathbf{R}$:

(a) $\{(-1,1,0), (-1,1,1), (-2,-1,1), (1,1,1)\}$.

(b) $\{(1,0,0), (1,1,0), (1,1,1)\}$.
(c) $\{(1,1,1), (1,2,3), (2,-1,1)\}$.

12 Let $\{u_i\}_{i=1}^n$ be a set of vectors of the form

$$u_i=(g_{1i}, g_{2i},\ldots, g_{ni}).$$

Show that the vectors are linearly dependent only if the determinant of the matrix of coefficients g_{ij} vanishes:

$$[G]=\begin{bmatrix} g_{11} & g_{12} & \cdots & g_{1n} \\ g_{21} & g_{22} & \cdots & g_{2n} \\ \vdots & \vdots & & \vdots \\ g_{n1} & g_{n2} & \cdots & g_{nn} \end{bmatrix}.$$

The converse can also be shown to be true.

13 Determine which of the following sets span R^3:
(a) $\{(2,1,0), (1,-1,2), (0,3,-4)\}$.
(b) $\{(1,1,1), (-2,-1,2), (-1,1,1), (-1,1,0)\}$.
(c) $\{(-1,1,0), (-1,1,1), (-2,-1,1), (1,1,1)\}$.
(d) $\{(1,0,0), (1,2,0), (1,2,3)\}$.

14 Let S_1 be the subspace spanned by the set $\{(1,2,3,6), (4,-1,3,6), (5,1,6,12), (1,-1,1,1)\}$. Determine the subspace S_1.

15 Let S_2 be the subspace spanned by $\{(1,-1,1,1), (2,-1,4,5)\}$. Determine the subspace S_2.

16 Show that the vector space P of all polynomials over the field of real numbers cannot be generated by a finite number of vectors.

17 Determine a basis for the linear vector space containing solutions of the differential equation

$$\frac{d^4u}{dx^4}-k^4u=0, \qquad k>0.$$

18 Let $S_1=\{(1,2,3,6), (4,-1,3,6), (5,1,6,12)\}$ and $S_2=\{(1,-1,1,1), (2,-1,4,5)\}$ be subspaces of R^4. Find a basis for S_1+S_2.

19 Show that $(1,1,1,0)$, $(2,1,0,1)$ spans the subspace of all solutions of the system of linear equations

$$3U_1-2U_2-U_3-4U_4=0$$

$$U_1+U_2-2U_3-3U_4=0.$$

20 Find the set of solutions of the following system of nonhomogeneous linear equations

$$U_1+3U_2+5U_3-2U_4=11$$

$$3U_1-2U_2-7U_3-5U_4=0$$

$$2U_1+U_2-0+U_4=7.$$

21 Find the dimension of and a basis for the solution space of the differential equation

$$y'''+6y''+11y'+66y=0.$$

22 Find the dimension of and a basis for the solution space of the algebraic equations

$$3x_1-2x_2-x_3-x_4=0$$

$$x_1-3x_2+x_3-2x_4=0$$

$$3x_1-x_2-2x_3-5x_4=0.$$

23 Show that the solutions of the equation $ax_1+bx_2+cx_3=0$ comprise a two-dimensional subspace of R^3. Describe the subspace in geometric terms.

2.5 NORMED AND INNER PRODUCT SPACES

Since a good deal of exact science is dominated by the idea of approximation, a means for measuring the difference (or distance) between the exact solution and an approximate solution to a given problem is desirable. We have seen in the preceding sections that there exist a number of ways the distance between two elements of a space can be measured. In most practical applications an element of a linear vector space can be endowed with, like an ordinary physical vector, a length or distance—called a *norm* of the element. For an element that is a continuous function of an independent variable, the norm can be defined, for example, by the *net* area under the function or by the maximum of all possible values the function could take. All of these measures should obey certain basic rules, called *norm axioms* (similar to the properties of a metric).

2.5.1 Norm of a Vector

Let V be a linear vector space over the field R. The value of a function n: $V\to\mathsf{R}$, mapping vectors into real numbers, is called a *norm* on V if and only if the following conditions are satisfied by $n(u), u \in V$:

1a $n(u) \geq 0$
1b $n(u)=0$ if $u=0$, positive
2 $n(\alpha u)=|\alpha| n(u)$, $\alpha \in \mathrm{R}$, homogeneous (2.37)
3 $n(u+v) \leq n(u)+n(v)$, $u, v \in V$, triangle inequality.

We shall use the standard notation $\|u\| \equiv n(u)$ to denote the norm of the vector u. If $n(u)$ satisfies (1a), (2), and (3) only, we call it the *seminorm* of the vector u, and denote it by $|u|$. Note that the axioms are similar to those satisfied by a metric, with the exception of the axiom (2), which is not required of a metric. Associated with each norm, we can define a metric

$$d(u, v) \equiv \|u-v\|. \tag{2.38}$$

It is easy to check that $d(u, v)$ defined above, called *natural metric induced by the norm*, satisfies the axioms of a metric.

Clearly, the axioms of a norm are satisfied by the length (or magnitude) of an ordinary vector $\mathbf{A}$:

$$\|\mathbf{A}\| \equiv |\mathbf{A}| = \sqrt{\mathbf{A} \cdot \mathbf{A}}\,.$$

Since a normed vector space is a linear space as well as a metric space, we can refer to its elements as *vectors* or *points*. This is consistent with the notion of ordinary vectors and points in three-dimensional geometry, where the space consists of points with the usual Euclidean distance, Eq. (2.20). On the other hand, each point is the terminal point of a vector emanating from a common origin, and therefore we can refer to them as vectors. In this case the length of a vector $\mathbf{x}$ is given by $d(\mathbf{x},0)=\|\mathbf{x}\|=|\mathbf{x}|$.

Many of the metrics defined in Section 2.3 are qualified as norms. The reader is asked to show that the sup-metric, and the L_2-metric satisfy the axioms of a norm. While every norm is a metric, the converse does not hold in general.

Example 2.24

1 (*Lebesque Norm*, $\|\cdot\|_p$) Consider the space $L_p[0,T]$, $1 \leq p < \infty$, consisting of all scalar-valued measurable functions (for an account of the Lebesque measure theory, see Ref. 14) $u(t)$ defined over $0 \leq t \leq T$ such that

$$\|u\|_p = \left[\int_0^T |u(t)|^p \, dt\right]^{1/p} < \infty. \tag{2.39}$$

Here it should be understood that $u=0$ if $\|u\|_p=0$ and $u=0$ if $u(t)=0$ almost everywhere. Using the Minkowski inequality for integrals, one can show that $\|u\|_p$ satisfies the triangle inequality. Other axioms of the norm can be easily verified and hence $\|u\|_p$ is a norm on $L_p[0,T]$ space.

2 (*Sobolev Norm*) Let $C^m(\Omega)$, where Ω is an open bounded set in R^n, denote the set of all real-valued functions with m continuous derivatives defined on Ω, and let $C^\infty(\Omega)$ denote the set of infinitely differentiable continuous functions. We define on $C^m(\Omega)$ the norm

$$\|u\|_{m,p}=\left[\int_\Omega \sum_{|\alpha|\le m} |D^\alpha u|^p\,dx\right]^{1/p} \tag{2.40a}$$

for $1\le p<\infty$ and for all $u\in C^m(\Omega)$ for which $\|u\|_{m,p}$ is finite. Here α denotes an n-tuple of nonnegative integers:

$$\alpha=(\alpha_1,\alpha_2,\ldots,\alpha_n),\qquad |\alpha|=\sum_i^n \alpha_i,\qquad \alpha_i\ge 0$$

$$D^\alpha=\left(\partial^{|\alpha|}/\partial x_1^{\alpha_1}\partial x_2^{\alpha_2}\cdots\partial x_n^{\alpha_n}\right). \tag{2.40b}$$

The triangle inequality can be verified by means of the Minkowski inequality for finite sums. The norm $\|u\|_{m,p}$ is called the *Sobolev norm*. Note also that for $m=0$, we have $\|u\|_{0,p}=\|u\|_p$, the Lebesque norm, and for $m=0$ and $p=2$ we get the L_2-norm. For $m=1$, $n=2$, and any $1\le p<\infty$, we have ($\alpha=(\alpha_1,\alpha_2), \alpha_1,\alpha_2=0,1$)

$$\|u\|_{1,p}=\left\{\int_{\Omega\subset\mathsf{R}^2}\left[|u|^p+\left|\frac{\partial u}{\partial x}\right|^p+\left|\frac{\partial u}{\partial y}\right|^p\right]dx\,dy\right\}^{1/p}. \quad \blacksquare \tag{2.41}$$

2.5.2 Normed Linear Spaces

A linear vector space on which a norm can be defined is called a *normed linear space*. A linear subspace S of a normed linear space V is a linear subspace equipped with the norm of V. Recall that a subspace is itself a vector space.

Two norms $\|\cdot\|_1$ and $\|\cdot\|_2$ on a normed linear space V are said to be *equivalent* if there exist positive numbers C_1 and C_2, independent of $u\in V$, such that the following double inequality holds:

$$C_1\|u\|_1\le\|u\|_2\le C_2\|u\|_1. \tag{2.42}$$

The notion of completeness discussed with respect to metric spaces can be carried to normed spaces with the metric $d(x,y)$ replaced by the norm $\|x-y\|$. Also, the concept of isometry carries onto normed linear spaces. Indeed the linear spaces U and V are said to be *isomorphic* if there exists a one-to-one linear mapping T (we shall discuss linear mappings shortly) of U onto V. We state without proof that two linear spaces are isomorphic if and only if $\dim(U)=\dim(V)$. As a corollary to this statement we remark that every n-dimensional linear space is isomorphic to the linear space of ordered

n-tuples of real numbers $\mathbf{R}^n$. These last two considerations have an important implication on finite-dimensional linear spaces. Before we prove that every finite-dimensional linear space is complete, we give the definition of a Banach space.

Banach Space A normed linear space which is complete in its natural metric is called a *Banach space*. A linear subspace of a Banach space is itself a Banach space if and only if the subspace is closed (hence complete).

Some examples of complete normed spaces are given below.

Example 2.25

1 The n-dimensional Euclidean space $\mathbf{R}^n$ is a Banach space with respect to the *Euclidean norm*:

$$\|\mathbf{x}\| \equiv \sqrt{\sum_{i=1}^{n} x_i^2}\,. \tag{2.43}$$

The completeness of $\mathbf{R}^n$ is proved in Example 2.29.

2 The space $C[0,1]$ with the sup-norm is a Banach space (see Example 2.10). It is a linear vector space with respect to the vector addition and scalar multiplication defined in Eq. (2.32). Further, it is complete with respect to the sup-norm:

$$\|x\|_\infty \equiv \max_{0 \le t \le 1} |x(t)|.$$

3 Consider once again the space $C[0,1]$ with the L_2-norm

$$\|u\|_0 \equiv \left\{ \int_0^1 |u(t)|^2\, dx \right\}^{1/2}.$$

The space $(C[0,1], \|\cdot\|_0)$ is not complete, as shown in Example 2.13. Let S be the two-dimensional subspace of all functions $u(x) \in C[0,1]$ of the form $u(x) = c_1 \sin 2\pi x + c_2 \cos 2\pi x$, where c_1 and c_2 are real numbers. We show that the subspace S is closed (or, equivalently, complete). This is done by showing that S is isometrically equivalent to the space $\mathbf{C}$ of complex numbers, considered as a two-dimensional real normed space, which is a complete metric space with respect to the norm

$$\|z\| = \|x + iy\| \equiv \frac{1}{\sqrt{2}} (x^2 + y^2)^{1/2}.$$

We have, for $u \in S$,

$$\begin{aligned}\|u\|_2^2 &= \int_0^1 \left(c_1^2 \sin^2 2\pi x + c_2^2 \cos^2 2\pi x + 2c_1 c_2 \sin 2\pi x \cdot \cos 2\pi x \right) dx \\ &= \frac{1}{2} \left(c_1^2 + c_2^2 \right).\end{aligned}$$

By identifying the isometry f: $S \to \mathbf{C}$ to be $f(u(x)) = c_1 + ic_2$, we conclude that S and $\mathbf{C}$ are isometric and that S is complete since $\mathbf{C}$ is complete.

4 (*Sobolev Space*, $W^{m,p}(\Omega)$) This is a continuation of part 2 of Example 2.24. The space $C^m(\Omega)$ is not complete with respect to the Sobolev norm $\|\cdot\|_{m,p}$. The completion of $C^m(\Omega)$ with respect to the norm $\|\cdot\|_{m,p}$ is called the *Sobolev space of order* (m, p), denoted by $W^{m,p}(\Omega)$. (Recall that the completion of $C(\Omega)$ is the $L_2(\Omega)$ space.) Hence the Sobolev space is a Banach space. Of course, the Lebesque space $L_p(\Omega)$ is a special case of $W^{m,p}$ for $m=0$, and of $L_2(\Omega)$ for $p=2$. ■

Product Spaces Recall from Section 2.3.3 that if (X, d_x) and (Y, d_y) were metric spaces, then many metrics [see Eq. (2.29)] involving the metrics d_x and d_y could be constructed for $X \times Y$. Analogously, if U and V are each normed linear spaces, we can define a norm on $U \times V$. First we must define addition of vectors and scalar multiplication of vectors from $U \times V$ so that $U \times V$ can qualify as a linear vector space. Let (u_1, v_1) and (u_2, v_2) be two vectors in $U \times V$ with $u_1, u_2 \in U$ and $v_1, v_2 \in V$. We define addition of two vectors by

$$(u_1, v_1) + (u_2, v_2) = (u_1 + u_2, v_1 + v_2) \tag{2.44a}$$

and multiplication by a scalar α by

$$\alpha(u, v) = (\alpha u, \alpha v). \tag{2.44b}$$

It is a simple matter to verify that $U \times V$ is a linear vector space with respect to these operations.

A norm on $U \times V$ can be defined in a number of ways. Some of them are given below. The reader is expected to verify that they satisfy the axioms of a norm:

$$\begin{aligned} &1 \quad \|(u, v)\| = \|u\|_U + \|v\|_V \\ &2 \quad \|(u, v)\| = (\|u\|_U^p + \|v\|_V^p)^{1/p}, \qquad p \geq 1 \\ &3 \quad \|(u, v)\| = \max(\|u\|_U, \|v\|_V). \end{aligned} \tag{2.45}$$

We will show that every finite-dimensional linear subspace of a normed linear space (that is not necessarily finite dimensional) is complete. We need the following result first. A proof of the following lemma can be found in Ref. 18.

Lemma 2.1 Let V be a finite-dimensional normed linear space and let $\{\phi_k\}_{k=1}^n$ be a basis for V. Then each coefficient α_i, $i = 1, 2, \ldots, n$, in the expansion

$$u = \alpha_1\phi_1 + \alpha_2\phi_2 + \cdots + \alpha_n\phi_n$$

is bounded in the sense that there exists a constant $M>0$ such that

$$|\alpha_i| < M\|u\|$$

for $i=1,2,\ldots,n$ and all $u \in V$. ■

Theorem 2.4 Any finite-dimensional normed linear space is complete (hence a Banach space).

Proof Let $\{\phi_i\}_{i=1}^n$ be any basis for V and let $\{u_k\}$ be any Cauchy sequence in V. We must show that the sequence $\{u_k\}$ is convergent in V. Since u_k is in V, we can express u_k as

$$u_k = \alpha_{k1}\phi_1 + \alpha_{k2}\phi_2 + \cdots + \alpha_{kn}\phi_n, \qquad k=1,2,\ldots,n.$$

From Lemma 2.1 we know that there exists an $M>0$ such that

$$|\alpha_{ki} - \alpha_{li}| \leq M\|u_k - u_l\|, \qquad i=1,2,\ldots,n.$$

This shows that each sequence of real numbers $\{\alpha_{ki}\}$ is a Cauchy sequence. Since the space of real numbers is complete, the sequence $\{\alpha_{ki}\}$ is convergent. Let α_{0i} be the limit of $\{\alpha_{kj}\}$ (i.e., $\alpha_{0j} = \lim_{k\to\infty}\alpha_{kj}$), $j=1,2,\ldots,n$, and let $u_0 = \alpha_{01}\phi_1 + \alpha_{02}\phi_2 + \cdots + \alpha_{0n}\phi_n$. Then

$$\begin{aligned}\|u_k - u_0\| &= \|(\alpha_{k1}-\alpha_{01})\phi_1 + (\alpha_{k2}-\alpha_{02})\phi_2 + \cdots + (\alpha_{kn}-\alpha_{0n})\phi_n\| \\ &\leq |\alpha_{k1}-\alpha_{01}|\|\phi_1\| + |\alpha_{k2}-\alpha_{02}|\|\phi_2\| + \cdots + |\alpha_{kn}-\alpha_{0n}|\|\phi_n\|,\end{aligned}$$

from which it follows that $u_k \to u_0$ as $k\to\infty$, that is, $\{u_k\}$ converges to u_0. This completes the proof of the theorem. ■

The next result is concerned with finite-dimensional subspaces of normed linear spaces.

Theorem 2.5 Every finite-dimensional subspace of a normed linear space is closed.

Proof The proof follows from Theorem 2.4 and the closed set theorem (see Exercise 2.2.24). ■

Note that in the above theorem the space is not assumed to be complete.

2.5.3 Inner Product and Inner Product Spaces

The notion of a norm provides us with a way of measuring the length of a vector, or the difference between two vectors of a linear vector space. In order to have a means of measuring the angle between two vectors, we need to introduce the notion of an inner product (analogous to the scalar product or dot product of ordinary physical vectors).

An *inner product* on a linear vector space V over $\mathbf{R}$ is a real-valued function that associates with each pair of vectors u and v in V a scalar, denoted (u, v), which satisfies the following axioms:

$$\begin{array}{lll} \mathbf{1} & (u,v)=(v,u), \quad \text{(symmetry)} & \\ \mathbf{2a} & (\alpha u, v)=\alpha(u,v), \quad \text{(homogeneous)} & \left.\right\} \text{(bilinear)} \\ \mathbf{2b} & (u_1+u_2,v)=(u_1,v)+(u_2,v), \quad \text{(additive)} & \\ \mathbf{3} & (u,u)\geq 0, \text{ and } (u,u)=0 \text{ if and only if } u=0, & \text{(positive definite)}, \end{array} \tag{2.46}$$

for every u, u_1, u_2, $v \in V$ and $\alpha \in \mathbf{R}$.

The dot product of vectors provides a familiar example of the inner product defined in Eq. (2.46).

Note that the square root of the inner product of a vector with itself satisfies the axioms of a norm. Consequently one can associate with every inner product a norm by defining

$$\|u\|=\sqrt{(u,u)}\,. \tag{2.47}$$

The norm thus obtained is called the *norm induced by the inner product*.

Inner Product Space A linear vector space on which an inner product can be defined is called an *inner product space*. A linear subspace S of an inner product space V is a subspace with the inner product of V.

Since we can associate with each inner product a norm, every inner product space is also a normed linear space. It should be obvious to the reader that the converse does not hold in general.

Cauchy-Schwarz Inequality A special case of the Hölder inequality, for $p=q=2$, is known as the Cauchy-Schwarz inequality. The inequality is proved here in terms of a norm. Let u and v be any elements of an inner product space V. Then the Cauchy-Schwarz inequality states that

$$|(u,v)|\leq\sqrt{(u,u)(v,v)}=\|u\|\,\|v\|. \tag{2.48}$$

Obviously the relation holds when, say, $v=0$. Suppose now that $v\neq 0$. Then for any scalar $\alpha \in \mathbf{R}$,

$$0\leq(u-\alpha v, u-\alpha v)=(u,u)+\alpha^2(v,v)-2\alpha(u,v).$$

Since α is arbitrary, we can pick α to be $(u,v)/(v,v)$. We have

$$0\leq(u,u)-\frac{(u,v)^2}{(v,v)} \qquad \text{or} \qquad |(u,v)|^2\leq(u,u)(v,v).$$

Orthogonality of Vectors The concept of orthogonality (or perpendicularity) from the ordinary physical vectors can be generalized to the elements of linear vector spaces with the aid of an inner product. Two vectors u and v in an inner product space are said to be *orthogonal* if

$$(u, v)=0. \tag{2.49}$$

A set of vectors $\{u_1, u_2, \dots\}$ is called an *orthogonal set* if each pair of the set is orthogonal: $(u_i, u_j)=0$ for $i \neq j$. The set $\{u_n\}$ is called *orthonormal* if

$$(u_i, u_j)=\delta_{ij} \tag{2.50}$$

where δ_{ij} is the Kronecker delta defined in (1.36b).

If two vectors u and v of an inner product space V are orthogonal, then the Pythagorean theorem holds even in function spaces:

$$\|(u+v)\|^2=(u+v, u+v)=(u,u)+2(u,v)+(v,v)$$
$$=\|u\|^2+\|v\|^2.$$

Theorem 2.6 An orthonormal set of nonzero vectors is linearly independent.

Proof Let $\{u_i\}$ be an orthonormal set. Consider the linear combination

$$\alpha_1 u_1+\alpha_2 u_2+\cdots+\alpha_n u_n=0.$$

We wish to show that the only solution of this equation is $\alpha_1=\alpha_2=\cdots=\alpha_n=0$. Indeed, we have

$$0=(0, u_1)=(\alpha_1 u_1+\alpha_2 u_2+\cdots+\alpha_n u_n, u_1)=\alpha_1(u_1, u_1)+\alpha_2(u_2, u_1)+\cdots.$$

Since $(u_i, u_1)=\delta_{i1}$, we have

$$0=\alpha_1(u_1, u_1) \qquad \text{or} \qquad \alpha_1=0 \qquad (\text{since } u_1 \neq 0).$$

Similarly, we get $\alpha_2=\alpha_3=\cdots=\alpha_n=0$. ■

It should be noted that any nonzero set of orthogonal vectors can be changed into an orthonormal set by replacing each vector u_i with $u_i/\|u_i\|$ (see Exercise 2.4.20).

Projections Once again we will find that the concept of the projection in ordinary physical vector case can be generalized to vectors in linear spaces. For any nonzero vector v the set $\{\alpha v\}$, for every $\alpha \in \mathsf{R}$, is called the *line generated* by v. One can view αv to be the amplification or stretching of the element v. The *projection of u on the line generated by v* is the vector $Pu=(u,\hat{e})\hat{e}$, where $\hat{e}$ is the unit vector lying on the line generated by v, $\hat{e}=v/\|v\|$. It will be shown

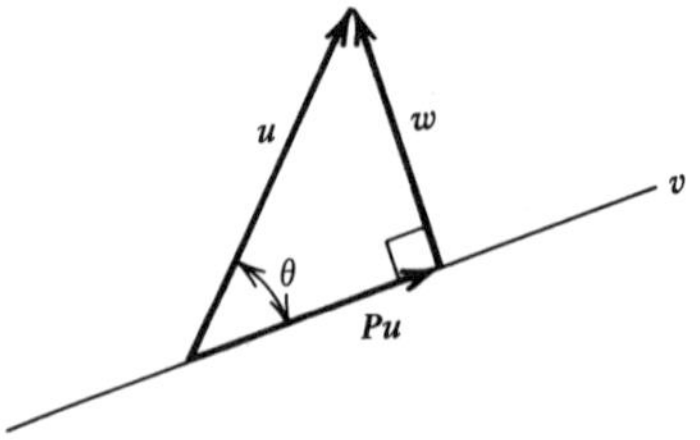

Figure 2.7 Projection of vector u along a line generated by v.

in the next section that P is an operator mapping V into itself. We can write the projection of an element $u \in V$ in the alternate form

$$Pu = \frac{(u,v)v}{\|v\|^2}. \tag{2.51}$$

Any nonzero vector u can be decomposed uniquely as the sum of two vectors, the first lying on the line generated by a given nonzero vector and the second perpendicular to this line. In other words, we can write an element u in V as (see Fig. 2.7)

$$u = Pu + w, \qquad w = u - \frac{(u,v)v}{\|v\|^2}, \qquad (w, Pu) = 0. \tag{2.52}$$

The angle θ, $0 \leq \theta \leq \pi$, between u and v (or Pu) is given by

$$\cos\theta = \frac{(u,\hat{e})}{\|u\|} = \frac{(u,v)}{\|u\|\,\|v\|}. \tag{2.53}$$

Example 2.26 Let $V = \mathsf{R}^n$ and define for $\mathbf{x} = (x_1, x_2, \ldots, x_n)$ and $\mathbf{y} = (y_1, y_2, \ldots, y_n)$ in V, called the *n-dimensional Euclidean space*, the inner product

$$(\mathbf{x},\mathbf{y}) = \sum_{i=1}^{n} x_i y_i. \tag{2.54a}$$

This is the usual inner product on R^n. Similarly, the inner product on the space $V = \mathsf{C}$ of complex numbers is defined by

$$(\mathbf{x},\mathbf{y}) = \sum_{i=1}^{n} x_i \bar{y}_i \tag{2.54b}$$

where $\bar{y}_i$ is the complex conjugate of y_i. ■

Example 2.27 Let Ω be an open bounded set (i.e., region) in R^3, and let $V = C^1(\Omega)$ be the space of real-valued functions that have continuous first partial derivatives in Ω. If $u(\mathbf{x}), v(\mathbf{x}) \in V, \mathbf{x}=(x_1, x_2, x_3) \in \Omega$, we define

$$(u, v)_1 = \int_\Omega \left(uv + \frac{\partial u}{\partial x_1}\frac{\partial v}{\partial x_1} + \frac{\partial u}{\partial x_2}\frac{\partial v}{\partial x_2} + \frac{\partial u}{\partial x_3}\frac{\partial v}{\partial x_3} \right) dx_1\, dx_2\, dx_3. \quad (2.55a)$$

Clearly, $(u, v)_1$ is linear in u and v, also symmetric, $(u, v)_1 = (v, u)_1$ and $(u, u)_1 \geq 0$. Furthermore, if $(u, u)_1 = 0$, then

$$\int_\Omega |u|^2\, dx_1\, dx_2\, dx_3 = 0, \qquad \int_\Omega \left| \frac{\partial u_1}{\partial x_1} \right|^2 dx_1\, dx_2\, dx_3 = 0, \dots . \quad (2.55b)$$

Since u is continuous, $\int_\Omega |u|^2\, dx_1\, dx_2\, dx_3 = 0$ implies that $u=0$. Therefore $(u, v)_1$ defined in Eq. (2.55a) is an inner product on V.

The *natural* norm is given by

$$\|u\|_1 = \sqrt{(u, u)_1} = \left(\int_\Omega (|u|^2 + |\nabla u|^2)\, dx_1\, dx_2\, dx_3 \right)^{1/2}, \quad (2.56)$$

where $|\nabla u|^2 = \nabla u \cdot \nabla u$. ■

Example 2.28

1 Let V be the set of ordinary (physical) vectors in R^3. An element $\mathbf{a} \in V$ has the form $\mathbf{a} = (a_1, a_2, a_3)$. We have already seen that V is a linear vector space with respect to the addition and scalar multiplication of vectors discussed in Chapter I. It is an inner product space with respect to the inner product,

$$(\mathbf{a}, \mathbf{b}) \equiv \mathbf{a} \cdot \mathbf{b} = a_1 b_1 + a_2 b_2 + a_3 b_3$$

$$\|\mathbf{a}\| = \sqrt{a_1^2 + a_2^2 + a_3^2}\,.$$

The angle θ between two vectors $\mathbf{a}$ and $\mathbf{b}$ is given by

$$\cos\theta = \frac{\mathbf{a} \cdot \mathbf{b}}{|\mathbf{a}|\,|\mathbf{b}|} = \frac{a_1 b_1 + a_2 b_2 + a_3 b_3}{\sqrt{a_1^2 + a_2^2 + a_3^2} \cdot \sqrt{b_1^2 + b_2^2 + b_3^2}}.$$

If two vectors $\mathbf{a}$ and $\mathbf{b}$ are orthogonal in V, then

$$\mathbf{a} \cdot \mathbf{b} = 0, \qquad \cos\theta = 0 \qquad \text{or} \qquad \theta = \pi/2.$$

This means vectors $\mathbf{a}$ and $\mathbf{b}$ are perpendicular to each other. The projection of the vector **a** on the line generated by **b** is given by

$$P\mathbf{a}=\frac{(\mathbf{a}\cdot\mathbf{b})\mathbf{b}}{|\mathbf{b}|^2}=(\mathbf{a}\cdot\hat{\mathbf{e}}_b)\hat{\mathbf{e}}_b, \qquad \hat{\mathbf{e}}_b=\frac{\mathbf{b}}{|\mathbf{b}|}$$

$$=|\mathbf{a}|\cos\theta\,\hat{\mathbf{e}}_b.$$

Note that the projection is a vector.

2 Let V be the two-dimensional linear subspace of $C[0,1]$ spanned by $\{\sin 2\pi x, \cos 2\pi x\}$. Define the inner product on $C[0,1]$ (hence on V) by

$$(u,v)=\int_0^1 u(x)v(x)\,dx.$$

The inner product of $u=a_1\sin 2\pi x+a_2\cos 2\pi x$ and $v=b_1\sin 2\pi x+b_2\cos 2\pi x$ is the real number

$$(u,v)=\tfrac{1}{2}(a_1b_1+a_2b_2).$$

An element $u=a_1\sin 2\pi x+a_2\cos 2\pi x$ is orthogonal to $v\in V$ if and only if v is of the form

$$v=-a_2\sin 2\pi x+a_1\cos 2\pi x.$$

The "angle" between u and v is given by

$$\cos\theta=\frac{(u,v)}{\|u\|\,\|v\|}=\frac{(a_1b_1+a_2b_2)}{\sqrt{a_1^2+a_2^2}\,\sqrt{b_1^2+b_2^2}}.$$

The projection of the vector u on the line generated by v is

$$Pu=\frac{(a_1b_1+a_2b_2)(b_1\sin 2\pi x+b_2\cos 2\pi x)}{(b_1^2+b_2^2)}.$$

The Schwarz inequality takes the form

$$|(a_1b_1+a_2b_2)|\le\sqrt{a_1^2+a_2^2}\cdot\sqrt{b_1^2+b_2^2}. \quad \blacksquare$$

The following lemma, referred to as the *fundamental lemma of variational calculus*, plays an important role in variational theory.

Lemma 2.2 Let V be an inner product space. If $(u,v)=0$ for all $v\in V$, then $u=0$.

Proof Since $(u, v)=0$ for all v, it must also hold for $v=u$. Then $(u, u)=0$ implies $u=0$. ■

Orthogonal Complement Let V be an inner product space and let M be any subset of V. The orthogonal complement of M, denoted by $M^{\perp}$, is defined by

$$M^{\perp}=\{u\in M: \quad (u, v)=0 \quad \text{for all} \quad v\in M\}. \tag{2.57}$$

That is, $M^{\perp}$ is made up of all vectors that are orthogonal to every vector in M. Obviously if $M=\varnothing$, then $M^{\perp}=V$. The orthogonal complement of a linear (closed or not) subspace M in V is a closed subspace itself.

From previous discussions and examples it is clear that the geometry of normed and inner product vector spaces is much like the familiar two- and three-dimensional Euclidean geometry. The geometry of Hilbert spaces is even closer to the Euclidean geometry.

Hilbert Space A complete (in its natural metric) inner product space is called *Hilbert space*.

We mention without proof the fact that every inner product space (hence a normed space) has a completion. The concept of *completion* for metric spaces was defined in Section 2.3.4.

Example 2.29

1 (*Euclidean Space* R^n) The n-dimensional Euclidean space is a Hilbert space. The inner product in R^n is defined by Eq. (2.54a). We prove that the space is complete. Let $\mathbf{x}^k=(x_1^k, x_2^k,\dots, x_n^k)$ be a Cauchy sequence in R^n. Then for any $\varepsilon>0$ there exists an N such that

$$d(\mathbf{x}^m,\mathbf{x}^p)\equiv(\mathbf{x}^m-\mathbf{x}^p,\mathbf{x}^m-\mathbf{x}^p)=\left[\sum_{i=1}^{n}\left(x_i^{(m)}-x_i^{(p)}\right)^2\right]^{1/2}\leq\varepsilon$$

whenever $m, p>N$. This implies that

$$|x_1^{(m)}-x_1^{(p)}|\leq\varepsilon, |x_2^{(m)}-x_2^{(p)}|\leq\varepsilon,\dots, |x_n^{(m)}-x_n^{(p)}|\leq\varepsilon, \qquad \text{for } m, p>N.$$

That is, each of the sequences $x_n^{(k)}$ must converge, since R is complete, as $k\to\infty$. Let $\lim_{k\to\infty}x_i^{(k)}=x_i$; then $\lim_{k\to\infty}\mathbf{x}^k=(x_1, x_2,\dots, x_n)$, and the space R^n is complete. Hence it is a Hilbert space. The Schwarz inequality becomes

$$\left[\sum_{k=1}^{n} x_k y_k\right]^2\leq\left[\sum_{k=1}^{n} x_k^2\right]\left[\sum_{k=1}^{n} y_k^2\right].$$

The completeness of R^n implies the completeness of all finite-dimensional linear vector spaces because every finite-dimensional linear space is isomorphic to R^n.

2 ($H^m(\Omega)$ *inner product.*) Consider the Sobolev space $W^{m,2}(\Omega) \equiv H^m(\Omega)$. For $u, v \in H^m(\Omega)$ define

$$(u, v)_m = \int_\Omega \sum_{|\alpha| \leq m} D^\alpha u \cdot D^\alpha v \, dx. \tag{2.58a}$$

This defines an inner product on $H^m(\Omega)$. Once again, we obtain the $L_2(\Omega)$ inner product as a special case from the $H^m(\Omega)$ inner product by setting $m=0$:

$$(u, v)_0 = \int_\Omega u \cdot v \, dx. \tag{2.58b}$$

For example, let $n=2$ and $m=1$ in (2.58a). Then the $H^1(\Omega)$ inner product becomes

$$(u, v)_1 = \int_\Omega \left[uv + \frac{\partial u}{\partial x}\frac{\partial v}{\partial x} + \frac{\partial u}{\partial y}\frac{\partial v}{\partial y} \right] dx\, dy. \tag{2.58c}$$

Similarly, we have

$$(u, v)_2 = (u, v)_1 + \int_\Omega \left[\frac{\partial^2 u}{\partial x \partial y}\frac{\partial^2 v}{\partial x \partial y} + \frac{\partial^2 u}{\partial x^2}\frac{\partial^2 v}{\partial x^2} + \frac{\partial^2 u}{\partial y^2}\frac{\partial^2 v}{\partial y^2} \right] dx\, dy. \tag{2.58d}$$

The norm induced by the inner product gives us back the Sobolev norm $\|\cdot\|_{m,2}$. However, there is no inner product associated with the Sobolev norm $\|\cdot\|_{m,p}$ (i.e., when $p \neq 2$).

3 (*Hilbert spaces* $H^m(\Omega)$.) This is a continuation of part 2 above. Since the Sobolev space $W^{m,p}(\Omega)$ is a Banach space, $W^{m,2}(\Omega) = H^m(\Omega)$ is an Hilbert space with the inner product defined in (2.58a). It should be pointed out once again that the Hilbert space $H^m(\Omega)$ is a special case of the Sobolev space $W^{m,p}(\Omega)$. Of course $W^{m,p}(\Omega)$ is not an Hilbert space for $p \neq 2$.

The Hilbert space $H_0^m(\Omega)$ is a linear subspace of functions from $H^m(\Omega)$ that vanish along with their derivatives up to order $m-1$ on the boundary of Ω. ■

Exercises 2.4

1 Consider the space $L_p[0,1]$, $0 < p < 1$, of all functions $u(x)$ with

$$\|u\| = \int_0^1 |u(x)|^p \, dx < \infty.$$

Show that $\|u\|$ is *not* a norm on $L_p[0,1]$. Show that $\|u - v\| \equiv d(u, v)$ is a metric on $L_p[0,1]$.

2 A real-valued function f defined on a linear space X is said to be *convex* if

$$f(\alpha x+\beta y)\leq \alpha f(x)+\beta f(y)$$

for all $x, y\in X$ and all scalars α, β such that $0\leq\alpha, \beta\leq 1, \alpha+\beta=1$. Show that a norm is a convex function.

3 Compute the L_2-norm and the sup-norm of the following functions in the interval indicated:

(a) $u(x)=\sin\pi x-x$, on $0\leq x\leq 1$.

(b) $u(x)=x^{1/3}$, on $0\leq x\leq 1$.

(c) $u(x)=\cos\pi x+2x-1$, on $0\leq x\leq 1$.

(d) $u(x)=\sqrt{1+x^2}$, on $0\leq x\leq 1$.

(e) $u(x)=\begin{cases}100\sin 100\pi x, & 0\leq x\leq 0.01\\ 0, & 0.01\leq x\leq 1.\end{cases}$

(f) $u(x, y)=\sin^2\pi x\sin^2\pi y$, on $0\leq x, y\leq 1$.

4 Two norms $\|\cdot\|_1$ and $\|\cdot\|_2$ in a vector space V are said to be *equivalent* if there exist positive numbers C_1 and C_2 such that

$$C_1\|u\|_1\leq\|u\|_2\leq C_2\|u\|_1$$

for all u in V. Show that on a finite-dimensional vector space all norms are equivalent.

5 Let $\Omega=(a, b)$ be an open bounded interval in $\mathbf{R}$ and let $C^1(\Omega)$ denote the collection of all real-valued functions defined on Ω with continuous first derivatives. Define, for $1\leq p<\infty$ and $u\in C^1(\Omega)$,

$$\|u\|_{1,p}=\left\{\int_\Omega\left[|u|^p+\left|\frac{du}{dt}\right|^p\right]dt\right\}^{1/p}$$

$$|u|_{1,p}=\left\{\int_\Omega\left|\frac{du}{dt}\right|^p dt\right\}^{1/p}.$$

Show that $\|\cdot\|_{1,p}$ is a norm and $|\cdot|_{1,p}$ is a seminorm on $C^1(\Omega)$. Also show that $|\cdot|_{1,p}$ is a norm on the subspace $C_0^1(\Omega)$

$$C_0^1(\Omega)=\{u\in C^1(\Omega):\quad u=0 \text{ on the boundary of } \Omega\}.$$

6 (*Continuation of Exercise* 5) Show that the norm $|\cdot|_{1,2}$ is *equivalent* to the norm $\|\cdot\|_{1,2}$ on $C_0^1(\Omega)$. In other words, show that there exist positive numbers C_1 and C_2 such that the following inequality holds:

$$C_1|u|_{1,\Omega}\leq\|u\|_{1,\Omega}\leq C_2|u|_{1,\Omega},\qquad \Omega\subset\mathbf{R}^2$$

for all $u \in C_0^1(\Omega)$. (*Hint*: Consider a square K of length a that encloses the open bounded region Ω, and extend $u \in C_0^1(\Omega)$ to $u \in C_0^1(K)$ by zero (i.e., $u(\mathbf{x}) \equiv 0$ for $\mathbf{x}$ in $K-\Omega$) so that u is continuously differentiable on K and vanishes on the boundary of K. Using the identity

$$u(x, y) = \int_0^x \frac{\partial u}{\partial \xi}(\xi, y)\, d\xi,$$

show that $\|u\|_{0,\Omega}^2 \leq a^2 \int_\Omega \left|\frac{\partial u}{\partial x}\right|^2 dx\, dy$, and

$$\|u\|_{0,\Omega} \leq C(\Omega)|u|_{1,\Omega}, \qquad C(\Omega) = \frac{a}{\sqrt{2}}.$$

7 Let $\vec{\vec{\sigma}} = \sigma_{ij}\hat{\mathbf{e}}_i\hat{\mathbf{e}}_j$ be a symmetric dyadic (stress tensor) where $\hat{\mathbf{e}}_i$, $i=1,2,3$, is a system of unit orthonormal basis vectors in R^3. The components σ_{ij} can be viewed as components of a 3×3 matrix (see Section 1.5.3). Let $\mathsf{L}_2(\Omega)$ be the linear vector space of six-tuples $(\sigma_{11}, \sigma_{22}, \sigma_{33}, \sigma_{13}, \sigma_{23}, \sigma_{12})$ of real-valued functions that are square integrable,

$$\int_\Omega |\sigma_{ij}|^2\, dx < \infty, \qquad \text{for } i, j = 1,2,3.$$

The vector addition and scalar multiplication are defined as in the case of a set of matrices. The open bounded domain Ω denotes the region occupied by the material (fluid or solid). Note also that $\mathsf{L}_2(\Omega)$ is the product space obtained by taking the product of $L_2(\Omega)$ with itself six times. Define for $\vec{\vec{\sigma}}$ in $\mathsf{L}_2(\Omega)$,

$$\|\vec{\vec{\sigma}}\|_s^2 = \int_\Omega |\text{trace}\, \vec{\vec{\sigma}} \cdot \vec{\vec{\sigma}}|\, dx = \int_\Omega \left| \sum_{i,j=1}^{3} \sigma_{ij}\sigma_{ij} \right| dx$$

where $dx = dx_1\, dx_2\, dx_3$. Show that $\|\cdot\|_s$ defines a norm on $\mathsf{L}_2(\Omega)$.

8 (*Continuation of Exercise* 7) Consider the following quadratic functional of σ_{ij}:

$$\Psi(\vec{\vec{\sigma}}) = \int_\Omega \sum_{i,j,k,l=1}^{3} C_{ijkl}\sigma_{ij}\sigma_{kl}\, dx,$$

where C_{ijkl} are positive real numbers such that

$$C_{ijkl} = C_{jikl} = C_{ijlk} = C_{klij},$$

and the Korn inequality (Ref. 12) holds:

$$\mu_1 \|\vec{\vec{\sigma}}\|_s \geq \Psi(\vec{\vec{\sigma}}) \geq \mu_2 \|\vec{\vec{\sigma}}\|_s.$$

Here the notation of Exercise 7 above is employed. Show that $\sqrt{\Psi(\vec{\sigma})}$ (complementary energy of an elastic body) satisfies the axioms of a norm, and thus is an equivalent norm on $\mathsf{L}_2(\Omega)$.

9 Show that the normed linear space $C[-\pi,\pi]$ with the norm

$$\|u\|=\left[\int_{-\pi}^{\pi}|u|^2\,dx\right]^{1/2}$$

is *not* complete, whereas $L_2[-\pi,\pi]$ is complete with respect to the above norm. More specifically, show that the series

$$\sum_{n=1}^{\infty}\frac{\sin nx}{n}$$

converges to the function

$$u_0(x)=\begin{cases}-\dfrac{\pi}{4}, & -\pi\leq x<0\\[2mm] \dfrac{\pi}{4}, & 0\leq x\leq\pi,\end{cases}$$

which, not being continuous, is not in $C[-\pi,\pi]$. However, $u_0(x)$ is in $L_2[-\pi,\pi]$.

10 Let $C^{\infty}(a,b)$ be the linear space of all real-valued functions continuous with their derivatives of all orders in (a,b), and let $C_0^{\infty}(a,b)$ denote the set of functions from $C^{\infty}(a,b)$ which are equal to zero in some neighborhood of the boundary of the domain. Prove that the linear space $C_0^{\infty}(a,b)$ is dense in $L_2(a,b)$. More precisely, prove that to every $u\in L_2(a,b)$ and to every $\varepsilon>0$ there exists a function $v\in C_0^{\infty}(a,b)$ such that $\|u-v\|_{L_2(a,b)}<\varepsilon$.

11 Prove the following relations in a real inner product space:

(a) Parallelogram law:

$$\|u+v\|^2+\|u-v\|^2=2(\|u\|^2+\|v\|^2).$$

(b) $(u,v)=\frac{1}{4}[\|u+v\|^2-\|u-v\|^2]$.

(c) $|\,\|u\|-\|v\|\,|\leq\|u-v\|$.

12 Compute the inner product of the following pairs of functions on the interval indicated. Use the L_2 inner product and the H^1 inner product.

(a) $u=x-x^2,\ v=\sin\pi x$, on $0\leq x\leq 1$.

(b) $u=(1+x),\ v=3x^2-1$, on $-1\leq x\leq 1$.

(c) $u=\sin\pi x,\ v=\cos\pi x$, on $0\leq x\leq 1$.

(d) $u=\sin\pi x,\ v=a+bx+cx^2$, on $0\leq x\leq 1$.

(e) $u=\sin\pi x\sin\pi y,\ v=(1-x^2-y^2)$, on $0\leq x,\ y\leq 1$.

13 Let V be the linear vector space of continuous complex-valued functions defined on an interval $[1,0]$. Define, for $u, v \in V$,

$$(u,v) \equiv \int_0^1 u(x)\bar{v}(x)\,dx$$

where $\bar{v}(x)$ is the complex conjugate of $v(x)$ (i.e., if $v=a+ib$, then $\bar{v}=a-ib$). Show that $(\cdot,\cdot)$ is an inner product on V. If S is the linear subspace of V spanned by $\{\sin \pi x\}$, find a function in S that is orthogonal to $f(x)=(a+ib)\sin \pi x$.

14 Consider the linear vector space V of continuous real-valued functions $u(x,t)$ on $\Omega=[0,1]\times[0,T]$. Define for any pair of elements $u, v \in V$,

$$(u,v)=\int_0^1\int_0^T u(x,T-t)v(x,t)\,dt\,dx.$$

Show that $(\cdot,\cdot)$ is an inner product on V. Also show that $(\partial u/\partial t, v)$ is symmetric.

15 If S_1 and S_2 are subspaces of a vector space V, such that

$$S_1+S_2=V \qquad \text{and} \qquad S_1\cap S_2=\{0\},$$

prove that for each vector $u\in V$ there exist *unique* vectors $v\in S_1$ and $w\in S_2$ such that $u=v+w$.

16 Determine whether the following functions belong to the $C[0,1]$ space with (i) sup-norm (ii) and L_2-norm:

(a) $u(x)=\dfrac{1}{\sqrt{x}}$.

(b) $u(x)=x^{-\frac{1}{3}}$.

(c) $u(x)=\dfrac{1}{(1+x)}$.

17 Determine C if the indicated pairs of vectors are orthogonal in R^4:

(a) $(1, C-1, 2, 1+C)$, $(2C, 4, C, 1)$.

(b) $(2, 1-C, 4, 3C)$, $(C, 1+C, 3C, -7)$.

18 Determine the subspace of vectors in R^3 that are orthogonal to:

(a) $(1,1,0)$.

(b) $(-1,2,1)$.

(c) $(1,2,3)$.

19 Determine a vector in the space P of polynomials of degree 2 such that the vector is orthogonal to the polynomials $p_1=1+x-2x^2$ and $p_2=-2+4x+x^2$ in $L_2\,(1,0)$.

20 (*Gram-Schmidt Orthonormalization*) Let $\{u_k\}_{k=1}^n$ be a linearly independent set in an inner product space V. Construct an orthonormal set $\{\phi_k\}_{k=1}^n$ from $\{u_k\}_{k=1}^n$. See Exercise 1.2.12.

21 Let V be an inner product space and let S be any subset of V. The *orthogonal complement* of S, denoted by $S^{\perp}$, is

$$S^{\perp}=\{u\in V:\quad (u,v)=0 \text{ for all } v\in S\}.$$

Prove the following properties of orthogonal complements. Here S_1 and S_2 denote nonempty subsets of V.

(a) If $S_1\subset S_2$, then $S_2^{\perp}\subset S_1^{\perp}$.

(b) If $u\in S_1\cap S_1^{\perp}$, then $u=0$.

(c) If S_1 is a dense subset of V, then $S_1^{\perp}=\{0\}$.

22 Find the orthogonal complement of the subspace $S\subset\mathbb{R}^3$ spanned by the set $\{(1,0,1),(0,2,3)\}$.

23 Discuss why an orthogonal decomposition of a linear vector space is a more restrictive type than a direct sum of subspaces.

24 Show that the condition that S_1 and S_2 are orthogonal complements in $\mathbb{R}^n$ implies that each is the solution space of a system of homogeneous linear equations.

25 Let $V=C[-1,1]$ be the space of all continuous functions on the interval $[-1,1]$ with the L_2 inner product. Let S be the subspace of "odd" functions $f(x)$ such that $f(-x)=-f(x)$. Determine the orthogonal complement of S.

26 Consider the linear vector space of all $n\times n$ matrices of real numbers.

(a) Using the analogy of the standard inner product in $\mathbb{R}^n$, show that the inner product (A,B) of two matrices A and B is equal to $\mathrm{tr}(AB^T)$.

(b) Determine the orthogonal complement of the subspace of diagonal matrices.

2.6 LINEAR TRANSFORMATIONS (OR OPERATORS) AND FUNCTIONALS

2.6.1 Linear Transformations

The notion of a function or transformation from one set into another can be extended to vector spaces. A transformation T from a linear vector space U into another linear vector space V (both vector spaces are defined on the same field of scalars) is a correspondence which assigns to each element u in U a unique element $v=Tu$ in V. We use the terms *transformation*, *mapping*, and *operator* (*not function*) interchangeably.

A transformation L of vector space U into a vector space V, where U and V have the same scalar field, is said to be *linear* if

1 $L(\alpha u)=\alpha L(u)$, for all $u\in U$, scalars α (homogeneous)
2 $L(u_1+u_2)=L(u_1)+L(u_2)$, for all $u_1, u_2\in U$, (additive). (2.59)

Otherwise it is said to be a *nonlinear transformation*. Conditions (1) and (2) can be combined into one: a transformation L is linear if

$$L(\alpha u_1+\beta u_2)=\alpha L(u_1)+\beta L(u_2) \tag{2.60}$$

for all $u_1, u_2\in U$ and scalars α and β. Roughly speaking, the image of the sum is the sum of images, and the image of the product (of a vector with a scalar) is the product of the image of the vector with the scalar (see the introduction at the beginning of Section 2.4).

If L_1: $U\to V$, and L_2: $V\to W$ are linear transformations, then their composition L_2L_1 is also linear:

$$\begin{aligned}(L_2L_1)(\alpha u_1+\beta u_2)&=L_2(\alpha L_1(u_1)+\beta L_1(u_2))\\&=\alpha L_2(L_1(u_1))+\beta L_2(L_1(u_2))\\&=\alpha(L_2L_1)(u_1)+\beta(L_2L_1)(u_2).\end{aligned}$$

A one-to-one transformation T [i.e., for $u\neq v$ it necessarily follows that $T(u)\neq T(v)$] is called a *monomorphism*. If U is a vector space and S is a subset of U, and T: $S\to V$, the set $T(S)$ denotes the collection of images of elements in S; $T(S)$ is called the *image* of T, denoted by $\mathrm{Im}(T)$. Obviously $\mathrm{Im}(T)\subset V$. If $\mathrm{Im}(T)=V$, the transformation is *onto* and is called an *epimorphism*. A transformation that is both a monomorphism and an epimorphism is called an *isomorphism*. Strictly speaking, a linear transformation must specify its domain space, its range space, and the definition of the mapping.

The *null space*, also known as the *kernel* of the transformation T: $U\to V$, denoted by $\mathfrak{N}(T)$, is defined by

$$\mathfrak{N}(T)=\{u:\ u\in U, Tu=0\}. \tag{2.61}$$

The null space is the subset of elements of U which have the zero image. It is easy to verify that this set is a linear vector space.

Let L: $U\to V$ be a linear transformation. If L is a monomorphism (i.e., one-to-one), then $u\in\mathfrak{N}(L)$ implies that $u=0$ (since $L0=0$ and $Lu=0$). Hence $\mathfrak{N}(L)$ is the null set $\{0\}$. Conversely, if $\mathfrak{N}(L)=\{0\}$, $Lu_1=Lu_2$ implies that (since L is linear)

$$L(u_1-u_2)=0 \quad\text{or}\quad u_1-u_2\in\mathfrak{N}(L)\Rightarrow u_1-u_2=0 \quad\text{or}\quad u_1=u_2.$$

Thus a linear transformation is one-to-one if and only if its null space is trivial, $\mathfrak{N}(L)=\{0\}$.

The dimension of the subspace $\mathfrak{R}(T)=\mathrm{Im}(T)\subset V$ is called the *rank* of the transformation, T: $U\to V$. The rank of T is smaller than the dimension of U as well as V. The dimension of the null space of T is called the *nullity* of T. The sum of the rank and nullity of the transformation is equal to the dimension of U (see Theorem 2.7).

Example 2.30

1 Let $U=V$ be the set of complex numbers (on the scalar field of complex numbers) $z=x+iy$, $i=\sqrt{-1}$, and let A be the operator that maps a complex number z into its conjugate, $\bar{z}=x-iy$:

$$A(z)=\bar{z}, \qquad A(z_1+z_2)=\bar{z}_1+\bar{z}_2=A(z_1)+A(z_2).$$

Hence A is additive. However, for a complex scalar $\alpha\neq 0$, we have

$$A(\alpha z)=\overline{\alpha z}=\bar{\alpha}\,\bar{z}\neq\alpha A(z).$$

Hence it is not homogeneous.

2 Let $U=\mathsf{R}\times\mathsf{R}=\mathsf{R}^2$ and $V=\mathsf{R}$ (over the field of real numbers), and let H be such that it maps an element (x, y) in U into V according to

$$H((x, y))=\begin{cases} x+y, & \text{if } xy>0\\ 0, & \text{if } xy\le 0.\end{cases}$$

Note that

$$H(\alpha(x, y))=H((\alpha x, \alpha y))=\alpha x+\alpha y=\alpha(x+y)=\alpha H((x, y)),$$

for *any* (x, y). Hence H is a homogeneous transformation. However, it is *not* additive, since for $x>0$, $y>0$, we have $(x, y)=(x,0)+(0, y)$, and $H((x, y))=x+y$, whereas $H((x,0))=0$, $H((0, y))=0$.

3 Let U be the space of functions that are integrable and twice differentiable over $[a, b]$, and V the set of continuous functions on $[a, b]$. The transformation from U into V, defined by

$$L(u)=u+\frac{d^2u}{dx^2}+\int_a^b K(x, y)u(y)\,dy,$$

is linear, Indeed, we have for any $u_1, u_2 \in U$ and real numbers α, β,

$$
\begin{aligned}
L(\alpha u_1+\beta u_2) &= (\alpha u_1+\beta u_2)+\frac{d^2}{dx^2}(\alpha u_1+\beta u_2) \\
&\quad + \int_a^b K(x,y)\left[\alpha u_1(y)+\beta u_2(y)\right] dy \\
&= \alpha u_1+\alpha\frac{d^2u_1}{dx^2}+\alpha\int_a^b K(x,y)u_1(y)\,dy+\beta u_2 \\
&\quad +\beta\frac{d^2u_2}{dx^2}+\beta\int_a^b K(x,y)u_2(y)\,dy \\
&= \alpha L(u_1)+\beta L(u_2).
\end{aligned}
$$

Note that the operator L is the sum of three linear operators, $L(u)=L_1(u)+L_2(u)+L_3(u)$:

$$
L_1(u)=u, \qquad L_2(u)=\frac{d^2u}{dx^2}, \qquad L_3(u)=\int_a^b K(x,y)u(y)\,dy.
$$

4 The operators $N_1(u)=c$, a constant, $N_2(u)=u\,du/dx$ are not additive, nor homogeneous:

$$
N_1(\alpha u)=c\neq \alpha N_1(u)
$$

$$
N_1(u_1+u_2)=c\neq N(u_1)+N(u_2)
$$

$$
N_2(\alpha u)=(\alpha u)\frac{d(\alpha u)}{dx}=\alpha^2u\frac{du}{dx}\neq\alpha\left(u\frac{du}{dx}\right)
$$

$$
N_2(u_1+u_2)=(u_1+u_2)\frac{d}{dx}(u_1+u_2)=u_1\frac{du_2}{dx}+u_2\frac{du_1}{dx}+u_2\frac{du_2}{dx}+u_1\frac{du_1}{dx}.
$$

However, a zero transformation is linear.

5 Consider a rectangular $m\times n$ matrix $[A]=[a_{ij}]$. The expression

$$
y_i=\sum_{j=1}^{n} a_{ij}x_j, \qquad i=1,2,\dots,m
$$

defines an operator A that transforms an element $\mathbf{x}=(x_1,x_2,\dots,x_n)$ of the n-dimensional Euclidean space R^n into an element $\mathbf{y}=(y_1,y_2,\dots,y_m)$ of the m-dimensional Euclidean space R^m. Clearly, A is a linear operator.

6 If V is an inner product space, the inner product $((u, v))$ is a mapping I from $V \times V$ into the real numbers R. The mapping is *not* linear, since

$$I(\alpha(u, v)) = I(\alpha u, \alpha v) = ((\alpha u, \alpha v)) = \alpha^2((u, v)) = \alpha^2 I(u, v). \quad \blacksquare$$

Example 2.31 Consider the spring and mass system (an idealization, for example, of the motion of a piston and piston rod in an automobile engine) shown in Fig. 2.8. We assume that there is viscous friction between the piston (mass) and the cylinder wall it slides on, and that the piston rod behaves elastically. In the mathematical modeling of the system we assume that the elastic and viscous responses are linear (i.e., the restoring force is linearly proportional to the displacement and the time rate of the displacement, respectively). We wish to model the motion of the system for times $t > 0$. At time $t = 0$ we assume that $x(0) = 0$ and $(dx/dt)(0) = 0$. The differential equation relating the displacement $x(t)$ and the applied force $f(t)$ is given by

$$m\frac{d^2x}{dt^2} + \eta\frac{dx}{dt} + kx = f \tag{a}$$

or

$$Tx = f.$$

For $f \in C[0, \infty)$, the equation describes a mapping T from $C[0, \infty)$ into $C[0, \infty)$. Clearly, T is linear.

Solving the differential equation with the initial conditions $x(0) = 0$ and $(dx/dt)(0) = 0$, we obtain

$$x(t) = \frac{1}{m(\lambda_1 - \lambda_2)} \int_0^t [e^{\lambda_1(t-\tau)} - e^{\lambda_2(t-\tau)}] f(\tau)\, d\tau \tag{b}$$

where λ_1 and λ_2 are the roots of the equation

$$m\lambda^2 + \eta\lambda + k = 0.$$

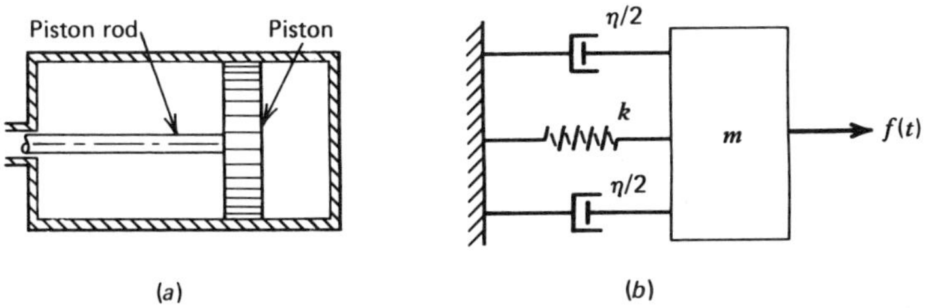

Figure 2.8 Model for a spring mass system. (*a*) Physical model. (*b*) Mathematical model.

We assume that m, η, and k are such that λ_i are real and distinct. Equation (b) describes a linear transformation L from $C[0,\infty)$ into $C[0,\infty)$:

$$L(f)=x.$$

It can be easily verified that $T(L(f))=f$ or $TL=I$. ■

When the domain and range spaces of a linear transformation are finite-dimensional spaces, the following result is very useful.

Theorem 2.7 Let T: $U\to V$ be a linear transformation where U and V are linear spaces. Then the dimensions of the kernel $\mathcal{N}(T)$, range $\mathcal{R}(T)$, and U are related by the formula

$$\dim\{\mathcal{N}(T)\}+\dim\{\mathcal{R}(T)\}=\dim\{U\}. \tag{2.62}$$

Proof See Exercise 2.5.8. ■

2.6.2 Continuous Linear Transformations

The concept of inverse transformation (or mapping) is the same as that of a function, which is a set-theoretic concept. A transformation T (linear or not) has an inverse defined on its range if and only if T is one-to-one. From the discussion above we see that if T is linear, then it is one-to-one if and only if the null space of the transformation is trivial. When the domain and range spaces of a transformation are equipped with norms, one can also speak of the continuity and norm of the transformation.

Let U and V be normed linear spaces. A transformation L is said to be a *continuous linear transformation* if it is (1) linear and (2) continuous. We say that L is bounded if there exists a real number $M\geq 0$ such that

$$\|Lu\|_v\leq M\|u\|_u, \qquad \text{for all } u\in U \tag{2.63}$$

where $\|\cdot\|_u$ is the norm in U and $\|\cdot\|_v$ is the norm in V.

As remarked in the case of continuous functions, a linear transformation L: $U\to V$ is continuous if and only if it is bounded. Indeed, if L is bounded, we have (since L is linear)

$$\|Lu_1-Lu_2\|_v=\|L(u_1-u_2)\|_v\leq M\|u_1-u_2\|_u.$$

By setting $\delta=\varepsilon/M$ we see that L is continuous. On the other hand, if L is continuous, it is continuous at 0. For $\varepsilon=1$ there is a $\delta>0$ such that $\|L\bar{u}\|_v\leq 1$ whenever $\|\bar{u}\|_u<\delta$. If u is any nonzero element in U, let $\bar{u}=\alpha u$, with

$\alpha = \delta / \|u\|_u$,

$$1 \geq \|L\bar{u}\|_v = |\alpha| \, \|Lu\|_v \qquad \text{or} \qquad \|Lu\|_v \leq \frac{1}{\delta} \|u\|_u.$$

Hence L is bounded.

The *norm* $\|L\|$ of a bounded linear transformation $L\colon\; U \to V$, where U and V are normed spaces, is defined by

$$\|L\| = \inf\{M\colon\; \|Lu\|_v \leq M \|u\|_u \quad \text{for all } u \in U\}. \tag{2.64}$$

Other equivalent operator norms are

$$\|L\| = \sup\{\|Lu\|_v\colon\; \|u\|_u \leq 1\} \tag{2.65}$$

$$\|L\| = \sup\left\{\frac{\|Lu\|_v}{\|u\|_u}\colon\; \|u\|_u \neq 0\right\}.$$

Physically, the number $\|Lu\|/\|u\|$ can be viewed as the "amplification" of L at the point u, and $\|L\|$ as the maximum amplification of a system.

Example 2.32

1 Consider the integration operator defined by

$$y(t) = \int_0^t x(\tau)\, d\tau$$

from $C[0,1]$ into itself. The operator is defined on the whole space. Continuity follows from the Schwarz inequality.

2 Now consider the differential operator $D \equiv d/dt$,

$$y(t) = \frac{dx(t)}{dt} \equiv Dx(t).$$

The operator is not defined on the whole of $C[0,1]$ and when $Dx(t)$ exists, then $y(t)$ not necessarily belongs to $C[0,1]$. However, if we take the domain of D to be the linear manifold $\mathfrak{D}$ of functions with continuous first derivatives, then the range is also contained in $C[0,1]$. The operator D is not continuous from $\mathfrak{D}$ into $C[0,1]$. For example, let $x_n(t) = \sin n\pi t$; then $\|x_n\|^2 = \frac{1}{2}$, where $\|\cdot\|$ is the norm induced by the L_2-metric in Eq. (2.30b). We have $\|Dx_n\|^2 = (n\pi)^2/2$, and $\|Dx_n\|/\|x_n\|$ is unbounded as $n \to \infty$.

3 The transformation defined in part 6 of Example 2.30 is continuous (*not* bounded). Consider

$$|I(u_1, v_1) - I(u_2, v_2)| = |((u_1, v_1)) - ((u_2, v_2))|.$$

Let $u = u_1 - u_2$ and $v = v_1 - v_2$. Then

$$\begin{aligned}|((u_1, v_1)) - ((u_2, v_2))| &= |((u + u_2, v + v_2)) - ((u_2, v_2))| \\ &= |((u_2, v)) + ((u, v_2)) + ((u, v))| \\ &\leq \|u_2\| \|v_1 - v_2\| + \|v_2\| \|u_1 - u_2\| \\ &\quad + \|v_1 - v_2\| \|u_1 - u_2\|\end{aligned}$$

where in the last step the Cauchy-Schwarz inequality is invoked. Thus the continuity of the transformation follows.

4 Consider the integral operator $Tx = y$ defined by

$$y(t) = Tx(t) \equiv \int_{-\infty}^{\infty} k(t-\tau) x(\tau)\, d\tau$$

where $t \in (-\infty, \infty)$. Here T defines a linear transformation from the space $L_1(-\infty, \infty)$ of integrable functions into itself. We show that T is a continuous transformation. We have

$$\begin{aligned}\int_{-\infty}^{\infty} |y_1(t) - y_2(t)|\, dt &= \int_{-\infty}^{\infty} \left| \int_{-\infty}^{\infty} k(t-\tau)[x_1(\tau) - x_2(\tau)]\, d\tau \right| dt \\ &\leq \int_{-\infty}^{\infty} \int_{-\infty}^{\infty} |k(t-\tau)| |x_1(\tau) - x_2(\tau)|\, d\tau\, dt \\ &\leq K \int_{-\infty}^{\infty} |x_1(\tau) - x_2(\tau)|\, d\tau.\end{aligned}$$

In arriving at the last step, we interchanged the order of integration and assumed that

$$\int_{-\infty}^{\infty} |k(t-\tau)|\, dt \equiv K < \infty.$$

Thus we have

$$\|y_1 - y_2\|_{L_1(-\infty,\infty)} \leq K \|x_1 - x_2\|_{L_1(-\infty,\infty)}.$$

Hence T is continuous. ■

In finite-dimensional normed linear spaces, *all* linear transformations are continuous (hence bounded). This result is stated in the following theorem.

Theorem 2.8 Let T: $U \to V$ be a linear transformation where U and V are normed linear spaces. If U is finite dimensional, then T is continuous.

Proof Let $\{\phi_1, \phi_2, \ldots, \phi_n\}$ be a basis for U. For any $u \in U$ we have

$$u = \alpha_1\phi_1 + \alpha_2\phi_2 + \cdots + \alpha_n\phi_n$$

$$Tu = \alpha_1 T(\phi_1) + \alpha_2 T(\phi_2) + \cdots + \alpha_n T(\phi_n)$$

$$\|Tu\| \leq |\alpha_1|\,\|T(\phi_1)\| + |\alpha_2|\,\|T(\phi_2)\| + \cdots + |\alpha_n|\,\|T(\phi_n)\|$$

$$\leq C_0(|\alpha_1| + |\alpha_2| + \cdots + |\alpha_n|)$$

where $C_0 = \max\{\|T\phi_i\|,\ i = 1,2,\ldots,n\}$. By Lemma 2.1 there exists a constant C_1 such that

$$|\alpha_1| + |\alpha_2| + \cdots + |\alpha_n| \leq C_1\|u\|,$$

which gives

$$\|Tu\| \leq C_0 C_1 \|u\| = M\|u\|.$$

Hence T is bounded. Note that the range space is not required to be finite dimensional. Note also that the finite dimensionality of U is used in the proof via Lemma 2.1. ■

Example 2.33

1 Consider the linear transformation T: $\mathsf{R}^3 \to \mathsf{R}^3$ defined by $T(\mathbf{x}) = (x_1 + x_3, x_2 - x_1, x_3)$. Using the Euclidean norm, we write

$$\|T(\mathbf{x})\|^2 = (x_1 + x_3)^2 + (x_2 - x_1)^2 + x_3^2$$

$$= 2x_1^2 + x_2^2 + 2x_3^2 + 2x_1x_3 - 2x_1x_2$$

$$\leq 4x_1^2 + 2x_2^2 + 3x_3^2$$

$$\leq 4(x_1^2 + x_2^2 + x_3^2) = 4\|\mathbf{x}\|^2.$$

Thus T is bounded (with $M = 2$).

2 Consider the linear transformation T: $\mathsf{R}^2 \to \mathsf{R}^2$ represented by the matrix

$$[T] = \begin{bmatrix} t_{11} & t_{12} \\ t_{12} & t_{22} \end{bmatrix}, \qquad t_{11} > 0, \qquad t_{22} > 0.$$

We have

$$[T]\{\mathbf{x}\}=\begin{Bmatrix} t_{11}x_1+t_{12}x_2 \\ t_{12}x_1+t_{22}x_2 \end{Bmatrix}$$

$$\|T\mathbf{x}\|=\left[\left(t_{11}^2+t_{12}^2\right)x_1^2+\left(t_{12}^2+t_{22}^2\right)x_2^2+2t_{12}(t_{11}+t_{22})x_1x_2\right]^{1/2}$$

$$\|T\|=\sup\frac{\|T\mathbf{x}\|}{\|\mathbf{x}\|}=\sup\frac{\left[\left(t_{11}^2+t_{12}^2\right)x_1^2+\left(t_{12}^2+t_{22}^2\right)x_2^2+2t_{12}(t_{11}+t_{22})x_1x_2\right]^{1/2}}{\left[x_1^2+x_2^2\right]^{1/2}}$$

To simplify this expression, we use the transformation $x_1=r\cos\theta$ and $x_2=r\sin\theta$.

$$\|T\|=\sup\left[\tfrac{1}{2}\left(t_{11}^2+2t_{12}^2+t_{22}^2\right)+\tfrac{1}{2}\left(t_{11}^2-t_{22}^2\right)\cos 2\theta+t_{12}(t_{11}+t_{22})\sin 2\theta\right]^{1/2}$$

$$\leq\left[\tfrac{1}{2}\left(t_{11}^2+2t_{12}^2+t_{22}^2\right)+\tfrac{1}{2}(t_{11}+t_{22})\sqrt{(t_{11}-t_{22})^2+4t_{12}^2}\right]^{1/2}$$

$$=\tfrac{1}{2}\left[t_{11}+t_{22}+\sqrt{(t_{11}-t_{22})^2+4t_{12}^2}\right].$$

It can be easily verified that this is precisely the maximum eigenvalue of the matrix (see Exercise 2.5.23). ■

Example 2.34 The matrix operator of Example 2.30, part 5 is continuous. The continuity follows from the Schwarz inequality

$$\sqrt{\sum_{i=1}^{m}\|y_i^{(k)}-y_i\|^2}\leq\sqrt{\sum_{i=1}^{m}\sum_{j=1}^{n}a_{ij}^2}\sqrt{\sum_{j=1}^{n}\left(x_j^{(k)}-x_j\right)^2},$$

where we assumed

$$\mathbf{x}^{(k)}=(x_1,x_2,\ldots,x_n),\qquad \mathbf{y}^{(k)}=(y_1,y_2,\ldots,y_m),\qquad \mathbf{x}^{(k)}\to\mathbf{x}.$$

It follows, for $m=n$, that

$$\|A\|\leq\sqrt{\sum_{i,j=1}^{n}|a_{ij}|^2}.\quad ■$$

Bounded Below Operators A linear operator L: $U\to V$ is said to be *bounded below* if there exists a constant $C>0$ such that

$$\|Lu\|_v\geq C\|u\|_u,\qquad u\in U. \tag{2.66}$$

The following theorem, which is needed in the proof of the Lax-Milgram Theorem 2.19, establishes that a bounded below operator has a continuous inverse defined on its range. Note that L is not required to be continuous.

Theorem 2.9 (*Bounded Inverse Theorem*) Let $L: \quad U \to V$ be a linear transformation, where U and V are normed linear spaces. If L is bounded below, then L has a continuous inverse L^{-1} defined on its range $\mathcal{R}(L)$. In other words, the inverse operator L^{-1} is bounded:

$$\|L^{-1}v\|_u \leq \frac{1}{C}\|v\|_v. \tag{2.67}$$

Proof To show that L^{-1} exists, we must show that L is one-to-one, or equivalently, that $\mathcal{N}(L)=\{0\}$. This follows immediately from $\|Lu\|_v \geq C\|u\|_u$, which implies that if $u \neq 0$, then $Lu \neq 0$. To show that $L^{-1}: \quad \mathcal{R}(L) \to U$ is continuous (or bounded), let $v \in \mathcal{R}(L)$. Then there is a $u \in U$ such that $Lu=v$ and $u=L^{-1}v$. We have

$$\|L^{-1}v\|_u = \|u\|_u \leq \frac{1}{C}\|Lu\|_v = \frac{1}{C}\|v\|_v$$

for all v in $\mathcal{R}(L)$. Hence L^{-1} is bounded. ∎

Example 2.35

1 The differential operator $D=d/dx$ was shown to be unbounded from $C[0,1]$ into itself. It is also unbounded from $C_0[0,1]$ (the space of functions from $C[0,1]$ which vanish on the boundary, that is, at $x=0$ and $x=1$) itself. However, the operator is bounded below on $C_0[0,1]$ with the sup-norm. For $u \in C_0[0,1]$ we have (since $u(y)=0$ at $y=0$),

$$u(x)=\int_0^x \frac{du}{dy}\,dy \leq \sup_{x\in(0,1)}\left|\frac{du}{dx}\right| = \|Du\|_\infty$$

$$\sup_{x\in(0,1)} |u(x)| \leq \|Du\|_\infty \qquad \text{or} \qquad \|u\|_\infty \leq \|Du\|_\infty. \tag{2.68}$$

2 The differential operator D is also bounded from $L_2[0,1]$ into itself. We have

$$u(x)=\int_0^x \frac{du}{dy}\,dy \leq x\left[\int_0^x \left|\frac{du}{dy}\right|^2 dy\right]^{1/2}$$

$$\int_0^1 |u(x)|^2\,dx \leq \int_0^1 \left|\frac{du}{dx}\right|^2 dx$$

or

$$\|u\|_0 \leq \|Du\|_0 \tag{2.69}$$

where $\|\cdot\|_0$ denotes the L_2-norm. ∎

2.6.3 Orthogonal Projection Operators

Projection A *linear* transformation P of a linear vector space into itself is said to be a *projection* if $P^2 = P$.

We hasten to point out that the requirement $P^2 = P$ *does not* imply that P is linear.

Example 2.36

1 Let $V = \mathsf{R}^2$, and S_1 and S_2 be the subspaces

$$S_1 = \{\mathbf{x} \in \mathsf{R}^2:\ x_1 = 0\}, \qquad S_2 = \{\mathbf{x} \in \mathsf{R}^2:\ x_2 = 0\}.$$

Here V is the x_1x_2 plane, S_1 is the x_2 axis, and S_2 is the x_1 axis. An arbitrary element (or point) $\mathbf{x}$ in V can be uniquely expressed in the form $\mathbf{x} = \mathbf{x}^1 + \mathbf{x}^2$, where $\mathbf{x}^1 = (0, x_2) \in S_1$ and $\mathbf{x}^2 = (x_1, 0) \in S_2$. Define the mapping P by $P(\mathbf{x}^1 + \mathbf{x}^2) = \mathbf{x}^1$. Clearly, P is linear since

$$\begin{aligned} P(\alpha\mathbf{x} + \beta\mathbf{y}) &= P(\alpha\mathbf{x}^1 + \alpha\mathbf{x}^2 + \beta\mathbf{y}^1 + \beta\mathbf{y}^2) \\ &= \alpha\mathbf{x}^1 + \beta\mathbf{y}^1 = \alpha P(\mathbf{x}^1 + \mathbf{x}^2) + \beta P(\mathbf{y}^1 + \mathbf{y}^2) \\ &= \alpha P(\mathbf{x}) + \beta P(\mathbf{y}) \end{aligned}$$

for all $\mathbf{x}, \mathbf{y} \in V$, $\mathbf{x}^1, \mathbf{y}^1 \in S_1$, $\mathbf{x}^2, \mathbf{y}^2 \in S_2$, and $\alpha, \beta \in \mathsf{R}$. We see also that $P^2 = P$. Hence P is a projection. Geometrically P projects the plane V onto the subspace S_1 along the subspace S_2. We note also that $\mathcal{R}(P) = S_1$ and $\mathcal{N}(P) = S_2$.

2 Consider $V = \mathsf{R}$, and define Q by

$$Q(x) = \begin{cases} 1, & x \geq 1 \\ 0, & -1 < x < 1 \\ -1, & x \leq -1 \end{cases}.$$

We note that $Q \cdot (Q(x)) = Q(x)$. However, Q is *not* linear. Therefore Q is *not* a projection.

3 Consider a partition of a linear space V into a collection of n disjoint linear subspaces V_i, $i = 1, 2, \ldots, n$:

$$V = V_1 + V_2 + \cdots + V_n, \qquad V_i \cap V_j = \{0\} \quad \text{for} \quad i \neq j.$$

Let P_j, $j = 1, 2, \ldots, n$, be the projection on V such that $P_j(V) = V_j$, or equivalently $\mathcal{R}(P_j) = V_j$, and $\mathcal{N}(P_j) = V_1 + V_2 + \cdots + V_{j-1} + V_{j+1} + \cdots + V_n$. Clearly, P_j defines a projection. The linear combination

$$T = \sum_{j=1}^{n} \lambda_j P_j,$$

where λ_j are scalars, defines a linear transformation of V into itself. The *restriction* of T to V_j, denoted by T_j, is a mapping of V_j into itself, hence $T_j = \lambda_j I_j$, where I_j is the identity operator on V_j. Note also that

$$P_i \cdot P_j(V) = P_i(V_j) = V_j \delta_{ij}, \qquad \text{no sum.}$$

Next consider

$$TT(V) = T(\lambda_1 V_1 + \lambda_2 V_2 + \cdots + \lambda_n V_n)$$
$$= \lambda_1 V_1 + \lambda_2 V_2 + \cdots + \lambda_n V_n.$$

Thus T is not a projection unless λ_j is either 0 or 1 for $j = 1, 2, \ldots, n$. ■

Proof of the following result is left as an exercise to the reader.

Theorem 2.10 Let V be a linear vector space and S_1 and S_2 be two disjoint linear subspaces of V such that $V = S_1 + S_2$. Then there exists a projection P on V such that $\mathcal{R}(P) = S_1$ and $\mathcal{N}(P) = S_2$. Equivalently,

$$V = \mathcal{R}(P) + \mathcal{N}(P), \mathcal{R}(P) \cap \mathcal{N}(P) = \{0\}. \quad ■$$

Orthogonal Projections A projection P on an inner product space V is said to be *orthogonal* if its range and null space are orthogonal, $\mathcal{R}(P) \perp \mathcal{N}(P)$:

$$(u, v) = 0, \text{ for all } u \in \mathcal{R}(P) \text{ and } v \in \mathcal{N}(P). \tag{2.70}$$

The following theorem establishes some of the properties of orthogonal projections on inner product spaces.

Theorem 2.11 Let V be an inner product space and let P be an orthogonal projection on V. Then

(i) $\mathcal{N}(P)$ and $\mathcal{R}(P)$ are closed linear subspaces of V.

(ii) $\mathcal{N}(P) = \mathcal{R}(P)^{\perp}$ and $\mathcal{R}(P) = \mathcal{N}(P)^{\perp}$.

(iii) Each element $w \in V$ can be written uniquely as $w = u + v$, where $u \in \mathcal{R}(P)$ and $v \in \mathcal{N}(P)$.

(iv) $\| w \|^2 = \| u \|^2 + \| v \|^2$.

Proof Part (iii) of the theorem follows from the fact that P is a projection [see eq. (2.52)].

To prove (i), we first show that P is a continuous operator. Since each $w \in V$ can be uniquely expressed as $w = u + v$ with $u \in \mathcal{R}(P)$ and $v \in \mathcal{N}(P)$, and that $u \perp v$, it follows from the Pythagorean theorem that part (iv) holds. Hence $\| Pw \|^2 = \| u \|^2 \leq \| w \|^2$. Note also that the operator $I - P$, where I is the identity operator, is continuous. The proof of (i) follows from the fact that $\mathfrak{N}(P)$ and $\mathfrak{R}(P)$ are the null spaces of the continuous operators P and $I-P$, respectively (see Exercise 2.5.22).

Proof of part (ii) is next. Since $\mathfrak{N}(P)\perp\mathfrak{R}(P)$, we have $\mathfrak{N}(P)\subset\mathfrak{R}(P)^{\perp}$. Consider an element w in $\mathfrak{R}(P)^{\perp}$. Then there exists a unique $u_0\in\mathfrak{R}(P)$ and $v_0\in\mathfrak{N}(P)$ such that $w=u_0+v_0$. Since $w\in\mathfrak{R}(P)^{\perp}$, $(w,u)=0$ for all $u\in\mathfrak{R}(P)$. Hence $0=(w,u)=(u_0,u)$ for all $u\in\mathfrak{R}(P)$. This implies that $u_0=0$ and $w=v_0\in\mathfrak{N}(P)$. Thus $\mathfrak{R}(P)^{\perp}\subset\mathfrak{N}(P)$, and therefore $\mathfrak{N}(P)=\mathfrak{R}(P)^{\perp}$. A similar argument proves that $\mathfrak{R}(P)=\mathfrak{N}(P)^{\perp}$. ■

In the next section we consider the topic of representing linear transformations on finite-dimensional spaces by matrices.

2.6.4 Use of Matrices to Represent Linear Transformations

Transformations on finite-dimensional vector spaces can be represented by matrices. Let U and V be finite dimensional, and let T: $U\to V$ be linear. Let $\{\phi_1,\phi_2,\ldots,\phi_n\}$ and $\{\psi_1,\psi_2,\ldots,\psi_m\}$ be bases for U and V, respectively. Then $u\in U$ and $v\in V$ can be expressed uniquely as

$$u=\sum_{i=1}^{n}\alpha_i\phi_i,\; v=\sum_{j=1}^{m}\beta_j\psi_j. \tag{2.71}$$

Thus for any $u\in U$, $T(u)=v$ can be written as

$$T(u)\equiv\sum_{i}^{n}\alpha_i T(\phi_i)=\sum_{j}^{m}\beta_j\psi_j. \tag{2.72}$$

Since $T(\phi_i)\in V$, we can write

$$T(\phi_i)=\sum_{j}^{m}t_{ji}\psi_j. \tag{2.73}$$

We have from Eqs. (2.71) and (2.72)

$$\sum_{j}^{m}\beta_j\psi_j-\sum_{i}^{n}\alpha_i\left(\sum_{j}^{m}t_{ji}\psi_j\right)=0$$

or, since ψ_j are linearly independent, we have

$$\left(\beta_j-\sum_{i}^{n}t_{ji}\alpha_i\right)=0. \tag{2.74}$$

In matrix form we have

$$\begin{Bmatrix} \beta_1 \\ \beta_2 \\ \vdots \\ \beta_m \end{Bmatrix} = \begin{bmatrix} t_{11} & t_{12} & \cdots & t_{1n} \\ t_{21} & t_{22} & \cdots & t_{2n} \\ \vdots & \vdots & & \vdots \\ t_{m1} & & & t_{mn} \end{bmatrix} \begin{Bmatrix} \alpha_1 \\ \alpha_2 \\ \vdots \\ \alpha_n \end{Bmatrix}. \tag{2.75}$$

The matrix $[t_{ij}]$ is said to represent the linear transformation relative to the bases $\{\phi_i\}^n$ and $\{\psi_i\}^m$.

Example 2.37 Let $U = P_3$ be the space of polynomials of degree 3, and let $V = P_2$ be the space of polynomials of degree 2. Let $D = d/dx$ be the linear operator. The matrix representing the transformation D can be represented with respect to any pair of bases in P_3 and P_2.

1 Let the bases in P_3 and P_2 be

$$\{\phi_i\}^4 = \{1, x, x^2, x^3\}, \qquad \{\psi_i\}^3 = \{1, x, x^2\}.$$

Then we have, from Eq. (2.73),

$$D(\phi_1) = 0 = 0 \cdot \psi_1 + 0 \cdot \psi_2 + 0 \cdot \psi_3 = \sum_j^3 d_{j1}\psi_j$$

$$D(\phi_2) = 1 = 1 \cdot \psi_1 + 0 \cdot \psi_2 + 0 \cdot \psi_3 = \sum_j^3 d_{j2}\psi_j$$

$$D(\phi_3) = 2x = 0 \cdot \psi_1 + 2 \cdot \psi_2 + 0 \cdot \psi_3 = \sum_j^3 d_{j3}\psi_j$$

$$D(\phi_4) = 3x^2 = 0 \cdot \psi_1 + 0 \cdot \psi_2 + 3 \cdot \psi_3 = \sum_j^3 d_{j4}\psi_j$$

or

$$\begin{Bmatrix} \beta_1 \\ \beta_2 \\ \beta_3 \end{Bmatrix} = \begin{bmatrix} 0 & 1 & 0 & 0 \\ 0 & 0 & 2 & 0 \\ 0 & 0 & 0 & 3 \end{bmatrix} \begin{Bmatrix} \alpha_1 \\ \alpha_2 \\ \alpha_3 \\ \alpha_4 \end{Bmatrix}, \qquad [d_{ij}] = \begin{bmatrix} 0 & 1 & 0 & 0 \\ 0 & 0 & 2 & 0 \\ 0 & 0 & 0 & 3 \end{bmatrix}.$$

The matrix $[d]$ represents the transformation relative to the bases $\{1, x, x^2, x^3\}$ and $\{1, x, x^2\}$.

2 If we select the bases for P_3 and P_2 to be $\{1, 1+t, 1+t+t^2, 1+t+t^2+t^3\}$ and $\{1, t, t^2\}$, we get, for $D = d/dt$,

$$D(\phi_1)=0, \qquad D(\phi_2)=\psi_1, \qquad D(\phi_3)=1+2t=\psi_1+2\psi_2,$$

$$D(\phi_4)=1+2t+3t^2=\psi_1+2\psi_2+3\psi_3.$$

Hence,

$$[d_{ij}]=\begin{bmatrix} 0 & 1 & 1 & 1 \\ 0 & 0 & 2 & 2 \\ 0 & 0 & 0 & 3 \end{bmatrix}.$$

3 Let $\{\phi_i\}=\{1, 1+t, 1+t+t^2, 1+t+t^2+t^3\}$, and $\{\psi_i\}=\{1, 1+t, t+t^2\}$. Then

$$D(\phi_1)=0, \qquad D(\phi_2)=\psi_1, \qquad D(\phi_3)=1+2t=-\psi_1+2\psi_2,$$

$$D(\phi_4)=1+2t+3t^2=2\psi_1-\psi_2+3\psi_3.$$

Therefore we have

$$[d_{ij}]=\begin{bmatrix} 0 & 1 & -1 & 2 \\ 0 & 0 & 2 & -1 \\ 0 & 0 & 0 & 3 \end{bmatrix}$$

4 Recall the rotation transformation of the plane through an angle θ, in which the point (x, y) is mapped into the point (x', y') (see Example 2.9). The transformation T: $\mathsf{R}^2 \to \mathsf{R}^2$ can be represented by

$$T(x, y)=(x\cos\theta - y\sin\theta, x\sin\theta + y\cos\theta).$$

Note that

$$T(\hat{\mathbf{e}}_1)\equiv T(1,0)=(\cos\theta, \sin\theta)=\cos\theta\hat{\mathbf{e}}_1+\sin\theta\hat{\mathbf{e}}_2$$

$$T(\hat{\mathbf{e}}_2)\equiv T(0,1)=(-\sin\theta, \cos\theta)=-\sin\theta\hat{\mathbf{e}}_1+\cos\theta\hat{\mathbf{e}}_2.$$

Hence the matrix T relative to the basis $(\hat{\mathbf{e}}_1, \hat{\mathbf{e}}_2)$ is given by

$$[t_{ij}]=\begin{bmatrix} \cos\theta & \sin\theta \\ -\sin\theta & \cos\theta \end{bmatrix}.$$

It turns out that the inverse of the matrix $[T]$ is equal to its transpose. ■

2.6.5 Linear Functionals, Bilinear Forms, and Quadratic Forms

Linear transformations that map a given linear vector space or a product of two linear vector spaces into the real numbers R are of considerable interest in the study of variational solutions of operator equations.

Linear Functional Let U be a linear vector space over the real number field R. A linear transformation l of U into R is called a *linear form* or *linear functional.*

Since linear functionals are a special case of linear transformations, the concepts and results given in the preceding pages remain valid for linear functionals. The set of all linear functionals on a linear vector space is itself a vector space, called the *dual* or *conjugate space* of U. The dual space is denoted by U'. When U is a finite-dimensional vector space of dimension n, then the dual space U' is a finite-dimensional vector space of dimension n.

The basis of the dual space U' of U has a special relation to the basis of U. Let $\{\phi_i\}$ be the basis of U. Define the linear functional ψ^i by

$$u=\sum_{i=1}^{n}\alpha^i\phi_i,\qquad \psi^i(u)=\alpha^i\in\mathsf{R}. \tag{2.76}$$

We shall call ψ^i the ith coordinate function. We can show that ψ^i is a linear functional. Indeed, for $u=\Sigma_i\alpha^i\phi_i$, $v=\Sigma_i\beta^i\phi_i$, and for any scalars μ and λ, we have

$$\psi^j(\mu u+\lambda v)=\psi^j\Big(\sum_i(\mu\alpha^i+\lambda\beta^i)\phi_i\Big)=\mu\alpha^j+\lambda\beta^j=\mu\psi^j(u)+\lambda\psi^j(v).$$

Note that the relationship between $\{\psi^i)$ and $\{\phi_i\}$ is characterized by

$$\psi^i(\phi_j)=\delta^i_j,\qquad \text{for all } i,\ j. \tag{2.77}$$

Since ϕ_j are linearly independent, ψ^i are also linearly independent. Therefore $\{\psi^i\}$ form a basis for the dual space U', and we shall call $\{\psi^i\}$ the *dual* (or *conjugate*) *basis.*

Bilinear Forms Let U and V be two vector spaces with the same field of scalars. The mapping B of pairs (u,v), $u\in U, v\in V$, into the field of scalars such that

$$B(\alpha u_1+\beta u_2,\mu v_1+\lambda v_2)=\alpha\mu B(u_1,v_1)$$
$$+\alpha\lambda B(u_1,v_2)+\beta\mu B(u_2,v_1)+\beta\lambda B(u_2,v_2) \tag{2.78}$$

for all u_1, $u_2\in U$, v_1, $v_2\in V$, and scalars α, β, μ, and λ is called a *bilinear form.* That is, $B\colon\ U\times V\to\mathsf{R}$.

A simple example of the bilinear form is the inner product, in which $U=V$ and $B(u,v)=(u,v)$.

Analogous to the representation of linear operators on finite-dimensional spaces, one can represent bilinear forms on finite-dimensional spaces by matrices. Consider a bilinear form $B(\cdot,\cdot)$ on finite-dimensional vector spaces U and V. Let U be an n-dimensional and V be an m-dimensional vector space. Let ϕ_i be a basis for U and let ψ_i be a basis for V. For any u in U and v in V, we have

$$u=\sum_{i=1}^{m}\alpha_i\phi_i, \qquad v=\sum_{i=1}^{n}\beta_i\psi_i \tag{2.79}$$

for $\alpha_i, \beta_i \in \mathrm{R}$. Then

$$\begin{aligned} B(u,v)&=B\left(\sum_i^m \alpha_i\phi_i, \sum_j^n \beta_j\psi_j\right)=\sum_i^m\sum_j^n \alpha_i\beta_j B(\phi_i,\psi_j)\\ &=\sum_i^m\sum_j^n\alpha_i\beta_j b_{ij}=\{\alpha\}^T[B]\{\beta\}, \qquad b_{ij}=B(\phi_i,\psi_j), \end{aligned} \tag{2.80}$$

The matrix $[B]=[b_{ij}]$ represents the bilinear form $B(u,v)$ with respect to the bases $\{\phi_i\}$ and $\{\psi_i\}$.

A bilinear form $B(\cdot,\cdot)$: $U\times U\to \mathrm{R}$ is said to be *symmetric* if $B(u,v)=B(v,u)$ for all $u,v\in U$. If $B(u,u)=0$ for all $u\in U$, we say that the bilinear form is *skew-symmetric*. Note that if $B(\cdot,\cdot)$ is symmetric, we have (see Section 1.3.7),

$$b_{ij}=B(\phi_i,\phi_j)=B(\phi_j,\phi_i)=b_{ji}\to[B]=[B]^T. \tag{2.81}$$

Similarly, if $B(\cdot,\cdot)$ is skew-symmetric, we have

$$0=B(u+v,u+v)=B(u,u)+B(u,v)+B(v,u)+B(v,v)$$

and hence $[B]^T=-[B]$.

Every bilinear form can be represented uniquely as a sum of a symmetric bilinear form and a skew-symmetric bilinear form (see Sec. 1.5.6),

$$B(u,v)=\tfrac{1}{2}[B(u,v)+B(v,u)]+\tfrac{1}{2}[B(u,v)-B(v,u)]=B_s(u,v)+B_{ss}(u,v). \tag{2.82}$$

Quadratic Form Let U be a vector space, and let $B(\cdot,\cdot)$ be a bilinear form on U. A *quadratic form* is a functional Q on U defined by setting

$$Q(u)=B(u,u). \tag{2.83}$$

Note that if $B(\cdot,\cdot)$ is represented as the sum of a symmetric and a skew-symmetric bilinear form, we can write

$$Q(u)=B(u,u)=B_s(u,u)+B_{ss}(u,u)=B_s(u,u). \qquad (2.84)$$

That is, $Q(u)$ is completely determined by the symmetric part of the bilinear form. Also, two different bilinear forms with the same symmetric part must generate the same quadratic form.

Example 2.38

1 The linear functionals in R^n are of the form,

$$f(\mathbf{x})=\sum_{i=1}^{n} f_i x_i,$$

where x_i are components of the vector $\mathbf{x}$ in R^n, and f_i are any numbers which can be thought of as components of a vector $\mathbf{f}$. Clearly, the scalar product of ordinary vectors is a special case of the above statement. Indeed, every bounded linear functional can be written as a scalar product,

$$f(\mathbf{x})=(\mathbf{f},\mathbf{x}).$$

2 Let $V=L_2(0,1)$ and define a linear functional on V by

$$l(v)=\int_0^1 f(\mathbf{x})v(\mathbf{x})\,dx$$

where f is an arbitrary function. The functional l is linear because it is homogeneous,

$$l(\alpha v)=\int_0^1 f(\mathbf{x})\alpha v(\mathbf{x})\,dx=\alpha\int_0^1 f(\mathbf{x})v(\mathbf{x})\,dx=\alpha l(v),$$

and additive,

$$l(v_1+v_2)=\int_0^1 f(\mathbf{x})(v_1+v_2)\,dx=\int_0^1 f(\mathbf{x})v_1(\mathbf{x})\,dx+\int_0^1 f(\mathbf{x})v_2(\mathbf{x})\,dx$$
$$=l(v_1)+l(v_2).$$

The functional is also continuous. If $\|v_n-v\|\to 0$, then

$$|l(v_n)-l(v)|=|l(v_n-v)|\leq\|f\|\,\|v_n-v\|.$$

Since $\|f\|$ is independent of v and v_n, it follows that l is continuous.

3 Consider the bilinear form defined by

$$B(u,v)=\int_\Omega (c_1 \operatorname{grad} u \cdot \operatorname{grad} v + c_0 uv)\,dx, \qquad c_1, c_0 = \text{constants}$$

on $H_0^1(\Omega)\times H_0^1(\Omega)$, where Ω is a bounded domain in R^3, and $H_0^1(\Omega)$ is the Hilbert space defined in Example 2.29. It is easy to verify that $B(\cdot,\cdot)$ is linear in its arguments. We claim that $B(\cdot,\cdot)$ is also continuous. Indeed, by appealing to the Schwarz inequality, we have

$$|B(u,v)| \le M\left[\int_\Omega (|\operatorname{grad} u|^2+|u|^2)\,dx\right]^{1/2}\left[\int_\Omega (|\operatorname{grad} v|^2+|v|^2)\,dx\right]^{1/2}$$

$$= M\|u\|_1\|v\|_1, \qquad M=\max(c_1,c_0)$$

where $\|\cdot\|_1$ denotes the $H^1(\Omega)$-norm in Eq. (2.56). Further, the associated quadratic form is bounded below (or positive definite):

$$Q(u)=B(u,u)=\int_\Omega (c_1|\operatorname{grad} u|^2+c_0|u|^2)\,dx$$

$$\ge \mu\|u\|_1^2, \qquad \mu=\min(c_1,c_0).$$

The continuity and positive-definite properties play a crucial role in the existence and uniqueness of solutions to functional equations, as will be seen in Section 2.8.4. ■

Exercises 2.5

1 Determine whether the following transformations are linear. (Check for homogeneity and additivity in each case.)

(a) T: $U=C[0,T]\to V=C[0,T]$,

$$Tu=\begin{cases} u_0, & u(t)\ge u_0 \\ u(t), & -u_0<u(t)<u_0 \\ -u_0, & u(t)\le -u_0 \end{cases}$$

where $u_0>0$ is a real number.

(b) T: $\mathsf{R}^3\to\mathsf{R}^3$, $\quad T(\mathbf{x})=(2+x_1-x_2, x_2-x_1, x_3)$.

(c) T: $\mathsf{R}^3\to\mathsf{R}^3$, $\quad T(\mathbf{x})=\lambda\mathbf{x}$, $\quad \lambda$ is a real number.

2 Describe the range of the transformations in **(a)** of Exercise 1. Is T onto and/or one-to-one?

3 Describe the linear transformation T of R^3 that maps the unit orthogonal vectors $\hat{\mathbf{e}}_i$ onto (1,0,0), (1,1,0), (1,1,1), respectively.

4 Is the transformation T: $C[0,\infty)\to C[0,\infty)$,

$$T(x(t))=\int_0^t K(t-s)x(s)\,ds, \qquad K(t)=e^{-t}-e^{-2t}$$

one-to-one and/or onto?

5 Describe the null spaces of transformations in Exercise 1.

6 Prove that the range and kernel of a linear transformation are linear vector spaces.

7 Find the null space of the linear transformation T: $\mathsf{R}^3\to\mathsf{R}^3$ described by the algebraic equations:

$$\begin{aligned} 3x_1-2x_2 \qquad &= y_1\\ -2x_1+5x_2-3x_3 &= y_2\\ -3x_2+4x_3 &= y_3. \end{aligned}$$

8 Let T: $U\to V$ be a linear transformation. Show that the dimensions of the null space $\mathfrak{N}(T)$, the range space $\mathfrak{R}(T)$, and the domain U are related by

$$\dim\{\mathfrak{N}(T)\}+\dim\{\mathfrak{R}(T)\}=\dim\{U\}.$$

9 Show that the dimension of the dual space V' of a finite-dimensional linear vector space V is the same as that of V.

10 Let T: $U\to V$ be a linear transformation of U and V, where both U and V are finite-dimensional linear spaces. Show that:

(a) T: $U\to V$ is *onto* if and only if $\dim\{\mathfrak{R}(T)\}=\dim\{V\}$.

(b) T: $U\to V$ is *one-to-one* if and only if $\dim\{\mathfrak{R}(T)\}=\dim\{U\}$.

(c) T: $U\to V$ is invertible if and only if T is one-to-one and onto.

11 If the rank of a linear transformation T of a linear space is 1, prove that $T(Tu)=\lambda(Tu)$ for any $u\in V$ and associated real number λ.

12 Let U and V be n-dimensional vector spaces. Then, if there exist linear transformations T: $U\to V$ and L: $V\to U$ such that $LT(u)=L(Tu)=u$ for any $u\in U$, prove that T and L are isomorphisms (i.e., one-to-one mappings) onto V and U, respectively.

13 Let T_1 and T_2 be linear operators defined on R^3 by $T_1(\mathbf{x})=(0, x_1-x_3, x_2+x_3)$, $T_2(\mathbf{x})=(x_2-x_3, 2x_1-x_2, x_1)$ determine:

(a) T_1+T_2.

(b) T_1T_2.

(c) T_2T_1.

14 Let $D = d/dt$ be the differential operator on the space P of polynomials in a real variable t. Suppose that the transformation T is defined on P by $Tp(t) = tp(t)$ for any $p(t) \in P$. Show that:

(a) $DT = TD + I$.

(b) $(TD)^2 = T^2D^2 + TD$.

Here I is the identity operator.

15 Find the matrix representation, relative to the Cartesian base vectors $\{\hat{e}_i\}$ of R^3, of each of the transformations in Exercise 13.

16 If

$$[T] = \begin{bmatrix} 1 & 2 & 1 \\ 0 & -1 & 2 \\ -1 & 2 & 3 \end{bmatrix}$$

is the matrix of a linear operator T on R^3 relative to $\{\hat{e}_i\}$, find $T(\mathbf{x})$ if

(a) $\mathbf{x} = (2, -2, 3)$.

(b) $\mathbf{x} = (-1, 2, 4)$.

(c) $\mathbf{x} = (1, -2, 0)$.

17 Show that the integral operator of Example 2.32 is a continuous linear transformation from $L_2(-\infty, \infty)$ into itself.

18 If T: $U \to V$ is a continuous linear transformation, then prove that the null space of the transformation is a closed subspace.

19 Prove that the following transformations T: $\mathsf{R}^2 \to \mathsf{R}^2$ are continuous:

(a) $T(\mathbf{x}) = (2x_1 + 3x_2, x_1 - 2x_2)$.

(b) $T(\mathbf{x}) = 2x_1 + 2x_2, 2x_1 + x_2)$.

(c) $T(\mathbf{x}) = (2x_1 + x_2, x_1 + 2x_2)$.

20 Find the operator norms of the linear operators in Exercise 19.

21 Let V be the normed linear space of all bounded continuous functions defined on $[0, \infty)$ with the sup-norm $\|u\|_\infty$. Let L be the operator defined by

$$(Lu)(x) = L(u(x)) = \frac{1}{x}\int_0^x u(t)\,dt.$$

Show that L is linear and continuous.

22 (*Closed Operators*) Let U and V be two Banach spaces, and let S be a linear subspace of U. A linear operator T: $U \to V$ is said to be *closed* if, whenever $u_n \to u$ in U and $v_n = Tu_n \to v$ in V, one has a $u \in S$ and $v = Tu$. Show that every bounded linear operator is closed. Also, show that the null space of every bounded linear operator is closed.

23 Show that the maximum eigenvalue of the matrix

$$\begin{bmatrix} t_{11} & t_{12} \\ t_{12} & t_{22} \end{bmatrix}, \qquad t_{11}>0, \qquad t_{22}>0$$

is given by

$$\lambda_{\max}=\frac{1}{2}\left[t_{11}+t_{22}+\sqrt{(t_{11}-t_{22})^2+4t_{12}^2}\,\right].$$

24 Compute the operator norm and the maximum eigenvalue of the matrix

$$\begin{bmatrix} t_{11} & t_{12} \\ t_{21} & t_{22} \end{bmatrix}, \qquad t_{11}>0, \qquad t_{22}>0, \qquad t_{12}\neq t_{21}.$$

Use the results to compute the norm and eigenvalues of the transformations in Exercise 13.

25 A linear transformation T on a Euclidean space V is said to be *orthogonal* if $(Tu, Tu)=(u, u)$ for any $u\in V$. That is, a linear operator is orthogonal if it preserves the length of every vector in space. Show that:

(a) Orthogonal transformation preserves inner products and also "angle" between any two vectors of the space.

(b) The set of orthogonal linear transformations on a vector space V is itself a linear space.

26 Let $V=L_2[-\pi,\pi]$. Show that

$$P(x)=\int_{-\pi}^{\pi} K(t,x)x(s)\,ds, \qquad K(t,s)=\frac{1}{2\pi}\sum_{n=-10}^{n=10} e^{in(t-s)}$$

represents a projection on V. Here $i=\sqrt{-1}$.

27 A projection P: $U\rightarrow S$, where U is a Hilbert space and S is a subspace of U, is said to be an *orthogonal projection* if, for all $u\in U$ and all $v\in S$,

$$(u-Pu, v)=0.$$

Prove that:

(a) The orthogonal projection P of a Hilbert space into a subspace is unique.

(b) The norm of the remainder, $\|u-Pu\|$, is the smallest distance from u to the subspace.

28 Let P be a projection operator on a linear space V.

(a) Show that $I-P$, where I is the identity operator, is also a projection.

(b) What is the kernel of $I-P$?

(c) Show that $P(I-P)=0$.

29 Let $\{P_1, P_2, \ldots, P_n\}$ be the collection of orthogonal projections with $P_iP_j=0$ for $i \neq j$. Show that

$$T=\sum_{i=1}^{m} P_i$$

is an orthogonal projection.

30 Let S be a linear subspace of $L_2[0,1]$ whose elements are all constants in the interval $[0,1]$. Decompose the function $u(x) \in L_2[0,1]$ into the sum $u=\alpha+v$ where $\alpha \in S$ and $v(x) \in S^{\perp}$. Take $u(x)$ to be:

(a) $u(x)=x^2$.

(b) $u(x)=1-x$.

(c) $u(x)=x(1-x)$.

31 Let $V=\mathsf{R}^3$, and consider the transformation $T(\mathbf{x})=(x_1, x_2, 0)$. Show that:

(a) T is a projection (of R^3 to R^2).

(b) $\mathcal{R}(T)=\mathsf{R}^2$.

(c) $\|\mathbf{x}_0-T\mathbf{x}_0\|<\|\mathbf{x}_0-\mathbf{x}\|$ for all $\mathbf{x} \neq T\mathbf{x}_0$ in R^2, where $\|\cdot\|$ denotes the Euclidean norm. Interpret the result geometrically.

32 Let $X=Y$ denote the space of all fourth-degree polynomials in t and define $T\colon\ X \to Y$ by

$$Tx=\frac{d^2x}{dt^2}+2\frac{dx}{dt}+x.$$

Represent T by a matrix with respect to the basis $\{1, t, t^2, t^3, t^4\}$. Represent T^2 by a matrix and verify that $[T^2]=[T][T]$.

33 Find the matrix representing the linear transformation from R^4 into R^3 with respect to the standard bases of R^4 and R^3:

$$T(x_1, x_2, x_3, x_4)=(x_1+2x_2, 2x_2+x_3, x_3+x_1).$$

34 Let D be the first-order differential operator on $V=\{a_0+a_1x+a_2x^2, a_i$ in $\mathsf{R}\}$. Find the matrix that represents D relative to the basis

$$p_1(x)=3-x, \qquad p_2(x)=3-x^2, \qquad p_3(x)=x+x^2-7.$$

35 Consider the functional

$$B(u,v)=\frac{1}{4}(\|u+v\|^2-\|u-v\|^2)$$

on an inner product space. Show that:

(a) $B(\cdot,\cdot)$ is bilinear in u and v.

(b) For fixed u, $f_u(v)=B(u,v)$ is a bounded linear functional.

(c) $B(u,v)$ satisfies the axioms of an inner product.

36 Consider the bilinear and linear forms

$$B(u,v)=\int_0^L \alpha \frac{d^2u}{dx^2}\frac{d^2v}{dx^2}\,dx, \qquad l(v)=\int_0^1 \beta v(x)\,dx.$$

Determine the associated matrices with respect to the basis

$$\phi_1=1-3\xi^2+2\xi^3, \qquad \phi_2=-\xi L(1-\xi)^2, \qquad \phi_3=3\xi^2-2\xi^3,$$

$$\phi_4=-\xi L(\xi^2-\xi), \qquad \xi=x/L.$$

37 Let V be a vector space and V' its (algebraic) conjugate with respect to the linear functional l. Show that if

$$\langle u,l\rangle=0 \qquad \text{for all } l\in V',$$

then $u=0$. Here $\langle u,l\rangle\equiv l(u)$.

38 (*Transpose of a Linear Transformation*) Let T: $U\to V$ be a linear transformation and U' and V' the algebraic conjugate of U and V, respectively. A linear transformation L: $V'\to U'$ such that

$$\langle u, Lv'\rangle_u=\langle Tu, v'\rangle_v$$

for all $u\in U$, $v'\in V'$ is called the transpose of T, and is denoted by $L=T^t$. Again $\langle\cdot,\cdot\rangle_x$ denotes the duality pairing of elements from X and X' (note the order). Show that the transpose of a linear operator is unique.

39 (*Continuation of Exercise* 38) Let $U\equiv \mathrm{R}^4$ and $V\equiv\mathrm{R}^3$, and define the linear transformation T by the following matrix equation:

$$\begin{Bmatrix} v_1\\ v_2\\ v_3\end{Bmatrix}=\begin{bmatrix}1&2&0&0\\0&2&1&0\\1&0&2&3\end{bmatrix}\begin{Bmatrix}u_1\\u_2\\u_3\\u_4\end{Bmatrix}.$$

Since the dimensions of a linear space and its algebraic conjugate are the same in the finite-dimensional case (see Exercise 9), we have $U'=\mathsf{R}^4$ and $V'=\mathsf{R}^3$. Select bases in U' and V' as follows. For $u=(u_1,u_2,u_3,u_4)\in U$,

$$\langle u,\psi_i\rangle \equiv \psi_i(u)=u_i, \qquad i=1,2,3,4,$$

and for $v=(v_1,v_2,v_3)\in V$,

$$\langle v,\phi_i\rangle \equiv \phi_i(v)=v_i, \qquad i=1,2,3.$$

Show that the transpose operator T' is indeed represented by the transpose of the matrix representing T.

40 Let V be a linear vector space. Then for fixed $u\in V$ prove that the following define linear functionals on V for all $v\in V$:

(a) $l_1(v)\equiv(u,v)$.

(b) $l_2(v)=(Tu,v)$.

Here $(\cdot,\cdot)$ is the inner product on V, and T is a fixed linear operator on V.

41 Determine which of the following operators represent bilinear forms:

(a) B: $\mathsf{R}^2\times\mathsf{R}^2\to\mathsf{R}$, $B(\mathbf{x},\mathbf{y})=(x_1+y_1)^2-(x_1-y_1)^2$.

(b) B: $\mathsf{R}^2\times\mathsf{R}^2\to\mathsf{R}$, $B(\mathbf{x},\mathbf{y})=x_1y_2-x_2y_1$.

(c) B: $C^1[0,1]\times C^1[0,1]\to\mathsf{R}$, $B(u,v)=\int_0^1 \frac{d}{dt}u(1-t)\cdot v(t)\,dt$. Is $B(u,v)$ symmetric?

(d) B: $U\times U\to\mathsf{R}$, $B(u,v)=(Tu,v)-(f,u)$, where T is a linear operator, T: $U\to U$, and f is a fixed element in U.

42 Represent the following bilinear forms on finite-dimensional vector spaces by matrices:

(a) B: $\mathsf{R}^2\times\mathsf{R}^3$, $B(\mathbf{x},\mathbf{y})=x_1y_1-3x_1y_3+2x_1y_2-x_2y_1+x_2y_2$.

(b) B: $\mathsf{R}^3\times\mathsf{R}^3$, $B(\mathbf{x},\mathbf{y})=2x_1y_1+4x_2y_2-3x_1y_3+5x_2y_3+x_3y_1$.

(c) B: $\mathsf{R}^2\times\mathsf{R}^2$, $B(\mathbf{x},\mathbf{y})=x_1^2+y_1^2+x_1y_2-x_2y_1$.

43 Express the bilinear forms associated with the following matrices. (Identify the spaces on which the bilinear form is defined.)

(a) $\begin{bmatrix} 1 & -1 & 2 \\ 1 & 1 & 1 \\ 2 & 0 & 1 \end{bmatrix}$.

(b) $\begin{bmatrix} 2 & 1 & 1 \\ 1 & -1 & 2 \\ 3 & 1 & 0 \end{bmatrix}$.

(c) $\begin{bmatrix} 1 & -1 & 1 \\ 2 & 1 & 2 \\ 3 & 2 & 1 \end{bmatrix}$.

44 Find the symmetric matrix representing each of the following quadratic forms:

(a) $q(x_1, x_2, x_3) = x_1^2 + 2x_1x_2 + 4x_1x_3 + 3x_2^2 + 6x_3^2 + x_2x_3$.

(b) $q(x_1, x_2) = 3x_1^2 + 4x_2^2 - 2x_1x_2$.

(c) $q(x_1, x_2) = ax_1^2 + bx_2^2 + cx_1x_2$.

45 Let B: $\mathsf{R}^n \times \mathsf{R}^n \to \mathsf{R}$. Under what conditions does it follow that $\{\mathbf{x}\}^T[B]\{\mathbf{x}\} = 0$ for all $\{\mathbf{x}\}$?

2.7 ADDITIONAL CONCEPTS FROM HILBERT SPACE THEORY

As pointed out in Section 2.5, Hilbert spaces resemble much the familiar two-and three-dimensional Euclidean spaces. Since most approximate methods seek solutions to operator equations in finite-dimensional linear spaces (which are complete), it is of interest to study certain additional properties of Hilbert spaces. Many of the results to be presented for infinite-dimensional Hilbert spaces, especially $L_2(\Omega)$, will apply to finite-dimensional spaces.

2.7.1 Projection Theorem

We first recall the definition of an orthogonal complement. Let V be an inner product space and let S be any subset of V. The orthogonal complement of S, denoted by $S^\perp$, is made up of all elements that are orthogonal to every element in S. If S is an empty subset then obviously $S^\perp = V$. If V is complete, then the orthogonal complement $S^\perp$ of any subset S of V is also a complete subspace of V. Note that $S^\perp$ is defined for any set S (whether a linear space or not), and it is a linear subspace of V. When S is a linear subspace of a Hilbert space H, the orthogonal complement $S^\perp$ possesses the following properties.

Theorem 2.12 Let H be a Hilbert space and S be a linear subspace of H. Then one has:

(i) $S^\perp = \{0\}$ if and only if S is dense in H.

(ii) If S is closed and $S^\perp = \{0\}$, then $S = H$.

(iii) $S^{\perp\perp} \equiv (S^\perp)^\perp = \bar{S}$, where $\bar{S}$ is the closure of S.

(iv) If S is closed, then $S^{\perp\perp} = S$.

Proof

(i) Let S be dense in H (see Exercise 2.4.21). For all $u \in S$ and $v \in S^\perp$ we have by the Pythagorean theorem

$$\|u - v\|^2 = \|u\|^2 + \|v\|^2 \geq \|v\|^2.$$

Since S is dense in H, we can choose u such that $\|u - v\|$ can be made arbitrarily small (for every $v \in S^{\perp} \subset H$). Hence $v = 0$ or $S^{\perp} = \{0\}$. The necessity can be proved by considering $S^{\perp} = \{0\}$. We have $S^{\perp\perp} = \{0\}^{\perp} = H$. By part (iii) (to be proved), $H = \overline{S}$, and hence S is closed.

(ii) Follows directly from **(i)**.

(iii) Since $M \subset \overline{M}$, $M \subset M^{\perp\perp}$, and $M^{\perp\perp}$ is closed, we have $\overline{M} \subset M^{\perp\perp}$. Now let us show that $M^{\perp\perp} \subset \overline{M}$. Let $w \in M^{\perp\perp}$. Then $w \perp (M^{\perp\perp})^{\perp} = M^{\perp}$. Since $\overline{M}^{\perp} \subset M^{\perp}$, we have $w \in \overline{M}^{\perp}$ and hence $w \in \overline{M}$. Since w is an arbitrary element in $M^{\perp\perp}$, it follows that $M^{\perp\perp} \subset \overline{M}$. Thus we have $\overline{M} = M^{\perp\perp}$.

(iv) Follows directly from **(iii)**. ■

We now state and prove the projection theorem, which plays a central role in the theory and applications of Hilbert spaces.

Theorem 2.13 (*The Projection Theorem*) Let S be any closed linear subspace of a Hilbert space H. Then the following statements are valid:

(i) $H = S + S^{\perp}$.

(ii) Each $u \in H$ can be expressed uniquely as $u = v + w$, where $v \in S$ and $w \in S^{\perp}$.

(iii) There is one and only one orthogonal projection P with $\mathcal{R}(P) = S$.

Proof

(i) First we prove that $W = S + S^{\perp}$ is a closed linear subspace of H. Let $\{w_n\}$ be a convergent sequence in W with the limit w. If we show that w is in W, then we have shown that W is closed (hence complete). By definition, w_n is of the form $w_n = u_n + v_n$, where $u_n \in S$ and $v_n \in S^{\perp}$. Since $S \perp S^{\perp}$, we also know that $u_n \perp v_n$, and

$$\| w_n - w_m \|^2 = \| u_n - u_m \|^2 + \| v_n - v_m \|^2 \geq \| u_n - u_m \|^2$$

by the Pythagorean theorem. Hence $\{u_n\}$ and $\{v_n\}$ are Cauchy sequences in S and $S^{\perp}$, respectively. Since S and $S^{\perp}$ are complete, the limits u and v of the sequences $\{u_n\}$ and $\{v_n\}$, respectively, exist and $u \in S$ and $v \in S^{\perp}$. Then it follows from the continuity of addition that $w = u + v$ and that $W = S + S^{\perp}$ is closed.

Since $S \subset W$ and $S^{\perp} \subset W$, we have, from Exercise 2.4.21, that $W^{\perp} \subset S^{\perp}$ and $W^{\perp} \subset S^{\perp\perp}$. This implies that $W^{\perp} \subset (S^{\perp} \cap S^{\perp\perp})$. Again by Exercise 2.4.21 it follows that $W^{\perp} = \{0\}$, proving that S and $S^{\perp}$ are disjoint. Part (i) follows immediately from part (ii) of Theorem 2.12.

(ii) Since S and $S^{\perp}$ are disjoint, each $w \in W = S + S^{\perp}$ can be expressed as $w = u + v$, where $u \in S$ and $v \in S^{\perp}$. To prove that this representation is unique, assume that

$$w=u_1+v_1=u_2+v_2, \qquad \text{for } u_1, u_2 \in S, \qquad v_1, v_2 \in S^{\perp}.$$

Then $u_1-u_2=v_2-v_1$, and since $u_1-u_2 \in S$, $v_2-v_1 \in S^{\perp}$, and $S \cap S^{\perp}=\{0\}$, it follows that $u_1-u_2=0$ and $v_1-v_2=0$. Hence the representation is unique. Also by the Pythagorean theorem one has

$$\|w\|^2=\|u\|^2+\|v\|^2. \tag{2.85}$$

(iii) Define the operator P: $H \to H (H=S+S^{\perp})$ by

$$P(u+v)=u$$

for all $u \in S$ and $v \in S^{\perp}$. It can be easily verified that P is an orthogonal projection with $\mathcal{R}(P)=S$. We wish to show that P is unique. If Q is another orthogonal projection with $\mathcal{R}(Q)=S$, then one has $\mathcal{N}(Q)=S^{\perp}$ and

$$P(u)=u=Q(u), \qquad u \in S$$

$$P(v)=0=Q(v), \qquad v \in S^{\perp}.$$

Therefore $P(u+v)=Q(u+v)$ for all $w=u+v \in H$. This implies that $P=Q$. This completes the proof of the theorem. ■

It should be pointed out that the completeness of H is essential in the above theorem. The completeness of H was used in proving that $S+S^{\perp}$ is a closed linear subspace of H. Note also that the name projection theorem is derived from the fact that the projection operator P partitions H into S and $S^{\perp}$. When H is not complete, there may not exist an orthogonal projection operator with the property $\mathcal{R}(P)=S$, where S is a closed subspace of H. On the other hand if S is not closed, then $S+S^{\perp}$ is not closed, and we cannot establish that $H=S+S^{\perp}$ and $S \cap S^{\perp}=\{0\}$. (Recall from Theorem 2.4 that every finite-dimensional linear subspace of a normed space is closed.)

Next we present a result that plays an important role in approximation theory.

Theorem 2.14 Let H be a Hilbert space H, and let S be a closed linear subspace of H. Further, let P be the orthogonal projection with $\mathcal{R}(P)=S$. Then for every $u_0 \in H$ the following inequality holds:

$$\|u_0-Pu_0\| \leq \|u_0-u\| \tag{2.86}$$

or, equivalently,

$$\|u_0-Pu_0\|=\inf\{\|u_0-u\|: \quad u \in S\}.$$

In other words, for every element $u_0 \in H$ the nearest element in S is given by Pu_0.

Proof Since $u_0 = Pu_0 + v$, where $Pu_0 \in S$ and $v \in S^{\perp}$, we have $v = u_0 - Pu_0 \in S^{\perp}$. For all $u \in S$ we have $Pu_0 - u \in S$ and therefore

$$(u_0 - Pu_0, Pu_0 - u)_H = 0.$$

Hence,

$$\begin{aligned}\|u_0 - u\|^2 &= \|u_0 - Pu_0 + Pu_0 - u\|^2 \\ &= \|u_0 - Pu_0\|^2 + \|Pu_0 - u\|^2 + 2(u_0 - Pu_0, Pu_0 - u) \\ &\geq \|Pu_0 - u_0\|^2.\end{aligned}$$

The inequality is a strict inequality except for $u = u_0 = Pu_0$. Note also that the projection operator P is continuous, we also have

$$\|u_0 - u\| \geq \|Pu_0 - u\| = \|P(u_0 - u)\|,$$

where we have used the fact that $P(u) = u$ for all $u \in S$. ■

2.7.2 Continuous Linear Functionals and Adjoint Operators in Hilbert Spaces

The concept of linear functionals on an inner product space was discussed in Section 2.6. Since a linear functional is a special kind of linear operator, the concept of continuity for linear functionals is the same as that for linear operators. In a normed linear vector space the concepts of continuity and boundedness are equivalent for linear functionals.

Bounded linear functionals on Hilbert spaces have simple representations. Consider an inner product space V and let v_0 be a fixed element in V. Consider the operator l defined by

$$l(u) = (v_0, u)$$

for every u in V. Clearly, l is a functional. It is linear, since

$$\begin{aligned}l(\alpha u + \beta v) &= (v_0, \alpha u + \beta v) = \alpha(v_0, u) + \beta(v_0, v) \\ &= \alpha l(u) + \beta l(v).\end{aligned}$$

Also, by the Schwarz inequality we get

$$|l(u)| = |(v_0, u)| \leq \|v_0\| \|u\| = M \|u\|$$

where $M=\|v_0\|$. Hence l is bounded. Further, note that $\|l\|\leq\|v_0\|$. However, we also have $|l(v_0)|=\|v_0\|^2$ or $\|l\|\geq\|v_0\|$. Thus the operator norm of l is given by $\|l\|=\|v_0\|$. From this discussion it is clear that we can associate a bounded linear functional with each element of an inner product space. However, the converse is not true in general. The following theorem, whose proof is not included here, states that the converse holds in Hilbert spaces.

Theorem 2.15 (*Reisz Representation Theorem*) Let l be a bounded linear functional on a Hilbert space H. Then there is a unique vector v_0 in H such that

$$l(u)=(v_0,u) \tag{2.87}$$

for all u in H. ■

The vector $v_0\in H$ is called the *representation* of l. However, it should be noted that l is a linear functional on H while v_0 is an element in H.

Next we discuss the notion of the adjoint operators.

Consider a bounded operator T on a Hilbert space H (i.e., T: $H\rightarrow H$). For a fixed element v in H, the inner product (Tu,v) in H can be regarded as a number that varies with u. Clearly, $(Tu,v)=l(u)$ is a linear functional on H. Since T is bounded, using the Schwarz inequality, we can show that $l(u)$ is bounded. Therefore, by the Reisz representation theorem there exists a unique element g_v in H such that

$$(Tu,v)=(u,g_v)$$

for all u in H. This implies that given a $v\in H$ there is a unique g_v associated with element v. In other words, there exists a mapping T^* of H into itself. We write $g_v=T^*v$ and call T^* the *adjoint* (not to be confused with the adjunct of a matrix) of T. We have

$$(Tu,v)=(u,T^*v) \tag{2.88}$$

for all u, $v\in H$. It can be easily verified that the adjoint of a linear bounded operator is also linear and bounded. Furthermore, the adjoint is uniquely defined.

The following properties of adjoint operators should be noted:

1 $(T_1+T_2)^*=T_1^*+T_2^*$.
2 $(\alpha T_1)^*=\alpha T_1^*$, α is a real number.
3 $(T_1T_2)^*=T_2^*T_1^*$. (2.89)
4 $(T^*)^*=T$.

Example 2.39 Let $H = L_2(a, b)$ be the space of square-integrable real-valued functions on the interval (a, b). Consider the integral operator T defined by

$$Tu = \int_a^b k(x, y)u(y)\,dy.$$

Here $k(x, y)$ is a given function defined over the area $a \leq x, y \leq b$, and is called the *kernel* of the operator. We assume that $k(x, y)$ satisfies the condition

$$\int_a^b \int_a^b |k(x, y)|^2\,dx\,dy \equiv C < \infty.$$

Then the operator T is bounded. In fact, by the Schwarz inequality we have

$$\|Tu\|_0^2 \leq \|u\|_0^2 \left[\int_a^b \int_a^b |k(x, y)|^2\,dx\,dy\right]$$

or $\|T\| \leq C$. We now wish to find the adjoint of T. Consider

$$(Tu, v) = \int_a^b (Tu)v\,dx = \int_a^b \left(\int_a^b k(x, y)u(y)\,dy\right) v\,dx$$

$$= \int_a^b \int_a^b k(x, y)u(y)v(x)\,dx\,dy$$

$$= \int_a^b \left(\int_a^b k(x, y)v(x)\,dx\right) u(y)\,dy \equiv (u, T^*v).$$

Thus the adjoint T^* is also an integral operator on $L_2(a, b)$.

$$T^*v = \int_a^b k(x, y)v(x)\,dx.$$

Note that $T^* \neq T$; T^* is a transpose of T. Indeed, if an orthonormal basis is used to represent the operators, it can be seen that $t_{ij} = t_{ji}^*$. ■

We now turn to unbounded operators on Hilbert spaces. Let T be a linear operator on a real Hilbert space H. That is, T is defined on domain $\mathfrak{D}_T$ dense in H. The adjoint operator T^* is defined so that Eq. (2.88) is valid. The domain of T^* is defined to be the space

$$\mathfrak{D}_T^* = \{v:\quad v \in H, (Tu, v) = (u, g_v) \text{ for some } g_v \text{ in } H \text{ and } u \in \mathfrak{D}_T\}$$

so that the mapping $v \to g_v$ is uniquely defined.

A linear operator T defined on dense subspace $\mathfrak{D}_T$ of a Hilbert space H is said to be *closed* if for every sequence $\{u_n\}$ in $\mathfrak{D}_T$, with the property that both limits $u = \lim u_n$ and $v = \lim Tu_n$ exist, one has $u \in \mathfrak{D}_T$ and $v = Tu$.

We remark that the domain $\mathfrak{D}_T^*$ of the adjoint T^* of a closed operator T is dense in H.

A linear operator T: $\mathfrak{D}_T \to H$ is said to be *self-adjoint* if $\mathfrak{D}_T = \mathfrak{D}_T^*$ and $Tu = T^*u$ for all $u \in \mathfrak{D}_T$. If T is bounded, $\mathfrak{D}_T = \mathfrak{D}_T^* = H$. A linear operator T is said to be *normal* if $T^*T = TT^*$. Clearly, every self-adjoint operator is normal.

Example 2.40

1 Let $\mathfrak{D}_T$ be the subspace of $L_2(0,1)$ consisting of continuously differentiable functions $u(x)$ on $0 \le x \le 1$ with $u(0)=0$, and let T be the first-order differential operator, $Tu = du/dx$. We wish to calculate T^*. We write

$$\int_0^1 Tu\,v\,dx = \int_0^1 \frac{du}{dx} v\,dx = \int_0^1 u\left(-\frac{dv}{dx}\right) dx + (uv)|_0^1.$$

Since $u(0)=0$, one of the boundary terms vanishes. The second boundary term vanishes [so that we establish Eq. (2.88)] if we define $\mathfrak{D}_T^*$ to be the subset of $L_2(0,1)$ consisting of continuously differentiable functions $v(x)$ on $0 \le x \le 1$ with $v(1)=0$. Then we have $T^* = -d/dx$. Since $T^* \ne T$ and $\mathfrak{D}_T^* \ne \mathfrak{D}_T$, T is *not* self-adjoint.

2 Let $\mathfrak{D}_T$ be the subspace of $H^1[0,1]$ (see Example 2.29) of functions $u(x)$ with continuous first derivatives on $0 \le x \le 1$ with $u(0)=u(1)=0$, and let $T = d^2/dx^2$. The adjoint of T is given by

$$(Tu,v) = \int_0^1 \left(\frac{du^2}{dx^2}\right) v\,dx = \int_0^1 \frac{du}{dx}\left(-\frac{dv}{dx}\right) dx + \left(\frac{du}{dx}v\right)\Big|_0^1$$

$$= \int_0^1 u\left(\frac{dv^2}{dx^2}\right) dx + \left[\frac{du}{dx}v - u\frac{dv}{dx}\right]_0^1.$$

If we define $\mathfrak{D}_T^*$ to be the subspace of $H^1(0,1)$ of functions $v(x)$ with continuous first derivatives on $0 \le x \le 1$ with $v(1) = v(0)=0$, we have

$$(Tu,v) = \int_0^1 u\frac{d^2v}{dx^2} dx = (u, T^*v)$$

or $T^* = d^2/dx^2 = T$. Hence T is self-adjoint. ■

The notion of self-adjoint operator can be extended to linear operators defined from one Hilbert space into another. Let T: $\mathfrak{D}_T \subset H_1 \to H_2$ be a linear operator, and let v be a fixed element in H_2. As u varies over H_1, $(Tu,v)_2$ takes on various numerical values, so that $(Tu,v)_2$ is a linear functional (not necessarily bounded) in u. The operator T^*: $H_2 \to H_1$ is the adjoint of T if the following relation holds:

$$(Tu,v)_2 = (u,T^*v)_1. \tag{2.90}$$

Here $(\cdot,\cdot)_1$ and $(\cdot,\cdot)_2$ denote the inner products in H_1 and H_2, respectively. We have to be more precise about the domain of T^*. We define $\mathfrak{D}_T^*$ as follows:

$$\mathfrak{D}_T^* = \{v:\quad v \in H_2, (Tu, v)_2 = (u, g_v)_1 \text{ for some } g_v \text{ in } H_1 \text{ and all } u \text{ in } \mathfrak{D}_T\}.$$

Example 2.41 Let $\mathfrak{D}_T$ be the subspace of $L_2(\Omega)$ consisting of continuously differentiable functions u on a bounded domain Ω in three-dimensional Euclidean space R^3. Let T be the gradient operator mapping u in $\mathfrak{D}_T$ into a continuous vector-valued function $\mathbf{v} = (v_1, v_2, v_3)$ in $H_2 = L_2(\Omega) \times L_2(\Omega) \times L_2(\Omega)$. We choose the inner product in H_2 to be the integral of the vector dot product:

$$(\mathbf{v}, \mathbf{w})_2 = \int_\Omega \mathbf{v} \cdot \mathbf{w}\, dx.$$

We wish to calculate the adjoint of $T = \text{grad}$. We have

$$\begin{aligned}(Tu, \mathbf{v})_2 &= \int_\Omega (\text{grad}\, u) \cdot \mathbf{v}\, dx \\ &= \int_\Omega [\text{div}(u\mathbf{v}) - u \,\text{div}\, \mathbf{v}]\, dx \\ &= \int_\Omega u(-\text{div}\, \mathbf{v})\, dx + \int_S \hat{\mathbf{n}} \cdot (u\mathbf{v})\, ds.\end{aligned}$$

Here $dx = dx_1\, dx_2\, dx_3$ and $\hat{\mathbf{n}}$ is the unit outward normal to the surface S. It follows that $\mathfrak{D}_T^*$ is the subspace of H_2 with $\hat{\mathbf{n}} \cdot \mathbf{v} = 0$ on S, and $T^* = -\text{div}$. The inner product $(\cdot,\cdot)_1$ in $\mathfrak{D}_T$ is taken to be the L_2 inner product. ■

An operator A, linear in its domain $\mathfrak{D}_A$, is called *symmetric* in $\mathfrak{D}_A$ if for every pair of elements u, v from $\mathfrak{D}_A$ we have

$$(Au, v) = (u, Av). \tag{2.91}$$

Clearly, every self-adjoint operator (i.e., $A = A^*$) is always a symmetric operator. A symmetric operator is said to be *positive in its domain* $\mathfrak{D}_A$ if for all u in $\mathfrak{D}_A$ the following relation holds:

$$(Au, u) \geq 0, \qquad \text{and} (Au, u) = 0 \text{ implies } u = 0 \text{ in } \mathfrak{D}_A. \tag{2.92}$$

Further, if we can find a constant $\gamma > 0$ such that for all u in $\mathfrak{D}_A$ the relation

$$(Au, u) \geq \gamma \|u\|^2 \tag{2.93}$$

holds, then the operator A is called *positive definite in* $\mathfrak{D}_A$.

Example 2.42 Let $H = L_2(\Omega)$, where $\Omega \subset R^2$ is a plane domain. Let $\mathfrak{D}_A$ be the set of functions from $C^2(\Omega)$ which vanish on the boundary $\partial\Omega$. Since $C^2(\Omega)$ is dense in $L_2(\Omega)$, $\mathfrak{D}_A$ is dense in $L_2(\Omega)$. Let A be the differential operator, $A \equiv -(\partial^2/\partial x^2 + \partial^2/\partial y^2) \equiv -\nabla^2$. We wish to show first that A is symmetric on $\mathfrak{D}_A$. Consider

$$(Au, v)_0 = \int_\Omega -\left(\frac{\partial^2 u}{\partial x^2} + \frac{\partial^2 u}{\partial y^2}\right) v\, dx\, dy = \int_\Omega (-\nabla^2 u) v\, dx\, dy$$

$$= \int_\Omega -[\nabla \cdot (\nabla uv) - \nabla u \cdot \nabla v]\, dx\, dy.$$

Using the divergence theorem (Eq. (1.235b))

$$\int_\Omega \nabla \cdot \mathbf{f}\, dx = \int_{\partial\Omega} \hat{\mathbf{n}} \cdot \mathbf{f}\, ds, \tag{2.94}$$

where $\hat{\mathbf{n}}$ is the unit outward normal, we obtain

$$(Au, v)_0 = \int_\Omega \nabla u \cdot \nabla v\, dx\, dy - \int_{\partial\Omega} (\hat{\mathbf{n}} \cdot \nabla u) v\, ds$$

$$= \int_\Omega \left(\frac{\partial u}{\partial x}\frac{\partial v}{\partial x} + \frac{\partial u}{\partial y}\frac{\partial v}{\partial y}\right) dx\, dy - \int_{\partial\Omega} \left(\frac{\partial u}{\partial x} n_x + \frac{\partial u}{\partial y} n_y\right) v\, ds. \tag{2.95}$$

Since $v = 0$ on $\partial\Omega$, the boundary term vanishes. Owing to the symmetry of the right side in u and v, we immediately have $(Au, v)_0 = (Av, u)_0 = (u, Av)_0$.

Next we prove that A is positive. For u in $\mathfrak{D}_A$ we have

$$(Au, u)_0 = \int_\Omega \left[\left(\frac{\partial u}{\partial x}\right)^2 + \left(\frac{\partial u}{\partial y}\right)^2\right] dx\, dy \geq 0.$$

If $(Au, u) = 0$, then it follows that (since $\partial u/\partial x = \partial u/\partial y = 0$) $u =$ constant. Since $u = 0$ on C it follows that this constant is zero, and $u = 0$. Thus $(Au, u) = 0$ implies $u = 0$. This proves that A is positive.

To prove that A is positive definite on $\mathfrak{D}_A$, we invoke the *Friedrichs inequality*. For u in $C^1(\Omega)$, the following inequality holds:

$$\int_\Omega u^2(x)\, dx \leq C_1 \sum_{k=1}^{2} \int_\Omega \left(\frac{\partial u}{\partial x_k}\right)^2 dx + C_2 \int_{\partial\Omega} u^2(s)\, ds \tag{2.96}$$

where C_1 and C_2 are nonnegative constants dependent on Ω, but independent of u. For u in $\mathfrak{D}_A$ we have $u=0$ on $\partial\Omega$, hence the second term on the right-hand side of Eq. (2.96) is zero. Consequently,

$$(Au,u)_0 \geq \frac{1}{C_1}\int_\Omega u^2\,dx = \frac{1}{C_1}\|u\|_0.$$

Thus A is positive definite on $\mathfrak{D}_A$. ■

2.7.3 Orthonormal Bases and Generalized Fourier Series

Orthonormal Bases Recall from Section 2.5 that a set of elements $\{\phi_i\}$ in an inner product space V is said to be orthonormal if

$$(\phi_i,\phi_j)=\delta_{ij}$$

for all i and j, where δ_{ij} is the Kronecker function. The index i may range over a finite or an infinite index set.

An orthonormal set $S=\{\phi_n\}$ in an inner product space V is called *maximal* if there is no unit vector ϕ_0 in V, such that $S\cup\{\phi_0\}$ is an orthonormal set. This is equivalent to saying that an orthonormal set $\{\phi_i\}$ in an inner product space V is maximal if and only if $(\phi,\phi_i)=0$ for all i implies that $\phi=0$. A maximal orthonormal set $\{\phi_i\}$ in a Hilbert space H is referred to as an *orthonormal basis* for H. Given any countable linearly independent set in an inner product space, it is possible to construct an orthonormal set by means of the Gram-Schmidt process (see Exercise 2.4.20).

Example 2.43

1 Let $H=L_2[0,1]$ be the Hilbert space on a complex number field with the usual inner product,

$$(u,v)_0=\int_0^1 u\bar{v}\,dx$$

where $\bar{v}$ is the complex conjugate of v. Consider the set $\{\phi_n\}$,

$$\phi_n=\cos 2\pi nx + i\sin 2\pi nx, \qquad i=\sqrt{-1}.$$

For $m\neq n$, we have

$$(\phi_m,\phi_n)_0=\int_0^1(\cos 2\pi mx + i\sin 2\pi mx)(\cos 2\pi nx - i\sin 2\pi nx)\,dx$$

$$=\int_0^1\left[\cos 2\pi(m-n)x + i\sin 2\pi(m-n)x\right]dx=0$$

and for $m=n$, we have

$$(\phi_n,\phi_n)_0=\|\phi_n\|_0^2=\int_0^1 1\cdot dx=1.$$

Thus $\{\phi_n\}$ is an orthonormal set in $L_2[0,1]$.

2 (*The Laguerre Functions*) The Laguerre functions $\phi_n(x)$ form an orthonormal basis in $L_2[0,\infty)$:

$$\phi_n(x)=\frac{1}{n!}e^{-x/2}L_n(x), \qquad n=0,1,\ldots \tag{2.97}$$

where $L_n(x)$ is the Laguerre polynomial:

$$L_n(x)=\sum_{i=1}^{n}(-1)^i\binom{n}{i}n(n-1)\cdots(i+1)x^i$$

$$=e^x\frac{d^n}{dx^n}(x^ne^{-x}). \tag{2.98}$$

3 (*The Hermite Functions*) The Hermite functions $\phi_n(x)$ form an orthonormal basis in $L_2(-\infty,\infty)=L_2(\mathbf{R})$:

$$\phi_n(x)=\left[2^nn!\sqrt{\pi}e^{x^2}\right]^{-1/2}H_n(x), \qquad n=0,1,2,\ldots \tag{2.99}$$

where $H_n(x)$ is the Hermite polynomial

$$H_n(x)=(-1)^ne^{x^2}\frac{d^n}{dx^n}(e^{-x^2}). \tag{2.100}$$

4 (*The Legendre Functions*) The Legendre functions $\phi_n(x)$ form an orthonormal basis in $L_2[-1,1]$:

$$\phi_n(x)=\left(\frac{2n+1}{2}\right)P_n(x), \qquad n=0,1,2,\ldots \tag{2.101}$$

where P_n are the Legendre polynomials:

$$P_n(x)=\frac{1}{2^nn!}\frac{d^n}{dx^n}\left[(x^2-1)^n\right]. \tag{2.102}$$

5 Consider the linear vector space $V=P_2$ of polynomials of degree less than or equal to 2 in real variable x in $[-1,1]$. We chose $\{1,x,x^2\}$ as a basis for P_2 and define the inner product in P_2 to be the L_2 inner product,

$$(p_1,p_2)_0=\int_{-1}^{1}p_1(x)p_2(x)\,dx.$$

We wish to use the Gram-Schmidt orthogonalization process to construct an orthonormal basis. If $\{\psi_i\}$ denotes the original basis and $\{\phi_i\}$ denotes the orthogonal basis, the Gram-Schmidt orthogonalization process gives the formula

$$\phi_{i+1}=\psi_{i+1}-\sum_{j=1}^{i}(\psi_{i+1},\hat{e}_j)\hat{e}_j,\ \hat{e}_j=\frac{\phi_j}{\|\phi_j\|}.$$

We set $\phi_1=\psi_1=1(\|\phi_1\|=\sqrt{2})$ and compute the remaining elements using the formula:

$$\phi_2=x-\sum_{j=1}^{2-1}(\psi_2,\hat{e}_j)\hat{e}_j=x-(\psi_2,\hat{e}_1)\hat{e}_1$$

$$=x-\frac{1}{2}\int_{-1}^{1}x\cdot 1\,dx=x-0$$

$$\phi_3=x^2-(\psi_3,\hat{e}_1)\hat{e}_1-(\psi_3,\hat{e}_2)\hat{e}_2$$

$$=x^2-\frac{1}{2}\int_{-1}^{1}x^2\cdot dx-\frac{3}{2}\left(\int_{-1}^{1}x^2\cdot x\,dx\right)x$$

$$=x^2-\frac{1}{3}.$$

Thus the orthogonal set is given by $\{1, x, x^2-\frac{1}{3}\}$. The set can be orthonormalized by replacing ϕ_i by $\phi_i/\|\phi_i\|$, we get

$$\left\{\frac{1}{\sqrt{2}},\sqrt{\frac{3}{2}}\,x,\sqrt{\frac{45}{8}}\left(x^2-\frac{1}{3}\right)\right\}$$ ■

The Bessel Inequality Let $\{\phi_n\}$ be an orthonormal set in an inner product space V. For any finite subset $\{\phi_1,\phi_2,\ldots,\phi_N\}$ from $\{\phi_n\}$, we have

$$0\le\left\|u-\sum_{i=1}^{N}(u,\phi_i)\phi_i\right\|^2=\left(u-\sum_{i}^{N}(u,\phi_i)\phi_i,\,u-\sum_{j}^{N}(u,\phi_j)\phi_j\right)$$

$$=\|u\|^2-2\sum_{i}^{N}(u,\phi_i)(u,\phi_i)+\sum_{i,j=1}^{N}(u,\phi_j)(u,\phi_j)(\phi_i,\phi_j)$$

$$=\|u\|^2-\sum_{i}^{N}|(u,\phi_i)|^2.$$

In arriving at the last step we have employed the orthonormal property of

$\{\phi_n\}$. Thus we have

$$\|u\|^2 \geq \sum_i^N |(u, \phi_i)|^2,$$

and since the left-hand side does not depend on N, the inequality, known as the Bessel inequality, holds for countable sums

$$\|u\|^2 \geq \sum_i |(u, \phi_i)|^2. \tag{2.103}$$

Next we study certain properties of the linear combinations of the form $\Sigma_i \alpha_i \phi_i$, where $\{\phi_i\}$ is an orthonormal set.

Lemma 2.3 Let $\{\phi_n\}$ be a countably infinite orthonormal set in a Hilbert space H. Then the following statements hold:

(i) The infinite series $\Sigma_{n=1}^{\infty} \alpha_n \phi_n$, where α_n are scalars, converges if and only if the series of real numbers $\Sigma_{n=1}^{\infty} |\alpha_n|^2$ converges.

(ii) If $\Sigma_{n=1}^{\infty} \alpha_n \phi_n$ converges and

$$u = \sum_{n=1}^{\infty} \alpha_n \phi_n = \sum_{n=1}^{\infty} \beta_n \phi_n,$$

then $\alpha_n = \beta_n$ for all n and $\|u\|^2 = \Sigma_{n=1}^{\infty} |\alpha_n|^2$.

Proof

(i) First consider the "if" part. Suppose that the series $\Sigma_{n=1}^{\infty} \alpha_n \phi_n$ is convergent and let $u = \Sigma_{n=1}^{\infty} \alpha_n \phi_n$. Equivalently,

$$\lim_{N \to \infty} \left\| u - \sum_{n=1}^{N} \alpha_n \phi_n \right\|^2 = 0.$$

We have, since the inner product is continuous,

$$(u, \phi_m) = \left(\sum_{n=1}^{\infty} \alpha_n \phi_n, \phi_m \right) = \sum_{n=1}^{\infty} \alpha_n (\phi_n, \phi_m) = \alpha_m, \qquad m = 1, 2, \ldots .$$

In view of the Bessel inequality we obtain

$$\sum_{m=1}^{\infty} |(u, \phi_m)|^2 = \sum_{m=1}^{\infty} |\alpha_m|^2 \leq \|u\|^2,$$

which shows that $\Sigma_{m=1}^{\infty}|\alpha_m|^2$ converges. It should be noted that the completeness of H is not used in this part of the proof.

Next consider the "only if" part. Suppose that $\Sigma_{n=1}^{\infty}|\alpha_n|^2$ converges. Consider the finite sum $S_n = \Sigma_{i=1}^{n}\alpha_i\phi_i$. We have

$$\|S_n - S_m\|^2 = \left(\sum_{i=m+1}^{n} \alpha_i\phi_i, \sum_{j=m+1}^{n} \alpha_j\phi_j \right) = \sum_{i=m+1}^{n} |\alpha_i|^2,$$

which shows that $\{S_n\}$ is a Cauchy sequence. Since H is complete, the sequence of partial sums $\{S_n\}$ is convergent in H, and therefore the series $\Sigma_n\alpha_n\phi_n$ converges.

(ii) We first prove that $\|u\|^2 = \Sigma_{n=1}^{\infty}|\alpha_n|^2$. Consider

$$\|u\|^2 - \sum_{n=1}^{N} |\alpha_n|^2 = (u,u) - \sum_{n=1}^{N}\sum_{m=1}^{N} (\alpha_n\phi_n, \alpha_m\phi_m)$$

$$= \left(u, u - \sum_{n=1}^{N} \alpha_n\phi_n \right) + \left(\sum_{n=1}^{N} \alpha_n\phi_n, u - \sum_{n=1}^{N} \alpha_n\phi_n \right)$$

$$\leq \left\| u - \sum_{n=1}^{N} \alpha_n\phi_n \right\| \left\{ \|u\| + \left\| \sum_{n=1}^{N} \alpha_n\phi_n \right\| \right\}.$$

Since $\Sigma_{n=1}^{\infty}\alpha_n\phi_n$ converges to u, we have that the right-hand side converges to zero, giving the desired result. If

$$u = \sum_{n=1}^{\infty} \alpha_n\phi_n = \sum_{n=1}^{\infty} \beta_n\phi_n$$

then

$$0 = \lim_{N\to\infty} \left[\sum_{n=1}^{N} (\alpha_n - \beta_n)\phi_n \right] \to 0^2 = \sum_{n=1}^{\infty} \|\alpha_n - \beta_n\|^2,$$

implying that $\alpha_n = \beta_n$ for all n. This completes the proof of the lemma. ■

In the following we characterize the orthogonal projections on a Hilbert space in terms of an orthonormal basis.

Lemma 2.4 Let V be an inner product space, and let S be the finite-dimensional linear subspace spanned by the finite orthonormal set $\{\phi_i\}_{i=1}^{n}$. Then the orthogonal projection of V into S is given by

$$Pu = \sum_{i=1}^{n} (u, \phi_i)\phi_i.$$

Proof First we note that P is linear. Next we prove that P is a projection:

$$P(Pu)=\sum_{i=1}^{n}(u,\phi_i)P\phi_i=\sum_{i=1}^{n}(u,\phi_i)\sum_{j=1}^{n}(\phi_i,\phi_j)\phi_j$$

$$=\sum_{i=1}^{n}(u,\phi_i)\phi_i=Pu.$$

Further, by definition, $\mathcal{R}(P)\subset S$. On the other hand, for $u=\sum_{i=1}^{n}\alpha_i\phi_i\in S$, we have $Pu=u$ [and $\alpha_i=(u,\phi_i)$], so $\mathcal{R}(P)=S$.

Now it remains to be shown that P is orthogonal. Let $v\in\mathcal{N}(P)$ and $u\in\mathcal{R}(P)$. Since $u\in\mathcal{R}(P)=S$ we have $Pu=u$ and

$$(v,u)=(v,Pu)=\left(v,\sum_{i=1}^{n}(u,\phi_i)\phi_i\right)$$

$$=\sum_{i=1}^{n}(v,\phi_i)(u,\phi_i)$$

$$=\left(\sum_{i=1}^{n}(v,\phi_i)\phi_i,u\right)=(Pv,u).$$

Since $v\in\mathcal{N}(P)$, we have $Pv=0$ and hence $\mathcal{N}(P)\perp\mathcal{R}(P)$. This proves that P is orthogonal (and self-adjoint). ■

We remark that the above result can be carried on to closed linear subspaces (not necessarily finite dimensional) spanned by a countable orthonormal set. We state this fact in the following lemma.

Lemma 2.5 Let S be the closed linear subspace generated by a countable orthonormal set $\{\phi_n\}$ in a Hilbert space H. Then every vector $u\in S$ can be written uniquely as

$$u=\sum_{n}(u,\phi_n)\phi_n. \tag{2.104}$$

Also, the operator P defined by

$$Pu=\sum_{n}(u,\phi_n)\phi_n \tag{2.105}$$

is the orthogonal projection of H into S.

Proof The uniqueness of Eq. (2.104) follows from Lemma 2.4. We wish to show that every vector $u\in S$ can be represented by Eq. (2.104). For $u\in S$ we have (since S is closed).

$$u=\lim_{N\to\infty}\sum_{n=1}^{M}\alpha_n\phi_n,\qquad M\geq N.$$

From Theorem 2.11 and Lemma 2.3 it follows that

$$\left\| u - \sum_{n=1}^{M} (u, \phi_n)\phi_n \right\| \leq \left\| u - \sum_{n=1}^{M} \alpha_n \phi_n \right\|$$

and, as $N \to \infty$, we get the desired result.

The fact that P is an orthogonal projection of H into S was already established in Lemma 2.4. ■

We now come to an important topic of this section, namely, the generalized Fourier representation of functions. Much of the preceding discussion may have already prompted the reader to compare the representations in Eqs. (2.104) and (2.105) with the Fourier series representations he is familiar with. The following theorem establishes the fundamental properties of orthonormal bases.

Theorem 2.16 (*Fourier-Series Theorem*) For any orthonormal set $\{\phi_n\}$ in a Hilbert space H, the following statements are *equivalent*:

(i) $\{\phi_n\}$ is an (maximal) orthonormal set.

(ii) (*Fourier Series Expansion*) Every u in H can be represented by

$$u = \sum_n (u, \phi_n)\phi_n. \tag{2.106}$$

(iii) (*Parseval Equality*) For any pair of vectors $u, v \in H$ one has

$$(u, v) = \sum_n (u, \phi_n)(v, \phi_n). \tag{2.107}$$

(iv) For any $u \in H$ one has

$$\|u\|^2 = \sum_n |(u, \phi_n)|^2. \tag{2.108}$$

(v) Any linear subspace S of H that contains $\{\phi_n\}$ is dense in H.

Proof

(i)→(ii) Let S be the linear subspace generated by the set $\{\phi_n\}$. If $u \in S^{\perp}$, then $(u, \phi_n) = 0$ for all n. Since $\{\phi_n\}$ is maximal in H, we have $u = 0$, implying that $S^{\perp} = \{0\}$ and $S = H$ (see Theorem 2.12, which requires the completeness of H). Hence the orthogonal projection is the identity transformation I, and we have

$$Iu = u = \sum_n (u, \phi_n)\phi_n$$

for every u in H.

(ii)→(iii) This follows immediately from part (ii) and the fact that $\{\phi_n\}$ is orthonormal.

(iii)→(iv) Follows from part (iii) by setting $v=u$.

(iv)→(i) If $\{\phi_n\}$ is not maximal, then there exists a unit vector u_0 such that $\{\phi_n\}\cup\{u_0\}$ is an orthonormal set. However, in view of part (iv) we have

$$1=\|u_0\|^2=\sum_n |(u_0,\phi_n)|^2=0,$$

which is a contradiction.

(ii)→(v) Statement (v) is equivalent to saying that the orthogonal projection onto $\bar{S}$, the closure of S, is the identity. In view of Lemma 2.5, statement (v) is equivalent to statement (ii). ■

The Fourier series representation (2.106) can be interpreted as the sum of projections of u along each vector ϕ_n in the orthonormal set $\{\phi_n\}$. The scalars (u,ϕ_n) are the *coordinates*, also known as the *Fourier coefficients*, of u with respect to the orthonormal basis $\{\phi_n\}$.

Example 2.44

1 (*Fourier Trigonometric Series*) Let $H=L_2[0,2\pi]$ be the Hilbert space with the L_2 inner product and associated norm. The set

$$B=\{1\}\cup\{\cos nx\}\cup\{\sin nx\}, \qquad n=1,2,\ldots \tag{2.109}$$

is a maximal (sometimes called *complete*) orthogonal set in the real vector space $L_2[0,\pi]$. The set can be made orthonormal by dividing each element by its norm. The associated orthonormal set is given by

$$B=\left\{\frac{1}{\sqrt{2\pi}},\frac{\cos x}{\sqrt{\pi}},\frac{\sin x}{\sqrt{\pi}},\frac{\cos 2x}{\sqrt{\pi}},\frac{\sin 2x}{\sqrt{\pi}},\ldots\right\}\equiv\{\phi_n\}_{n=0}^{\infty}. \tag{2.110}$$

The Fourier series theorem states that any element $f(x)$ in $L_2[0,2\pi]$ can be expressed in the form

$$f(x)=\alpha_0\phi_0+\sum_{n=1}^{\infty}(\alpha_{2n-1}\phi_{2n-1}+\alpha_{2n}\phi_{2n})$$

$$=\frac{a_0}{\sqrt{2\pi}}+\frac{1}{\sqrt{\pi}}\sum_{n=1}^{\infty}(a_n\cos nx+b_n\sin nx) \tag{2.111}$$

where $a_0=\alpha_0$, $a_n=\alpha_{2n-1}$, and $b_n=\alpha_{2n}$ are given by

$$a_0=(f,\phi_0)_0, \qquad a_n=(f,\phi_{2n-1})_0, \qquad b_n=(f,\phi_{2n})_0. \tag{2.112}$$

Let us now compute the L_2-norm of f.

$$\|f\|_0^2=(f,f)_0=a_0^2+\sum_{n=1}^{\infty}\left(a_n^2+b_n^2\right). \tag{2.113}$$

It is easy to verify that this is equal to $\Sigma_{n=0}^{\infty}|(f,\phi_n)_0|^2$, and therefore $\{\phi_n\}_{n=0}^{\infty}$ is maximal.

2 Next consider the linear subspace $S\subset L_2[0,2\pi]$ spanned by the set

$$C\equiv\left\{\frac{1}{\sqrt{2\pi}},\frac{\cos x}{\sqrt{\pi}},\frac{\cos 2x}{\sqrt{\pi}},\ldots\right\}.$$

Obviously, $\sin nx$ for any $n\geq 1$ is not in S, for it cannot be represented by a linear combination of elements in set C. Of course, $\sin nx$ is in $L_2[0,2\pi]$, but the set C is *not* maximal in $L_2[0,2\pi]$. Conversely, if we wish to determine whether a given element is in S, we look at its Fourier representation by the orthonormal set generating S. For example, consider the function $f(x)=\frac{1}{2}(2\pi-x)\sin x$. We compute the Fourier coefficients

$$a_0=\int_0^{2\pi}\frac{1}{\sqrt{2\pi}}\frac{1}{2}(2\pi-x)\sin x\,dx=\frac{\pi}{2}$$

$$a_1=\int_0^{2\pi}\frac{1}{\sqrt{\pi}}\cos x\cdot\frac{1}{2}(2\pi-x)\sin x\,dx=\frac{\sqrt{\pi}}{4}$$

$$a_n=\int_0^{2\pi}\frac{1}{\sqrt{\pi}}\cos nx\cdot\frac{1}{2}(2\pi-x)\sin x\,dx=\frac{\sqrt{\pi}}{1-n^2},\qquad n=2,3,\ldots.$$

Thus $f(x)=\frac{1}{2}(2\pi-x)\sin x$ can be expressed in terms of the set C:

$$\tfrac{1}{2}(2\pi-x)\sin x=\tfrac{1}{2}+\tfrac{1}{4}\cos x-\tfrac{1}{3}\cos 2x-\tfrac{1}{8}\cos 3x-\cdots$$

Hence the element $\frac{1}{2}(2\pi-x)\sin x$ is in S.

3 Consider the linear vector space P_2 of polynomials of degree less than or equal to 2 defined in the interval $[-1,1]$. As we have from Example 2.43, the set

$$\left\{\frac{1}{\sqrt{2}},\sqrt{\frac{3}{2}}\,x,\sqrt{\frac{45}{8}}\left(x^2-\frac{1}{3}\right)\right\}$$

forms an orthonormal basis for P_2 with respect to the L_2 inner product. An arbitrary element $p \in P_2$ can be expressed in the form

$$p = \sum_{i=1}^{3} (p, \phi_i)\phi_i = \sum_{i=1}^{3} \alpha_i \phi_i .$$

Let $p = a_0 + a_1 x + a_2 x^2$. Then we have

$$\alpha_1 = \int_{-1}^{1} (a_0 + a_1 x + a_2 x^2) \frac{1}{\sqrt{2}} dx = \frac{1}{\sqrt{2}} \left(2a_0 + \frac{2a_2}{3} \right) = \sqrt{2} \left(a_0 + \frac{a_2}{3} \right)$$

$$\alpha_2 = \int_{-1}^{1} (a_0 + a_1 x + a_2 x^2) \frac{3x}{2} dx = \sqrt{\frac{3}{2}} \left(\frac{2a_1}{3} \right) = \sqrt{\frac{2}{3}}\, a_1$$

$$\alpha_3 = \int_{-1}^{1} (a_0 + a_1 x + a_2 x^2) \frac{45}{8} \left(x^2 - \frac{1}{3} \right) dx = \sqrt{\frac{45}{8}} \left(\frac{8}{45} a_2 \right) = \sqrt{\frac{8}{45}}\, a_2$$

as expected. The coordinates α_i will be different for different sets of orthonormal bases (e.g., first three Legendre polynomials) in P_2. ■

2.7.4 Separable Hilbert Spaces and the Least-Squares Approximation

Separable Hilbert Space In the preceding sections we have seen that the Hilbert space theory is much like the Euclidean spaces. The so-called separable Hilbert spaces are even closer to the Euclidean spaces. The separability ensures that the infinite-dimensional Hilbert space is not too "large." In other words, separable Hilbert spaces are Hilbert spaces with countable orthonormal bases. We give the precise definition of separable Hilbert spaces below.

Let H be a Hilbert space, and let S be a set of vectors in H. The set $\mathfrak{L}$ of all finite linear combinations of vectors from the set S is itself a linear subspace, which is in general not closed. If $\mathfrak{L}(S)$ contains only a finite number of linearly independent vectors, then $\mathfrak{L}(S)$ is a finite-dimensional linear subspace and hence closed. If $\mathfrak{L}(S)$ is infinite dimensional, $\mathfrak{L}(S)$ and its closure $\bar{\mathfrak{L}}(S)$ differ. For example, the set $S = \{1, x, x^2, \dots\}$ generates the space P of all polynomials (which is an infinite-dimensional linear space), and the closure of P is the whole of $L_2[a, b]$.

A Hilbert space H is said to be *separable* if there exists a countable set of elements $\{\psi_n\}$ in H whose finite linear combinations are dense in H. In other words, given an element u in H and an $\varepsilon > 0$, there exist an index N (which usually depends on ε) and scalars $\alpha_1, \alpha_2, \dots, \alpha_N$ such that

$$\left\| u - \sum_{i}^{N} \alpha_i \psi_i \right\| < \varepsilon. \tag{2.114}$$

The set $\{\psi_n\}$ is called the *spanning set*. A separable Hilbert space contains a linearly independent spanning set of elements, which can be converted, suing the Gram-Schmidt process, to an orthonormal set. For this reason some authors define a separable Hilbert space to be a Hilbert space with a countable orthonormal basis.

A finite-dimensional normed linear space is separable since any basis serves as a countable (finite) spanning set. The Hilbert spaces $H^m(\Omega)$, $m=0,1,2,\ldots,$ are separable since the countable set $\{1, x, x^2, \ldots\}$ is a spanning set.

One must note that Eq. (2.114) does not guarantee that u can be expanded in an infinite series, unless it is possible to choose the scalars α_i in Eq. (2.114) independently of ε.

Least-Squares Approximation Suppose that we are required to seek an approximation to an element u in a Hilbert space H as a linear combination of a given finite orthonormal set $\{\phi_i\}_{i=1}^N$ such that the approximation $\Sigma_i^n \alpha_i \phi_i$ is closest to u in the metric of H. Since the set of linear combinations of $\{\phi_i\}_{i=1}^N$ is an N-dimensional linear vector space S_N that is closed, the projection theorem, Theorem 2.13, and the approximation result of Theorem 2.14 are valid. In other words, the *best approximation* to u is given by its projection into S_N,

$$\left\| u - \sum_{i=1}^{N} (u, \phi_i)\phi_i \right\| \leq \|u - v\|$$

for all v in S_N. The approximation is the best possible one in the space S_N, the error being measured in the natural metric of H.

The same result can be obtained by seeking the minimum of the error between the elements and its approximation:

$$e \equiv \left\| u - \sum_{i}^{N} \alpha_i \phi_i \right\|.$$

We have

$$e^2 = \|u\|^2 + \sum_{i}^{N} [(u, \phi_i) - \alpha_i]^2 - \sum_{i}^{N} |(u, \phi_i)|^2.$$

Clearly, the error e attains its minimum when $\alpha_i = (u, \phi_i)$. Alternately, we can view e^2 as an ordinary function of the parameters α_i. The necessary condition for the minimum of a function of several variables is that its first derivative with respect to each of the variables must be zero. Thus,

$$\frac{d}{d\alpha_j}(e^2) \equiv 0 + 2\sum_{i}^{N} [(u, \phi_i) - \alpha_i](-\delta_{ij}) - 0 = 0$$

or $\alpha_i = (u, \phi_i)$. The technique of minimizing the square of the error in an approximation is known as the *least-squares method*. The uniquely determined best approximation to u in S_N is the *Fourier sum*

$$u_N = \sum_{i=1}^{N} \alpha_i \phi_i = \sum_{i=1}^{N} (u, \phi_i)\phi_i. \tag{2.115}$$

It is expected that the approximation to u improves as the space S_N is made larger (i.e., as the set $\{\phi_i\}$ is made larger). Note that there is a definite advantage of using an orthonormal basis over a merely linearly independent basis that is not orthogonal. The use of the orthonormal basis simplifies the calculation of the coefficients $\alpha_i = (u, \phi_i)$. Furthermore, with the addition of a new element to the orthonormal set, the Fourier sum becomes

$$u_{N+1} = \sum_{i}^{N+1} (u, \phi_i)\phi_i = \sum_{i}^{N} (u, \phi_i)\phi_i + (u, \phi_{N+1})\phi_{N+1},$$

which indicates that previously computed Fourier coefficients remain unchanged, and one is required to compute only the coefficient corresponding to the new element ϕ_{N+1}.

We close this section with a discussion of the least-squares approximation of a linear operator equation. Let A be a linear operator such that the operator equation

$$Au = f, \qquad u \in \mathfrak{D}(A) \tag{2.116}$$

has a unique solution. We wish to determine an approximation to $u \in \mathfrak{D}(A)$ in the form

$$u_N = \sum_{i}^{N} \alpha_i \phi_i \tag{2.117}$$

where $\{\phi_i\}_i^N$ are linearly independent (not necessarily orthogonal) elements, by minimizing the norm of the residual, $R = \| Au_N - f \|$ (*not* of the error e). We get

$$\begin{aligned} R^2 &= \| Au_N - f \|^2 = (Au_N - f, Au_N - f) \\ &= \| Au_N \|^2 - 2(Au_N, f) + \| f \|^2 \\ &= \left\| A\left(\sum_{i}^{N} \alpha_i \phi_i\right) \right\|^2 - 2\left(A\left(\sum_{i}^{N} \alpha_i \phi_i\right), f\right) + \| f \|^2 \end{aligned}$$

and

$$\frac{\partial R^2}{\partial \alpha_j} = 2\left(A\left(\sum_{i}^{N} \alpha_i \phi_i\right), A\phi_j\right) - 2(A\phi_j, f) = 0.$$

This gives, since A is linear,

$$\sum_i^N (A\phi_j, A\phi_i)\alpha_i - (A\phi_j, f) = 0 \rightarrow [A]\{\alpha\} = \{F\} \tag{2.118}$$

where

$$A_{ij} = (A\phi_i, A\phi_j), \qquad F_i = (A\phi_i, f). \tag{2.119}$$

Thus the solution $\{\alpha\}$ is obtained by solving the matrix equation. Note that $B(\phi_i, \phi_j) = A_{ij}$ is bilinear (and symmetric) in ϕ's, hence is a bilinear form on $\mathfrak{D}(A)$. Similarly, F_i is a linear form on $\mathfrak{D}(A)$. Existence and uniqueness of solutions to variational problems, that is, to determine u in a Hilbert space H such that

$$B(u, v) = F(v), \qquad \text{for all } v \in H, \tag{2.120}$$

is the subject of the next section.

Exercises 2.6

1 Let S_1 and S_2 be linear subspaces of a Hilbert space H.
- **(a)** If $S_1 \perp S_2$, show that $S_1^{\perp\perp} \perp S_2^{\perp\perp}$.
- **(b)** If S_i, $i = 1, 2, \ldots, n$, are closed subspaces of H with $S_i \perp S_j$ for $i \neq j$, show that $S = \Sigma_i^n S_i$ is a closed linear subspace of H.
- **(c)** Prove or disprove: (i) $S_1^{\perp} \perp S_2^{\perp}$. (ii) $S_1^{\perp\perp\perp} \perp S_2^{\perp\perp\perp}$.

2 Use geometric language to explain the results of Theorems 2.13 and 2.14. (*Hint*: Take $H = \mathsf{R}^3$ and $S = \mathsf{R}^2$, and identify the projection operator.)

3 Prove that for every linear functional l, defined on a finite-dimensional space V, there exists a unique element u_l in V such that $l(u) = (u, u_l)$ for all $u \in V$ (a special case of the Riesz representation theorem, Theorem 2.15). (*Hint*: Assume that $\{\phi_i\}$ is an orthonormal basis for V and choose $u = \Sigma_i^n l(\phi_i)\phi_i$.)

4 Consider the set S of all constant functions in $L_2[0,1]$.
- **(a)** Show that S is a complete linear subspace of $L_2[0,1]$.
- **(b)** Decompose the function $u = x^2 \in L_2[0,1]$ into the sum $u = v + w$, where $v \in S$ and $w \in S^{\perp}$. Characterize the space S.
- **(c)** Repeat part (b) with $u(x) = \sin \pi x$.

5 Determine whether the operator $T \equiv -\nabla^2$: $H \rightarrow H$, $H = H_0^1(\Omega) \cap H^2(\Omega)$ and $\Omega \subset \mathsf{R}^2$, is bounded. Is the operator bounded below?

6 Prove the following identities:
- **(a)** $(A + B)^* = A^* + B^*$.
- **(b)** $(AB)^* = B^* A^*$.
- **(c)** $\|AA^*\| = \|A\|^2$, A is a bounded operator.

7 Let A be a linear operator mapping R^n into itself: $\mathbf{y}=A\mathbf{x}$, where $\mathbf{x}=(x_1, x_2, \ldots, x_n)$ and $\mathbf{y}=(y_1, y_2, \ldots, y_n)$ in R^n. Show that the matrix representing A^* is the transpose of the matrix representing A.

8 Find the adjoint of the Laplace operator on appropriate spaces. Use the L_2-inner product. (Exercise 1.7.5 may prove to be useful.)

9 Find the adjoint of the first-order differential operator with respect to the scalar product

$$(u, v) \equiv \int_0^T u(t) v(T-t)\, dt.$$

10 Show that every self-adjoint linear operator $A=A^*$ can be represented uniquely as a product of a linear operator T with its adjoint T^*:

$$A = T^*T = TT^*.$$

11 Prove that any linear operator on a finite-dimensional vector space can be expressed in the form $T=T_1+T_2$, where $T_1^*=T_1$ and $T_2^*=-T_2$. (*Hint*: Note that $(T+T^*)$ is self-adjoint.)

12 Describe T^* if T is the linear operator described below:

(a) T: $\mathsf{R}^3 \to \mathsf{R}^3$, T is a rotation about the z axis through an angle of θ counterclockwise.

(b) T: $\mathsf{R}^3 \to \mathsf{R}^2$, T is a projection onto the xy plane.

13 (*Solvability Conditions*) Consider the problem of determining the solution to the nonhomogeneous equation

$$Au = f, \qquad u \in \mathfrak{D}(A) \subset H \tag{i}$$

where A is a linear operator. Next consider the adjoint homogeneous equation

$$A^*v = 0, \qquad v \in \mathfrak{D}(A^*). \tag{ii}$$

Show that:

(a) A necessary condition for Eq. (i) to have solution(s) is that

$$(f, v) = 0$$

for all v satisfying Eq. (ii).

(b) The orthogonal complement of the range of A is the null space of the adjoint, $\mathfrak{R}(A)^\perp = \mathfrak{N}(A^*)$ and $\overline{\mathfrak{R}}(A) = \mathfrak{N}(A^*)^\perp$.

14 Prove that the biharmonic operator ∇^4 is self-adjoint and positive definite in its domain. Identify the domain and range spaces of the operator (see Example 1.15).

15 (*Elastic Bending of a Clamped Beam*) Consider the fourth-order differential equation

$$\frac{d^2}{dx^2}\left(b(x)\frac{d^2w}{dx^2}\right)=f(x), \qquad 0<x<L$$

which governs the transverse deflection w of an elastic beam of flexural rigidity $b(x)$ and subjected to distributed loading $f(x)$. Assume that the boundary conditions of a clamped beam are,

$$w=\frac{dw}{dx}=0 \qquad \text{at } x=0 \text{ and } x=L.$$

(a) Identify the domain and the range space of the operator,

$$T=\frac{d^2}{dx^2}\left(b\frac{d^2}{dx^2}\right).$$

(b) Show that the operator is self-adjoint and positive definite.

16 Prove that a continuous projection P on a Hilbert space H is orthogonal if and only if it is self-adjoint.

17 Let H be a Hilbert space, and let S be a closed linear subspace of H. If P_1 and P_2 are the orthonormal projections of H onto S and $S^{\perp}$, respectively, show that the operator $T=\lambda_1 P_1+\lambda_2 P_2$ is self-adjoint if and only if λ_1 and λ_2 are real numbers.

18 (*Normal Operators*) A linear operator T is called a normal operator if it commutes with its adjoint: $T^*T=TT^*$. Show that the operator $T=\lambda_1 P_1+\lambda_2 P_2$ defined in the above exercise is a normal operator.

19 Prove that the eigenvalues of a self-adjoint operator T: $H\to H$ are real.

20 Consider the differential operator of order $2m$,

$$Lu=\sum_{0\le|\alpha|,|\beta|\le m}(-1)^{|\alpha|}D^{\alpha}\left[a^{\alpha\beta}(\mathbf{x})\,D^{\beta}u\right],$$

where the coefficients $a^{\alpha\beta}$ are assumed to be sufficiently differentiable in a region Ω in $\mathbf{R}^n$ [say that $a^{\alpha\beta}$ belongs to $C^m(\Omega)$]. If $u, v\in C_0^{2m}(\Omega)$ (space of functions that are $2m$ times continuously differentiable and vanish in the neighborhood of the boundary $\partial\Omega$ of Ω), compute the adjoint of L on $C_0^{2m}(\Omega)$ with respect to the L_2 inner product [see Eq. (2.40b) for the notation].

21 (*Strongly Elliptic Operators*) Let L be the differential operator defined in the above exercise with $a^{\alpha\beta}$ as constants. The operator L is called *strongly* *elliptic* if there is a positive constant μ_0 such that (the real part)

$$(-1)^m \operatorname{Re}\left(\sum_{|\boldsymbol{\alpha}|=|\boldsymbol{\beta}|=m} a^{\alpha\beta}\mathbf{x}^{\alpha}\mathbf{x}^{\beta}\right) \geq \mu_0 |\mathbf{x}|^{2m}, \ \mathbf{x} \in \mathbf{R}^n$$

where $\mathbf{x}^{\alpha} = x_1^{\alpha_1} x_2^{\alpha_2} \cdots x_n^{\alpha_n}$, $|\mathbf{x}|^2 = x_1^2 + x_2^2 + \cdots + x_n^2$, and

$$|\mathbf{x}|^{2m} = \sum_{|\boldsymbol{\alpha}|=|\boldsymbol{\beta}|=m} \delta^{\alpha\beta}\mathbf{x}^{\alpha}\mathbf{x}^{\beta}.$$

Show that the Laplace operator and the biharmonic operator are strongly elliptic.

22 (*Classification of Partial Differential Operators*) The classification of partial differential equations of second order is based on the analogy of classifying the equation of a conic section:

$$ax^2 + bxy + cy^2 + dx + ey + f = 0. \tag{i}$$

The equation represents

(a) An ellipse if $b^2 - 4ac < 0$.

(b) A parabola if $b^2 - 4ac = 0$. (ii)

(c) A hyperbola if $b^2 - 4ac > 0$.

We now consider the general partial differential equation of second order in two dimensions (note the analogy between this and the equation of a conic section given above):

$$a\frac{\partial^2 u}{\partial x^2} + b\frac{\partial^2 u}{\partial x \partial y} + c\frac{\partial^2 u}{\partial y^2} + d\frac{\partial u}{\partial x} + e\frac{\partial u}{\partial y} + f = 0. \tag{iii}$$

The partial differential Eq. (iii) is classified as elliptic, parabolic, and hyperbolic according to Eq. (ii). Show how the strong ellipticity defined in Exercise 21 is related to the ellipticity defined here.

23 Classify the following differential operators into elliptic, strongly elliptic, parabolic, or hyperbolic:

(a) $A = \dfrac{\partial^4}{\partial x^4} + \dfrac{\partial^4}{\partial y^4} - \dfrac{\partial^4}{\partial z^4} + i\dfrac{\partial^2}{\partial z^2}\left(\dfrac{\partial^2}{\partial x^2} + \dfrac{\partial^2}{\partial y^2}\right), \qquad i = \sqrt{-1}.$

(b) $A = \dfrac{\partial^2}{\partial x^2} - \dfrac{\partial^2}{\partial y^2}.$

(c) $A = \dfrac{\partial}{\partial x} + \dfrac{\partial}{\partial y}.$

24 Show that a necessary and sufficient condition for an orthonormal set $\{\phi_n\}$ to be a basis for a Hilbert space H is that the Parseval equality holds:

$$(u,v)=\sum_{n=1}^{\infty}(u,\phi_n)(v,\phi_n), \qquad \text{for all } u,v\in H.$$

25 Let $\{\phi_n\}$ be an orthonormal set in $L_2[a,b]$. Show that:

(a) $\{\phi_n\}$ is a basis if and only if

$$\sum_{n=1}^{\infty}\left|\int_a^x \phi_n(y)\,dy\right|^2=x-a.$$

(b) $\{\phi_n\}$ is a basis if and only if

$$\sum_{n=1}^{\infty}\left|\int_a^b\int_a^x \phi_n(y)\,dy\,dx\right|^2=\frac{(b-a)^2}{2}.$$

Hint: Show that $\|u\|_0^2=\sum_{n=1}^{\infty}|(u,\phi_n)|^2$ for all step functions u.

26 Show that the set $\{\phi_n\}$, where

$$\phi_n(x)=\frac{1}{\sqrt{b-a}}\exp\left(2\pi in\frac{x-a}{b-a}\right), \qquad n=0,\pm1,\pm2,\ldots$$

forms an orthonormal basis for $L_2[a,b]$.

27 Show that the following sets are maximal orthonormal sets in $L_2(\Omega)$ (note the interval):

(a) $\sqrt{\dfrac{2}{\pi}}\,\{\frac{1}{2},\cos x,\cos 2x,\ldots\}, \qquad \Omega=(0,\pi).$

(b) $\sqrt{\dfrac{2}{\pi}}\,\{\sin nx\}_{n=1}^{\infty}, \qquad \Omega=(0,\pi).$

(c) $\dfrac{2}{\sqrt{ab}}\left\{\sin\left(\dfrac{m\pi x}{a}\right)\sin\left(\dfrac{n\pi y}{b}\right)\right\}_{m,n=1}^{\infty},$
$\Omega=\{(x,y):\ 0<x<a, 0<y<b\}.$

28 Use the Gram-Schmidt orthonormalization process to convert the set $\{1,x,x^2x,\ldots\}$ to the following orthonormal set in $L_2[-1,1]$:

$$\left\{\frac{\sqrt{2}}{2},\sqrt{\frac{3}{2}}\,x,\sqrt{\frac{5}{8}}\,(3x^2-1),\ldots\right\},$$

which differ from the Legendre polynomials in multiplicative constraints only. (*Hint*: Obtain the fourth and fifth entries in the series.)

29 Repeat the above exercise (for the first three terms) in $H^1[-1,1]$.

30 Expand the following functions in a Fourier trigonometric series in $L_2[0,\pi]$:

(a) $f(x)=\pi x(\pi-x)$.

(b) $f(x)=\pi(\pi-2x)(\pi^2+2\pi x-2x^2)$.

$\left(\textit{Hint}: \int_0^\pi f(x)\cos nx\,dx=\frac{1}{n^4}\int_0^\pi (d^4f/dx^4)\cos nx\,dx \text{ for all } f(x) \text{ such that } f'(0)=f'(\pi)=0.\right)$

(c) $f(x)=\cos x+\frac{2}{\pi}x-1$.

(d) $f(x)=\sin^2 x$.

(e) $f(x)=\cos^2 x$.

31 Let S be the linear subspace of $L_2[-1,1]$ generated by the set

$$\left\{\frac{\sqrt{2}}{2}, \sqrt{\frac{3}{2}}\,x, \sqrt{\frac{5}{8}}\,(3x^2-1)\right\}.$$

(a) Characterize the orthogonal projection of $L_2[-1,1]$ onto S.

(b) Expand the following functions (using the results of Lemma 2.5) in terms of the orthonormal basis in S:
(i) $\sin[(1+x)\pi/2]$. **(ii)** $a_0+a_1x+a_2x^2+a_3x^3$.

32 Let $A=L_2[0,1]\to L_2[0,1]$ be a linear operator, and consider the nonhomogeneous equation

$$Au=f, \qquad f\in L_2[0,1]. \tag{i}$$

The domain $\mathfrak{D}(A)$ of operator A consists of twice differentiable functions that are square integrable along with their derivatives. Determine an approximate solution to u, in the form (the Galerkin method)

$$u=\sum_{i=1}^{N}\alpha_i\phi_i \tag{ii}$$

in the N-dimensional linear subspace S_N spanned by the linearly independent set $\{\phi_i\}_{i=1}^N$ such that the error in the equation (i.e., the residual) is made orthogonal to S_N. In particular, show that the coefficients α_i are given by solving the equation

$$\sum_{j=1}^{N}K_{ij}\alpha_j=F_i$$

where

$$K_{ij}=(A\phi_i,\phi_j)_0, \qquad F_j=(f,\phi_j)_0.$$

Show that when $A = T^*T$, where T is a linear operator, the elements K_{ij} are given by

$$K_{ij} = (T\phi_i, T\phi_j)_0$$

33 Solve the following differential equation using (a) Eq. (2.118) or the least-squares method, and (b) the Galerkin method described in the above exercise, on the N-dimensional subspace spanned by $\{\sqrt{(2/\pi)}\sin nx\}_{n=1}^{n=N}$ in $L_2[0,\pi]$:

$$-\frac{d^2u}{dx^2} = \cos x, \qquad 0 < x < \pi$$

$$u(0) = u(\pi) = 0.$$

34 Solve the differential equation $-d^2u/dx^2 = 1$, $u(0) = u(1) = 0$, in the one-dimensional subspace spanned by $\{x(1-x)\}$ in $L_2[0,1]$. Use both the least-squares method and the Galerkin method. Compute the exact solution to the differential equation and compare with the approximate solutions obtained here.

2.8 EXISTENCE AND UNIQUENESS OF SOLUTIONS

2.8.1 Vector Form of Systems of Linear Algebraic Equations

Here we show the relationship between linear operator equations and systems of linear equations and thereby give the vector interpretation of systems of linear equations. Let $f^1(u) \equiv b^1, f^2(u) \equiv b^2, \ldots, f^m(u) \equiv b^m$ be m given linear functionals defined over an n-dimensional vector space U_n. Let $\{\phi_i\}_{i=1}^n$ be the basis of U_n. Then using eq. (2.115), we can write

$$f^k(u) = \sum_{i=1}^{n} \alpha^i f^k(\phi_i) \equiv \sum_{i=1}^{n} a_i^k \alpha^i = b^k \qquad k = 1, 2, \ldots, m. \tag{2.121}$$

Equation (2.121) contains m algebraic equations in n unknowns α^i. Solving this system is equivalent to finding the components α^i, $i = 1, 2, \ldots, n$, such that the m linear functionals f^k assume the m given values b^k. The same problem can be cast into a vector form. Let $\{\psi^j\}_{j=1}^m$ be a basis of m-dimensional vector space V_m and let

$$A_i = \sum_{k=1}^{m} a_i^k \psi_k, \qquad i = 1, 2, \ldots, n, \qquad B = \sum_{k=1}^{m} b^k \psi_k.$$

Then the system of Eqs. (2.121) reduces to the vector equation

$$B=\sum_{i=1}^{n}\alpha^i A_i \qquad \text{or} \qquad \{B\}=[A]\{\alpha\}, \tag{2.122}$$

which amounts to finding all possible expressions for the vector B as a linear combination of the vectors $A_1, A_2, \ldots, A_n$. The rectangular matrix associated with the set $(A_1, A_2, \ldots, A_m)$ is the array a_i^k, which represents the linear operator with respect to the bases $\{\phi_i\}$ and $\{\psi_k\}$.

Any system of linear equations in which all b^k are zero is called *homogeneous*.

2.8.2 Conditions for the Existence and Uniqueness of Solutions of Linear Equations

First consider the case $m=n$ ($V_m=U_n$). The set of eqs. (2.122) has a unique solution only if the vectors $(A_1, A_2, \ldots, A_m)$ are linearly independent (i.e., the rows or columns of the matrix a_i^k are linearly independent). This follows from the fact that every vector B in V_n can be expressed as a unique linear combination of the vectors A_i, which, being linearly independent, form the basis of $U_n=V_n$.

Next consider the case when $U_n \subset V_m$ with vectors A_i and the vector B belong to the vector space V_m. Recall from Section 2.4.3 that the elements of V_m which can be expressed as a linear combination of n arbitrary vectors A_i of V_m form a subspace of V_m. Let r be the rank of the set of vectors A_i, and let V_r be the corresponding subspace. Since r of the vectors A_i are linearly independent, the other $(n-r)$ vectors of the set can be expressed by

$$A_{r+k}=\sum_{i=1}^{r} c_{r+k}^i A_i, \qquad (k=1,2,\ldots,n-r) \tag{2.123a}$$

where c_{r+k}^i are arbitrary. For a given vector B there exist two alternatives:

1 The vector B does not belong to V_r, and the system then has no solution.
2 The vector B belongs to V_r, and there exists a unique set of numbers α_0^i such that

$$B=\sum_{i=1}^{r}\alpha_0^i A_i. \tag{2.123b}$$

In the second case we have, from Eqs. (2.123a) and (2.123b),

$$B = \sum_{i=1}^{r} \alpha^i A_i + \sum_{k=1}^{n-r} \alpha^{r+k} A_{r+k}$$

$$= \sum_{i=1}^{r} \left(\alpha_i + \sum_{k=1}^{n-r} c^i_{r+k} \alpha^{r+k} \right) A_i .$$

Comparing the above equation with Eq. (2.123b), we see that a necessary and sufficient condition for the numbers α^i to be a solution of Eq. (2.121) is that they satisfy the equations

$$\alpha^i + \sum_{k=1}^{n-r} c^i_{r+k} \alpha^{r+k} = \alpha^i_0, \qquad i=1,2,\ldots,r. \tag{2.124}$$

The remaining $(n-r)$ unknowns α^{r+k} can be chosen arbitrarily. We summarize these results in the following theorem (cf. Theorem 1.1).

Theorem 2.17 If V_r is the subspace of V_m generated by the vectors A_i associated with the given linear system (2.121), then one of the following two alternatives holds:

(i) Either vector B does not belong to V_r and the linear system has no solution.

(ii) Or vector B belongs to V_r and the linear system has solutions (not unique): the values of $(n-r)$ of the unknowns can be chosen arbitrarily, and the values of the remaining r unknowns are determined by Eq. (2.124). ■

We now consider a special case of Eq. (2.121), namely, a system of n linear algebraic equations in n unknowns (i.e., $m=n$):

$$\sum_{j=1}^{n} a_{ij}\alpha_j = b_i, \qquad i=1,2,\ldots,n \tag{2.125a}$$

In operator form we can write Eq. (2.125a) in the form

$$A\alpha = B \tag{2.125b}$$

where A is the $n \times n$ matrix operator with real entries a_{ij}, and α and B are column vectors (i.e., n-tuples) of real numbers. Consider the following related equations:

$$A\alpha = 0, \qquad \text{homogeneous equation} \tag{2.126}$$

$$A^*\beta = 0, \qquad \text{adjoint homogeneous equation} \tag{2.127}$$

where the adjoint matrix A^* of A is the transpose A^T of A.

In light of Theorem 2.17, we can make the following observations concerning the solutions of eqs. (2.125)–(1.127):

1 If Eq. (2.126) has only the trivial (i.e., zero) solution, it follows that $\det A \neq 0$ (otherwise the trivial solution cannot be determined), and hence $\det A^* \neq 0$. Therefore the adjoint homogeneous Eq. (2.127) also has only the trivial solution. Moreover the *solvability conditions* (see Exercise 2.6.13) are automatically satisfied for any B [since the only solution of Eq. (2.127) is $\beta = 0$], and the nonhomogeneous Eq. (2.125) has one and only one solution, $\alpha = A^{-1}B$, where A^{-1} is the inverse of the matrix A.

2 If Eq. (2.126) has nontrivial solutions, then $\det A = 0$. This in turn implies that the rows (or columns) of A are linearly dependent. If these linear dependencies are also reflected in the column vector B (i.e., for example, if the third row of A is the sum of the first and second rows, we must have $B_3 = B_1 + B_2$ in order to have any solutions), then there is a hope of having a solution to the system. If there are a number of $r(\leq n)$ independent solutions to Eq. (2.125), A is said to have an *r-dimensional null space* (i.e., nullity of A is r). It can be shown that A^* also has an r-dimensional null space, which is in general different from that of A. A necessary and sufficient condition for Eq. (2.125) to have solutions is provided by the solvability condition $(B, \beta) \equiv \Sigma_{i=1}^{n} b_i \beta_i = 0$.

Example 2.45

1 Consider the following pair of equations in two unknowns α_1 and α_2:

$$3\alpha_1 - 2\alpha_2 = 4$$

$$2\alpha_1 + \alpha_2 = 5$$

or

$$\begin{bmatrix} 3 & -2 \\ 2 & 1 \end{bmatrix} \begin{Bmatrix} \alpha_1 \\ \alpha_2 \end{Bmatrix} = \begin{Bmatrix} 4 \\ 5 \end{Bmatrix}, \qquad (A\alpha = B).$$

We note that $\det A = 3 + 4 = 7 \neq 0$. The solution is then given by

$$\begin{Bmatrix} \alpha_1 \\ \alpha_2 \end{Bmatrix} = \begin{bmatrix} \frac{1}{7} & \frac{2}{7} \\ -\frac{2}{7} & \frac{3}{7} \end{bmatrix} \begin{Bmatrix} 4 \\ 5 \end{Bmatrix} = \begin{Bmatrix} 2 \\ 1 \end{Bmatrix}.$$

2 Next consider the pair of equations

$$6\alpha_1 + 4\alpha_2 = 4$$

$$3\alpha_1 + 2\alpha_2 = 2$$

or

$$\begin{bmatrix} 6 & 4 \\ 3 & 2 \end{bmatrix} \begin{Bmatrix} \alpha_1 \\ \alpha_2 \end{Bmatrix} = \begin{Bmatrix} 4 \\ 2 \end{Bmatrix}, \qquad A\alpha = B.$$

We have $\det A = 0$, because row 1 (R_1) is equal to two times row 2 (R_2). However, we also have $2B_2 = B_1$. Consequently we have one linearly independent solution, say $\boldsymbol{\alpha}^{(1)}$, and the other depends on $\boldsymbol{\alpha}^{(1)}$:

$$\boldsymbol{\alpha}^{(1)} = (2, -2).$$

Note that there are many dependent solutions to the pair. For example, $(2,-2)$, $(4,-5)$, $(-2,4), \ldots$ are solutions of $A\alpha = B$. The solution to the adjoint homogeneous equation

$$\begin{bmatrix} 6 & 3 \\ 4 & 2 \end{bmatrix} \begin{Bmatrix} \beta_1 \\ \beta_2 \end{Bmatrix} = \begin{Bmatrix} 0 \\ 0 \end{Bmatrix}$$

is given by $\beta_2 = -2\beta_1$. Note that $(B, \beta) \equiv B_1\beta_1 + B_2\beta_2 = 4(-\frac{1}{2}\beta_2) + 2\beta_2 = 0$. Hence the solvability condition is satisfied.

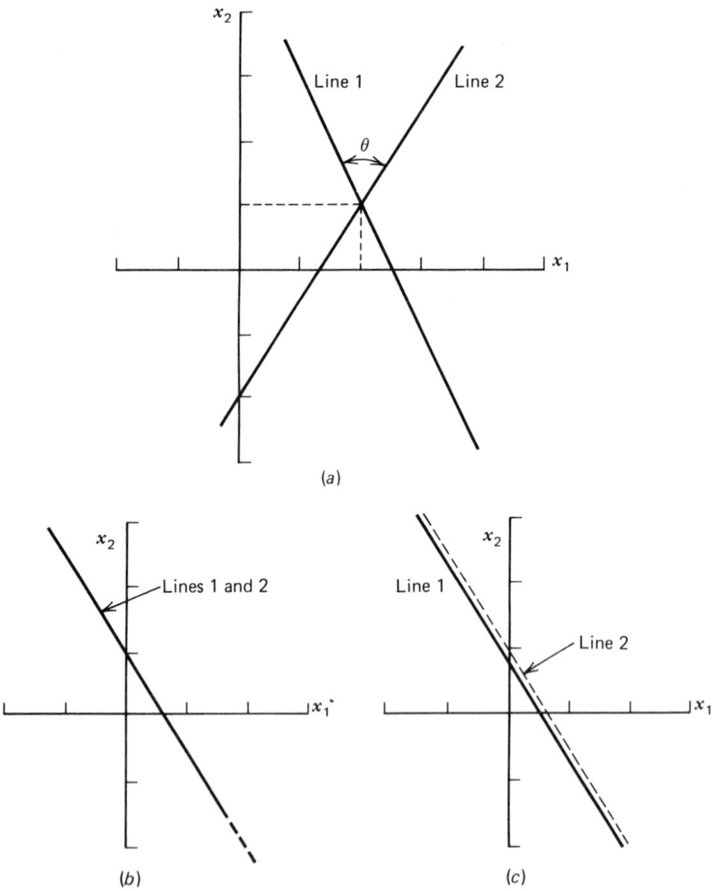

Figure 2.9 Geometric interpretation of a pair of algebraic equations. (a) Unique solution. (b) Many solutions. (c) No solution.

3 Finally consider the pair of equations

$$6\alpha_1+4\alpha_2=3$$

$$3\alpha_1+2\alpha_2=2$$

or

$$\begin{bmatrix}6 & 4\\ 3 & 2\end{bmatrix}\begin{Bmatrix}\alpha_1\\ \alpha_2\end{Bmatrix}=\begin{Bmatrix}3\\ 2\end{Bmatrix}.$$

We note that $\det A=0$ because $2R_2=R_1$. However, $2B_2\neq B_1$. Hence the pair of equations is inconsistent, and therefore no solutions exist.

Geometrically we can interpret these three pairs of equations as pairs of straight lines in R^2 with $\alpha_i=x_i$, $i=1,2$ (see Fig. 2.9). In Fig. 2.9*a* the lines represented by the two equations intersect at the point $(\alpha_1,\alpha_2)=(2,1)$. In Fig. 2.9*b* the lines coincide, or intersect at an infinite number of points, whereas in Fig. 2.9*c* the lines do not intersect at all. From this geometric interpretation one can see that if the lines are nearly parallel (i.e., the angle θ is nearly zero), the determinant of A is nearly zero [because $\tan\theta=(a_{11}a_{22}-a_{12}a_{21})/(a_{11}a_{21}+a_{12}a_{22})$], And therefore it is difficult to obtain an accurate numerical solution. In such cases the system of equations is said to be *ill-conditioned*. While these observations can be generalized to a system of n equations, the geometric interpretation becomes complicated. ■

Example 2.46

1 Consider the following set of three equations in three unknowns:

$$\alpha_1+\alpha_2+\alpha_3=2$$

$$\alpha_1-\alpha_2-3\alpha_3=3$$

$$3\alpha_1+\alpha_2-\alpha_3=1$$

or

$$A\alpha=B.$$

The adjoint homogeneous equations become

$$\beta_1+\beta_2+3\beta_3=0$$

$$\beta_1-\beta_2+\beta_3=0$$

$$\beta_1-3\beta_2-\beta_3=0$$

or

$$A^*\beta=0.$$

Solving for β, we obtain $\beta_1 = 2\beta_2 = -2\beta_3$. Hence the null space of A^* is defined by

$$\mathcal{N}(A^*) = \{(2a, a, -a), a \text{ is a real number}\}.$$

Clearly, $\mathcal{N}(A^*)$ is one-dimensional. The null space of A is given by

$$\mathcal{N}(A) = \{(a, -2a, a), a \text{ is a real number}\}.$$

Note that $\mathcal{N}(A^*) \neq \mathcal{N}(A)$, but their dimensions are the same. Clearly, $(2, 1, -1)$ is a solution of $A^*\beta = 0$, while $(1, -2, 1)$ is a solution of $A\alpha = 0$. The solvability condition gives

$$(B, \beta) = 2 \times 2 + 3 \times 1 + 1 \times (-1) \neq 0,$$

and therefore $A\alpha = B$ has *no* solution.

2 Reconsider the above linear equations with $B = \{-1, 3, 1\}^T$. Then the solvability condition is clearly satisfied. Hence there is one linearly independent solution to $A\alpha = B$ (note that $-2R_1 + R_3 = R_2$ and $-2B_1 + B_3 = B_2$):

$$\alpha \equiv (\alpha_1, \alpha_2, \alpha_3) = (1, -2, 0).$$

Only one of the three α's is arbitrary (not determined), and the remaining two α's are given in terms of the arbitrary α. For example, if α_1 is arbitrary, we have

$$\alpha_2 = -2\alpha_1 \qquad \text{and} \qquad \alpha_3 = \alpha_1 - 1.$$

3 Now consider the following set of linear equations:

$$\alpha_1 + \alpha_2 + \alpha_3 = 1$$

$$\alpha_1 + \alpha_2 - 3\alpha_3 = 2$$

$$3\alpha_1 + \alpha_2 - \alpha_3 = 3$$

or

$$A\alpha = B.$$

We have $\det A = -8 \neq 0$. It can be easily verified that $\mathcal{N}(A) = \mathcal{N}(A^*) = \{(0, 0, 0)\}$. The unique solution to $A\alpha = -B$ is given by

$$\alpha_1 = \tfrac{3}{4}, \qquad \alpha_2 = \tfrac{1}{2}, \qquad \alpha_3 = -\tfrac{1}{4}. \quad \blacksquare$$

2.8.3 Solvability (or Compatibility) Conditions for Linear Operator Equations

In the preceding section we have discussed the solvability conditions (i.e., conditions for the existence and uniqueness of solutions) to operator equations on finite-dimensional linear spaces. Since finite-dimensional linear vector spaces are closed (hence complete), one may suspect that the results presented do not carry over to infinite-dimensional inner product spaces without some restrictions on the operator as well as the associated vector spaces. This section is devoted to the discussion of the solvability conditions for abstract operator equations.

Let A be a linear operator on a domain $\mathfrak{D}(A)$ dense in a Hilbert space H. We wish to know for what *data* (or source terms) f can we solve the nonhomogeneous equation

$$Au=f, \qquad u\in\mathfrak{D}(A). \tag{2.128}$$

The existence of solutions to Eq. (2.128) largely depends on the solution of the homogeneous adjoint equation

$$A^*v=0. \tag{2.129}$$

Solvability Conditions Let u be the solution of Eq. (2.128) and v the solution of Eq. (2.129). Taking the inner product of Eq. (2.128) with v, we get

$$(Au,v)=(f,v).$$

By the definition of the adjoint and Eq. (2.129) it follows that a *necessary* condition for Eq. (2.128) to have solutions is that the data f should be orthogonal to the null space of A^*:

$$(f,v)=0 \qquad \text{for all } v\in\mathfrak{N}(A^*). \tag{2.130}$$

Clearly, the condition is identically satisfied for $v=0\in\mathfrak{N}(A^*)$. We wish to know whether there are any nonzero elements in $\mathfrak{N}(A^*)$. In particular we ask ourselves the following questions. Is Eq. (2.130) sufficient for the solvability of Eq. (2.128)? If Eq. (2.128) has solutions, how many are there?

The following theorem establishes the sufficiency of Eq. (2.130) for closed bounded below linear operator.

Theorem 2.18 Let A be a closed, bounded below linear operator on a Hilbert space H. Then Eq. (2.128) has solutions if and only if

$$\mathfrak{R}(A)=\mathfrak{N}(A^*)^{\perp}. \tag{2.131}$$

Proof We first prove that the range of a closed bounded below operator is closed [i.e., $\overline{\mathfrak{R}(A)}=\mathfrak{R}(A)$]. Let $f_n\in\mathfrak{R}(A)$ with $f_n\to f$ and $Av_n=f_n$. We wish to show that $f \in \mathcal{R}(A)$. Define $u_n = v_n - Pv_n$, where P is the projection operator from $\mathcal{D}(A)$ onto $\mathcal{N}(A)$. Then $u_n \in \mathcal{D}(A) \cap \mathcal{N}(A)^{\perp}$ and $Au_n = f_n$. We have, since A is bounded below,

$$\|u_n - u_m\| \leq \frac{1}{c}\|A(u_n - u_m)\| = \frac{1}{c}\|f_n - f_m\|,$$

which shows that $\{u_n\}$ is a Cauchy sequence in $\mathcal{D}(A)$ and therefore $u_n \to u$ in H. Since A is closed, $Au_n \to f_n$ and $u_n \to u$ imply that $u \in \mathcal{D}(A)$ and $Au = f$. Hence $\mathcal{R}(A)$ is closed.

Next we show that condition (2.131) is equivalent to Eq. (2.130). Since $\mathcal{R}(A)$ is closed $\mathcal{R}(A)^{\perp\perp} = \mathcal{R}(A) = \mathcal{N}(A^*)^{\perp}$ or $\mathcal{R}(A)^{\perp} = \mathcal{N}(A^*)$. This implies that $(Au, v) = (f, v) = 0$ for all $u \in \mathcal{D}(A)$ and $v \in \mathcal{N}(A^*)$.

Sufficiency (if) Let $f \in \mathcal{R}(A)$. Then there exists an element $u \in \mathcal{D}(A)$ such that $Au = f$. Since A is bounded below, A has a continuous inverse A^{-1} defined on its range, and therefore $u = A^{-1}f$. Moreover we have

$$\|u\| = \|A^{-1}f\| \leq C\|f\|. \tag{2.132}$$

Necessity (only if) Let $u = A^{-1}f$. Hence $Au = f$ and $(Au, v) = (f, v)$ for any $v \in \mathcal{R}(A)$. Clearly, $(Au, v) = (u, A^*v) = 0$ for any $v \in \mathcal{N}(A^*)$. In other words $u \in \mathcal{N}(A^*)^{\perp}$. This completes the proof of the theorem. ■

Example 2.47

1 (*Transverse Deflection of a String*) Consider the second-order differential equation

$$Au \equiv -\frac{d}{dx}\left(a\frac{du}{dx}\right) = f, \qquad 0 < x < L$$

$$u(0) = 0, \qquad u(L) = 0.$$

Here $a = a(x)$ and $f = f(x)$ are given functions of x. We have $\mathcal{D}(A) \subset L_2(0, L)$. More precisely, $\mathcal{D}(A) = H_0^1(0, L) \cap H^2(0, L)$. Since A is self-adjoint (with respect to the L_2 inner product), it is continuous (hence closed). A is also bounded below. The homogeneous adjoint equation becomes

$$-\frac{d}{dx}\left(a\frac{dv}{dx}\right) = 0, \qquad v \in \mathcal{D}(A^*)$$

where $\mathcal{D}(A^*) = \mathcal{D}(A)$. For constant values of a the solution v to the homogeneous adjoint equation is of the form $v = C_1 + C_2 x$. However, since $v \in H_0^1(\Omega)$, we have $v = 0$, so that $\mathcal{N}(A^*) = \{0\}$. Hence the differential equation has a unique solution.

2 (*Axial deformation of a bar*) Let us consider the same differential equation as in the previous example, with the following boundary conditions:

$$u(0)=0, \qquad \left(a\frac{du}{dx}\right)\bigg|_{x=L}=0.$$

The domains of A and A^* are given by $\mathfrak{D}(A)=H_*^1(0,L)\cap H^2(0,L)$, where

$$H_*^1(0,L)=\left\{u:\quad u\in H^1(0,L), u(0)=0, a\frac{du}{dx}(L)=0\right\},$$

and $\mathfrak{D}(A^*)=\mathfrak{D}(A)$. Again the null space of A^* is trivial and hence the solution is unique.

3 We now look at the same differential equation as in part 1 of this example, but with the boundary conditions

$$\left(a\frac{du}{dx}\right)\bigg|_{x=0}=-P_1, \qquad \left(a\frac{du}{dx}\right)\bigg|_{x=L}=P_2.$$

The domains $\mathfrak{D}(A)$ and $\mathfrak{D}(A^*)$ can be easily identified:

$$\mathfrak{D}(A)=\left\{u:\quad u\in H^1(0,L), a\frac{du}{dx}(0)=-P_1, a\frac{du}{dx}(L)=P_2\right\}\cap H^2(0,L)$$

and

$$\mathfrak{D}(A^*)=\left\{v:\quad v\in H^1(0,L), a\frac{dv}{dx}(0)=a\frac{dv}{dx}(L)=0\right\}\cap H^2(0,L).$$

The null space $\mathfrak{N}(A^*)$ contains the constant element, hence the solvability condition becomes

$$\int_0^L f\,dx+P_1+P_2=0.$$

Note that this equation represents nothing but the equilibrium of the forces. ■

It should be pointed out that the domain $\mathfrak{D}(A)$ defined in part 3 of the above example is not a linear subspace. It would be a linear space if $P_1=P_2=0$. It is possible to reformulate the problem so that the boundary conditions of a linear operator equation become homogeneous. For example, consider the operator equation $Au = f$ in Ω with the nonhomogeneous boundary condition $u = g$ on the boundary Γ of Ω. Introduce $v \in H$ such that $v = g$ on Γ. Then the problem $Aw = A(u - v) = f - Av \equiv h$ has homogeneous boundary condition $w = 0$ on Γ. However, finding the function v is not always easy in practice (see Exercise 2.7.12).

2.8.4 Conditions for the Existence and Uniqueness of Solutions of Variational Problems

In Section 2.8.2 we studied conditions for the existence and uniqueness of solutions of a system of a finite number of equations. Such sets of equations are encountered in problems with a finite number of degrees of freedom. All approximate methods seek solutions of continuum problems (i.e., problems with infinite degrees of freedom) in finite-dimensional subspaces of vector spaces in which the actual solution lies. Recall that in the least-squares approximation we approximated the solution $u \in H$ by representing it in a finite Fourier series (i.e., the index i in Eq. (2.115) takes only a finite value N). Thus all approximate methods lead ultimately (as will be seen in Chapter 3) to systems of algebraic equations. Whenever these equations are linear, the existence and uniqueness of solutions are governed by conditions discussed in Section 2.8.2.

This section is devoted to the study of the existence and uniqueness of solutions of continuum problems which lie in some infinite-dimensional vector spaces. More specifically, we state and prove the well-known Lax-Milgram theorem. The theorem can also be used for problems defined on finite-dimensional spaces, and therefore it is useful in the approximation theory. In the latter case the Lax-Milgram theorem should, as one might suspect, require the same conditions as Theorem 2.17.

Regularity of the Weak Solution Before we embark on the theory of the existence and uniqueness of solutions to the weak (or variational) problem Eq. (2.120), associated with the (differential) operator Eq. (2.116), we must impose certain smoothness conditions on the data f and the operator A so that if the weak solution exists, it is also the classical solution [i.e., the solution of Eq. (2.116)]. Here we shall not attempt to study the regularity theory, but only point out to the reader that certain smoothness conditions are implied in the existence results to be given shortly. The interested reader can consult Ref. 16 and 21 for details of the regularity theory. If A is the $2m$th order differential operator (see Exercises 2.6.20 and 2.6.21), we assume that the coefficients $a^{\alpha\beta}(x)$ in the operator A belong to $C^m(\Omega)$ and $f \in C(\Omega)$, and that the domain Ω and its boundary $\partial\Omega$ are sufficiently smooth (i.e., the so-called Lipschitz boundary—boundary without geometric singularity). Note that the bilinear form $B(\cdot, \cdot)$ associated with the operator A is well defined for $u, v \in C^m(\Omega)$, but we require $u, v \in C^{2m}(\Omega)$ in order that the weak solution coincides with the classical solution of the problem. However, in the approximate solution of the weak problem, Eq. (2.120), we seek solutions in $H^m(\Omega)$ only. With these comments we now turn to the Lax-Milgram theorem.

As has always been the case in this chapter, the Lax-Milgram theorem is given for an abstract variational (or weak) problem so that it holds for any problem that can be cast into the general form. To fix the ideas, consider the variational problem of finding u in a Hilbert space (i.e. a complete normed linear vector space) such that

$$B(u,v)=l(v), \qquad \text{for all } v \text{ in } H. \tag{2.133}$$

Here $B(\cdot,\cdot)$ is a bilinear form on the product space $H\times H$, and $l(v)$ is a linear functional on H. Many of the important physical problems can be cast into this form (indeed, one of the objectives of Chapter 3 is to formulate the variational problem associated with a given problem). We now give the conditions under which the variational Eq. (2.133) has a unique solution. The theorem is valid only for symmetric bilinear forms. A generalization of this theorem to unsymmetric bilinear forms can be found in the references given at the end of this chapter (see Ref. 20 and 21).

Theorem 2.19 (*The Lax-Milgram Theorem*) *Let H be a Hilbert space, and let $B(\cdot,\cdot)$: $H\times H\to\mathbf{R}$ be a bilinear form on $H\times H$, with the following properties*:

(i) Continuity of $B(\cdot,\cdot)$: $|B(u,v)|\le M\|u\|\|v\|$, $0<M<\infty$ (2.134)

(ii) Positive-definiteness of $B(\cdot,\cdot)$: $|B(u,u)|\ge\mu\|u\|^2$, $\mu>0$

for all $u,v\in H$. Then for any continuous linear functional l: $H\to\mathbf{R}$ on H there exists a unique vector u_0 in H such that

$$B(u_0,v)=l(v), \qquad \text{for every } v\in H. \tag{2.135}$$

Proof:

Existence For each $u\in H$, $B(u,v)$ defines a linear form $l_u(v)$ on H (i.e., l_u is an element of the dual space H'),

$$l_u(v)=B(u,v), \qquad v\in H.$$

Using the continuity of $B(\cdot,\cdot)$, we have

$$\|u\|_{H'}=\sup_{v\in H}\frac{|l_u(v)|}{\|v\|}=\sup_{v\in H}\frac{|B(u,v)|}{\|v\|}\le M\|u\|.$$

By the Reisz representation theorem, Theorem 2.15, there exists a unique element $u_l=Au$ such that

$$B(u,v)=l_u(v)=(u_l,v)=(Au,v) \tag{2.136}$$

where $(\cdot,\cdot)$ is the inner product on H, and A is a linear operator, $A\colon\ H\to H$. The operator A is continuous,

$$\|Au\|_H=\|l_u\|\le M\|u\| \qquad (\text{or } \|A\|\le M)$$

and bounded below,

$$(Au,u)=B(u,u)\ge\mu\|u\|^2$$

or

$$\|Au\|\ge\mu\|u\|, \qquad \text{for all } u\in H. \tag{2.137}$$

Since A is bounded below (see Theorem 2.9), A has a continuous inverse A^{-1}: $\mathfrak{R}(A)\to H$,

$$\|A^{-1}u\|\le\frac{1}{\mu}\|u\|. \tag{2.138}$$

Further, the range of $A, \mathfrak{R}(A)$, is the whole space H. To prove this, first we show that $\mathfrak{R}(A)$ is a closed subspace of H. Let $\{Au_n\}$ be a Cauchy sequence in H. Since A is linear and bounded below, we have

$$\begin{aligned}0=\lim_{m,n\to\infty}\|Au_n-Au_m\|&=\lim_{m,n\to\infty}\|A(u_n-u_m)\|\\&\ge\mu\lim_{m,n\to\infty}\|u_m-u_n\|,\end{aligned}$$

which shows that $\{u_n\}$ is a Cauchy sequence. Since A is continuous, $\mathfrak{R}(A)$ is closed in H (see Exercise 2.5.22).

Now suppose that $\mathfrak{R}(A)\ne H$. Then, since $\mathfrak{R}(A)$ is a closed subspace of H, there exists a nonzero element $u_0\in\mathfrak{R}(A)^\perp$ [the orthogonal complement of $\mathfrak{R}(A)$] such that

$$(Au,u_0)=0, \qquad \text{for every } u\in H.$$

In particular, we have

$$(Au_0,u_0)=B(u_0,u_0)=0, \qquad u_0\ne 0,$$

which contradicts assumption (ii) in Eq. (2.134). Hence $\mathfrak{R}(A)=H$.

Now let u^* be the unique element (by the Reisz representation theorem, Theorem 2.15) in H corresponding to $l(u)$,

$$l(u)=(u^*,u), \qquad \|l\|=\|u^*\|. \tag{2.139}$$

Hence

$$0=B(u,v)-l(v)=(Au,v)-(u^*,v)=(Au-u^*,v), \qquad \text{for every } v\in H.$$

Hence a solution to Eq. (2.135) does exist, and is given by

$$u=A^{-1}u^*. \tag{2.140}$$

Uniqueness Suppose that the problem (2.135) has two solutions, u_1 and u_2. This implies that

$$0=B(u_1-u_2,v)=(A(u_1-u_2),v), \qquad \text{for every } v\in H.$$

By choosing $v=A(u_1-u_2)$, we obtain $\|A(u_1-u_2)\|=0$. This implies that $u_1=u_2$, hence the solution is unique. ■

Note that the variational problem (2.135) can be written in terms of the operator A as

$$0=B(u,v)-l(v)=(Au,v)-l(v).$$

If the linear functional l is of the form

$$l(v)=(f,v), \tag{2.141}$$

where f is a fixed element in H, known as the *data*, we have

$$(Au-f,v)=0, \qquad \text{for every } v \text{ in } H. \tag{2.142}$$

This implies that

$$Au=f \qquad \text{in } H. \tag{2.143}$$

Thus Eq. (2.142) is the variational (or weak) problem associated with the operator Eq. (2.143). In practice we know the operator equation, and we wish to know whether the problem has a solution.

Example 2.48 Consider a boundary-value problem described by the operator equation

$$Au\equiv-\nabla^2u+c_0u=f \qquad \text{in } \Omega\subset\mathbf{R}^2 \tag{2.144}$$

$$u=0 \qquad \text{on } \Gamma$$

where Γ is the boundary of the two-dimensional region Ω. The solution to the above problem belongs to the space (with the H^1 inner product)

$$H(\Omega)=H_0^1(\Omega)\cap H^2(\Omega), \qquad (u,v)=\int_\Omega\left(uv+\frac{\partial u}{\partial x_1}\frac{\partial v}{\partial x_1}+\frac{\partial u}{\partial x_2}\frac{\partial v}{\partial x_2}\right)dx_1dx_2 \tag{2.145}$$

The bilinear and linear forms are given by

$$B(u,v)\equiv(Au,v)=\int_\Omega(-\nabla^2u+c_0u)v\,dx_1dx_2=\int_\Omega(\nabla u\cdot\nabla v+c_0uv)\,dx_1dx_2 \tag{2.146}$$

$$l(v)=\int_\Omega fv\,dx_1dx_2, \qquad u,v\in H_0^1(\Omega).$$

A quadratic form (or functional) for the problem can be defined on $H_0^1(\Omega)$ by:

$$I(u)\equiv\frac{1}{2}B(u,u)-l(u)=\int_\Omega\left\{\frac{1}{2}\left[\left(\frac{\partial u}{\partial x_1}\right)^2+\left(\frac{\partial u}{\partial x_2}\right)^2+c_0u^2\right]-fu\right\}dx_1dx_2. \tag{2.147}$$

Note that $B(\cdot,\cdot)$ is symmetric, continuous, and positive definite. Indeed, the continuity follows from the Schwarz inequality

$$\begin{aligned}|B(u,v)|\le&\left[\int_\Omega\left(\left|\frac{\partial u}{\partial x_1}\right|^2+\left|\frac{\partial u}{\partial x_2}\right|^2\right)dx_1dx_2\right]^{1/2}\left[\int_\Omega\left(\left|\frac{\partial v}{\partial x_1}\right|^2+\left|\frac{\partial v}{\partial x_2}\right|^2\right)dx_1dx_2\right]^{1/2}\\&+c_0\left(\int_\Omega|u|^2\,dx_1dx_2\right)^{1/2}\cdot\left(\int_\Omega|v|^2\,dx_1dx_2\right)^{1/2}\\&\le M\|u\|\|v\|, \qquad M=\max(1,c_0).\end{aligned} \tag{2.148}$$

The positive definiteness follows from Eq. (2.146) by setting $u=v$:

$$\begin{aligned}B(u,u)&=\int_\Omega(|\nabla u|^2+c_0u^2)\,dx_1dx_2\ge\mu\int_\Omega(|\nabla u|^2+|u|^2)\,dx_1dx_2\\&=\mu\|u\|^2, \qquad \mu=\min(1,c_0).\end{aligned} \tag{2.149}$$

Since the linear form $l(v)$ is continuous, it follows from the Lax-Milgram theorem that the problem

$$B(u,v)=l(v) \tag{2.150}$$

has a unique solution in $H_0^1(\Omega)$.

Now suppose that we wish to find an approximate solution to the variational problem in Eq. (2.150). Let S be a finite-dimensional subspace of $H_0^1(\Omega)$, and let $\{\phi_i\}_{i=1}^N$ be the basis of S. Then a typical element U in S is of the form

$$U=\sum_{i=1}^{N}\phi_i\alpha_i \approx u \tag{2.151}$$

where α_i, $i=1,2,\dots,N$, are (components of U) constants to be determined such that the problem (2.150) holds on the finite-dimensional space. Thus the approximate problem associated with Eq. (2.150) is to find U in S such that

$$B(U,V)=l(V), \qquad \text{for every } V \text{ in } S. \tag{2.152}$$

In view of Eq. (2.151), the problem can be stated as one of determining α_i such that

$$\sum_{i=1}^{N} B(\phi_i,\phi_k)\alpha_i=l(\phi_k), \qquad k=1,2,\dots,N. \tag{2.153}$$

Note that Eq. (2.153) is of the same form as Eq. (2.121),

$$a_i^k=B(\phi_i,\phi_k), \qquad \alpha^i=\alpha_i, \qquad l(\phi_k)=b^k.$$

Thus the discrete problem involves solving a system of algebraic equations. The existence and uniqueness of solutions to such problems are guaranteed by the conditions of Theorem 2.17. The positive-definiteness condition (2.134) on $B(\cdot,\cdot)$ is equivalent to the positive definiteness of the matrix $[a_i^k]$, and continuity (2.134ii) is satisfied by all bilinear forms on finite-dimensional spaces. Thus Theorem 2.19 and Theorem 2.17 are the same for finite-dimensional spaces.

Example 2.49 Consider the following bilinear form associated with plane elasticity problems: Let $\Lambda=(u,v)$ and

$$B(\Lambda,\bar{\Lambda})=\int_{\Omega}\left\{\lambda\left(\frac{\partial u}{\partial x}+\frac{\partial v}{\partial y}\right)\left(\frac{\partial \bar{u}}{\partial x}+\frac{\partial \bar{v}}{\partial y}\right)+\mu\left[\left(\frac{\partial u}{\partial y}+\frac{\partial v}{\partial x}\right)\left(\frac{\partial \bar{u}}{\partial y}+\frac{\partial \bar{v}}{\partial x}\right)\right.\right.$$
$$\left.\left.+2\frac{\partial u}{\partial x}\frac{\partial \bar{u}}{\partial x}+2\frac{\partial v}{\partial y}\frac{\partial \bar{v}}{\partial y}\right]\right\}dx\,dy \tag{2.154}$$

$$l(\bar{\Lambda})=\int_{\Omega}(f_x\bar{u}+f_y\bar{v})\,dx\,dy+\int_{\partial\Omega_2}(\hat{t}_x\bar{u}+\hat{t}_y\bar{v})\,ds$$

where u and v are the displacements, f_x and f_y are the body forces, $\hat{t}_x$ and $\hat{t}_y$ are specified tractions on the portion $\partial\Omega_2$ of the boundary, and μ, $\lambda>0$ are the *Lamé constants*. We assume that the boundary $\partial\Omega$ is *Lipschitz continuous* (in order to apply *Korn's inequality*, Ref. 12).

We introduce the following space:

$$\mathbf{H}(\Omega)=\{\Lambda=(u,v):\quad (u,v)\in H^1(\Omega)\times H^1(\Omega),\, u=0, v=0 \text{ on } \partial\Omega_1\} \tag{2.155}$$

equipped with the norm

$$\|\Lambda\|_{\mathbf{H}(\Omega)}=\left(\|u\|_{1,\Omega}^2+\|v\|_{1,\Omega}^2\right)^{1/2} \tag{2.156}$$

It can be shown that $\mathbf{H}(\Omega)$ is a closed subspace of $H^1(\Omega)\times H^1(\Omega)$. We note that the bilinear form $B(\cdot,\cdot)$ is symmetric on $\mathbf{H}(\Omega)$. The quadratic form $I(\Lambda)\equiv\frac{1}{2}B(\Lambda,\Lambda)-l(\Lambda)$ is usually referred to as the *total potential energy* of the elastic continuum Ω (assumed to be isotropic). We have

$$I(\Lambda)=\int_\Omega\left\{\frac{\lambda}{2}\left(\frac{\partial u}{\partial x}+\frac{\partial v}{\partial y}\right)^2+\mu\left[\frac{1}{2}\left(\frac{\partial u}{\partial y}+\frac{\partial v}{\partial x}\right)^2\right.\right.$$
$$\left.\left.+\left(\frac{\partial u}{\partial x}\right)^2+\left(\frac{\partial v}{\partial y}\right)^2\right]-f_xu-f_yv\right\}dx\,dy-\int_{\partial\Omega_2}\left(u\hat{t}_x+v\hat{t}_y\right)ds. \tag{2.157}$$

The problem of minimizing the total potential energy in Eq. (2.157) is equivalent to seeking the solution $(u,v)=\Lambda\in\mathbf{H}(\Omega)$ to the weak problem

$$B(\Lambda,\bar{\Lambda})=l(\bar{\Lambda}),\qquad \text{for every } \bar{\Lambda}\in\mathbf{H}(\Omega). \tag{2.158}$$

In order to show that there exists a unique solution to the problem (2.158), we verify the conditions of Theorem 2.19.

1 (*Continuity*) Using the Hölder inequality for sums

$$\sum_{i=1}^{n}|x_iy_i|\le\left(\sum_{i=1}^{n}|x_i|\right)^{1/p}\left(\sum_{i=1}^{n}|y_i|\right)^{1/q},\qquad \frac{1}{p}+\frac{1}{q}=1 \tag{2.159}$$

we write

$$|B(\Lambda,\bar{\Lambda})|\le M|\Lambda|_{\mathbf{H}(\Omega)}|\bar{\Lambda}|_{\mathbf{H}(\Omega)}\le M\|\Lambda\|_{\mathbf{H}(\Omega)}\|\bar{\Lambda}\|_{\mathbf{H}(\Omega)}$$

where $M=\max(\lambda,2\mu)$. Continuity of $l(\bar{\Lambda})$ is obvious.

2 (**H**(Ω) *Ellipticity*) Here we make use of Korn's inequality. For $\Lambda=(u,v)\in$ **H**(Ω) and $\partial\Omega$ Lipschitz continuous, there exists a constant $C(\Omega)$ such that

$$\|\Lambda\|_{\mathbf{H}(\Omega)}\leq C(\Omega)\left\{\int_{\Omega}\left[2\left(\frac{\partial u}{\partial x}\right)^2+2\left(\frac{\partial v}{\partial y}\right)^2+\frac{1}{2}\left(\frac{\partial u}{\partial y}+\frac{\partial v}{\partial x}\right)^2\right]dx\,dy\right\}^{1/2}. \tag{2.160}$$

Then we have **H** ellipticity of $B(\cdot,\cdot)$,

$$B(\Lambda,\Lambda)=\int_{\Omega}\left\{\lambda\left(\frac{\partial u}{\partial x}+\frac{\partial v}{\partial y}\right)^2+\mu\left[\left(\frac{\partial u}{\partial y}+\frac{\partial v}{\partial x}\right)^2+2\left(\frac{\partial u}{\partial x}\right)^2+2\left(\frac{\partial v}{\partial y}\right)^2\right]\right\}dx\,dy$$

$$\geq\left[\frac{\mu}{C^2(\Omega)}\right]\|\Lambda\|^2_{\mathbf{H}(\Omega)}=\mu_0\|\Lambda\|^2_{\mathbf{H}(\Omega)}. \tag{2.161}$$

Therefore the variational problem (2.158) has a unique solution. ■

Exercises 2.7

1 For the following sets of linear algebraic equations, check whether the solvability condition is satisfied. Also, discuss whether the system has solutions. If so, give the linearly independent solution.

(a) $x_1+x_2+x_3=1$
$2x_1-x_2+x_3=2$
$x_1-2x_2+3x_3=3.$

(b) $x_1+2x_2+x_3=1$
$x_1-x_2-2x_3=3$
$x_1+x_2=1.$

(c) $3x_1-x_2+2x_3=2$
$2x_1+x_2+x_3=-1$
$x_1+3x_3=2.$

(d) $10x_1+10x_2+9x_3=25$
$10x_1+9x_2+8x_3=24$
$9x_1+8x_2+7x_3=23.$

(e) $2x_1+3x_2-x_3=4$
$x_1-x_2+2x_3=2$
$x_1+2x_2-x_3=1.$

(f) $-x_1+5x_2+4x_3=0$
$x_1-x_2+2x_3=2$
$x_1+2x_2-x_3=1.$

2 Show that the system of equations in part (d) of Exercise 1 is ill-conditioned. Specifically, show that there are at least two solutions to the system that differ only in the second decimal point.

3 (*An Application of the Contraction Mapping Theorem*) Show that the system

$$\left.\begin{aligned} x_1 &= \tfrac{1}{5}x_1 + \tfrac{1}{2}x_2 + \tfrac{1}{4}x_3 - 1 \\ x_2 &= \tfrac{1}{3}x_1 - \tfrac{1}{3}x_2 - \tfrac{1}{4}x_3 + 2 \\ x_3 &= \tfrac{1}{4}x_1 + \tfrac{2}{15}x_2 + \tfrac{1}{4}x_3 + 3 \end{aligned}\right\} \Leftrightarrow x = Ax$$

has a unique solution by using the contraction mapping theorem, Theorem 2.3.

4 Consider the fourth-order differential equation governing the static equilibrium of an elastic beam:

$$\frac{d^2}{dx^2}\left[b(x)\frac{d^2w}{dx^2}\right] + f(x)=0, \qquad b(x)>0, \quad 0<x<L$$

$$w = \frac{dw}{dx} = 0 \qquad \text{at} \quad x=0 \text{ and } L.$$

Here b and f are given functions of x.

(a) Identify the domains of the operator A and its adjoint A^*.

(b) Show that A is bounded below on its domain.

(c) Determine whether the operator equation has a unique solution.

5 Repeat Exercise 3 for the boundary conditions

$$w = \frac{dw}{dx} = 0 \qquad \text{at } x=0, \quad \frac{d}{dx}\left(b\frac{d^2w}{dx^2}\right)\bigg|_{x=L} = 0, \quad \left(b\frac{d^2w}{dx^2}\right)\bigg|_{x=L} = 0.$$

6 Consider the operator equations (Dirichlet problem for the Poisson equation) in Eq. (2.144) with $c_0 = 0$. Prove that there exists a unique solution to the problem (*Hint*: Use the equivalence of norm and seminorm on $H_0^1(\Omega)$).

7 Consider the boundary-value problem (Newton's problem for the Poisson equation) described by

$$-\nabla^2 u = f \text{ in } \Omega, \qquad \frac{\partial u}{\partial n} + \alpha u = g \text{ on } \Gamma, \qquad \alpha > 0$$

where α is a constant, $\partial/\partial n$ denotes the normal derivative, and Γ is the boundary of a two-dimensional domain Ω. Assuming the fact that the mapping of $H^1(\Omega)$ into $L_2(\Gamma)$ is bounded,

$$\|u\|_{L_2(\Gamma)} \le C\|u\|_{H^1(\Omega)}, \qquad C>0,$$

show that the bilinear form and linear forms

$$B(u,v)=\int_{\Omega}\operatorname{grad} u\cdot\operatorname{grad} v\,dx\,dy+\int_{\Gamma}\alpha uv\,ds$$

$$l(v)=\int_{\Omega} fv\,dx\,dy+\int_{\Gamma} gv\,ds$$

satisfy the conditions of the Lax-Milgram theorem.

8 Consider the Newmann problem for the Poisson equation,

$$-\nabla^2 u=f \text{ in } \Omega, \qquad \frac{\partial u}{\partial n}=g \text{ on } \Gamma.$$

Show that the problem does not have a unique solution. What conditions of the Lax-Milgram theorem are violated?

9 Given the bilinear form $B(\cdot,\cdot)$: $H_0^1(\Omega)\times H_0^1(\Omega)\to \mathbf{R}$, $H_0^1(\Omega)$ as defined in Example 2.38 with $\Omega\subset\mathbf{R}^2$

$$B((u,v),(\bar{u},\bar{v}))=\int_{\Omega}\Big\{(1-\nu)(u_x\bar{u}_x+v_y\bar{v}_y)+\nu(u_x+v_y)(\bar{u}_x+\bar{v}_y)$$

$$+\frac{1-\nu}{2}(u_y+v_x)(\bar{u}_y+\bar{v}_x)\Big\}\,dx\,dy$$

where $u_x=\partial u/\partial x$, etc. Show that $B(\cdot,\cdot)$ satisfies the conditions of the Lax-Milgram theorem. Use the product norm in $H_0^1(\Omega)\times H_0^1(\Omega)$:

$$\|(u,v)\|=\sqrt{\|u\|_1^2+\|v\|_1^2}\,.$$

(*Hint*: To prove the continuity of $B(\cdot,\cdot)$ use the Hölder inequality, and to prove the positive-definite property, use the Korn inequality (Ref. 12):

$$C_1\int_{\Omega}\left(u_x^2+u_y^2+v_x^2+v_y^2\right)dx\,dy\le\int_{\Omega}\left[u_x^2+v_y^2+(u_y+v_x)^2\right]dx\,dy$$

$$\le C_2\int_{\Omega}\left(u_x^2+u_y^2+v_x^2+v_y^2\right)dx\,dy$$

where C_1 and C_2 are constraints independent of u and v. Also, the equivalence of norm and seminorm on $H_0^1(\Omega)$ must be used.)

10 (*Bending of Thin Elastic Plates*) Consider the biharmonic equation

$$Au\equiv\nabla^2\nabla^2 u=f \qquad \text{in } \Omega\subset\mathbf{R}^2$$

with the boundary conditions

$$u=0, \qquad \frac{\partial u}{\partial n}=0 \qquad \text{on } \partial\Omega.$$

Define, on the domain $\mathfrak{D}$ of ∇^4, the bilinear form by

$$B(u,v)\equiv(Au,v)=\int_\Omega\left(\frac{\partial^2 u}{\partial x^2}\frac{\partial^2 v}{\partial x^2}+2\frac{\partial^2 u}{\partial x\partial y}\frac{\partial^2 v}{\partial x\partial y}+\frac{\partial^2 u}{\partial y^2}\frac{\partial^2 v}{\partial y^2}\right)dx\,dy.$$

(a) Identify the domain $\mathfrak{D}(A)$ of the operator $A=\nabla^4$.

(b) Show that $B(u,v)$ satisfies the conditions of the Lax-Milgram theorem.

Hint: Since $u=0$, $\partial u/\partial x=0$, and $\partial u/\partial y=0$, the *Friedrichs inequality* (2.96) gives

$$\int_\Omega u^2\,dx\,dy\le C_1\int_\Omega\left[\left(\frac{\partial u}{\partial x}\right)^2+\left(\frac{\partial u}{\partial y}\right)^2\right]dx\,dy$$

$$\int_\Omega\left(\frac{\partial u}{\partial x}\right)^2dx\,dy\le C_1\int_\Omega\left[\left(\frac{\partial^2 u}{\partial x^2}\right)^2+\left(\frac{\partial^2 u}{\partial x\partial y}\right)^2\right]dx\,dy$$

$$\int_\Omega\left(\frac{\partial u}{\partial y}\right)^2dx\,dy\le C_1\int_\Omega\left[\left(\frac{\partial^2 u}{\partial x\partial y}\right)^2+\left(\frac{\partial^2 u}{\partial y^2}\right)^2\right]dx\,dy.$$

11 Consider the weak (or variational) problem: Find $u\in H\equiv H^1(\Omega)$ such that

$$B(u,v)=l(v), \qquad \text{for all } v\in H^1(\Omega)$$

where

$$B(u,v)=\int_\Omega\nabla u\cdot\nabla v\,dx\,dy, \qquad l(v)=\int_\Omega fv\,dx\,dy+\int_{\partial\Omega}gv\,dx.$$

Here f and g are known functions. Assume that a solution $u(x)$ to the problem exists. Since an arbitrary real number C also belongs to $H^1(\Omega)$, show that the necessary conditions for the variational problem to have a solution is that (*compatibility condition*)

$$\int_\Omega f\,dx\,dy+\int_{\partial\Omega}g\,ds=0.$$

12 (*Nonhomogeneous Boundary Conditions*) Consider the operator equation

$$Au = f \qquad \text{in } \Omega, \qquad f \in L_2(\Omega)$$

subjected to the nonhomogeneous boundary conditions

$$Bu = g \qquad \text{on } \Gamma.$$

The domain of the operator $\mathfrak{D}_A = \{u \in L_2(\Omega): \quad Au \in L_2(\Omega),\ Bu = g \text{ on } \Gamma\}$ is not a linear subspace of $L_2(\Omega)$. Then it is meaningless to talk about the positive definiteness, and so on, of the operator A on $\mathfrak{D}_A$. However, it is possible to convert the nonhomogeneous boundary conditions to homogeneous boundary conditions for any linear boundary-value problem so that we can define a linear subspace as the domain of A.

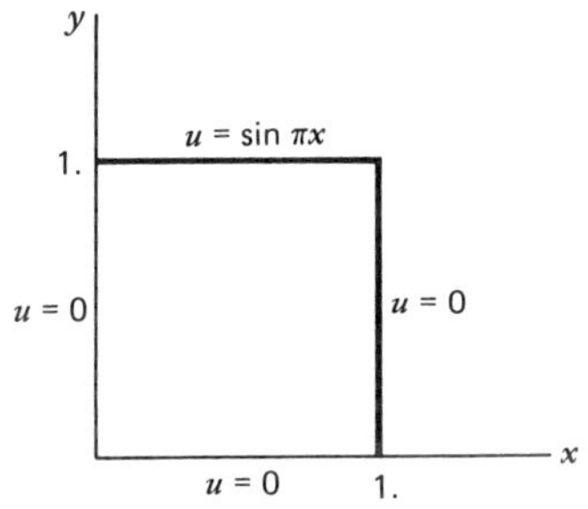

Let v be a function such that $Bv = g$ on Γ, and $Av \in L_2(\Omega)$. Then the function $w \equiv u - v$ satisfies

$$Aw = f - Av \qquad \text{in } \Omega \qquad \text{and} \qquad Bw = 0 \qquad \text{on } \Gamma.$$

Consider the boundary-value problem

$$Au \equiv -\nabla^2 u \text{ in } \Omega = \{(x, y): \quad 0 < x, y < 1\}$$

with the boundary conditions shown in the figure. Convert the problem to one with homogeneous boundary conditions, i.e., find the function v.

REFERENCES AND ADDITIONAL READING

1 Agmon, S., *Lectures on Elliptic Boundary Value Problems*, Van Nostrand, Princeton, NJ, 1965.

2 Akhiezer, N. I., and I. M. Glazman, *Theory of Linear Operators in Hilbert Space*, Frederick Unger, New York, 1961.

3 Aubin, J. P., *Approximation of Elliptic Boundary-Value Problems*, Wiley-Interscience, New York, 1972.

4 Aubin, J. P., *Applied Functional Analysis*, Wiley, New York, 1979.

5 Bachman, G., and L. Narici, *Functional Analysis*, Academic, New York, 1966.

6 Berezanskii, Ju. M., *Expansions in Eigenfunctions of Self-Adjoint Operators*, Translations of Mathematical Monographs, vol. 17, American Mathematical Society, Providence, RI, 1968.

7 Curtain, R. F., *Functional Analysis in Modern Applied Mathematics*, Academic, New York, 1977.

8 Edwards, R. E., *Functional Analysis*, Holt, Rinehart and Winston, New York, 1965.

9 Goffman, C., and G. Pedrick, *First Course in Functional Analysis*, Prentice-Hall, Englewood Cliffs, NJ, 1965.

10 Halmos, P., *Finite Dimensional Vector Spaces*, Van Nostrand Reinhold, New York, 1958.

11 Halmos, P., *Introduction to Hilbert Space*, Chelsea, New York, 1964.

12 Hlavacek, I., and J. Necas, "On Inequalities of Korn's Type: I. Boundary-Value Problems for Elliptic Systems of Partial Differential Equations; II. Applications to Linear Elasticity," *Archives of Rational Mechanics and Analysis*, vol. 36, pp. 305–311,312–334, 1970.

13 Kantorovich, L. V., and G. P. Akilov, *Functional Analysis in Normed Spaces*, Pergamon, New York, 1964.

14 Kolmogorov, A. N., and S. V. Fomin, *Elements of the Theory of Functions and Functional Analysis*, vols. 1 and 2, Graylock, Albany, NY, 1957 and 1961.

15 Kreyszig, E., *Introductory Functional Analysis with Applications*, Wiley, New York, 1978.

16 Lions, J. L., and E. Magenes, *Non-homogeneous Boundary-Value Problems and Applications*, vol. 1, Springer-Verlag, New York, 1972.

17 Liusternik, L. A., and V. J. Sobolev, *Elements of Functional Analysis*, Frederick Unger, New York, 1961.

18 Naylor, A. W., and G. R. Sell, *Linear Operator Theory in Engineering and Science*, Holt, Rinehart, and Winston, New York, 1971.

19 Nering, E. D., *Linear Algebra and Matrix Theory*, Wiley, New York, 1963.

20 Oden, J. T., *Applied Functional Analysis*, Prentice-Hall, Englewood Cliffs, NJ, 1979.

21 Oden, J. T., and J. N. Reddy, *An Introduction to the Mathematical Theory of Finite Elements*, Wiley-Interscience, New York, 1976 (for Hilbert space theory and variational theory of elliptic boundary-value problems).

22 Reed, M., and B. Simon, *Methods of Mathematical Physics*, vol. 1: *Functional Analysis*, Academic, New York, 1972.

23 Schechter, M., *Principles of Functional Analysis*, Academic, New York, 1971.

24 Showalter, R. E., *Hilbert Space Methods in Partial Differential Equations*, Pitman, London, 1977.

25 Smirnov, V. I., *A Course of Higher Mathematics*, vol. 5 *Integration and Functional Analysis*, Pergamon, New York, 1964.

26 Sobolev, S. L., *Applications of Functional Analysis in Mathematical Physics*, Translations of Mathematical Monographs, vol. 7, American Mathematical Society, Providence, RI, 1963.

27 Taylor, A. E., *Advanced Calculus*, Blaisdell, Waltham, MA, 1955.

28 Taylor, A. E., and D. C. Lay, *Introduction to Functional Analysis*, 2nd ed., Wiley, New York, 1980.

29 Yosida, K., *Functional Analysis*, 5th ed., Springer-Verlag, Berlin, 1978.

30 Wouk, A., *A Course of Applied Functional Analysis*, Wiley, New York, 1979.

Chapter Three

CALCULUS OF VARIATIONS AND VARIATIONAL METHODS

3.1 INTRODUCTION

The *calculus of variations* is a discipline that is concerned with a general theory of extreme values of functions and functions of functions. This subject, in a sense, is an extension of ordinary calculus, being concerned primarily with the theory of maxima and minima. The functions to be extremized may be functions of points, arcs, surfaces, energies, or other geometrical or physical entities. We are here concerned with functions of functions, or *functionals*.

From its inception, the calculus of variations has been closely associated with applications to mechanics, physics, engineering, numerical methods of analysis, and other fields. The purpose of this chapter is to study some of the problems that can be described by variational methods and to present some of the simpler applications of the theory. The application of the calculus of variations can be divided into three related roles. First, there are numerous problems that are posed in terms of maxima and minima and thus by their nature are formulated in terms of variational principles. Second, there are problems that can be formulated by other means, such as by vector mechanics, but that can also be formulated by means of variational principles. This leads to greater insight into the problems. Third, variational properties are not only useful for obtaining properties of solutions to practical problems, but form a powerful basis for obtaining approximate numerical answers to these problems, many of which are intractable otherwise.

Traditionally the term *variational principle* is used to imply the existence of a quadratic functional. That is, a set of partial differential equations is said to have a variational principle if there exists a quadratic functional such that the vanishing of its first variation generates the partial differential equations (called *Euler equations* of the functional). In recent times, however, the term *variational formulation* is used in a more general sense to include the variational principle as well as *weak* or *generalized formulations* of partial differential equations. Whereas the existence of a quadratic functional is not necessary for the construction of a variational approximation of equations, it is convenient for the study of certain mathematical properties such as the existence of solutions and the stability and accuracy of approximate solutions of the equations. The quadratic functional associated with a set of equations (governing a physical problem) includes all of the intrinsic features of the problem, such as the governing equations, boundary and initial conditions, constraint conditions, and even jump (discontinuity) conditions. Hamilton's principle and the formulation of the dynamical laws of motion in terms of the Lagrangian functional constitute an alternative means of formulating problems in mechanics to those of vectorial methods. Variational formulations suggest new theories, provide a means for studying mathematical properties of solutions, and provide a natural and powerful means of approximation.

The use of approximate methods based on variational statements of physical problems dates back to the work of Lord Rayleigh in the late nineteenth century. The well-known Rayleigh-Ritz method grew out of Rayleigh's method and the extensions proposed by W. Ritz in 1908. To such variationally based techniques one might add Galerkin's method, introduced by B. G. Galerkin in 1915, and a number of other closely related techniques, such as the method of Kantorovich, the collocation method, the least-squares method, or the general methods of weighted residuals. All of these methods have a common denominator: they seek to approximate variational solutions of operator (i.e., differential or integral) equations. They differ principally in the way weight functions are dealt with and in the degree of continuity required of approximation functions.

Following this introduction we shall review briefly the maxima and minima of functions. Then to give some indication of what "variational formulations" are all about, we shall discuss a couple of historical variational problems and then introduce the variational technique to determine the necessary conditions for the existence of extreme values. The Euler-Lagrange equations and Hamilton's principle shall be discussed, and the inverse procedure of constructing a variational statement of a given set of equations shall be presented. Then several methods of approximation based on the variational statements of physical problems shall be presented. These include the Ritz method, the Galerkin method, the least-squares method, and the collocation method. Numerous examples illustrating the selection of the coordinate functions in the variational methods shall be given.

3.2 MAXIMA AND MINIMA OF FUNCTIONS

3.2.1 Unconstrained Minimization

In elementary calculus we encounter the problem of finding the value of x, $a<x<b$, for which a differentiable function $f(x)$ attains a maximum or minimum value in the open interval $S=(a, b)$. A necessary condition at a

point x_0 in S at which f is either a maximum or a minimum is

$$f'(x_0) \equiv \left.\frac{df}{dx}\right|_{x=x_0} = 0 \tag{3.1a}$$

where $f'(x_0)$ is the derivative of $f(x)$ at x_0,

$$\lim_{h \to 0} \frac{f(x_0+h) - f(x_0)}{h} = f'(x_0). \tag{3.1b}$$

Note that the end points are not included in the intervals. Hence $f(x_0)$ must be compared against $f(a)$ and $f(b)$ to determine the absolute maximum or minimum value. It is also necessary in order for f to attain its maximum (or minimum) that

$$\frac{d^2f}{dx^2} < 0 \qquad \left(\frac{d^2f}{dx^2} > 0\right). \tag{3.2}$$

An example of the minimum and maximum of a function is shown in Fig. 3.1. Maximum and minimum points of f on S are called *extreme* (or *critical*) points of f on S and $f(x_0)$ is termed a *critical value* of f. Since a maximum point of f is a minimum point of $-f$, we restrict our discussion to the study of minimum points.

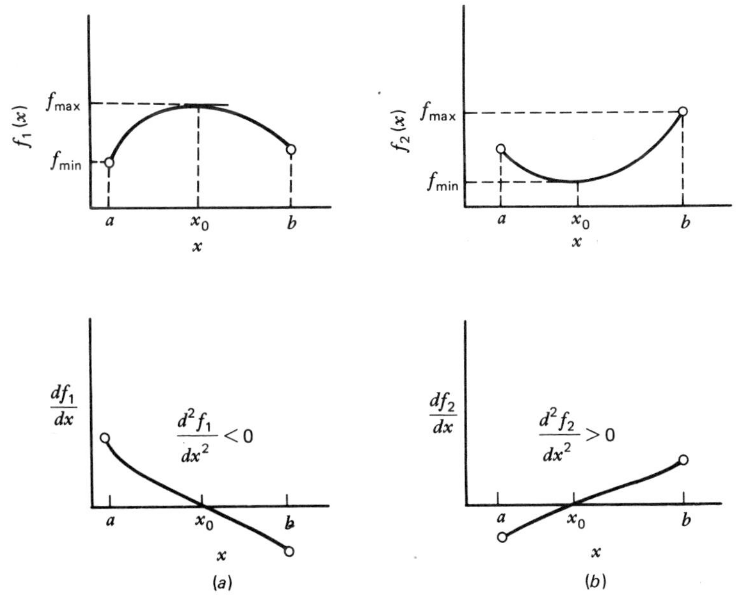

Figure 3.1 Maxima and minima of functions. (*a*) Convex function. (*b*) Concave function.

In the notation of Chapter 2 we write the statement, "a point x_0 in S minimizes f on S if and only if $f(x) \geq f(x_0)$ for all x in S" as

$$f(x_0) = \inf f(x) \qquad \text{on } S,$$

where inf denotes the infimum (see Chapter 2).

Similarly, if f is a differentiable function of x and y, and if x and y are independent, a necessary condition at a point (x_0, y_0) at which f is a (local) minimum is that the *total* derivative of f be zero at $(x, y) = (x_0, y_0)$:

$$df\bigg|_{(x_0, y_0)} = \left(\frac{\partial f}{\partial x} dx + \frac{\partial f}{\partial y} dy \right)\bigg|_{(x_0, y_0)} = d\mathbf{r} \cdot \nabla f\,|_{(x_0, y_0)} = 0. \tag{3.3}$$

Since x and y are independent variables (by hypothesis), the variations dx and dy are independent, and we must have

$$\left(\frac{\partial f}{\partial x} \right)\bigg|_{(x_0, y_0)} = 0 \qquad \text{and} \qquad \left(\frac{\partial f}{\partial y} \right)\bigg|_{(x_0, y_0)} = 0. \tag{3.4}$$

To determine whether f is minimum at (x_0, y_0), higher order derivatives must be evaluated. Similar arguments apply to a function in three-dimensional space.

3.2.2 Constrained Minimization

In the previous section we were concerned with the problem of minimizing a function f without constraints. Here we seek to minimize f on a manifold S defined by the equation of the form

$$G(x, y) = 0. \tag{3.5}$$

Equation (3.5) implies that y is dependent on x (or vice versa), and any arbitrary variation (or change) in x will affect an associated variation in y. Thus x and y are constrained by the relation (3.5). Hence Eq. (3.5) is called the constraint. To find the minima (or maxima) of a function $f(x, y)$ subject to a constraint, say, of the form of Eq. (3.5), there exist several alternate methods. For further details on the Lagrange multiplier rule and penalty methods the reader can consult the bibliography at the end of the chapter.

Direct Method The direct method is applicable to situations in which the constraint condition (3.5) can be used to write y as a function of x explicitly, that is,

$$y = g(x), \qquad dy = \frac{dg}{dx} dx. \tag{3.6}$$

Then we have

$$f(x, y)=f(x, g(x)), \qquad df=\frac{\partial f}{\partial x}dx+\frac{\partial f}{\partial g}\frac{dg}{dx}dx \tag{3.7}$$

where x is the only independent variable. The condition that $df=0$ implies, since the change dx is arbitrary,

$$\frac{\partial f}{\partial x}+\frac{\partial f}{\partial g}\cdot\frac{dg}{dx}=0. \tag{3.8}$$

Equations (3.6) and (3.8) can be used to solve for the point (x, y) at which f attains its maxima or minima.

Lagrange Multiplier Method This method is very general and frequently used in constrained optimization theory. It does not require $G(x, y)$ to be decomposable to the form of Eq. (3.6). From Eq. (3.5) we have

$$\frac{\partial G}{\partial x}dx+\frac{\partial G}{\partial y}dy=d\mathbf{r}\cdot\nabla G=0. \tag{3.9}$$

Multiplying Eq. (3.9) with a quantity λ to be specified (arbitrary for the moment), called the *Lagrange multiplier*, and adding to Eq. (3.3) (which amounts to adding zero), we have

$$d\mathbf{r}\cdot\nabla(f+\lambda G)=0. \tag{3.10}$$

Since λ is arbitrary, we can choose it in such a way that the coefficient of dy vanishes:

$$\frac{\partial f}{\partial y}+\lambda\frac{\partial G}{\partial y}=0. \tag{3.11a}$$

Then we must have, since dx is arbitrary,

$$\frac{\partial f}{\partial x}+\lambda\frac{\partial G}{\partial x}=0. \tag{3.11b}$$

Equations (3.5) and (3.11) determine x, y, and λ.

Alternately we introduce a new function $F_L(x, y, \lambda)$ with no constraints,

$$F_L(x, y, \lambda)\equiv f(x, y)+\lambda G(x, y) \tag{3.12}$$

and set its total derivative to zero (since $df=0$ and $dG=0$),

$$dF_L=\frac{\partial F_L}{\partial x}dx+\frac{\partial F_L}{\partial y}dy+\frac{\partial F_L}{\partial \lambda}d\lambda=0.$$

Setting the coefficients of dx, dy, and $d\lambda$ to zero, we obtain once again Eqs. (3.11b), (3.11a), and (3.5) respectively.

When the constraint in Eq. (3.5) is only in terms of one of the variables, say x, then the constraint equation can be used to solve (generally a nonlinear equation) for x and can be substituted into $f(x, y)$ to yield a function of y alone.

Penalty Method In 1941 Courant suggested a novel method that can be used for transforming a given constrained minimization problem into a sequence of unconstrained minimization problems. The method, now known as the *penalty function method*, seeks the minimum of a modified functional obtained from the original functional by adding terms corresponding to the constraints which vanish for the actual solution. The method, by definition, is an approximate one and yields the true solution only in the limit as a certain parameter (called the *penalty parameter*) in the modified functional approaches the value of infinity. However, in practice the method yields acceptable solutions only for finite values of the penalty parameter. The method enjoys computational simplicity over the Lagrange multiplier method owing to the fact that no additional variables, such as the Lagrange multiplier, are introduced into the problem. The fact that the extremum nature of the original functional is retained, coupled with the computational simplicity, makes the method an attractive one from the computational point of view. We describe the technique for the problem at hand.

Instead of seeking the minimum of f subject to the constraint $G(x, y)=0$, the penalty function method seeks to minimize the augmented functional,

$$F_p(x, y, \varepsilon)=f(x, y)+\frac{\varepsilon}{2}\left[G(x, y)\right]^2 \tag{3.13}$$

where ε is a preselected parameter, called the *penalty parameter*. Clearly, the pair (x, y) is sensitive to the value of ε. For sufficiently large values of ε the solution $(x_\varepsilon, y_\varepsilon)$ that minimizes F_p is sufficiently close to the actual solution (x_0, y_0). Thus the penalty method seeks a minimum of f on a larger (compact) set N which contains the subset S on which f attains its actual minimum. We assume that $G(x, y)\geq 0$ on N and that the set S of points x in N having $G^2(x, y)=0$ is not empty. These assumptions imply the existence of a point (x_0, y_0) in N which minimizes F_p on S. The following theorem establishes that the minimum point of the augmented functional approaches the actual solution.

Theorem 3.1 Let $\mathbf{x}=(x, y)$ be the minimum point of the augmented function F_p in Eq. (3.13) on N. Then

$$\lim_{\varepsilon\to\infty} \mathbf{x}(\varepsilon)=\mathbf{x}_0, \qquad \lim_{\varepsilon\to\infty} \varepsilon\left[G(\mathbf{x})\right]^2=0 \tag{3.14}$$

where $\mathbf{x}_0$ is the minimum point of f on S. Further, if $0 \le \varepsilon < \varepsilon_0$, we have

$$F_p[\mathbf{x}(\varepsilon), \varepsilon] \le F_p[\mathbf{x}(\varepsilon_0), \varepsilon_0] \le f(\mathbf{x}_0)$$
$$f[\mathbf{x}(\varepsilon)] \le f[\mathbf{x}(\varepsilon_0)] \le f(\mathbf{x}_0) \tag{3.15}$$
$$0 \le G^2(\mathbf{x}(\varepsilon_0)) \le G^2(\mathbf{x}(\varepsilon)).$$

Proof Since $\mathbf{x}(\varepsilon)$ minimizes F_p on N, and $G^2(\mathbf{x}_0)=0$, we have

$$F_p(\mathbf{x}(\varepsilon), \varepsilon) = f(\mathbf{x}(\varepsilon)) + \frac{\varepsilon}{2} G^2(\mathbf{x}(\varepsilon)) \le F_p(\mathbf{x}_0, \varepsilon) = f(\mathbf{x}_0), \tag{3.16a}$$

which in turn implies that

$$f(\mathbf{x}(\varepsilon)) \le f(\mathbf{x}_0). \tag{3.16b}$$

Now suppose that $\varepsilon < \varepsilon_0$. Then

$$0 \le F_p[\mathbf{x}(\varepsilon_0), \varepsilon] - F_p[\mathbf{x}(\varepsilon), \varepsilon]$$
$$= f(\mathbf{x}(\varepsilon_0)) - f(\mathbf{x}(\varepsilon)) + \frac{\varepsilon}{2}\left[G^2(\mathbf{x}(\varepsilon_0)) - G^2(\mathbf{x}(\varepsilon))\right]$$
$$0 \le F_p[\mathbf{x}(\varepsilon), \varepsilon_0] - F_p[\mathbf{x}(\varepsilon_0), \varepsilon_0]$$
$$= f(\mathbf{x}(\varepsilon)) - f(\mathbf{x}(\varepsilon_0)) + \frac{\varepsilon_0}{2}\left[G^2(\mathbf{x}(\varepsilon)) - G^2(\mathbf{x}(\varepsilon_0))\right]. \tag{3.17a}$$

Adding the above two inequalities, we find that

$$0 \le (\varepsilon_0 - \varepsilon)\left[G^2(\mathbf{x}(\varepsilon)) - G^2(\mathbf{x}(\varepsilon_0))\right], \tag{3.17b}$$

which implies that $G^2(\mathbf{x}(\varepsilon)) \ge G^2(\mathbf{x}(\varepsilon_0))$. Using this result in the first of Eqs. (3.17a), we obtain Eq. (3.15) for $\varepsilon < \varepsilon_0$.

Consider a sequence $\{\varepsilon_n\}$ tending to $+\infty$, and set $\mathbf{x}_n = \mathbf{x}(\varepsilon_n)$. Since $G^2(\mathbf{x}) \ge 0$, Eq. (3.16a) implies that

$$\lim_{n\to\infty} f(\mathbf{x}_n) \le f(\mathbf{x}_0), \qquad \lim_{n\to\infty} \varepsilon_n G^2(\mathbf{x}_n) = 0.$$

Therefore any limit $\bar{\mathbf{x}} = \mathbf{x}(\varepsilon_0)$ of a convergent subsequence of $\{\mathbf{x}_n\}$ must have $f(\bar{\mathbf{x}}) \le f(\mathbf{x}_0)$, $G^2(\bar{\mathbf{x}}) = 0$. Since N is compact, $\bar{\mathbf{x}}$ is in N, and since $\mathbf{x}_0$ is the unique minimum of $f(\mathbf{x}_0)$ on N subject to $G^2 = 0$, it follows that $\bar{\mathbf{x}} = \mathbf{x}_0$. Since $\{\mathbf{x}_n\}$ is bounded and every convergent subsequence of $\{\mathbf{x}_n\}$ converges to $\mathbf{x}_0$, we must have

$$\lim_{n\to\infty} \mathbf{x}_n = \lim_{n\to\infty} \mathbf{x}(\varepsilon_n) = \mathbf{x}_0.$$

From this it follows that relations (3.14) hold. This completes the proof of the theorem. ■

Returning to the functional F_p in Eq. (3.13), we set the total derivative of F_p to zero,

$$dF_p = 0 = \frac{\partial F_p}{\partial x} dx + \frac{\partial F_p}{\partial y} dy = 0 \tag{3.18}$$

We obtain

$$\frac{\partial f}{\partial x} + \varepsilon G \frac{\partial G}{\partial x} = 0, \qquad \frac{\partial f}{\partial y} + \varepsilon G \frac{\partial G}{\partial y} = 0. \tag{3.19}$$

Equations (3.19) can be used to solve for $(x_\varepsilon, y_\varepsilon)$ for any ε.

Example 3.1 Find the stationary point (i.e., critical point) of the quadratic function

$$f(x, y) = 2x^2 + y^2 - 8x + y + 1 \tag{i}$$

subject to the constraint

$$2x - y = 0. \tag{ii}$$

We will determine the stationary point by all three methods.

1 (*Direct Method*) Substituting for $y = 2x$ from (ii) into (i), we get

$$f(x) = 6x^2 - 6x + 1.$$

We have

$$\frac{df}{dx} = 0 = 12x - 6 \qquad \text{or} \qquad x = 0.5, \qquad y = 1.0. \tag{iii}$$

2 (*Lagrange Multiplier Method*) Introducing the constraint (ii) via the Lagrange multiplier λ produces the new functional F_L given by

$$F_L(x, y, \lambda) = (2x^2 + y^2 - 8x + y + 1) + \lambda(2x - y). \tag{iv}$$

The stationary point of this new function is determined by setting $dF_L = 0$:

$$\frac{\partial F_L}{\partial x} = 4x - 8 + 2\lambda = 0,$$

$$\frac{\partial F_L}{\partial y} = 2y + 1 - \lambda = 0, \qquad \frac{\partial F_L}{\partial \lambda} = 2x - y = 0. \tag{v}$$

Solving for x, y, and λ, we obtain

$$x = 0.5, \qquad y = 1.0, \qquad \lambda = 3.0. \tag{vi}$$

3 (*Penalty Method*) The penalty function F_p is given by

$$F_p(x, y) = (2x^2 + y^2 - 8x + y + 1) + \frac{\varepsilon}{2}(2x - y)^2. \tag{vii}$$

By setting $\partial F_p / \partial x = 0$, and $\partial F_p / \partial y = 0$, we obtain

$$4x - 8 + 2\varepsilon(2x - y) = 0, \qquad 2y + 1 - \varepsilon(2x - y) = 0 \tag{viii}$$

whose solution is given by

$$x_\varepsilon = \frac{8 + 3\varepsilon}{4 + 6\varepsilon}, \qquad y_\varepsilon = \frac{3\varepsilon - 1}{2 + 3\varepsilon}. \tag{ix}$$

Clearly, as $\varepsilon \to \infty$, we have

$$\lim_{\varepsilon \to \infty} x_\varepsilon = 0.5 = x, \qquad \lim_{\varepsilon \to \infty} y_\varepsilon = 1.0 = y.$$

Table 3.1 shows the convergence of $(x_\varepsilon, y_\varepsilon)$ to (x, y) for increasing values of ε. Figure 3.2*a* shows the plot of x_ε and y_ε against ε, while Fig. 3.2*b* shows the rate of convergence. For $\log \varepsilon > 2$, the rate of convergence is 1. ■

Table 3.1 Convergence of the penalty solution in Example 3.1

ε	1.0	2.0	6.0	10.0	25.0	50.0	100.0	1000.0
x_ε	1.1	0.875	0.650	0.5937	0.5389	0.5197	0.5099	0.5010
y_ε	0.4	0.625	0.850	0.9063	0.9610	0.9803	0.9901	0.9990
$2x_\varepsilon - y_\varepsilon$	1.8	1.125	0.450	0.2812	0.1167	0.0591	0.0298	0.0030

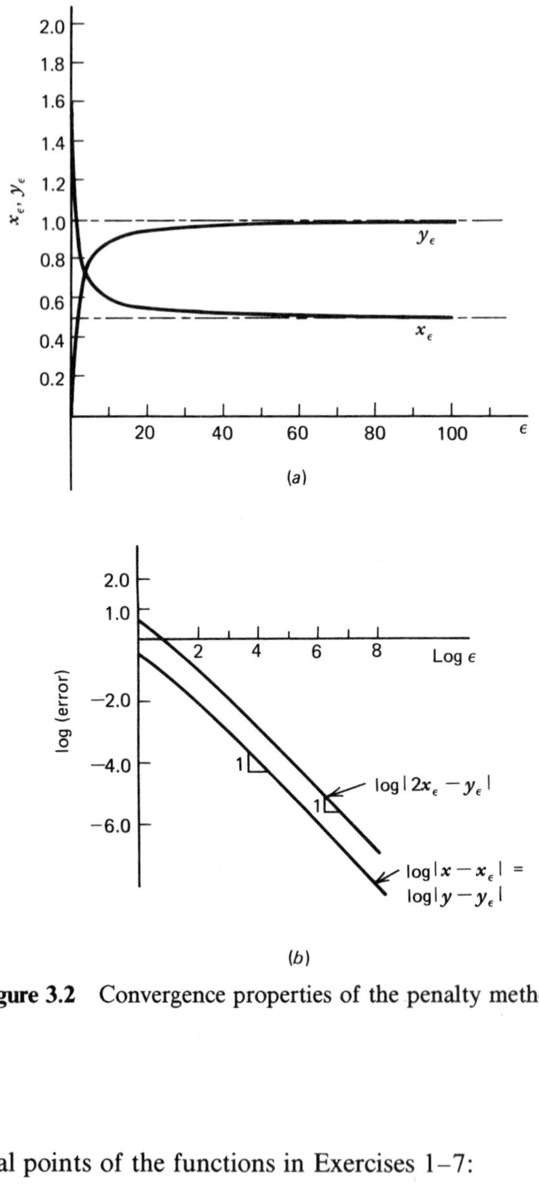

Figure 3.2 Convergence properties of the penalty method.

Exercises 3.1

Find the critical points of the functions in Exercises 1–7:

1 $f(x, y) = 5x^2 + y^2 - 4xy + 2(x - y) - 7$.

2 $f(x, y) = 10x^2 + 8xy + 2y^2 - 4(3x + y) + 6$.

3 $f(x, y) = x^4 - 8x^2 + y^2 + 16$.

4 $f(x, y) = 3xy^2(3x + 6y - 2)$.

5 $f(x, y, z) = -x^2 + y^2 + z^2 - 4z + 16$.

6 $f(x, y, z)=x^2+2y^2-3z^2+5.$

7 $f(x, y, z)=10x^2+2y^2+2z^2+6xy+4xz+2yz.$

Determine the critical points of the functions in Exercises 8–12 under the specified constraints, $G(\mathbf{x})=0$:

8 $f(x, y)=6x^2+4y^2+2xy-x+5y, \qquad G(x, y)=x+y-1.$

9 $f(x, y, z)=x^2+y^2+z^2-4z+16, \qquad G(x, z)=x^2-z.$

10 $f(x, y)=-2x^2+4xy+y^2, \qquad G(x, y)=x^2+y^2-1.$

11 $f(x, y, z)=6x^2+15y^2+33z^2+60xy-12xz+2yz,$
$G(x, y, z)=x^2+y^2+z^2-1.$

12 $f(x, y, z)=x^2+y^2+z^2, \qquad G(x, y, z)=z-xy-5.$

Find the values of x and y at which the functionals in Exercises 13–15 attain their minima. Use direct, Lagrange multiplier, and penalty methods to incorporate the constraints.

13 $f(x, y)=2x+6xy+y^2+3x^2+8, \qquad x+2y=0.$

14 $f(x, y)=5+3x-2y+6x^2-5y^2+xy, \qquad x-\sqrt{y}=0.$

15 $f(x, y)=-3+x^2-xy-y^2+6x-y, \qquad x^2+y^2=1.$

3.3 MAXIMA AND MINIMA OF FUNCTIONALS AND THE EULER EQUATIONS

Thus far we have confined our discussion to the minima or maxima of a function. In the formulation and solution of many engineering problems we encounter situations where we are required to seek the extremum (minima or maxima) of an integral expression involving the unknown function and its derivatives. The integral expression (as well as the integrand) is called a *functional*, which assumes a real value for every value of the required function. Thus, roughly speaking, functionals are "functions of functions." In most physical problems one is primarily interested in the necessary condition for a functional to achieve an extremum. This necessary condition on the functional leads to a (necessary) condition (generally in the form of a differential equation with boundary conditions) on the required function. Sufficient conditions for the functional to attain an extremum are beyond the scope of this study. Moreover, in the actual solution of physical problems one can avoid such additional considerations. To see how such problems arise naturally in practice, we shall list several historical examples here.

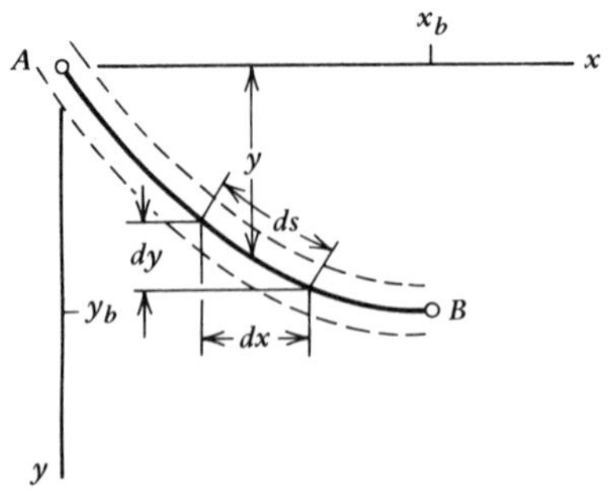

Figure 3.3 The brachistochrone problem.

3.3.1 Some Classical Variational Problems

The Brachistochrone Problem The so-called *brachistochrone* problem, posed by John Bernoulli in 1696, can be stated in engineering terms as follows. Design a chute between two points A: $(0,0)$ and B: (x_b, y_b) in a vertical plane such that a material particle, sliding without friction under its own weight, travels from point A to point B along the chute in the shortest time (see Fig. 3.3). The required curve (or geometry of the chute) is called a brachistochrone (shortest time). The problem can be formulated mathematically. Let the equation of the curve joining A to B be $y=u(x)$. At some instant of time t, let the moving particle be at a vertical distance y from point A. Then the velocity v is given by $v=\sqrt{2gy}=\sqrt{2gu}$,* where g is the acceleration due to gravity. If s represents the distance on $y=u(x)$, measured from A, we have

$$v=\frac{ds}{dt}=\sqrt{1+\left(\frac{du}{dx}\right)^2}\,\frac{dx}{dt}$$

and

$$v\,dt=\sqrt{1+\left(\frac{du}{dx}\right)^2}\,dx.$$

The total time taken by a particle in going from point A to point B is given by

$$T=\frac{1}{\sqrt{2g}}\int_0^{x_b}\sqrt{\frac{1+(du/dx)^2}{u}}\,dx\equiv I_1(u). \tag{3.20a}$$

The brachistochrone problem is then reduced to the following. Find a function

*Since the motion is, by assumption, frictionless with zero initial velocity and is influenced only by the gravitational force, we have, from the principle of the conservation of energy, $\frac{1}{2}mv^2=mgu$ or $v=\sqrt{2gu}$.

$y=u(x)$ within the class of functions with a continuous derivative [i.e., $u \in C^1(0, x_b)$] which satisfies the end conditions

$$u(0)=0, \qquad u(x_b)=y_b \tag{3.20b}$$

and which minimizes the definite integral in Eq. (3.20a). Conditions (3.20b) are called the *end conditions* or *boundary conditions*. The solution of the problem turns out to be an arc of a cycloid.

Problem of Minimal Surfaces of Revolution Given two points A: (a, y_a) and B: (b, y_b) is the plane, find a curve $y=u(x)$ with a continuous derivative such that the surface which is generated by rotation of this curve about the x axis has the smallest possible area (see Fig. 3.4).

We formulate the problem mathematically. Let S denote the area of surface generated by rotation of the curve $y=u(x)$ about the x axis, with the end conditions

$$y(a)=y_a, \qquad y(b)=y_b. \tag{3.21a}$$

Then the surface of revolution S is given by

$$S=2\pi\int_a^b y\,ds=2\pi\int_a^b u(x)\sqrt{1+\left(\frac{du}{dx}\right)^2}\,dx \equiv I_2(u), \tag{3.21b}$$

which is to be minimized by an appropriate choice of $u(x)$. The solution of this problem is a catenoid (i.e., a surface of revolution generated by a catenary).

Geodesic Problem This is one of the simplest problems of the calculus of variations. The problem consists of finding the curve of minimum length joining the two given points A: (a, y_a) and B: (b, y_b). Mathematically this amounts to minimizing the integral,

$$L=\int_a^b\sqrt{1+\left(\frac{dy}{dx}\right)^2}\,dx \equiv I_3(y). \tag{3.22}$$

Equivalently, among all curves $y=y(x)$ with a continuous derivative for which

$$y(a)=y_a, \qquad y(b)=y_b$$

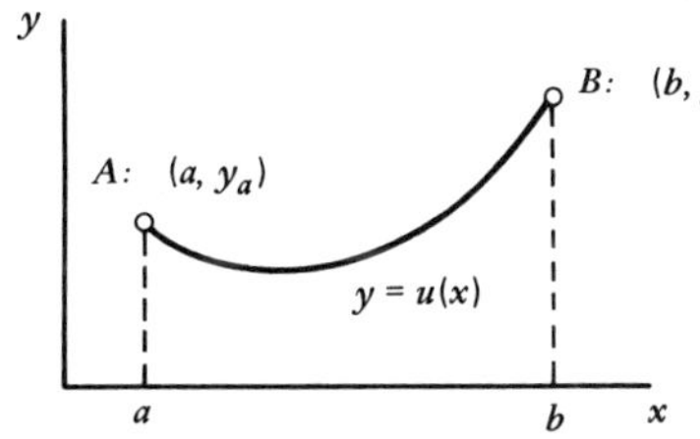

Figure 3.4 Problem of minimal surface of revolution.

find the one for which L is a minimum. The solution is a straight-line segment connecting the two points.

Isoperimetric Problem This problem is also a classic. Among all curves with a continuous derivative that join two given points A: (a, y_a) and B: (b, y_b) and have the given length L, find the one that encompasses the largest possible area. Mathematically, among all curves $y = y(x)$ for which

$$y(a) = y_a, \qquad y(b) = y_b, \qquad L = \int_a^b \sqrt{1 + \left(\frac{dy}{dx}\right)^2}\, dx \tag{3.23}$$

find the one for which

$$I_4(y) = \int_a^b y(x)\, dx \tag{3.24}$$

yields the largest possible value. A circular arc turns out to be the solution of the problem (see Exercise 3.2.4).

The isoperimetric problem differs from the other three problems in that an additional condition in the form of an integral constraint (3.23) is imposed on the class of "competing functions." The nature of this additional constraint makes the problem less easily accessible to mathematical analysis than the other three problems. As we shall see in the following sections, the Lagrange multiplier and the penalty function methods can be used to incorporate the constraints into variational problems.

In all of the examples discussed above, I_i, $i = 1, 2, 3$, is a functional because in each case a subspace of $H^1(a, b)$ is mapped into the real numbers. However, I_i is not a linear functional. (Of course, I_4 is a linear functional.) As in the case of functions the necessary condition for the existence of an extremum of a (nonlinear) functional involves its differentials. Therefore it is essential for one to understand the meaning of the differential of a functional. The next section is devoted to the discussion of the differentiation of functionals.

3.3.2 Differentiation of Functionals

Consider the following problem from the variational calculus. Given a functional I: $H \to \mathbf{R}$, where H is a Hilbert space, of the form

$$I(u) = \int_a^b F\left(x, u, \frac{du}{dx}\right) dx, \tag{3.25}$$

for what function $u(x)$ is the value of this functional a maximum? The integrand F is a known expression in u and its derivative. We will show that this problem can be reduced to the determination of the extreme value of an ordinary function.

In order for the functional $I(u)$ to make sense, we must impose certain restrictions on the choice of the functions $u(x)$ and on the integrand $F(x, u, du/dx)$. First we require that u satisfy the end conditions

$$u(a) = u_a, \qquad u(b) = u_b \tag{3.26}$$

where u_a and u_b are specified values. Next we assume that (it will be apparent in the sequel) u belongs to $H = H^2(\Omega)$, where Ω is the domain (or interval). The set $\mathcal{C} \subset H^2(\Omega)$ of all functions that satisfy these two requirements is called the *set of competing functions*. Note that $\mathcal{C}$ is not a linear space. The set $\mathcal{C}$ can be viewed as a family of smooth curves passing through points (a, u_a) and (b, u_b) in the x, y plane. Concerning the integrand F, we assume that it is twice differentiable with respect to x in some specified interval.

Suppose that u is the actual minimizer of I. Then $\bar{u} \equiv u + \alpha\eta$, where α is a small real number, and $\eta \in H^2(\Omega)$ and takes zero values at $x = a$ and b, is a neighboring function to the required function. For different values of α we get a different neighboring function, but all of these functions pass through the points (a, u_a) and (b, u_b). We note that $\bar{u}$ coincides with u for $\alpha = 0$. Upon substituting $\bar{u} \equiv u + \alpha\eta$ into the functional I, the functional becomes a continuous function of α, say $\bar{I}(\alpha)$. Consequently the differentiation of a functional is transformed into the differentiation of a function. We have

$$\bar{I}'(0) \equiv \lim_{\alpha \to 0} \frac{I(u + \alpha\eta) - I(u)}{\alpha} \equiv \frac{d}{d\alpha}(I(u + \alpha\eta))_{\alpha = 0}. \tag{3.27}$$

It should be noted that $\bar{I}'$ denotes the differential of I with respect to α (not with respect to the independent variable x). To avoid confusion we denote $\alpha\bar{I}'(0)$ by δI, called the *first variation* of I with respect to u. The *increment* $\alpha\eta$ in $\bar{u}$ is called the *variation* of u. Thus $\alpha\bar{I}'(0) = \delta I(u)$ represents the differential of I with respect to the dependent variable u.

A comment is in order concerning the requirement that u be in $H^2(a, b)$ as opposed to $H^1(a, b)$. The functional in Eq. (3.25) is well defined for u in $H^1(a, b)$. However, the necessary condition for the extremum of I leads, as we shall see shortly, to differential expressions involving the second derivatives of u. Keeping this in view, the requirement $u \in H^2(a, b)$ was imposed. When one is not interested in writing the necessary condition in differential equation form, as is the case in the variational methods of approximation, the continuity requirement on u can be weakened to $u \in H^1(a, b)$.

Example 3.2 Consider the functional on $H^4[a, b]$:

$$I(u) = \int_a^b \left\{ \frac{c_0}{2}u^2 + \frac{c_1}{4}\left(\frac{du}{dx}\right)^4 + \frac{c_2}{2}\left(\frac{d^2u}{dx^2}\right)^2 + fu \right\} dx. \tag{i}$$

Then

$$I(u+\alpha\eta)=\int_a^b\left\{\frac{c_0}{2}(u+\alpha\eta)^2+\frac{c_1}{4}\left(\frac{du}{dx}+\frac{\alpha d\eta}{dx}\right)^4\right.$$

$$\left.+\frac{c_2}{2}\left(\frac{d^2u}{dx^2}+\frac{\alpha d^2\eta}{dx^2}\right)^2+f(u+\alpha\eta)\right\}dx \tag{ii}$$

is defined for all $u,\eta\in H^4[a,b]$. We obtain

$$\delta I(u)=\alpha\left[\frac{d}{d\alpha}[I(u+\alpha\eta)]\right]_{\alpha=0}$$

$$=\alpha\left[\int_a^b\left\{c_0(u+\alpha\eta)\eta+c_1\left(\frac{du}{dx}+\frac{\alpha d\eta}{dx}\right)^3\frac{d\eta}{dx}\right.\right.$$

$$\left.\left.+c_2\left(\frac{d^2u}{dx^2}+\frac{\alpha d^2\eta}{dx^2}\right)\frac{d^2\eta}{dx^2}+f\eta\right\}dx\right]_{\alpha=0}$$

$$=\alpha\int_a^b\left\{c_0u\eta+c_1\left(\frac{du}{dx}\right)^3\frac{d\eta}{dx}+c_2\frac{d^2u}{dx^2}\frac{d^2\eta}{dx^2}+f\eta\right\}dx. \quad \blacksquare \tag{iii}$$

In the following sections frequent use of the terms *boundary* and *initial-value problems* and *homogeneous boundary conditions* is made. A differential equation is said to describe a *boundary-value problem* if the dependent variable and possibly its derivatives are required to take specified values on the boundary. An *initial-value problem* is one in which the dependent variable and possibly its derivatives are specified initially. Clearly, initial-value problems are generally time-dependent problems in which the initial values (i.e., time, $t=0$) of the dependent variable and its time derivatives are to be specified. Examples of a boundary-value problem and a boundary–initial-value problem are given by the following:

1a $$-\frac{d}{dx}\left(a\frac{du}{dx}\right)=f, \qquad 0<x<1$$

1b $$u(0)=d_0, \qquad \left.\left(a\frac{du}{dx}\right)\right|_{x=1}=g_0.$$

2a $$\rho\frac{d^2u}{dt^2}+\alpha u=f, \qquad 0<t\leq T_0$$

2b $$u(0)=u_0, \qquad \frac{du}{dt}(0)=V_0.$$

3a $$-\frac{\partial}{\partial x}\left(a\frac{\partial u}{\partial x}\right)+\rho\frac{\partial u}{\partial t}=f(x,t), \qquad 0<x<1, \quad 0<t\leq T_0$$

3b $$u(t,0)=d_0(t), \quad \left.\left(a\frac{\partial u}{\partial x}\right)\right|_{x=1}=g_0(t), \quad u(0,x)=u_0(x).$$

Conditions 1b are called *boundary conditions*, and conditions 2b are called *initial conditions*. When the specified values (i.e., d_0, g_0, u_0, and V_0) are nonzero, the conditions are said to be *nonhomogeneous*; otherwise they are called *homogeneous*. For example, $u(0)=d_0$ is a nonhomogeneous boundary condition, and the *associated* homogeneous boundary condition is $u(0)=0$.

The set of specified quantities (e.g., a, f, d_0, and g_0) is called the *data* of the problem. Differential equations in which the right-hand side f is zero are called *homogeneous differential equations*. The problem of determining the values of the constant λ such that

4a $$-\frac{d}{dx}\left(a\frac{du}{dx}\right)=\lambda u, \qquad 0<x<1$$

4b $$u(0)=0, \qquad \left.a\left(\frac{du}{dx}\right)\right|_{x=1}=0,$$

is called the *eigenvalue problem* associated with the differential equation in 1a. The values of λ are called the *eigenvalues*.

By *classical* (or exact) *solution* of a differential equation we mean the function that identically satisfies the differential equation and the specified boundary and/or initial conditions.

By *variational solution* of the operator Eq. (2.128) we mean the solution of the variational problem in Eq. (2.133). In other words, the solution $u\in H$ of Eq. (2.133) is not differentiable enough to satisfy the operator Eq. (2.128) but differentiable enough (as required in the bilinear form) to satisfy the variational Equation (2.133).

3.3.3 The Variational Symbol

Let F be a function of the dependent variable $u=u(x)$ and its derivative with respect to the independent variable x:

$$F=F(x,u,u'). \tag{3.28}$$

Then the first variation of F at u is given according to Eq. (3.27) by

$$\begin{aligned}\delta F &= \alpha\left[\frac{d}{d\alpha}F(x,u+\alpha\eta,u'+\alpha\eta')\right]_{\alpha=0}\\ &= \alpha\left[\frac{d}{d\alpha}F(x,\bar{u}(\alpha),\bar{u}'(\alpha))\right]_{\alpha=0}\\ &= \alpha\left[\left(\frac{\partial F}{\partial \bar{u}}\frac{d\bar{u}}{d\alpha}+\frac{\partial F}{\partial \bar{u}'}\frac{d\bar{u}'}{d\alpha}\right)\right]_{\alpha=0}\\ &= \alpha\left(\frac{\partial F}{\partial u}\eta+\frac{\partial F}{\partial u'}\eta'\right) \qquad (3.29)\end{aligned}$$

where $\bar{u}(\alpha)=u+\alpha\eta$ and $d\bar{u}/d\alpha=\eta$.

We now establish an analogy between the calculus of variations and the differential calculus. The use of the variational symbol avoids the intermediate algebra between the first and the last lines of Eq. (3.29).

Consider the function (of function u) F in Eq. (3.28). For a *fixed* value of x, F depends on u and its derivative. The change $\alpha\eta$ in u is called the *variation* of $\bar{u}$ and is denoted by δu,

$$\delta u \equiv \alpha\eta. \qquad (3.30)$$

Associated with this change in $u(x)$, there is a change ΔF in F,

$$\Delta F = F(x,u+\alpha\eta,u'+\alpha\eta') - F(x,u,u').$$

Expanding in powers of α gives (treating $u+\alpha\eta$ and $u'+\alpha\eta'$ as independent functions)

$$\begin{aligned}\Delta F &= F(x,u,u')+\alpha\eta\frac{\partial F}{\partial u}+\alpha\eta'\frac{\partial F}{\partial u'}+\frac{(\alpha\eta)^2}{2!}\frac{\partial^2 F}{\partial u^2}\\ &\quad+\frac{(\alpha\eta')(\alpha\eta)}{2!}\frac{\partial^2 F}{\partial u\,\partial u'}+\frac{(\alpha\eta')^2}{2!}\frac{\partial^2 F}{\partial u'^2}+\cdots-F(x,u,u')\\ &= \alpha\eta\frac{\partial F}{\partial u}+\alpha\eta'\frac{\partial F}{\partial u'}+\alpha R_1(\alpha) \qquad (3.31)\end{aligned}$$

where $\lim_{\alpha\to 0}R_1(\alpha)=0$. In analogy with the total differential of a function of two variables, the first two terms in Eq. (3.31) are defined to be the *first variation* of F,

$$\delta F = \frac{\partial F}{\partial u}\alpha\eta+\frac{\partial F}{\partial u'}\alpha\eta'. \qquad (3.32)$$

Note that in the special case when $F=u$, Eq. (3.32) gives Eq. (3.30). Further, when $F=u'$, we have

$$\alpha\eta'=\delta u'. \tag{3.33}$$

Consequently the first variation of F can be written in terms of the variations of the dependent variable u and its derivatives,

$$\delta F=\frac{\partial F}{\partial u}\delta u+\frac{\partial F}{\partial u'}\delta u', \tag{3.34}$$

which is equivalent to Eq. (3.29). Once again, note the analogy between the first variation, Eq. (3.34), and the total differential

$$dF=\frac{\partial F}{\partial x}dx+\frac{\partial F}{\partial u}du+\frac{\partial F}{\partial u'}du'. \tag{3.35}$$

Since x is *not varied* (during the variation of u to $u+\delta u$), $dx=0$, and the analogy between δF and dF becomes apparent. That is, δ *acts as a differential operator with respect to dependent variables.* It can be easily verified that the laws of variation of sums, products, ratios, powers, and so forth, are completely analogous to the corresponding laws of differentiation. For example, if $F_1=F_1(u)$ and $F_2=F_2(u)$, then

$$\delta(F_1F_1)=F_1\delta F_2+F_1\delta F_2$$

$$\delta\left(\frac{F_1}{F_2}\right)=\frac{F_2\,\delta F_1-F_1\,\delta F_2}{F_2^2} \tag{3.36}$$

Further, the variational operator can be interchanged with differential operators. From Eqs. (3.30) and (3.33) it follows that

$$\frac{d}{dx}(\delta u)=\frac{d}{dx}(\alpha\eta)=\alpha\frac{d\eta}{dx}=\alpha\eta'=\delta u'=\delta\left(\frac{du}{dx}\right). \tag{3.37}$$

All of the above discussion can be extended to two dimensions and to functions F which depend on more than one dependent variable. Let

$$F=F(x,y,u,v,u_x,v_x,u_y,v_y)$$

where $u=u(x,y)$ and $v=v(x,y)$ are dependent variables, and $u_x=\partial u/\partial x$, $u_y=\partial u/\partial y$, and so on. The first variation of F is given by

$$\delta F=\frac{\partial F}{\partial u}\delta u+\frac{\partial F}{\partial v}\delta v+\frac{\partial F}{\partial u_x}\delta u_x+\frac{\partial F}{\partial v_x}\delta v_x+\frac{\partial F}{\partial u_y}\delta u_y+\frac{\partial F}{\partial v_y}\delta v_y \tag{3.38}$$

and

$$\delta(F_1(u,v)F_2(u,v)) = \delta F_1(u,v)F_2 + \delta F_2(u,v)F_1$$

$$= \left(\frac{\partial F_1}{\partial u}\delta u + \frac{\partial F_1}{\partial v}\delta v\right)F_2 + \left(\frac{\partial F_2}{\partial u}\delta u + \frac{\partial F_2}{\partial v}\delta v\right)F_1. \tag{3.39}$$

Example 3.3 Consider again the functional of Example 3.2. Here we have

$$F = \left\{\frac{c_0 u^2}{2} + \frac{c_1}{4}\left(\frac{du}{dx}\right)^4 + \frac{c_2}{2}\left(\frac{d^2u}{dx^2}\right)^2 + fu\right\} \tag{i}$$

and

$$I(u) = \int_a^b F(u, u', u'')\, dx. \tag{ii}$$

The first variation of $I(u)$ for fixed end points is given by

$$\delta I = \int_a^b \delta F(u, u', u'')\, dx,$$

$$= \int_a^b \left(\frac{\partial F}{\partial u}\delta u + \frac{\partial F}{\partial u'}\delta u' + \frac{\partial F}{\partial u''}\delta u''\right) dx. \tag{iii}$$

From (i) we have

$$\frac{\partial F}{\partial u} = c_0 u + f, \qquad \frac{\partial F}{\partial u'} = c_1\left(\frac{du}{dx}\right)^3, \qquad \frac{\partial F}{\partial u''} = c_2\frac{d^2u}{dx^2}. \tag{iv}$$

Therefore

$$\delta I(u) = \int_a^b \left\{(c_0 u + f)\,\delta u + c_1\left(\frac{du}{dx}\right)^3 \delta u' + c_2\frac{d^2u}{dx^2}\delta u''\right\} dx. \tag{v}$$

Since $\delta u = \alpha\eta$, we get Eq. (iii) of Example 3.2. ■

3.3.4 The Space of Admissible Variations

Consider a simple, but typical, variational problem. Find a function $u = u(x) \in H^2[a, b]$ such that $u(a) = u_a$ and $u(b) = u_b$, and

$$I(u) = \int_a^b F(x, u(x), u'(x))\, dx \tag{3.40}$$

is a minimum.

In analyzing the problem we are not interested in all functions $u \in H^2[a,b]$, but only in those functions from $H^2[a,b]$ that satisfy the stated boundary (or end) conditions. The set of all such functions is called, for obvious reasons, the *set of competing functions*. We shall denote the set by $\mathcal{C}$. For the problem at hand $\mathcal{C}$ is given by

$$\mathcal{C} = \{u: \quad u \in H^2[a,b], u(a) = u_a, u(b) = u_b\}. \tag{3.41}$$

Note that the set is *not* a linear vector space unless the boundary conditions are homogeneous, that is $u_a = 0$ and $u_b = 0$. The problem is to seek an element u from $\mathcal{C}$ which renders I a minimum (or maximum).

If $u_0 \in \mathcal{C}$, then $(u_0 + \alpha\eta) \in \mathcal{C}$ for every η satisfying the conditions $\eta \in H^2[a,b], \eta(a) = \eta(b) = 0$. The space of all such elements is called the *space of admissible variations*,

$$\mathcal{H} = \{\eta: \quad \eta \in H^2[a,b], \eta(a) = \eta(b) = 0\}. \tag{3.42}$$

Although in general the space of admissible variations is not a linear vector space, it is a normed linear vector space in the present case (see Fig. 3.5). It is normed since $\mathcal{H} \subset H^2[a,b]$, and it is linear because for any $\eta_1, \eta_2 \in \mathcal{H}$, we have $\alpha\eta_1 + \beta\eta_2 \in \mathcal{H}, \alpha, \beta \in \mathrm{R}$.

We now define the set of competing functions and the space of admissible variations for a general case. Let I be a functional defined on the normed linear vector space V [more specifically, on a dense linear subspace $\mathfrak{D}(I)$ of V]. Then the *set of competing functions* $\mathcal{C}$ is a subset of V whose elements satisfy the boundary conditions and any constraint conditions of the problem. For a given set of competing functions $\mathcal{C} \subset V, \mathcal{H} \subset V'$ (the dual space of V) is called a *space of admissible variations* of $\mathcal{C}$ if, for all $u \in \mathcal{C}, \eta \in \mathcal{H}, (u + \alpha\eta) \in \mathcal{C}$. If the set of competing functions is a linear space, then the space of admissible variations is also a linear space. In most of the problems of interest to us, the vector space V is the $L_2(\Omega)$ space [which is self-dual, i.e., $L'_2(\Omega) = L_2(\Omega)$] and

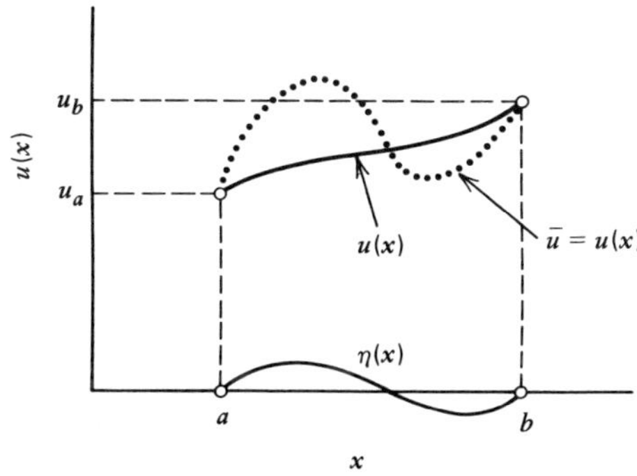

Figure 3.5 The variation of $u(x)$.

the domain $\mathfrak{D}(I)$ of the functional I is some dense subspace of $L_2(\Omega)$ [for example, $H^2(\Omega)$].

For example, for the isoperimetric problem discussed in Section 3.3.1, the space of competing functions is given by

$$\mathcal{C}=\left\{u:\quad u\in H^2[a,b], u(a)=y_a, u(b)=y_b, \int_a^b \sqrt{1+(u')^2}\,dx=L\right\}, \tag{3.43a}$$

which is not a linear vector space (even in the case in which $y_a=y_b=0$). The space of admissible variations is given by

$$\mathcal{H}=\{\eta:\quad \eta\in H^2[a,b], \eta(a)=\eta(b)=0, (u+\alpha\eta)\in\mathcal{C}\qquad \text{for all } u\in\mathcal{C}\}. \tag{3.43b}$$

3.3.5 Necessary Conditions for the Existence of a Minimum of a Functional

Let $I(u)$, $u\in V$, be a differentiable functional defined on a normed linear space V, and let $\mathcal{C}\in V$ denote the space of competing functions. We assume that $\mathcal{C}$ admits a linear space of admissible variations $\mathcal{H}$. Then an element $u_0\in\mathcal{C}$ is said to yield a *relative minimum* (*maximum*) for $I(u)$ in $\mathcal{C}$ if

$$I(u)-I(u_0)\geq 0 \qquad (\leq 0). \tag{3.44}$$

If $I(u)$ assumes a relative minimum (maximum) at $u_0\in\mathcal{C}$ relative to elements $u\in\mathcal{C}$, then it follows from the definition of the space of admissible variations and Eq. (3.43) that

$$I(u_0+\alpha\eta)-I(u_0)\geq 0 \qquad (\leq 0) \tag{3.45}$$

for all $\eta\in\mathcal{H}$, $\|\eta\|<\varepsilon$, and $\alpha\in\mathbf{R}$. Since u_0 is the minimizer, any other function $u\in\mathcal{C}$ is of the form $u=u_0+\alpha\eta$, and the actual minimizer is determined by setting $\alpha=0$. Once $u_0(x)$ and $\eta(x)$ are assigned, $I(u)$ is a *function* of α alone, say $\bar{I}(\alpha)$. Now a necessary condition for $I(u)=\bar{I}(\alpha)$ to attain a minimum is that

$$\frac{d\bar{I}(\alpha)}{d\alpha}=\frac{d}{d\alpha}[I(u_0+\alpha\eta)]=0. \tag{3.46}$$

On the other hand, $I(u)$ attains its minimum at u_0, that is, at $\alpha=0$. These two conditions together imply $(d\bar{I}(\alpha)/d\alpha)|_{\alpha=0}=0$, which is nothing but

$$\delta I(u_0)=0. \tag{3.47}$$

Analogous to the second necessary condition for ordinary functions, the second necessary condition for a functional to assume a relative minimum (maximum) is that the second variation $\delta^2 I(u)$ is greater (less) than zero. The second variation $\delta^2 I(u)$ of a functional $I(u)$ is given by

$$\delta^2 I(u) \equiv \frac{\alpha^2}{2}\left[\frac{d^2}{d\alpha^2} I(u+\alpha\eta)\right]_{\alpha=0} \tag{3.48}$$

for all $\eta \in V$ and $\alpha \in \mathbf{R}$.

3.3.6 Euler Equations: Natural and Essential Boundary Conditions

We now return to the problem of finding a minimum of the functional

$$I(u) = \int_{x_1}^{x_2} F\left(x, u, \frac{du}{dx}\right) dx \tag{3.49}$$

subject to the end conditions

$$u(x_1) = u_1, \qquad u(x_2) = u_2. \tag{3.50}$$

It is clear that any candidate for the minimizing function should satisfy the end conditions and be differentiable a sufficient number of times (twice in the present case, as will be seen later) with respect to the independent variable x. Thus the *set of competing functions* $\mathcal{C}$, for the present case, consists of functions that are differentiable twice and satisfy the end conditions (3.50). Functions from the admissible set can be viewed as smooth (i.e., differentiable twice) functions passing through points (x_1, u_1) and (x_2, u_2), as shown in Fig. 3.5. Assuming that for each admissible function $\bar{u}$ in $\mathcal{C}$, $F(x, \bar{u}, d\bar{u}/dx)$ exists and is continuously differentiable with respect to its arguments, and $I(\bar{u})$ takes one and only one real value, we seek the particular function $u(x)$ of the set $\mathcal{C}$ which makes the integral a minimum.

Suppose that $u(x)$ is the function that minimizes I in Eq. (3.49). Then every function $\bar{u}$ in the set $\mathcal{C}$ can be represented in the form

$$\bar{u}(x) = u(x) + \alpha\eta(x) \tag{3.51}$$

where α is a small number and $\eta(x)$ is a continuously differentiable function. Note that, since $\bar{u}$ and $u(x)$ satisfy the end conditions, $\eta(x)$ *vanishes* at the end points (i.e., satisfies the *homogeneous* boundary conditions):

$$\eta(x_1) = 0, \qquad \eta(x_2) = 0. \tag{3.52}$$

That is, $\eta(x)$ belongs to the set of admissible variations $\mathcal{H}$.

Now let us calculate δI as defined in Eq. (3.47) for the integral in Eq. (3.49):

$$\bar{I}(\alpha)=\int_{x_1}^{x_2} F(x,\bar{u},\bar{u}')\,dx, \qquad \bar{u}=u+\alpha\eta, \qquad \bar{u}'=\frac{du}{dx}+\alpha\frac{d\eta}{dx}.$$

Then

$$\frac{d\bar{I}(\alpha)}{d\alpha}=\int_{x_1}^{x_2}\frac{d}{d\alpha}F(x,\bar{u},\bar{u}')\,dx=\int_{x_1}^{x_2}\left(\frac{\partial F}{\partial \bar{u}}\frac{d\bar{u}}{d\alpha}+\frac{\partial F}{\partial \bar{u}'}\frac{d\bar{u}'}{d\alpha}\right)dx$$

$$=\int_{x_1}^{x_2}\left(\frac{\partial F}{\partial \bar{u}}\eta+\frac{\partial F}{\partial \bar{u}'}\eta'\right)dx$$

or

$$\delta I=\alpha\left[\frac{dI(\alpha)}{d\alpha}\right]_{\alpha=0}=\alpha\int_{x_1}^{x_2}\left(\frac{\partial F}{\partial u}\eta+\frac{\partial F}{\partial u'}\eta'\right)dx$$

$$=\alpha\int_{x_1}^{x_2}\left[\frac{\partial F}{\partial u}-\frac{d}{dx}\left(\frac{\partial F}{\partial u'}\right)\right]\eta\,dx+\alpha\left(\frac{\partial F}{\partial u'}\eta\right)\Bigg|_{x=x_1}^{x=x_2}=0 \qquad (3.53)$$

where integration by parts is used to arrive at the last step. Since η vanishes at the end points, the second term on the right-hand side of Eq. (3.53) becomes zero.

Note that $\delta I(u)$ is a linear functional on $L_2(a,b)$. It is also continuous, since $(\delta I(u;\eta)\equiv l_u(\eta))$

$$\|l_u(\eta)\|\leq\|T(u)\|\|\eta\|, \qquad Tu\equiv\frac{\partial F}{\partial u}-\frac{d}{dx}\left(\frac{\partial F}{\partial u'}\right) \qquad (3.54)$$

where $\|T(u)\|$ depends only on u. Thus Eq. (3.53) can be viewed as the representation (see Theorem 2.15) of the linear functional l_u:

$$l_u(\eta)=(T(u),\eta)_0.$$

The operator T: $H^2(a,b)\to L_2(a,b)$ can be determined explicitly once the function F is known. In the variational calculus $T(u)$ is referred to as the *gradient of the functional $I(u)$*.

In order for $I(u)$ to be an extremum, $\bar{I}'(\alpha)$ must vanish at $\alpha=0$. Since η is arbitrary in the interval (x_1,x_2), the integrand in Eq. (3.53) must vanish, and we have by the fundamental lemma of the calculus of variations, Lemma 2.2,

$$T(u)=\frac{\partial F}{\partial u}-\frac{d}{dx}\left(\frac{\partial F}{\partial u'}\right)=0. \qquad (3.55)$$

Thus in order for the functional I to attain its minimum at u, it is necessary that u satisfy Eq. (3.55). Equation (3.55) [i.e., grad $I(u)=0$] is called the *Euler* (or *Euler-Lagrange*) *equation*.

Next consider the problem of finding $(u,v)\in H^2(\Omega)\times H^2(\Omega)$, defined on a two-dimensional region Ω, such that the following functional is minimized:

$$I(u,v)=\int_\Omega F(x,y,u,v,u_x,v_x,u_y,v_y)\,dxdy \tag{3.56}$$

where $u_x=\partial u/\partial x$, $u_y=\partial u/\partial y$, and so on. For the moment we assume that u and v are arbitrary on the boundary Γ of Ω. The vanishing of the first variation of $I(u,v)$ is written

$$\delta I(u,v)=\delta_u I(u,v)+\delta_v I(u,v)=0.$$

Here δ_u and δ_v denote (partial) variations with respect to u and v, respectively. We have [see also Eq. (3.38)]

$$\delta I=\int_\Omega\left\{\frac{\partial F}{\partial u}\delta u+\frac{\partial F}{\partial u_x}\delta u_x+\frac{\partial F}{\partial u_y}\delta u_y+\frac{\partial F}{\partial v}\delta v+\frac{\partial F}{\partial v_x}\delta v_x+\frac{\partial F}{\partial v_y}\delta v_y\right\}dxdy. \tag{3.57}$$

The next step in the development involves the use of integration by parts, or the gradient theorem on the second, third, fifth, and sixth terms in Eq. (3.57). Consider the second term:

$$\int_\Omega\frac{\partial F}{\partial u_x}\delta u_x\,dxdy=\int_\Omega\left[\frac{\partial}{\partial x}\left(\frac{\partial F}{\partial u_x}\delta u\right)-\frac{\partial}{\partial x}\left(\frac{\partial F}{\partial u_x}\right)\delta u\right]dxdy$$

$$=\oint_\Gamma\frac{\partial F}{\partial u_x}\delta u n_x\,ds-\int_\Omega\frac{\partial}{\partial x}\left(\frac{\partial F}{\partial u_x}\right)\delta u\,dxdy. \tag{3.58}$$

Using a similar procedure on the other terms, we obtain

$$0=\int_\Omega\left\{\left[\frac{\partial F}{\partial u}-\frac{\partial}{\partial x}\left(\frac{\partial F}{\partial u_x}\right)-\frac{\partial}{\partial y}\left(\frac{\partial F}{\partial u_y}\right)\right]\delta u\right.$$

$$\left.+\left[\frac{\partial F}{\partial v}-\frac{\partial}{\partial x}\left(\frac{\partial F}{\partial v_x}\right)-\frac{\partial}{\partial y}\left(\frac{\partial F}{\partial v_y}\right)\right]\delta v\right\}dxdy$$

$$+\oint_\Gamma\left\{\left[\frac{\partial F}{\partial u_x}n_x+\frac{\partial F}{\partial u_y}n_y\right]\delta u+\left[\frac{\partial F}{\partial v_x}n_x+\frac{\partial F}{\partial v_y}n_y\right]\delta v\right\}ds. \tag{3.59}$$

Once again we have a continuous linear functional on $H^2(\Omega)\times H^2(\Omega)$:

$$l_\mu(\lambda)=(T(\mu),\lambda)_\Omega+(B(\mu),\lambda)_\Gamma$$
$$\equiv(T_1(\mu),\delta u)+(T_2(\mu),\delta v)+(B_1(\mu),\delta u)+(B_2(\mu),\delta v) \quad (3.60)$$

where $\mu=(u,v)$ and $\lambda=(\delta u,\delta v)$, and

$$T(\mu)=\begin{Bmatrix} T_1(\mu) \\ T_2(\mu) \end{Bmatrix}\equiv\begin{Bmatrix} \dfrac{\partial F}{\partial u}-\dfrac{\partial}{\partial x}\left(\dfrac{\partial F}{\partial u_x}\right)-\dfrac{\partial}{\partial y}\left(\dfrac{\partial F}{\partial u_y}\right) \\ \dfrac{\partial F}{\partial v}-\dfrac{\partial}{\partial x}\left(\dfrac{\partial F}{\partial v_x}\right)-\dfrac{\partial}{\partial y}\left(\dfrac{\partial F}{\partial v_y}\right) \end{Bmatrix}_\Omega \quad (3.61a)$$

$$B(\mu)=\begin{Bmatrix} B_1(\mu) \\ B_2(\mu) \end{Bmatrix}=\begin{Bmatrix} \dfrac{\partial F}{\partial u_x}n_x+\dfrac{\partial F}{\partial u_y}n_y \\ \dfrac{\partial F}{\partial v_x}n_x+\dfrac{\partial F}{\partial v_y}n_y \end{Bmatrix}_\Gamma. \quad (3.61b)$$

If λ is arbitrary in Ω and on Γ separately, we conclude from Lemma 2.2 that

$$T_1(\mu)=0, \qquad T_2(\mu)=0 \text{ in } \Omega \quad (3.62a)$$

$$B_1(\mu)=0, \qquad B_2(\mu)=0 \text{ on } \Gamma. \quad (3.62b)$$

Equations (3.62a) and (3.62b) are then the Euler equations for the functional $I(u,v)$. On the other hand, if μ is specified on the boundary Γ of Ω, then the variation λ of μ is identically zero on Γ, and therefore the boundary terms vanish identically [and we do not have Eq. (3.62b)]. The specification of μ on Γ is called the *essential boundary condition*. Specification of $B(\mu)=0$ is called the *natural boundary condition* of the problem.

From Eq. (3.60) it is easy to identify the natural and essential boundary conditions of the problem: in each of the pairings on boundary Γ, specifying the first element (which contains no variations of the dependent variables) constitutes the natural boundary condition, and vanishing of the second element (or, equivalently, specifying the quantity in front of the variational operator) constitutes the essential boundary condition. Thus we have

$$\begin{aligned} u&=\text{specified}, \hat{u} \quad \text{on} \quad \Gamma \\ v&=\text{specified}, \hat{v} \quad \text{on} \quad \Gamma \end{aligned} \quad (3.63a)$$

or

$$\frac{\partial F}{\partial u_x} n_x + \frac{\partial F}{\partial u_y} n_y = 0 \qquad \text{on} \quad \Gamma$$

$$\frac{\partial F}{\partial v_x} n_x + \frac{\partial F}{\partial v_y} n_y = 0 \qquad \text{on} \quad \Gamma. \tag{3.63b}$$

Note that either the natural or the essential boundary condition must be known on the boundary Γ (or on a portion of it). In all physical problems only one of the two types of boundary conditions is specified at a point on the boundary. Thus there are four possible combinations of natural and essential boundary conditions for the problem under discussion.

All of the foregoing discussion can be applied to the general case of a functional involving p dependent variables with mth-order partial derivatives with respect to n independent variables. In this case there will be p Euler equations involving $2m$th-order derivatives in n independent variables. We now consider some illustrative examples.

Example 3.4 Consider the functional

$$\Pi(u,w) = \int_0^L \left\{ \frac{EA}{2} \left[\frac{du}{dx} + \frac{1}{2} \left(\frac{dw}{dx} \right)^2 \right]^2 + \frac{EI}{2} \left(\frac{d^2 w}{dx^2} \right)^2 + qw \right\} dx$$

$$- Pu(L) - M_0 \frac{dw}{dx}(L) \tag{3.64}$$

defined on $\mathcal{C} = H_1(0, L) \times H_2(0, L)$, where

$$H_1(0, L) = \{u: \quad u \in H^2(0, L), u(0) = 0\}$$

$$H_2(0, L) = \left\{w: \quad w \in H^4(0, L), w(0) = \frac{dw}{dx}(0) = 0\right\}. \tag{3.65}$$

Note that the boundary (or end) conditions $u(0) = w(0) = (dw/dx)(0) = 0$ are included in the specification of the vector spaces. As will be clear in the sequel, these are essential boundary conditions. The space of admissible variation in this case coincides with the space of competing functions.

The functional in Eq. (3.64) represents the total potential energy associated with the large deflection bending and stretching of a beam of length L, the area of cross section A, the modulus of elasticity E, the moment of inertia I, and subjected to distributed transverse load q, axial load P, and bending moment M_0 (see Fig. 3.6). Here u denotes the axial displacement, and w denotes the transverse deflection of the beam. Calculating the first variation

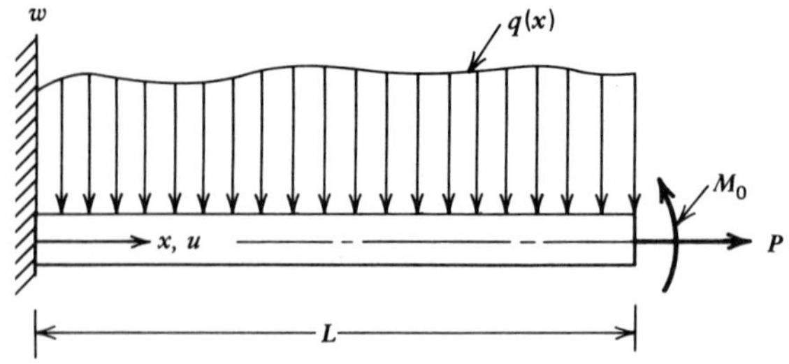

Figure 3.6 Bending and stretching of a cantilevered beam.

gives

$$\delta\Pi(u,w)=\int_0^L\left\{EA\left[\frac{du}{dx}+\frac{1}{2}\left(\frac{dw}{dx}\right)^2\right]\left(\frac{d\delta u}{dx}+\frac{dw}{dx}\frac{d\delta w}{dx}\right)\right.$$

$$\left.+EI\frac{d^2w}{dx^2}\frac{d^2\delta w}{dx^2}+q\delta w\right\}dx-P\delta u\Big|_{x=L}-M_0\frac{d\delta w}{dx}\Big|_{x=L}=0.$$

Integrating the first two terms by parts, we have

$$\delta\Pi=\int_0^L\left\{-\frac{d}{dx}\left[EA\left(\frac{du}{dx}+\frac{1}{2}\left(\frac{dw}{dx}\right)^2\right)\right]\delta u-\frac{d}{dx}\left[EA\frac{dw}{dx}\left(\frac{du}{dx}+\frac{1}{2}\left(\frac{dw}{dx}\right)^2\right)\right]\delta w\right.$$

$$\left.+\frac{d^2}{dx^2}\left(EI\frac{d^2w}{dx^2}\right)\delta w+q\delta w\right\}dx+\left[EA\left(\frac{du}{dx}+\frac{1}{2}\left(\frac{dw}{dx}\right)^2\right)\underline{\delta u}\right]_{x=0}^{x=L}$$

$$+\left[EA\frac{dw}{dx}\left(\frac{du}{dx}+\frac{1}{2}\left(\frac{dw}{dx}\right)^2\right)\underline{\delta w}\right]_{x=0}^{x=L}$$

$$+\left(EI\frac{d^2w}{dx^2}\underline{\frac{d\delta w}{dx}}-\frac{d}{dx}\left(EI\frac{d^2w}{dx^2}\right)\underline{\delta w}\right)\Bigg|_{x=0}^{x=L}-P\underline{\delta u}(L)-M_0\underline{\frac{d\delta w}{dx}}(L)=0.$$

It is clear that the essential boundary conditions for the functional involve (from the underlined terms) specification of the quantities

$$u,w,\frac{dw}{dx}.$$

However, for the problem under consideration the essential boundary conditions are specified only at $x=0$. Therefore the natural boundary conditions must be known at $x=L$.

Collecting the coefficients of δu and δw in $(0, L)$, and of $\delta u, \delta w$, and $\delta(dw/dx)$ at $x = L$, we obtain the Euler equations

$$\delta u: \quad -\frac{d}{dx}\left\{EA\left[\frac{du}{dx}+\frac{1}{2}\left(\frac{dw}{dx}\right)^2\right]\right\}=0, \qquad 0<x<L$$

$$\delta w: \quad \frac{d^2}{dx^2}\left(EI\frac{d^2w}{dx^2}\right)-\frac{d}{dx}\left\{EA\frac{dw}{dx}\left[\frac{du}{dx}+\frac{1}{2}\left(\frac{dw}{dx}\right)^2\right]\right\}+q=0, \qquad 0<x<L$$

(3.66)

and the *natural* boundary conditions

$$\delta u: \quad EA\left[\frac{du}{dx}+\frac{1}{2}\left(\frac{dw}{dx}\right)^2\right]-P=0, \qquad \text{at } x=L$$

$$\delta w: \quad \frac{d}{dx}\left(EI\frac{d^2w}{dx^2}\right)-\left\{EA\frac{dw}{dx}\left[\frac{du}{dx}+\frac{1}{2}\left(\frac{dw}{dx}\right)^2\right]\right\}=0, \qquad \text{at } x=L$$

$$\delta\left(\frac{dw}{dx}\right): \quad \left(EI\frac{d^2w}{dx^2}-M_0\right)=0, \qquad \text{at } x=L.$$

(3.67)

Note that, without regard to the applied loads, for a cantilever beam (i.e., a beam with one end fixed and the other end free) the boundary conditions are $u=w=dw/dx=0$ at $x=0$. These are dependent only on the geometry of support, and are sometimes called *geometric* or *static* boundary conditions. The natural boundary conditions depend on the loading, and can be changed accordingly. Hence the natural boundary conditions are also called *dynamic* conditions. ■

Example 3.5 We wish to minimize the functional

$$I(T)=\int_R\left\{\frac{1}{2}\left[\operatorname{grad} T\cdot\overleftrightarrow{\mathbf{k}}\cdot\operatorname{grad} T\right]+QT\right\}dxdy$$

$$-\int_{C_q}\hat{q}T\,ds+\int_{C_h}h\left(\frac{1}{2}T-T_\infty\right)T\,ds \tag{3.68}$$

on $\mathcal{C}$, where $\overleftrightarrow{\mathbf{k}}=k_1\hat{\mathbf{i}}\hat{\mathbf{i}}+k_2\hat{\mathbf{j}}\hat{\mathbf{j}}$ and

$$\mathcal{C}=\{T:\quad T\in H^2(R), T=\hat{T}\text{ on } C_T\}=\mathcal{H}. \tag{3.69}$$

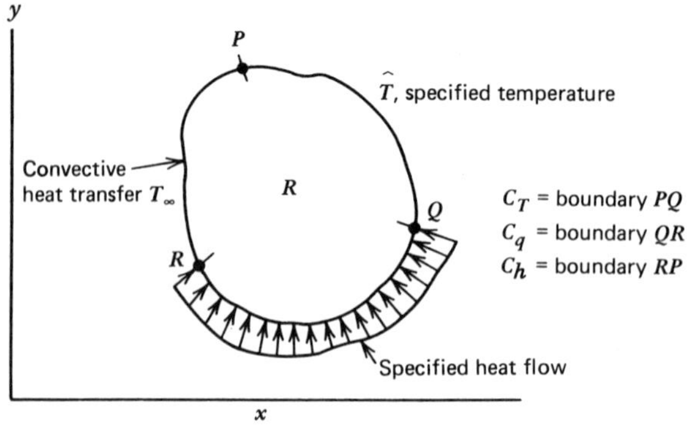

Figure 3.7 Two-dimensional conduction and convective heat transfer.

This functional arises in connection with heat-transfer problems. Here T is the temperature, Q is the internal heat generation (given as a function of x and y), k_1 and k_2 are the thermal conductivities in the principal directions (given, in general, as functions of x and y), $\hat{q}$ is a specified heat flow due to conduction on boundary C_q, h is the convective heat transfer coefficient (given), and T_∞ is the ambient temperature $[h(T-T_\infty)$ represents heat flux due to convection on boundary $C_h]$. The total boundary C is the sum of the disjoint boundary segments C_T, C_q, and C_h (see Fig. 3.7). The fact that $T=\hat{T}$ on C_T is an essential boundary condition will be clear when we take the variation of I. Using the variational method, we obtain

$$\delta I(T)=\int_R\{\operatorname{grad}\delta T\cdot\overleftrightarrow{\mathbf{k}}\cdot\operatorname{grad}T+Q\delta T\}\,dxdy-\int_{C_q}\hat{q}\delta T\,ds+\int_{C_h}h(T-T_\infty)\delta T\,ds \tag{3.70}$$

where we used the fact $\delta(\nabla T)=\nabla\delta T$. Using the divergence theorem on the first two terms, we have

$$\delta I(T)=\int_R\{-\operatorname{div}(\overleftrightarrow{\mathbf{k}}\cdot\operatorname{grad}T)+Q\}\delta T\,dxdy+\oint_C[\hat{\mathbf{n}}\cdot\overleftrightarrow{\mathbf{k}}\cdot\operatorname{grad}T]\delta T\,ds$$

$$-\int_{C_q}\hat{q}\delta T\,ds+\int_{C_h}h(T-T_\infty)\delta T\,ds=0.$$

Since $\oint_C ds=\int_{C_T}ds+\int_{C_q}ds+\int_{C_h}ds$, we can split the boundary integral on C

into three parts and combine them with the appropriate boundary integrals:

$$\delta I(T)=\int_R\{-\operatorname{div}(\overleftrightarrow{\mathbf{k}}\cdot\operatorname{grad}T)+Q\}\delta T\,dxdy$$

$$+\int_{C_T}(\hat{\mathbf{n}}\cdot\overleftrightarrow{\mathbf{k}}\cdot\operatorname{grad}T)\delta T\,ds+\int_{C_q}\left[(\hat{\mathbf{n}}\cdot\overleftrightarrow{\mathbf{k}}\cdot\operatorname{grad}T)-\hat{q}\right]\delta T\,ds$$

$$+\int_{C_h}\left[\hat{\mathbf{n}}\cdot\overleftrightarrow{\mathbf{k}}\cdot\operatorname{grad}T+h(T-T_\infty)\right]\delta T\,ds=0.$$

Since T is specified on C_T, $\delta T=0$; and since T is arbitrary elsewhere, we have the following Euler equation and natural boundary conditions:

$$\delta T:\quad -\operatorname{div}(\overleftrightarrow{\mathbf{k}}\cdot\operatorname{grad}T)+Q=0\qquad \text{in }\Omega \tag{3.71}$$

$$\delta T:\quad \hat{\mathbf{n}}\cdot\overleftrightarrow{\mathbf{k}}\cdot\operatorname{grad}T-\hat{q}=0\qquad \text{on }C_q$$

$$\delta T:\hat{\mathbf{n}}\cdot\overleftrightarrow{\mathbf{k}}\cdot\operatorname{grad}T+h(T-T_\infty)=0\qquad \text{on }C_h.\ \blacksquare \tag{3.72}$$

Example 3.6 Consider the problem of minimizing the functional

$$I(u,v)=\int_R\left\{\nu\left[u_x^2+v_y^2+\tfrac{1}{2}(u_y+v_x)^2\right]+f_1u+f_2v\right\}dxdy$$

$$-\int_{C_{1t}}\hat{t}_1u\,ds-\int_{C_{2t}}\hat{t}_2v\,ds \tag{3.73}$$

on $\mathcal{C}$,

$$\mathcal{C}=\mathcal{H}=\left\{(u,v):\quad u,v\in H^2(R),u=0\text{ on }C_{1d},v=0\text{ on }C_{2d}\right\}, \tag{3.74}$$

which arises in connection with slow viscous incompressible ($\nabla\cdot\mathbf{V}=0$) flow. Here ν is a constant (viscosity), f_1 and f_2 are given functions (body forces) of (x, y) in R, and $\hat{t}_1$ and $\hat{t}_2$ are specified surface stresses on portions C_{1t} and C_{2t}, respectively, of the boundary C,

$$C=C_{1t}+C_{1d}=C_{2t}+C_{2d}$$

and C_{1t} and C_{1d}, and C_{2t} and C_{2d} have no points in common. This is to say that if the velocity components u, v are specified on a portion of the boundary, associated components of the boundary forces (or tractions) cannot be specified on the same portion of the boundary. This should be intuitively obvious, since every force induces an associated velocity (and vice versa) and one cannot specify both. The first variation of I is

$$\delta I=\int_R \{\nu[2u_x\delta u_x+2v_y\delta v_y+(u_y+v_x)(\delta u_y+\delta v_x)]+f_1\delta u+f_2\delta v\}\,dx\,dy$$

$$-\int_{C_{1t}}\hat{t}_1\delta u\,ds-\int_{C_{2t}}\hat{t}_2\delta v\,ds$$

$$=0\equiv B((u,v),(\delta u,\delta v))-l((\delta u,\delta v)).$$

Use of the divergence theorem gives

$$0=\int_R\{-\nu[2u_{xx}+(u_y+v_x)_y]+f_1\}\delta u\,dx\,dy$$

$$+\int_R\{-\nu[(u_y+v_x)_x+2v_{yy}]+f_2\}\delta v\,dx\,dy$$

$$+\oint_C \nu\{[2u_xn_x+(u_y+v_x)n_y]\delta u+[(u_y+v_x)n_x+2v_yn_y]\delta v\}\,ds$$

$$-\int_{C_{1t}}\hat{t}_1\delta u\,ds-\int_{C_{2t}}\hat{t}_2\delta v\,ds$$

where n_x and n_y are the components of the unit normal $\hat{\mathbf{n}}$. We have

$$\delta u:\quad -\nu\left[2\frac{\partial^2 u}{\partial x^2}+\frac{\partial}{\partial y}\left(\frac{\partial u}{\partial y}+\frac{\partial v}{\partial x}\right)\right]+f_1=0\qquad \text{in } R$$

$$\delta v:\quad -\nu\left[\frac{\partial}{\partial x}\left(\frac{\partial u}{\partial y}+\frac{\partial v}{\partial x}\right)+2\frac{\partial^2 v}{\partial y^2}\right]+f_2=0\qquad \text{in } R. \tag{3.75}$$

Since

$$\oint_C ds=\int_{C_{1t}}ds+\int_{C_{1d}}ds=\int_{C_{2t}}ds+\int_{C_{2d}}ds,$$

we can write the boundary integrals as

$$0=\int_{C_{1t}}\{\nu[2u_xn_x+(u_y+v_x)n_y]-\hat{t}_1\}\delta u\,ds+\int_{C_{1d}}\nu[2u_xn_x+(u_y+v_x)n_y]\delta u\,ds$$

$$+\int_{C_{2t}}\{\nu[(u_y+v_x)n_x+2v_yn_y]-\hat{t}_2\}\delta v\,ds$$

$$+\int_{C_{2d}}\nu[(u_y+v_x)n_x+2v_yn_y]\delta v\,ds.$$

Since $(\delta u, \delta v)\in \mathcal{H}$, the second and fourth integrals in the above expression are zero and we have the following natural boundary conditions on C_{1t} and C_{2t} (since δu and δv are arbitrary there):

$$\delta u:\quad \nu\left[2\frac{\partial u}{\partial x}n_x+\left(\frac{\partial u}{\partial y}+\frac{\partial v}{\partial x}\right)n_y\right]-\hat{t}_1=0 \qquad \text{on } C_{1t}$$

$$\delta v:\quad \nu\left[\left(\frac{\partial u}{\partial y}+\frac{\partial v}{\partial x}\right)n_x+2\frac{\partial v}{\partial y}n_y\right]-\hat{t}_2=0 \qquad \text{on } C_{2t}. \tag{3.76}$$

Note that specifications of u and v on C_{1d} and C_{2d} constitute the essential boundary conditions. ■

3.3.7 Variable End Points: Transversality Conditions

In the preceding discussions it was assumed that the boundary of the domain of a problem is known a priori and that the extremizing function satisfies the specified conditions on the boundary. However, there arises an interesting situation where the boundary of the region of interest is not completely specified but is to be determined together with the solution. To fix the ideas, consider the problem of minimizing the functional

$$I(u)=\int_a^{x_0} F(x,u,u')\,dx \tag{3.77}$$

such that

$$u(a)=g \tag{3.78}$$

where a is fixed, g is given, and the point $(x_0, u_0=u(x_0))$ is required to lie on the curve $h(x)$.

Since the solution $u(x)$ depends on the variable end point x_0 (and vice versa), the first variation of I is given by

$$\delta I(u)=\alpha\left[\frac{d}{d\alpha}\int_a^{x_0(u+\alpha\eta)}F(x,u+\alpha\eta,u'+\alpha\eta')\,dx\right]_{\alpha=0}.$$

Using the *Liebnitz rule*, we obtain

$$\delta I(u)=\int_a^{x_0}\delta F(x,u,u')\,dx+F(x_0,u(x_0),u'(x_0))\delta x_0$$

$$=\alpha\int_a^{x_0}\left[-\frac{d}{dx}\left(\frac{\partial F}{\partial u'}\right)+\frac{\partial F}{\partial u}\right]\eta\,dx+\alpha\left[\frac{\partial F}{\partial u'}\eta\right]_a^{x_0}+(F)|_{x=x_0}\delta x_0.$$

Since $u(a)=g$, we have $\eta(a)=0$ and

$$0=\alpha\int_a^{x_0}\left[-\frac{d}{dx}\left(\frac{\partial F}{\partial u'}\right)+\frac{\partial F}{\partial u}\right]\eta\,dx+\left.\left(\frac{\partial F}{\partial u'}\alpha\eta+F\delta x_0\right)\right|_{x=x_0} \tag{3.79}$$

from which we obtain the usual Euler equation.

Now consider the boundary terms in the last expression. Suppose that Δx_0 is the change in x_0 corresponding to the change $\alpha\eta\equiv\delta u$ in u. Then since $u(x_0)=h(x_0)$, we must have

$$u(x_0+\Delta x_0)+\alpha\eta(x_0+\Delta x_0)=h(x_0+\Delta x_0)$$

$$u(x_0+\Delta x_0)-u(x_0)+\alpha\eta(x_0+\Delta x_0)=h(x_0+\Delta x_0)-h(x_0).$$

Using a Taylor series expansion, we write

$$0=\Delta x_0\frac{\partial u(x_0)}{\partial x_0}+\alpha\left[\eta(x_0)+\Delta x_0\frac{\partial\eta(x_0)}{\partial x_0}\right]-\Delta x_0\frac{\partial h(x_0)}{\partial x_0}$$

$$+(\text{higher-order terms in } \alpha \text{ and } \Delta x_0)$$

or

$$\Delta x_0=\frac{-\alpha\eta(x_0)}{u'(x_0)-h'(x_0)+\alpha\eta'(x_0)}+(\text{higher-order terms in } \alpha \text{ and } \Delta x_0).$$

The variation in x_0, δx_0, corresponds to the limiting value of Δx_0 as α goes to zero. Hence we have

$$\delta x_0=\frac{-\alpha\eta(x_0)}{u'(x_0)-h'(x_0)}.$$

Substituting into the boundary terms, we have

$$\left[\frac{\partial F}{\partial u'}\eta-\frac{F\eta}{u'-h'}\right]_{x=x_0}=0.$$

Since η is arbitrary at $x=x_0$, we have

$$\frac{\partial F}{\partial u'}(h'-u')+F=0 \qquad \text{at } x=x_0. \tag{3.80}$$

Equation (3.80) is known as the *transversality condition*. It is a boundary condition to be imposed on the function u at the upper limit $x=x_0$, corresponding to a *variable end point*. It replaces the boundary condition $u(b)=u_b$ which holds for a *fixed end point*.

3.3.8 Minimization of Functionals Subjected to Constraints

Recall from Section 3.3.6 that the function u which minimizes the integral in Eq. (3.49) is required to satisfy the fixed-end conditions (3.50). Other than that, u is not subject to any additional constraints. In some cases, such as in the case of the isoperimetric problem, the minimizing function is subject to additional constraints. In such cases the elements in the set of competing functions are also required to satisfy the constraint condition. Further, the variation δu of u is not arbitrary but must be consistent with the constraint on u.

To fix the ideas, consider the isoperimetric problem: Find the maximum of $I_4(u)$ in Eq. (3.24) subject to the integral constraint

$$G(u') \equiv \int_a^b \sqrt{1+(u')^2}\, dx - L = 0. \tag{3.81}$$

Stated in other words, among the elements of the domain of definition of $I_4(u)$, satisfying the constraint equation (3.81), find the element that assigns to the functional $I_4(u)$ its largest value. As the domain of definition of $I_4(u)$, we can take the set of those functions of $L_2[a, b]$ that vanish for $x = a$ and $x = b$. As the domain of definition of $G(u')$ we take the set of functions of $H^2[a, b]$ which satisfy the constraint condition (3.81). Clearly, the solution u belongs to the intersection of the domains of definition of $I_2(u)$ and $G(u')$, which we denote by $\mathcal{C}$. The constraint can be introduced into the variational problem by means of the Lagrange multiplier method or by the penalty method (see Section 3.2.2).

Lagrange Multiplier Method In the multiplier method the constrained minimization problem is replaced by the unconstrained stationary problem. Applied to the isoperimetric problem, the method seeks to find the stationary value of the modified functional

$$\begin{aligned} I(u) &= I_4(u) + \lambda G(u') \\ &= \int_a^b \left\{ u + \lambda\left(\sqrt{1+u'^2} - \frac{L}{b-a} \right) \right\} dx, \end{aligned} \tag{3.82}$$

$$u(a) = u(b) = 0. \tag{3.83}$$

The quantity λ is called the Lagrange multiplier and is a constant to be determined from the Euler equation of (3.82) and the constraint (3.83). Existence of such a constant is guaranteed by Euler's theorem, which states that when $\delta G(u', \eta') \neq 0$, the constant exists. The Euler equation of the modified functional is obtained by setting $\delta_u I = \delta_\lambda I = 0$:

$$\delta u:\quad 1 - \frac{d}{dx}\left[\frac{\lambda u'}{\sqrt{1+u'^2}}\right] = 0$$
$$\delta\lambda:\quad G(u') = 0. \tag{3.84}$$

Equations (3.84) and (3.81) together provide the solution (u_0, λ) to the isoperimetric problem. Indeed, integrating Eq. (3.84) successively, we arrive at the equation of a circle of radius λ:

$$(x - x_0)^2 + (u - u_0)^2 = \lambda^2 \tag{3.85}$$

where x_0 and u_0 are integration constants, which give the center of the circle. Thus the solution is an arc of a circle. The radius λ and center (x_0, u_0) are obtained by Eqs. (3.81) and (3.83).

Penalty Function Method The penalty function concept of Courant (1956) involves the reduction of variational problems, which are posed as conditional extremum problems to variational problems without constraints, by the introduction of a penalty functional associated with the constraints. For the problem at hand, the method involves constructing a modified functional:

$$I_p(u_\varepsilon) = I_4(u) + \frac{\varepsilon}{2}\left[\int_a^b \sqrt{1+(u')^2}\, dx - L\right]^2 \tag{3.86}$$

where ε is the penalty parameter (preassigned). The necessary condition for a minimum of I_p gives

$$1 - \frac{d}{dx}\left\{\frac{\varepsilon\left[\int_a^b \sqrt{1+(u')^2}\, dx - L\right] u'}{\sqrt{1+(u')^2}}\right\} = 0. \tag{3.87}$$

Comparing Eq. (3.87) with Eq. (3.84), one finds that

$$\lambda \sim \varepsilon\left[\int_a^b \sqrt{1+u'^2}\, dx - L\right]. \tag{3.88}$$

The discussion given above for the isoperimetric problem can be extended to a general class of problems that require minimization of a functional subjected to (differential) equality constraints. We consider the problem of minimizing a functional of two dependent variables (i.e., a functional on a product space) in two dimensions: Find $(u, v) \in H(\Omega) \times H(\Omega)$, where $H(\Omega) =$

$H^2(\Omega) \cap H_0^1(\Omega)$, such that

$$G(u, v, u_x, u_y, v_x, v_y) = 0 \quad \text{in } \Omega \subset \mathsf{R}^2, \qquad \text{constraint} \tag{3.89}$$

and

$$I_0(u, v) = \int_\Omega F(x, y, u, v, u_x, v_x, u_y, v_y)\, dx\, dy \tag{3.90}$$

is a minimum. It is understood that G: $H(\Omega) \to \mathsf{L}_2(\Omega) \equiv \mathsf{L}_2(\Omega) \times \mathsf{L}_2(\Omega)$ and I: $H(\Omega) \to \mathsf{R}$.

Lagrange Multiplier Method The necessary condition for I_0 to achieve its minimum requires us to compute the first variation of $I_0(u, v)$:

$$\delta I_0(u, v) = \int_\Omega \left\{ \left[\frac{\partial F}{\partial u} - \frac{\partial}{\partial x}\left(\frac{\partial F}{\partial u_x}\right) - \frac{\partial}{\partial y}\left(\frac{\partial F}{\partial u_y}\right) \right] \delta u \right.$$
$$\left. + \left[\frac{\partial F}{\partial v} - \frac{\partial}{\partial x}\left(\frac{\partial F}{\partial v_x}\right) - \frac{\partial}{\partial y}\left(\frac{\partial F}{\partial v_y}\right) \right] \delta v \right\} dx\, dy \tag{3.91}$$

where the boundary terms obtained by the integration by parts vanished because $u, v \in H_0^1(\Omega)$. Since u and v are constrained by the relation (3.89), the variations δu and δv are also constrained by relation (3.89) and cannot *both* be assigned arbitrarily in Ω. Hence we cannot make the usual arguments to set the coefficients of δu and δv to zero. From Eq. (3.89) we note that δu and δv are related according to

$$\delta G \equiv \frac{\partial G}{\partial u}\delta u + \frac{\partial G}{\partial u_x}\delta u_x + \frac{\partial G}{\partial u_y}\delta u_y + \frac{\partial G}{\partial v}\delta v + \frac{\partial G}{\partial v_x}\delta v_x + \frac{\partial G}{\partial v_y}\delta v_y = 0. \tag{3.92}$$

That is, if δv is arbitrarily chosen, then δu can be computed in terms of δv using Eq. (3.92). Multiplying Eq. (3.92) with a quantity $\lambda \in L_2(\Omega)$, arbitrary for the moment, integrating over Ω, and adding the result to Eq. (3.91), we obtain

$$\int_\Omega \left\{ \left[\frac{\partial F}{\partial u} - \frac{\partial}{\partial x}\left(\frac{\partial F}{\partial u_x}\right) - \frac{\partial}{\partial y}\left(\frac{\partial F}{\partial u_y}\right) + \lambda\frac{\partial G}{\partial u} - \frac{\partial}{\partial x}\left(\lambda\frac{\partial G}{\partial u_x}\right) - \frac{\partial}{\partial y}\left(\lambda\frac{\partial G}{\partial u_y}\right) \right] \delta u \right.$$
$$+ \left[\frac{\partial F}{\partial v} - \frac{\partial}{\partial x}\left(\frac{\partial F}{\partial v_x}\right) - \frac{\partial}{\partial y}\left(\frac{\partial F}{\partial v_y}\right) + \lambda\frac{\partial G}{\partial v} - \frac{\partial}{\partial x}\left(\lambda\frac{\partial G}{\partial v_x}\right) \right.$$
$$\left. \left. - \frac{\partial}{\partial y}\left(\lambda\frac{\partial G}{\partial v_y}\right) \right] \delta v \right\} dx\, dy = 0. \tag{3.93}$$

Since this must hold for any $\lambda \in L_2(\Omega)$, in particular it must also hold for the λ that makes the coefficient of δu vanish. Since we can choose δv arbitrarily in Ω, its coefficient must also vanish:

$$\frac{\partial F}{\partial u}-\frac{\partial}{\partial x}\left(\frac{\partial F}{\partial u_x}\right)-\frac{\partial}{\partial y}\left(\frac{\partial F}{\partial u_y}\right)+\lambda\frac{\partial G}{\partial u}-\frac{\partial}{\partial x}\left(\lambda\frac{\partial G}{\partial u_x}\right)-\frac{\partial}{\partial y}\left(\lambda\frac{\partial G}{\partial u_y}\right)=0$$

$$\frac{\partial F}{\partial v}-\frac{\partial}{\partial x}\left(\frac{\partial F}{\partial v_x}\right)-\frac{\partial}{\partial y}\left(\frac{\partial F}{\partial v_y}\right)+\lambda\frac{\partial G}{\partial v}-\frac{\partial}{\partial x}\left(\lambda\frac{\partial G}{\partial u_x}\right)-\frac{\partial}{\partial y}\left(\lambda\frac{\partial G}{\partial v_y}\right)=0. \tag{3.94}$$

Equations (3.89) and (3.94) together give three equations in the three unknown dependent variables u, v, and λ.

Alternately one can interpret that equations (3.89) and (3.94) are the three Euler equations obtained from the following functional by setting the coefficients of $\delta\lambda$, δu, and δv, respectively, to zero:

$$I_L(u,v,\lambda)=I_0(u,v)+\int_\Omega \lambda G(u,v,u_x,u_y,v_x,v_y)\,dx\,dy. \tag{3.95}$$

It should be noted that in the Lagrange multiplier methods an additional variable is introduced with each constraint.

Penalty Method As pointed out earlier, the penalty method is an approximate method in which the approximate solution converges to the exact solution in the limit as a certain parameter, called the *penalty parameter*, goes to infinity. The basic philosophy behind the penalty method is that a solution of a constrained minimum problem can be obtained as a limit of suitably chosen unconstrained minimum problems. The procedure involves the construction of a nonnegative functional $P(u,v)$ on $H(\Omega)$ such that the functions satisfying the constraint in Eq. (3.89) are given by the solutions of $P(u,v)=0$. Next the *penalty term* $\varepsilon P(u,v)$ is added to the functional $I_0(u,v)$ to be minimized under the constraint (3.89). Then we seek the stationary point $(u(\varepsilon), v(\varepsilon)) \in H(\Omega)$ of the augmented functional

$$I_p(u_\varepsilon,v_\varepsilon)=I_0(u_\varepsilon,v_\varepsilon)+\varepsilon P(u_\varepsilon,v_\varepsilon). \tag{3.96}$$

We hope that the limit $(u,v)=\lim_{\varepsilon\to\infty}(u_\varepsilon,v_\varepsilon)$ exists and is the critical point of $I_0(u,v)$ subject to the given constraint.

Obviously the selection of $P(u,v)$ is a crucial one. However, the following considerations prove to be useful in the selection. For most physical problems the functional to be minimized is a quadratic form. Hence to retain the quadratic nature of the functional, and at the same time to add a nonnegative penalty functional to it, it is convenient to choose $P(u,v)$ to be

$$P(u,v)=\frac{1}{2}\int_{\Omega}[G]^2\,dx\,dy \qquad \text{or} \qquad \delta P=\int_{\Omega}G\cdot\delta G\,dx\,dy. \tag{3.97}$$

In actual computations the penalty parameter ε is a finitely large number, and the solution $(u_\varepsilon, v_\varepsilon)$ differs from the actual solution. A formal proof of the convergence of $(u_\varepsilon, v_\varepsilon)$ to the actual solution can be found in the references listed at the end of this chapter.

Example 3.7 Consider the functional (3.73), and suppose that the dependent variables u and v are required to satisfy the constraint

$$G(u_x, v_y)\equiv\frac{\partial u}{\partial x}+\frac{\partial v}{\partial y}=0 \tag{3.98}$$

where G: $\mathsf{H}(\Omega)\to L_2(\Omega)$, $\mathsf{H}(\Omega)=H_*^2(\Omega)\times H_*^2(\Omega)$, and $H_*^2=\{u\colon\ u\in H^2(\Omega), u=0 \text{ on } \Gamma_2\}$.This constraint results from the conservation of the mass principle applied to an incompressible fluid. In the Lagrange multiplier method we construct a new functional on $\mathsf{H}(\Omega)\times L_2(\Omega)$:

$$I_L(u,v,\lambda)=I_0(u,v)+\int_{\Omega}\lambda\left(\frac{\partial u}{\partial x}+\frac{\partial v}{\partial y}\right)dx\,dy. \tag{3.99}$$

Setting the first variation of I_L to zero,

$$\delta I_L(u,v,\lambda)\equiv\delta I_0(u,v)+\int_{\Omega}\left[\lambda\left(\frac{\partial\delta u}{\partial x}+\frac{\partial\delta v}{\partial y}\right)+\delta\lambda\left(\frac{\partial u}{\partial x}+\frac{\partial v}{\partial y}\right)\right]dx\,dy=0,$$

one obtains the following Euler equations:

$$\begin{aligned}
\delta u&: \quad -\nu\left[2\frac{\partial^2 u}{\partial x^2}+\frac{\partial}{\partial y}\left(\frac{\partial u}{\partial y}+\frac{\partial v}{\partial x}\right)\right]-\frac{\partial\lambda}{\partial x}+f_1=0\\
\delta v&: \quad -\nu\left[\frac{\partial}{\partial x}\left(\frac{\partial u}{\partial y}+\frac{\partial v}{\partial x}\right)+2\frac{\partial^2 v}{\partial y^2}\right]-\frac{\partial\lambda}{\partial y}+f_2=0\\
\delta\lambda&: \quad \frac{\partial u}{\partial x}+\frac{\partial v}{\partial y}=0
\end{aligned} \tag{3.100}$$

in Ω, and the natural boundary conditions

$$\begin{aligned}
\delta u&: \quad \left(2\nu\frac{\partial u}{\partial x}+\lambda\right)n_x+\nu\left(\frac{\partial u}{\partial y}+\frac{\partial v}{\partial x}\right)n_y-t_1=0 \text{ on } \Gamma_1.\\
\delta v&: \quad \nu\left(\frac{\partial u}{\partial y}+\frac{\partial v}{\partial x}\right)n_x+\left(2\nu\frac{\partial v}{\partial y}+\lambda\right)n_y-t_2=0 \text{ on } \Gamma_1.
\end{aligned} \tag{3.101}$$

A comparison of Eqs. (3.100) and (3.101) with the equations of the Stokes flow shows that λ is indeed the negative of the hydrostatic pressure (or mean stress in the case of linear incompressible elasticity): $\lambda = -P$.

In the penalty method we construct the modified functional

$$I_p(u,v) = I_0(u,v) + \frac{\varepsilon}{2}\int_\Omega \left(\frac{\partial u}{\partial x} + \frac{\partial v}{\partial y}\right)^2 dx\,dy. \tag{3.102}$$

The Euler equations associated with I_p are [with $u = u(\varepsilon)$, $v = v(\varepsilon)$]

$$\begin{aligned}
\delta u:&\quad \nu\left[2\frac{\partial^2 u}{\partial x^2} + \frac{\partial}{\partial y}\left(\frac{\partial u}{\partial y} + \frac{\partial v}{\partial x}\right)\right] + \varepsilon\frac{\partial}{\partial x}\left(\frac{\partial u}{\partial x} + \frac{\partial v}{\partial y}\right) = f_1 \qquad \text{in } \Omega\\
\delta v:&\quad \nu\left[\frac{\partial}{\partial x}\left(\frac{\partial u}{\partial y} + \frac{\partial v}{\partial x}\right) + 2\frac{\partial^2 v}{\partial y^2}\right] + \varepsilon\frac{\partial}{\partial y}\left(\frac{\partial u}{\partial x} + \frac{\partial v}{\partial y}\right) = f_2 \qquad \text{in } \Omega
\end{aligned} \tag{3.103}$$

$$\begin{aligned}
\delta u:&\quad \left[2\nu\frac{\partial u}{\partial x} + \varepsilon\left(\frac{\partial u}{\partial x} + \frac{\partial v}{\partial y}\right)\right]n_x + \nu\left(\frac{\partial u}{\partial y} + \frac{\partial v}{\partial x}\right)n_y = t_1 \qquad \text{on } \Gamma_1\\
\delta v:&\quad \nu\left(\frac{\partial u}{\partial y} + \frac{\partial v}{\partial x}\right)n_x + \left[2\nu\frac{\partial v}{\partial y} + \varepsilon\left(\frac{\partial u}{\partial x} + \frac{\partial v}{\partial y}\right)\right]n_y = t_2 \qquad \text{on } \Gamma_1.
\end{aligned} \tag{3.104}$$

Comparison of Eqs. (3.103) and (3.104) with Eqs. (3.100) and (3.101), respectively, shows that

$$\varepsilon\left(\frac{\partial u}{\partial x} + \frac{\partial v}{\partial y}\right) \equiv \lambda_\varepsilon. \tag{3.105}$$

Thus in the penalty method the Lagrange multiplier λ_ε, which is presumably approximation to λ, can be computed using Eq. (3.105). ■

With the exception of the isoperimetric problem, in all other cases it was assumed that we were required to find the (conditions of) minimum of a *given* functional. We have seen in Section 3.3.1 how some of the functionals are constructed. All of the examples cited in Section 3.3.1 have a common denominator: all of the functionals were arrived at naturally from the mathematical statement of the problems there. The necessary condition for a minimum lead us to the associated differential equations. In many engineering problems of interest we already know the differential equations associated with the problem. Then if we wish to know the corresponding functional (in order to solve the equations by variational methods of approximation), we must have an *inverse* procedure to obtain the functional. Thus it is clear that the variational formulations (i.e., the functional form) of physical problems can be used (1) to derive the associated (differential) equations, and (2) to obtain an approximate solution to a given differential equation by means of a variational method. The latter, of course, requires the variational formulation of a given differential equation. Section 3.4 is devoted to the variational formulation of discrete and continuous systems using Hamilton's principle. In Section 3.5 we study the inverse problem of constructing variational statements from given differential equations. If the reader is already familiar with Hamilton's principle and the Lagrangian dynamics, Section 3.4 can be skipped without loss of continuity.

Exercises 3.2

1 (*A Problem of Navigation*) Consider the problem of crossing a river in a boat with a constant speed in the shortest possible time. Assume that the banks of the river are straight and parallel. Take the y axis to coincide with one of the banks and the x axis across the river (see the accompanying figure). The velocity vector $\mathbf{v}=(u, v)$ at any point (x, y) in the system

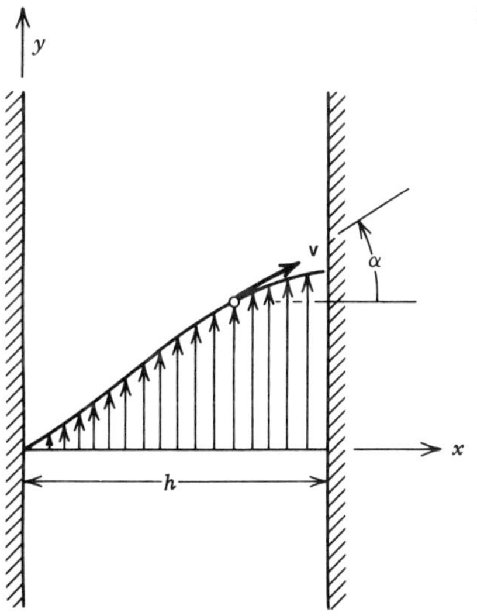

is given by

$$u=0, \qquad v=v(x).$$

Formulate the problem variationally by writing the time required to cross the river. Follow the procedure given below.

(a) Assume that the speed (constant) of the boat is v_b, and use the point $(0,0)$ as point of departure. Let α be the angle that depends on the course of the boat, as shown in the figure.

(b) Write the actual velocity of the boat in the river:

$$\frac{dx}{dt}=v_b\cos\alpha,\qquad \frac{dy}{dt}=v(x)+v_b\sin\alpha.$$

Express $\cos\alpha$ in terms of $dy/dx[y=y(x)$ is the path on which the boat moves] by solving the equation

$$y'=\frac{v\pm v_b\sqrt{(1-\cos^2\alpha)}}{v_b\cos\alpha}.$$

(c) Obtain the time functional in the form

$$t(y)=\int_0^h\frac{\sqrt{v_b^2[1+(y')^2]-v^2}-vy'}{v_b^2-v^2}dx,\qquad y(0)=0$$

where $v(x)$ is a given function (stream velocity) of x.

2 (*An Optimal Control Problem*) Consider the problem of finding the path of minimal travel time of a rocket under the influence of gravity and constant thrust force. The end conditions are as follows: The rocket is to be fired from a given point with a given initial direction and to arrive at another given point with a given terminal direction. Thus minimize the flight time T subject to the end conditions

$$x(0)=a,\qquad y(0)=y_a,\qquad x(T)=b,\qquad y(T)=y_b$$

where (x, y) are the coordinates of a point in space. Formulate the variational problem following the steps described below.

(a) Let F_0 be the magnitude and $\alpha=\alpha(t)$ the direction (measured counterclockwise from the positive x axis) of the thrust force $F(t)$ at time t, and write the equations of motion in component form:

$$\frac{d^2x}{dt^2}=F_0\cos\alpha(t)$$

$$\frac{d^2y}{dt^2}=F_0\sin\alpha(t)-g.$$

(b) Introduce the variables $u=dx/dt$ and $v=dy/dt$ to reduce the equations in step (a) to first-order equations of motion:

$$\frac{dx}{dt}=u,\qquad \frac{dy}{dt}=v$$

$$\frac{du}{dt}=F_0\cos\alpha(t),\qquad \frac{dv}{dt}=F_0\sin\alpha(t)-g$$

with the boundary conditions

$$x(0)=a, \qquad x(T)=b, \qquad y(0)=y_a, \qquad y(T)=y_b$$

$$u(0)=1, \qquad u(T)=1. \qquad v(0)=\dot{y}_a, \qquad v(T)=\dot{y}_b.$$

(c) Cast the differential equations (there are five unknowns x, y, u, v, and α, but only four equations—an underdetermined system) into an integral expression.

3 (*Newton's Problem*) One of the first problems that involved the essential ideas of the calculus of variations was *Newton's problem* of the solid of revolution of minimum resistance. The problem can be stated as follows: Find a curve $y=y(x)$ joining point A: (a, y_a) and point B: (b, y_b) such that the solid of revolution obtained by rotating the curve about the x axis shall suffer the least possible resistance when it moves to the left through air at a steady speed. If ρ is the density of the air, V is the speed of the projectile, $y'(=\tan\alpha)$ is the slope of the curve, and $p=\rho V^2\sin^2\alpha$ is the pressure on the surface, show that the functional to be minimized is given by

$$I(y)=2\pi\rho V^2\int_a^b \frac{y(y')^3}{1+(y')^2}\,dx.$$

4 (*Isoperimetric Problem*) Show that among all piecewise smooth curves of length L in the plane, the circle encloses the maximum area. Follow the steps described below:

(a) Use the parametric representation of the curve $x=x(s)$, $y=y(s)$, $0\le s\le L$. Let $t=s/L$ be the nondimensionalized parameter, and expand x and y in Fourier series expansions,

$$x(t)=a_0+\sqrt{2}\sum_{n=1}^{\infty}\left(a_n\cos 2\pi nt+b_n\sin 2\pi nt\right)$$

$$y(t)=c_0+\sqrt{2}\sum_{n=1}^{\infty}\left(c_n\cos 2\pi nt+d_n\sin 2\pi nt\right)$$

in the interval $0\le t\le 1$. Show that the constraint $(\int_0^L ds)^2=L^2$ can be expressed in the form

$$L^2=\sum_{n=1}^{\infty}4\pi^2n^2\left(a_n^2+b_n^2+c_n^2+d_n^2\right).$$

(b) Show that the area A is given by

$$A=\int_{y(0)}^{y(1)}x\,dy=\int_0^1 x\frac{dy}{dt}\,dt=\sum_{n=1}^{\infty}2\pi n\left(a_nd_n-c_nb_n\right).$$

(c) Show that $L^2 - 4\pi A \geq 0$. Then note that the equality holds if and only if $a_1 = d_1$, $b_1 = -c_1$, and $a_n = b_n = c_n = d_n = 0$ for all $n = 2, 3\ldots$. Show that the nonzero constraints give the equation of a circle.

5 Consider the functional

$$I(u) = \begin{cases} \dfrac{u_1 u_2^2}{u_1^2 + u_2^2}, & \text{if } u = (u_1, u_2) \neq 0 \\ 0, & \text{if } u = (u_1, u_2) = 0. \end{cases}$$

Compute the variation at $u = u_0$.

Calculate the first variation of the functionals in Exercises 6–10 at the values of y indicated:

6 $I(y) = \int_0^1 \sqrt{1 + (y')^2}\, dx$.

7 $I(y) = \int_a^b y\sqrt{1 + (y')^2}\, dx$.

8 $I(y) = \int_0^1 \sqrt{\dfrac{1 + (y')^2}{y}}\, dx$.

9 $I(u, v) = \int_0^1 \left(\dfrac{du}{dx}\dfrac{dv}{dx} + \dfrac{v^2}{2} - qu \right) dx, \qquad q = \text{constant}.$

10 $$I(w) = \frac{D}{2} \int_R \left\{ (\nabla^2 w)^2 + 2(1-\nu)\left[\left(\frac{\partial^2 w}{\partial x\, \partial y} \right)^2 - \left(\frac{\partial^2 w}{\partial x^2} \right)\left(\frac{\partial^2 w}{\partial y^2} \right) \right] \right\} dx\, dy - \int_R qw\, dx\, dy$$

where D and ν are constants and $q = q(x, y)$ is a known function.

Define the space of competing functions and the space of admissible variations for the following variational problems 11–14, where $u' = du/dx$, $u_x = \partial u / \partial x$, $u_y = \partial u / \partial y$, and so on.

11 $\min I(u) = \int_a^b F(x, u, u')\, dx, \qquad u(a) = u_a,\ u(b) = u_b.$

12 $\min I(u) = \int_R F(x, y, u, u_x, u_y)\, dx\, dy, \qquad u = g$ on C_1, $\dfrac{\partial u}{\partial n} = 0$ on C_2, and $C = C_1 \cup C_2, C_1 \cap C_2 = \varnothing$

13 $\min I(u, v) = \int_R F(x, y, u, v, u_x, v_x, u_y, v_y)\, dx\, dy, \qquad u = \hat{u}, v = \hat{v}$ on C.

14 $\min I(w) = \int_R F(x, y, w, w_x, w_y, w_{xx}, w_{xy}, w_{yy})\, dx\, dy,$

$w = \hat{w}$ and $\dfrac{\partial w}{\partial n} = \hat{w}_n$ on C.

Derive the Euler equations and the associated natural and essential boundary conditions for the functionals in Exercise 15–20.

15 $I(u)=\frac{1}{2}\int_a^b\{(u'')^2+2u'u''+(u')^2+u\}\,dx.$

16 $I(u,v)=\frac{1}{2}\int_R\{u_x^2+2u_xv_x+2u_yv_y+v_y^2+fu+gv\}\,dx\,dy.$

17 $I(u,\mathbf{v})=\int_R\left\{\frac{1}{2k}\mathbf{v}\cdot\mathbf{v}+\mathbf{v}\cdot\nabla u+Qu\right\}dx\,dy-\int_{C_1}qu\,ds.$

18
$$I(u_1,u_2,P)=\int_R\left\{\frac{\mu}{2}(u_{i,j}+u_{j,i})u_{i,j}-Pu_{i,i}\right\}dx_1\,dx_2$$
$$-\int_R f_iu_i\,dx_1\,dx_2+\int_C\hat{t}_iu_i\,ds$$

where index notation with summation on repeated indices is used, and $u_{i,j}=\partial u_i/\partial x_j$, μ is a constant, and f_i and $\hat{t}_i$ are known functions of position.

19
$$I(w,M_1,M_2,M_3)=\int_R\left\{\frac{S}{2}\left[M_1^2+M_2^2-2\nu M_1M_2+2(1+\nu)M_3^2\right]\right.$$
$$\left.-\frac{\partial w}{\partial x}\left(\frac{\partial M_1}{\partial x}+\frac{\partial M_3}{\partial y}\right)-\frac{\partial w}{\partial y}\left(\frac{\partial M_3}{\partial x}+\frac{\partial M_2}{\partial y}\right)+Pw\right\}dx\,dy$$
$$+\int_{C_2}\hat{V}w\,ds+\int_{C_3}\hat{\theta}_nM_n\,ds+\oint_C\frac{\partial w}{\partial s}M_s\,ds$$

where s and ν are constants, $\hat{V}$ and θ_n are known functions defined on the portions C_2 and C_3, respectively, of the boundary C,

$$C=C_1+C_2=C_3+C_4,\qquad C_1\cap C_2=\varnothing,\qquad C_3\cap C_4=\varnothing,$$

$w\in C^1(R)$, and w is specified to be $\hat{w}$ on C_1. Here M_n and M_s are defined by

$$M_n=M_1n_x^2-M_2n_y^2+2M_3n_xn_y,\qquad M_s=(M_1-M_2)n_xn_y+M_3\left(n_x^2-n_y^2\right)$$

and $\partial/\partial s$ is the tangential derivative

$$\frac{\partial}{\partial s}=n_x\frac{\partial}{\partial y}-n_y\frac{\partial}{\partial x}.$$

20 $$I(\mathbf{u}, \vec{\varepsilon}, \vec{\sigma}, \mathbf{T})=\int_R \{[\tfrac{1}{2}(u_{i,j}+u_{j,i}+u_{m,i}u_{m,j})-\varepsilon_{ij}]\sigma_{ij}+E_{ijkl}\varepsilon_{ij}\varepsilon_{kl}$$
$$-\rho f_m u_m\}\,dx_1\,dx_2\,dx_3-\int_{C_\sigma}\hat{T}_m u_m\,ds-\int_{C_u}T_m(u_m-\hat{u}_m)\,ds$$

where index notation with summation on repeated indices is used; ρf_m, $\hat{T}_m$, and u_m are known components of vectors $\rho\mathbf{f}$, $\hat{\mathbf{T}}$, and $\mathbf{u}$, respectively. The tensor $\vec{\sigma}$ is assumed to be symmetric, $\sigma_{ij}=\sigma_{ji}$. The boundary portions C_σ and C_u are the disjoint subsets whose union is the total boundary C.

21 Derive the Euler equations and identify the associated natural and essential boundary conditions of the functional

$$I=\int_{x_1}^{x_2}\int_{y_1}^{y_2}F(x, y, u, u_x, u_y, u_{xx}, u_{xy}, u_{yy})\,dx\,dy.$$

22 (*Bending of Isotropic Rectangular Plates*) Derive the Euler equations and the natural and essential boundary conditions of the functional

$$I(w)=\frac{D}{2}\int_0^a\int_0^b\left\{(\nabla^2 w)^2+2(1-\nu)\left[\left(\frac{\partial^2 w}{\partial x\,\partial y}\right)^2-\frac{\partial^2 w}{\partial x^2}\frac{\partial^2 w}{\partial y^2}\right]\right\}dx\,dy$$
$$-\int_0^a\int_0^b q_0 w\,dx\,dy$$

where D (flexural rigidity of the plate) and ν (Poisson's ratio) are constants and q_0 (distributed transverse load) is a given function of x and y. (*Hint*: Use the results of Exercise 21.)

23 (*Bending of Orthotropic Annular Plates*) Obtain the Euler equations and the associated boundary conditions for the following functional:

$$I(w)=\frac{\pi}{2}\int_{r_0}^{r_1}\left\{D_{11}\left(\frac{d^2w}{dr^2}\right)^2+\frac{2D_{12}}{r}\frac{dw}{dr}\frac{d^2w}{dr^2}+d_{22}\left(\frac{1}{r}\frac{dw}{dr}\right)^2\right\}r\,dr.$$

24 Minimize the functional

$$I(u)=\int_0^1\left[\frac{1}{48}\left(\frac{du}{dx}\right)^4+\left(\frac{du}{dx}\right)^2+u^6-6u\right]dx$$

on a set of functions from $H^2(0,1)$ that satisfy the end conditions $u(0)=1$ and $u(1)=0$.

25 Identify the linear and bilinear forms of the problems considered in Examples 3.5 and 3.6.

26 Consider the problem of minimizing the functional

$$I(\psi, w) = \int_0^L \left[\frac{EI}{2} \left(\frac{d\psi}{dx} \right)^2 + fw \right] dx$$

$\psi \in H_0^2(0, L) \cap H_0^1(0, L)$ and $w \in H_0^1(\Omega)$, subject to the constraint $dw/dx - \psi = 0$. Include the constraint in the variational problem. Find the Euler equations in each case.

(a) Use the Lagrange multiplier method.

(b) Use the penalty method.

27 Consider the problem of minimizing the functional

$$I(u, v, w) = \frac{D}{2} \int_R \left\{ \left(\frac{\partial u}{\partial x} \right)^2 + \left(\frac{\partial v}{\partial y} \right)^2 + 2\nu \frac{\partial u}{\partial x} \frac{\partial v}{\partial y} \right.$$

$$\left. + \frac{1-\nu}{2} \left(\frac{\partial u}{\partial y} + \frac{\partial v}{\partial x} \right)^2 \right\} dx\, dy - \int_R qw\, dx\, dy$$

subjected to the constraints

$$\frac{\partial w}{\partial x} - u = 0, \qquad \frac{\partial w}{\partial y} - v = 0.$$

Derive the Euler equations and associated natural and essential boundary conditions.

(a) Use the Lagrange multiplier method.

(b) Use the penalty method.

3.4 VARIATIONAL FORMULATION VIA HAMILTON'S PRINCIPLE

3.4.1 Introduction

Hamilton's principle provides an alternate formulation of the laws of mechanics. This formulation is based on energy concepts, and proves to be very helpful, particularly in cases where many degrees of freedom are involved. Hamilton's principle assumes that the system under consideration is characterized by two energy functions: a *kinetic energy* T and a *potential energy* V. When the system under consideration is *discrete* (i.e., can be described in terms of a finite number of degrees of freedom—analogous to a finite basis), these energies may be described in terms of a finite number of coordinates, called *generalized coordinates*, and their derivatives with respect to time t. When the system under consideration is *continuous* (i.e., systems that cannot be described by a finite set of generalized coordinates), the energies are expressed in terms of the dependent variables (which assume different values at different points in the system) of the problem.

Most of this section is devoted to the application of Hamilton's principle to discrete systems. Section 3.4.14 deals with continuous systems. All of the discussion that follows is for dynamical systems (i.e., systems whose behavior is time dependent). Hamilton's principle becomes the *total potential energy principle* for systems that are in static equilibrium.

3.4.2 Hamilton's Principle for a Single Particle

We now consider the variational formulation of the fundamental problem of a single point particle of mass m moving under the influence of a force field $\mathbf{F}=\mathbf{F}(\mathbf{r})$. The path $\mathbf{r}(t)$ followed by the point particle is governed by Newton's second law of motion, which we write as

$$m\frac{d^2\mathbf{r}}{dt^2}-\mathbf{F}(\mathbf{r})=0. \qquad (3.106)$$

A path that differs from the actual particle motion can be expressed as $\mathbf{r}+\delta\mathbf{r}$, where $\delta\mathbf{r}$ is the variation of the path. We suppose that the actual path and the varied path differ except at two distinct times t_1 and t_2, that is, $\delta\mathbf{r}(t_1)=\delta\mathbf{r}(t_2)=\mathbf{0}$. Let us form the scalar product of the variation $\delta\mathbf{r}$ and Eq. (3.106), and then integrate with time between t_1 and t_2. We then obtain

$$\int_{t_1}^{t_2}\left[m\frac{d^2\mathbf{r}}{dt^2}\cdot\delta\mathbf{r}-\mathbf{F}\cdot\delta\mathbf{r}\right]dt=0. \qquad (3.107)$$

The first term can be integrated by parts, and we find that

$$\int_{t_1}^{t_2}\frac{d^2\mathbf{r}}{dt^2}\cdot\delta\mathbf{r}\,dt=\left[\frac{d\mathbf{r}}{dt}\cdot\delta\mathbf{r}\right]_{t_1}^{t_2}-\int_{t_1}^{t_2}\frac{d\mathbf{r}}{dt}\cdot\delta\frac{d\mathbf{r}}{dt}\,dt. \qquad (3.108)$$

The integrated terms vanish since $\delta\mathbf{r}$ vanishes at the ends of the time interval. Further we note that

$$m\frac{d\mathbf{r}}{dt}\cdot\delta\frac{d\mathbf{r}}{dt}\equiv\delta\left[\frac{m}{2}\left(\frac{d\mathbf{r}}{dt}\right)^2\right]\equiv\delta T \qquad (3.109)$$

where

$$T\equiv\frac{m}{2}\left(\frac{d\mathbf{r}}{dt}\right)^2 \qquad (3.110)$$

is the kinetic energy of the particle. Equation (3.107) can now be written as

$$\int_{t_1}^{t_2} [\delta T + \mathbf{F} \cdot \delta \mathbf{r}]\, dt = 0. \tag{3.111}$$

This expression is the general form of what is known as *Hamilton's principle* for a single particle.

The generalized Hamilton's principle as represented by Eq. (3.111) has not yet been cast into the proper form of a variational problem since the variational operation has not been brought outside the integral. To do this, we need to deal further with the force term.

3.4.3 Conservative Forces

We consider the case where the force **F** is irrotational, such that it can be replaced by the gradient of a potential:

$$\mathbf{F} = -\nabla V \tag{3.112}$$

where $V = V(\mathbf{r})$ is the *potential energy* of the particle. Since for this system the sum of the potential and kinetic energies is conserved, the forces described by Eq. (3.112) are said to be *conservative*. With this requirement on **F**, we observe that

$$\mathbf{F} \cdot \delta \mathbf{r} = -\nabla V \cdot \delta \mathbf{r} = -\delta V \tag{3.113}$$

because, as we noted before, the variational operator behaves the same as the differential operator. The variational operator can now be brought out of the integral in Eq. (3.111), and we have

$$\delta \int_{t_1}^{t_2} (T - V)\, dt = 0. \tag{3.114}$$

The *difference* between the kinetic and potential energies is called the *Lagrangian function*:

$$L \equiv T - V. \tag{3.115}$$

We can state explicitly Hamilton's principle for the conservative motion of a particle: *The motion of a point particle acted on by conservative forces between two arbitrary instants of time t_1 and t_2 is such that the line integral over the Lagrangian function*

$$I = \int_{t_1}^{t_2} L\, dt \tag{3.116}$$

is an extremum for the path of motion. Thus out of all possible paths which the particle could travel from its position at time t_1 to its position at time t_2, its actual path will be the one for which the integral I is an extremum, whether a minimum, a maximum, or an inflection.

3.4.4 Generalized Coordinates

To perceive more clearly the functional dependence of the Lagrangian function on the coordinates of the problem, let us introduce a general orthogonal curvilinear coordinate system, such that q_1, q_2, q_3 are the generalized coordinates, and

$$\mathbf{r}=\mathbf{r}(q_1, q_2, q_3)$$

$$d\mathbf{r}=h_1\, dq_1\hat{\mathbf{e}}_1+h_2\, dq_2\hat{\mathbf{e}}_2+h_3\, dq_3\hat{\mathbf{e}}_3 \tag{3.117}$$

$$\frac{d\mathbf{r}}{dt}\equiv\dot{\mathbf{r}}=h_1\dot{q}_1\hat{\mathbf{e}}_1+h_2\dot{q}_2\hat{\mathbf{e}}_2+h_3\dot{q}_3\hat{\mathbf{e}}_3$$

where $h_1=h_1(q_1, q_2, q_3)$, $h_2=h_2(q_1, q_2, q_3)$, and $h_3=h_3(q_1, q_2, q_3)$ are the curvilinear scale factors, $\hat{\mathbf{e}}_1$, $\hat{\mathbf{e}}_2$, and $\hat{\mathbf{e}}_3$ are the orthonormal curvilinear basis vectors, and $\dot{q}_1\equiv dq_1/dt$, $\dot{q}_2\equiv dq_2/dt$, and $\dot{q}_3\equiv dq_3/dt$ are the generalized velocities. The kinetic energy is thus given by

$$T=\frac{m}{2}\left[(h_1\dot{q}_1)^2+(h_2\dot{q}_2)^2+(h_3\dot{q}_3)^2\right]. \tag{3.118}$$

Thus, in general, the kinetic energy depends functionally on both the generalized coordinates q_1, q_2, q_3 and the generalized velocities $\dot{q}_1, \dot{q}_2, \dot{q}_3$; that is,

$$T=T(q_1, q_2, q_3, \dot{q}_1, \dot{q}_2, \dot{q}_3). \tag{3.119}$$

The potential energy, on the other hand, depends only on the generalized coordinates:

$$V=V(q_1, q_2, q_3). \tag{3.120}$$

Consequently it follows that the Lagrangian function depends on both the generalized coordinates and the generalized velocities:

$$L=L(q_1, q_2, q_3, \dot{q}_1, \dot{q}_2, \dot{q}_3). \tag{3.121}$$

The variables involved in the functional I can now be written explicitly:

$$I=\int_{t_1}^{t_2}L(q_1, q_2, q_3, \dot{q}_1, \dot{q}_2, \dot{q}_3)\, dt. \tag{3.122}$$

The Lagrangian function depends on two physical vectors, the position vector $\mathbf{r}$ and the velocity vector $\dot{\mathbf{r}}$. Although general orthogonal curvilinear coordinates were used to describe these vectors, we could have used *any* coordinate system, and the functional dependencies obtained would be the same accordingly. It is not necessary that generalized coordinates should be conventional position coordinates; all sorts of quantities can be used for generalized coordinates. The Hamilton formulation is thus invariant to the coordinate system used to express the Lagrangian. The three coordinates of $\mathbf{r}$ comprise what is called the physical space or the *configuration* space. The three coordinates $(\dot{q}_1, \dot{q}_2, \dot{q}_3)$ of the velocity vector $\dot{\mathbf{r}}$ comprise what is called the *velocity* space. The six components of the configuration space and the velocity space taken together comprise a 6-space, for a single particle, which is called the *phase* space. At any given instant the dynamical state of the point mass is described by a point in the phase space, and this point determines the value of the Lagrangian at that instant.

3.4.5 Lagrangian Equations

Hamilton's principle amounts to saying that the variation of I is zero, that is,

$$\delta I = \delta \int_{t_1}^{t_2} L(q_1, q_2, q_3, \dot{q}_1, \dot{q}_2, \dot{q}_3)\, dt = 0. \tag{3.123}$$

The straightforward operational variation of this integral with fixed end points, interchange of the variation operation and the time derivative for the variation of the generalized velocities, and subsequent integration by parts lead to

$$\int_{t_1}^{t_2} \left[\sum_{i=1}^{3} \left(\frac{\partial L}{\partial q_i} - \frac{d}{dt}\frac{\partial L}{\partial \dot{q}_i} \right) \delta q_i \right] dt = 0 \tag{3.124}$$

where the integrated parts vanish by virtue of the conditions $\delta q_1 = \delta q_2 = \delta q_3 = 0$ at t_1 and t_2. When there are no constraint conditions, the coordinates are all independent, and the variations δq_1, δq_2, and δq_3 are all independent for $t \in (t_1, t_2)$. Thus for arbitrary end points the vanishing of the integral in Eq. (3.124) requires that the coefficients of the variations δq_1, δq_2, δq_3 vanish separately. Consequently we have the three equations:

$$\frac{d}{dt}\frac{\partial L}{\partial \dot{q}_i} - \frac{\partial L}{\partial q_i} = 0, \qquad i = 1, 2, 3. \tag{3.125}$$

These equations are called the *Lagrangian equations of motion*. They are the same as the Euler equations for the formal variational problem, and thus sometimes they are referred to as the Euler-Lagrange equations.

Example 3.8 (*General Conservative Motion of a Point Particle*) Consider the conservative motion of a particle of mass m described in spherical coordinates. Then we have

$$\begin{aligned} q_1 &= r, & \dot{q}_1 &= \dot{r} \\ q_2 &= \theta, & \dot{q}_2 &= \dot{\theta} \\ q_3 &= \phi, & \dot{q}_3 &= \dot{\phi}. \end{aligned} \tag{3.126}$$

The kinetic energy is

$$T = \frac{m}{2}\left[\dot{r}^2 + (r\dot{\theta})^2 + (r\sin\theta\dot{\phi})^2\right] \tag{3.127}$$

and the potential energy is given by $V = V(r, \theta, \phi)$. The Lagrangian equations of motion (3.125) become

$$\frac{d}{dt}(m\dot{r}) - mr(\dot{\theta}^2 + \sin^2\theta\dot{\phi}^2) = -\frac{\partial V}{\partial r} \tag{3.128a}$$

$$\frac{d}{dt}(mr^2\dot{\theta}) - mr^2(\sin\theta\cos\theta\dot{\phi}^2) = -\frac{\partial V}{\partial \theta} \tag{3.128b}$$

$$\frac{d}{dt}(mr^2\sin^2\theta\dot{\phi}) = -\frac{\partial V}{\partial \phi}. \tag{3.128c}$$

These equations can be placed in a form identical to the equations obtained from resolution of the vector equation

$$m\ddot{\mathbf{r}} = -\nabla V \tag{3.129}$$

into its component forms (see Exercises 1.5.2). ■

Example 3.9 (*Motion of a Pendulum in a Plane*) We imagine a pendulum described ideally as the motion of a mass m attached at the end of a rigid massless rod of length l that pivots about a fixed point O, as shown in Fig 3.8. Further we assume that the motion of the pendulum occurs in a fixed plane.

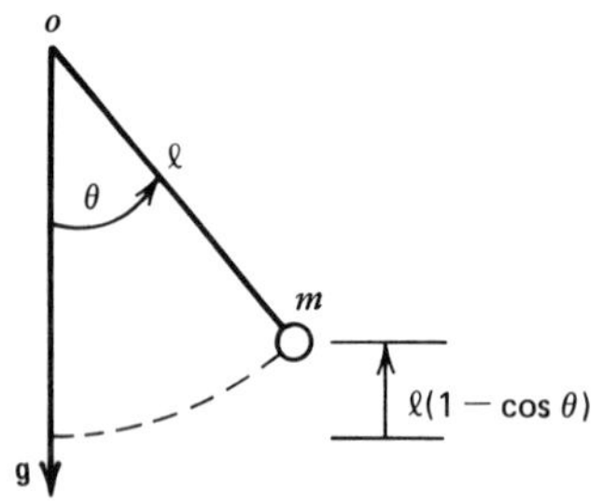

Figure 3.8 Configuration for the pendulum.

The motion is therefore a special case of the previous example in which we have

$$r = l = \text{constant}$$
$$\phi = \phi_0 = \text{constant}. \tag{3.130}$$

It follows that $\dot{r} = \dot{\phi} = 0$.

The acceleration of gravity **g** points downward, and θ is measured from this direction. The potential energy of the pendulum point bob is the work required to raise the bob above its lowest level:

$$V(\theta) = mgl(1 - \cos\theta). \tag{3.131}$$

We shall neglect the resistance of the point bob in the medium through which it moves.

The restrictions in the general motion that the pendulum bob remains a fixed distance away from its pivot point, $r = l$, and that the motion is contained in a plane, $\phi = \phi_0$, are called constraints. As we shall see, Eqs. (3.130) can be incorporated into the Lagrangian formulation as general constraint conditions by means of Lagrange multipliers. Alternatively, these particular constraints, which are *holonomic*, can be inserted directly into the Lagrangian function to effectively reduce the number of generalized coordinates. When this is done, we obtain for the pendulum problem

$$L = L(\theta, \dot{\theta}) = \frac{m}{2}(l^2\dot{\theta}^2) - mgl(1 - \cos\theta). \tag{3.132}$$

We thus have only one Lagrangian equation, Eq. (3.128b), which reduces to

$$\ddot{\theta} + \left(\frac{g}{l}\right)\sin\theta = 0 \tag{3.133}$$

This is the classical equation for a pendulum oscillating in a plane. Its solution involves elliptic functions (see Exercise 3.3.11). ■

3.4.6 Nonconservative Forces

When the forces are not conservative, then we must deal with Hamilton's principle directly in the form of Eq. (3.111). We first note that $\mathbf{F} \cdot \delta\mathbf{r}$ denotes the work done by the forces undergoing the variational displacement, that is,

$$\delta W = \mathbf{F} \cdot \delta\mathbf{r}. \tag{3.134}$$

The variational work δW is not in the form of an exact differential, except in the special case when **F** is given by the gradient of a potential. Thus for nonconservative forces Hamilton's principle must be written in the form

$$\delta\int_{t_1}^{t_2} T\,dt + \int_{t_1}^{t_2} \delta W\,dt = 0. \tag{3.135}$$

In this case there is no functional that must be an extremum. For nonconservative forces Hamilton's principle thus states that *the variation of the line integral of the kinetic energy between two time instants plus the line integral of the varied work over the same interval vanish.*

Now let us express the force in terms of components in a general orthogonal curvilinear coordinate system. Then we have

$$\begin{aligned}\delta W &\equiv \mathbf{F}\cdot\delta\mathbf{r} = F_1h_1\delta q_1 + F_2h_2\delta q_2 + F_3h_3\delta q_3 \\ &= Q_1\delta q_1 + Q_2\delta q_2 + Q_3\delta q_3\end{aligned} \tag{3.136}$$

where

$$Q_i \equiv F_ih_i, \qquad i=1,2,3. \tag{3.137}$$

The variables Q_1, Q_2, and Q_3 are called the *generalized forces* but, depending on the dimensions of the scale factors, they actually may not be forces but some other physical quantities such as moments.

The kinetic energy can be expressed functionally by Eq. (3.119) as before. Let us now take the variational operation on the line integral involving the kinetic energy for fixed end points, interchange the variation operation and the time derivative of the generalized velocities, and integrate by parts just as for the case of conservative forces. The integrated parts vanish since $\delta q_1 = \delta q_2 = \delta q_3 = 0$ at t_1 and t_2. Then substituting expression (3.136) for δW, we rewrite Eq. (3.135) as

$$\int_{t_1}^{t_2}\left[\sum_{i=1}^{3}\left(\frac{\partial T}{\partial q_i} - \frac{d}{dt}\frac{\partial T}{\partial \dot{q}_i} + Q_i\right)\delta q_i\right]dt = 0. \tag{3.138}$$

When all the generalized coordinates are independent, that is, when there are no constraints imposed on the motion, the integral will vanish for arbitrary end points only when the coefficients of the independent variations δq_1, δq_2, and δq_3 vanish separately. Thus we have

$$\frac{d}{dt}\frac{\partial T}{\partial \dot{q}_i} - \frac{\partial T}{\partial q_i} = Q_i, \qquad i=1,2,3. \tag{3.139}$$

These are the Lagrangian equations of motion for a point particle for *nonconservative forces.*

When the forces can be divided into a conservative part and a nonconservative part, then we have

$$\mathbf{F} = -\nabla V + \mathbf{F}^*, \tag{3.140}$$

where $\mathbf{F}^*$ is the nonconservative part. The Lagrangian operations can now be written in terms of a Lagrangian function, $L \equiv T - V$:

$$\frac{d}{dt}\frac{\partial L}{\partial \dot{q}_i} - \frac{\partial L}{\partial q_i} = Q_i^*, \qquad i = 1,2,3 \tag{3.141}$$

where

$$Q_i^* \equiv F_i^* h_i, \qquad i = 1,2,3 \tag{3.142}$$

and Q_i^* is the nonconservative part of the generalized force.

Example 3.10 (*Damped Motion of a Pendulum in a Plane*) We now assume that the pendulum bob in Example 3.9 experiences a resistance force proportional to its speed. Let μ be the viscosity of the surrounding medium, and let the radius of the spherical pendulum bob be a. Then the force of resistance on the bob is given by Stokes's law:

$$\mathbf{F}^* = -6\pi\mu a l \dot{\theta} \hat{\mathbf{e}}_\theta. \tag{3.143a}$$

The generalized forces are thus

$$Q_1^* = Q_3^* = 0$$

$$Q_2^* = -6\pi\mu a l^2 \dot{\theta}. \tag{3.143b}$$

This force of resistance is physically valid when the pertinent Reynolds number for the motion, $R_e \equiv \rho a \sqrt{gl}/\mu$, is sufficiently small, where ρ is the mass density of the surrounding medium. The resistance of the massless rod supporting the pendulum bob is neglected. The equation of motion of the pendulum is determined from the Lagrangian Eq. (3.141) for $i = 2$:

$$\ddot{\theta} + \left(\frac{g}{l}\right)\sin\theta = -\left(\frac{6\pi a\mu}{m}\right)\dot{\theta}. \tag{3.144}$$

The coefficient of $\dot{\theta}$ is called the damping coefficient c:

$$c = \frac{6\pi a\mu}{m}. \quad \blacksquare \tag{3.145}$$

3.4.7 Constraints

Kinematic relations that restrict a motion are called *constraint relations*, examples of which we have already encountered. When the relations are between the generalized coordinates, and possibly the time, in the form

$$G(q_1, q_2, q_3, t) = 0, \tag{3.146}$$

they are called *holonomic constraints*. When the constraints are *not* of this form, for instance, when they involve inequalities, they are called *nonholonomic constraints*. We shall first consider holonomic constraints.

For the simple problem of the motion of a single point mass, there are only three generalized coordinates. Hence there can be at most only two constraint equations. Thus for holonomic constraints we write

$$\begin{aligned} G_1(q_1, q_2, q_3, t) &= 0 \\ G_2(q_1, q_2, q_3, t) &= 0. \end{aligned} \tag{3.147}$$

The variation of these equations yields

$$\begin{aligned} \frac{\partial G_1}{\partial q_1}\delta q_1 + \frac{\partial G_1}{\partial q_2}\delta q_2 + \frac{\partial G_1}{\partial q_3}\delta q_3 &= 0 \\ \frac{\partial G_2}{\partial q_1}\delta q_1 + \frac{\partial G_2}{\partial q_2}\delta q_2 + \frac{\partial G_2}{\partial q_3}\delta q_3 &= 0. \end{aligned} \tag{3.148}$$

Clearly, only one of the variations δq_1, δq_2, or δq_3 is independent. We now multiply the first of Eq. (3.148) by the Lagrange multiplier λ_1 and the second by λ_2, integrate each equation over the time interval, and add the results to Eq. (3.138). The result is

$$\int_{t_1}^{t_2}\left[\sum_{i=1}^{3}\left(\frac{\partial T}{\partial q_i} - \frac{d}{dt}\frac{\partial T}{\partial \dot{q}_i} + Q_i + \lambda_1\frac{\partial G_1}{\partial q_i} + \lambda_2\frac{\partial G_2}{\partial q_i}\right)\delta q_i\right]dt = 0. \tag{3.149}$$

Since the variations δq_1, δq_2, δq_3 are not all independent, the conditions for the identical vanishing of the integral, for arbitrary end points, are to be obtained in the following way. The coefficients of two of the variations out of δq_1, δq_2, δq_3 are required to vanish, which essentially yields two equations for λ_1 and λ_2. Since the variation of one generalized coordinate remains, its coefficient is set to zero. By this procedure no generalized coordinate is given preferential treatment, and thus we obtain

$$\frac{d}{dt}\frac{\partial T}{\partial \dot{q}_i} - \frac{\partial T}{\partial q_i} = Q_i + \lambda_1\frac{\partial G_1}{\partial q_i} + \lambda_2\frac{\partial G_2}{\partial q_i}, \qquad i = 1, 2, 3. \tag{3.150}$$

These are the generalized Lagrangian equations for a single particle acted on by nonconservative forces and holonomic constraints.

The Lagrangian Eqs. (3.150) constitute three equations for the five unknowns q_1, q_2, q_3, $\lambda_1(t)$, and $\lambda_2(t)$. Together with the two constraint Eqs. (3.147), they comprise a closed set of five equations for the five unknowns. The Lagrange multipliers are themselves functions of time in general.

In the general Lagrangian Eqs. (3.150), the combinations

$$\lambda_1 \frac{\partial G_1}{\partial q_i} \qquad \text{and} \qquad \lambda_2 \frac{\partial G_2}{\partial q_i} \tag{3.151}$$

play the role of generalized forces. Thus the Lagrange multipliers are related physically to the forces or moments required to enforce the constraints. In many problems this information is desired explicitly.

Example 3.11 We shall consider the damped motion of a pendulum in a plane accounting for the constraints by means of the Lagrange multipliers. The constraint conditions are

$$\begin{aligned} G_1(r) &\equiv r - l = 0 \\ G_2(\phi) &= \phi - \phi_0 = 0. \end{aligned} \tag{3.152}$$

The kinetic energy is given by Eq. (3.127), and the generalized forces are determined from

$$\mathbf{F} = mg(-\sin\theta\hat{\mathbf{e}}_\theta + \cos\theta\hat{\mathbf{e}}_r) - 6\pi\mu a r\dot{\theta}\hat{\mathbf{e}}_\theta. \tag{3.153}$$

The set of governing equations, the Lagrange equations and the constraint equations, become

$$\frac{d}{dt}(m\dot{r}) - mr(\dot{\theta}^2 + \sin^2\theta\dot{\phi}^2) = mg\cos\theta + \lambda_1 \tag{3.154a}$$

$$\frac{d}{dt}(mr^2\dot{\theta}) - mr^2(\sin\theta\cos\theta\dot{\phi}^2) = -mgr\sin\theta - 6\pi\mu a r^2\dot{\theta} \tag{3.154b}$$

$$\frac{d}{dt}(mr^2\sin^2\theta\dot{\phi}) = \lambda_2 \tag{3.154c}$$

$$r - l = 0 \tag{3.154d}$$

$$\phi - \phi_0 = 0. \tag{3.154e}$$

The constraint equations lead to $\dot{r} = \dot{\phi} = 0$, and thus Eqs. (3.154a) and (3.154c) yield

$$\begin{aligned} \lambda_1 &= -mg\cos\theta \\ \lambda_2 &= 0. \end{aligned} \tag{3.155}$$

We can interpret λ_1 as the force exerted on the pendulum bob by the massless rod, and thus also the negative of the tension force in the rod. It is the force necessary to oppose gravity and maintain the motion of the bob in a circular arc. The result $\lambda_2=0$ is a statement of the fact that a zero side force is required to maintain the motion of the bob in a given plane. Equation (3.154b) reduces to Eq. (3.144), which was obtained earlier. ■

3.4.8 Nonholonomic Constraints

Nonholonomic constraints involve relations between the generalized velocities as well as the generalized coordinates. For a single particle a quasilinear nonholonomic constraint (linear in the generalized velocities) would be of the form

$$g_1(q_1,q_2,q_3,t)\dot{q}_1+g_2(q_1,q_2,q_3,t)\dot{q}_2 \\ +g_3(q_1,q_2,q_3,t)\dot{q}_3+g_4(q_1,q_2,q_3,t)=0. \qquad (3.156)$$

In a differential and summation form, this expression can be written as

$$\sum_{i=1}^{3} g_i(q_1,q_2,q_3,t)\,dq_i+g_4(q_1,q_2,q_3,t)\,dt=0. \qquad (3.157)$$

According to Hamilton's principle, the time is held constant for each point on the path which is varied. Hence the variational form of Eq. (3.157) is

$$\sum_{i=1}^{3} g_i(q_1,q_2,q_3,t)\delta q_i=0. \qquad (3.158)$$

This form is similar to that obtained by the variational differential of the holonomic constraint (3.146). The difference is that the nonholonomic constraint (3.158) does not have an exact differential and thus cannot be integrated and written in the form $G(q_1,q_2,q_3,t)=0$. Thus for nonholonomic constraints the dependent generalized coordinates cannot be solved for in terms of the independent generalized coordinates and thus eliminated directly in the kinetic energy or Lagrangian functions. A Lagrange multiplier method is thus a necessity. On the other hand, the total time derivative of the holonomic constraint (3.146) produces an equation with the same form as Eq. (3.156). Thus the holonomic constraint $G(q_1,q_2,q_3,t)=0$ can be regarded as a special case of the particular nonholonomic constraint Eq. (3.156).

For a single particle, which has three generalized coordinates, we may in general have a maximum of two constraint equations:

$$\sum_{i=1}^{3} g_{1i}\dot{q}_i + g_{14} = 0$$

$$\sum_{i=1}^{3} g_{2i}\dot{q}_i + g_{24} = 0. \tag{3.159}$$

The variational constraint conditions are therefore

$$\sum_{i=1}^{3} g_{1i}\delta q_i = 0$$

$$\sum_{i=1}^{3} g_{2i}\delta q_i = 0. \tag{3.160}$$

These equations have precisely the same form as for the holonomic conditions (3.148), with the identification $g_{ki} \to \partial G_k / \partial q_i$. Thus by multiplying Eqs. (3.160) by the Lagrange multipliers $\lambda_1(t)$ and $\lambda_2(t)$ respectively, and proceeding as for the holonomic constraint case, we obtain the following Lagrangian equations:

$$\frac{d}{dt}\frac{\partial T}{\partial \dot{q}_i} - \frac{\partial T}{\partial q_i} = Q_i + \lambda_1 g_{1i} + \lambda_2 g_{2i}, \qquad i = 1,2,3. \tag{3.161}$$

The closed set of equations for the problem thus consists of the three Eqs. (3.161) and the two constraint Eqs. (3.159) for the unknown functions q_1, q_2, q_3, λ_1, and λ_2. Clearly, it is possible to have one holonomic constraint and one nonholonomic constraint in this formulation.

Example 3.12 (*Spherical Pendulum with Prescribed Azimuthal Velocity*) We consider the motion of a pendulum which is not restricted to a plane and which has an azimuthal velocity that is prescribed. The constraints are

$$G_1(r) = r - l = 0 \tag{3.162a}$$

$$r \sin\theta\dot{\phi} - v(t) = 0 \tag{3.162b}$$

where $v(t)$ is the prescribed azimuthal velocity. The constraint (3.162a) is holonomic, and the constraint (3.162b) is nonholonomic. The holonomic constraint can be differentiated and placed in the form

$$\dot{r} = 0 \tag{3.163}$$

which renders it in the same form as the nonholonomic constraint. The force of resistance is now proportional to the total velocity, and thus the generalized forces are obtained from

$$\mathbf{F}=mg(\cos\theta\hat{\mathbf{e}}_r-\sin\theta\hat{\mathbf{e}}_\theta)-6\pi\mu a\left(\dot{r}\hat{\mathbf{e}}_r+r\dot{\theta}\hat{\mathbf{e}}_\theta+r\sin\theta\dot{\phi}\hat{\mathbf{e}}_\phi\right). \quad (3.164)$$

The generalized Lagrangian equations of motion (3.161) are thus found to be

$$\frac{d}{dt}(m\dot{r})-mr(\dot{\theta}^2+\sin^2\theta\dot{\phi}^2)=mg\cos\theta-6\pi\mu a\dot{r}+\lambda_1 \quad (3.165a)$$

$$\frac{d}{dt}(mr^2\dot{\theta})-mr^2(\sin\theta\cos\theta\dot{\phi}^2)=-mgr\sin\theta-6\pi\mu ar^2\dot{\theta} \quad (3.165b)$$

$$\frac{d}{dt}(mr^2\sin^2\theta\dot{\phi})=-6\pi\mu ar^2\sin^2\theta\dot{\phi}+\lambda_2 r\sin\theta. \quad (3.165c)$$

When the constraint equations are taken into account, these equations become

$$-ml\dot{\theta}^2-\frac{mv^2}{l}=mg\cos\theta+\lambda_1 \quad (3.166a)$$

$$\ddot{\theta}+\frac{g}{l}\sin\theta-\frac{v^2}{l^2}\cot\theta=\frac{-6\pi\mu a}{m}\dot{\theta} \quad (3.166b)$$

$$m\frac{d}{dt}(v\sin\theta)=-6\pi\mu a\sin\theta v+\lambda_2\sin\theta. \quad (3.166c)$$

When the azimuthal velocity $v=v(t)$ is prescribed explicitly, Eq. (3.166b) constitutes an ordinary differential equation for $\theta(t)$. When this is solved, Eq. (3.166a) is an equation for determining λ_1, which is the negative of the tension force in the massless pendulum rod. Likewise Eq. (3.166c) yields the value of λ_2, which is the force in the azimuthal direction required to impose the prescribed azimuthal velocity. ■

3.4.9 Hamilton's Principle for a System of *N* Particles

The derivation of Hamilton's principle and the Lagrangian equations for a system of *N* particles is analogous to that for a single particle. Newton's second law of motion applies for each particle, and we have

$$m_i\frac{d^2\mathbf{r}_i}{dt^2}-\mathbf{F}_i=0, \qquad i=1,2,3,\ldots,N. \quad (3.167)$$

Scalar multiplication with $\delta\mathbf{r}_i$, summation over all the particles, and integration over the time interval $t \in (t_1, t_2)$ then yield

$$\int_{t_1}^{t_2} \left[\sum_{i=1}^{N} \left(m_i \frac{d^2\mathbf{r}_i}{dt^2} \cdot \delta\mathbf{r}_i - \mathbf{F}_i \cdot \delta\mathbf{r}_i \right) \right] dt = 0. \tag{3.168}$$

When the acceleration terms are integrated by parts with the end conditions that $\delta\mathbf{r}_i = 0$ at t_1 and t_2 for all particles, the integrated parts vanish, and we have

$$\int_{t_1}^{t_2} (\delta T + \delta W)\, dt = 0 \tag{3.169a}$$

where

$$T \equiv \sum_{i=1}^{N} \frac{m_i}{2} \left(\frac{d\mathbf{r}_i}{dt} \right)^2 \tag{3.169b}$$

$$\delta W \equiv \sum_{i=1}^{N} \mathbf{F}_i \cdot \delta\mathbf{r}_i. \tag{3.169c}$$

We thus have Hamilton's principle for a system of N particles.

Since the position of each particle can be designated by three position coordinates, the spatial configuration of the system as a whole can be described by $3N$ position coordinates. This description can be fashioned in numerous ways, and in general we can describe the spacial configurations of the system by $3N$ generalized coordinates: $q_1, q_2, q_3, \ldots, q_{3N}$. These coordinates describe a *configuration space* of $3N$ dimensions. Correspondingly the velocities of the N particles will involve the time derivatives of the generalized coordinates, and thus also we will have $3N$ generalized velocities: $\dot{q}_1, \dot{q}_2, \dot{q}_3, \ldots, \dot{q}_{3N}$. The generalized velocities form a *velocity space* of $3N$ dimensions. The generalized coordinates and the generalized velocities taken together form a $6N$-space, called *phase space*. Since the state of the system as a whole is described by the positions of all the particles and the velocities of all the particles, the state of the system is denoted by a single point in phase space. As time changes, this point moves in phase space and traces out a curve called the *phase trajectory*. The kinetic energy of the system can be expressed functionally as

$$T = T(q_1, q_2, \ldots, q_{3N}, \dot{q}_1, \dot{q}_2, \ldots, \dot{q}_{3N}, t). \tag{3.170}$$

Frequently in physical problems, as we have seen, there are kinematic constraints on the variables. If we have k holonomic constraints (where $k < 3N$), then it may be possible to eliminate k generalized coordinates and reduce the problem to $3N-k$ independent generalized coordinates. With this method the constraints are thus contained implicitly, and the forces of constraint are not a direct outcome of the solution to the problem. If the k constraints are nonholonomic, then k dependent variables cannot be eliminated directly, and some other means must be utilized to accommodate the constraint conditions, such as Lagrange multipliers. At any rate, if there are k constraint equations, then the system is said to have $3N-k$ *degrees of freedom*.

We now assume that we have k nonholonomic constraint equations of the form (which may include holonomic constraints as a special case)

$$\sum_{i=1}^{3N} g_{ji}\dot{q}_i + g_j = 0, \qquad j = 1,2,\ldots,k \tag{3.171a}$$

where the coefficients g_{ji} and g_j are functions of q_i, $\dot{q}_i$, and t in general. The variational form is

$$\sum_{i=1}^{3N} g_{ji}\delta q_i = 0, \qquad j = 1,2,\ldots,k. \tag{3.171b}$$

By our previously established methods we can now obtain the generalized Lagrange equations for a system of N particles:

$$\frac{d}{dt}\frac{\partial T}{\partial \dot{q}_i} - \frac{\partial T}{\partial q_i} = Q_i + \sum_{j=1}^{k} \lambda_j g_{ji}, \qquad i = 1,2,\ldots,3N. \tag{3.172}$$

Here the Q_i and λ_j represent the generalized forces and Lagrange multipliers, as before. If the forces can be resolved into a conservative part and a nonconservative part, then the Eqs. (3.172) can be written in terms of a Lagrangian function, $L \equiv T - V$:

$$\frac{d}{dt}\frac{\partial L}{\partial \dot{q}_i} - \frac{\partial L}{\partial q_i} = Q_i^* + \sum_{j=1}^{k} \lambda_j g_{ji}, \qquad i = 1,2,\ldots,3N \tag{3.173}$$

where Q_i^* is the nonconservative part of the general forces.

Example 3.13 (*Atwood's Machine*) Two masses m_1 and m_2 are attached to the ends of an inextensible cord that is suspended over a frictionless stationary pulley, as shown in Fig. 3.9. We wish to find the motion of the two masses as they are affected by gravity. For this problem six appropriate generalized coordinates are x_1, y_1, z_1 and x_2, y_2, z_2. The motion is taken in a plane, and thus four constraints are $y_1 = z_1 = y_2 = z_2 = 0$, leaving x_1 and x_2 as the pertinent coordinates for m_1 and m_2. The kinetic energy is

$$T = \frac{1}{2}m_1\dot{x}_1^2 + \frac{1}{2}m_2\dot{x}_2^2 \tag{3.174}$$

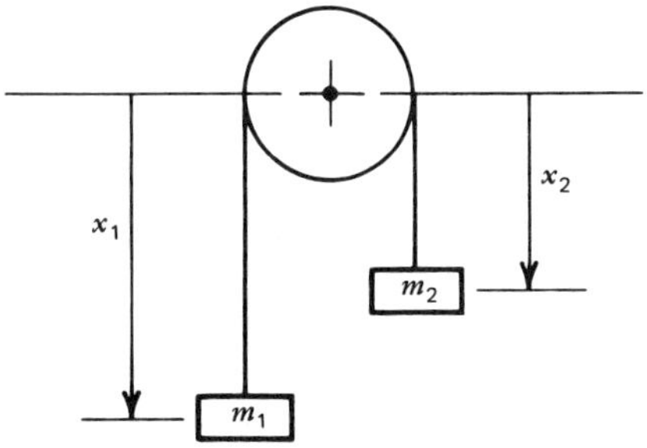

Figure 3.9 Atwood's machine.

and the potential energy is

$$V = -m_1 g x_1 - m_2 g x_2. \tag{3.175}$$

Since the cord is inextensible, the constraint (which is holonomic) is

$$x_1 + x_2 = l = \text{constant}. \tag{3.176}$$

We can now proceed by two means. We can eliminate x_2 in favor of x_1 directly in T and V, and thus reduce the problem to a single independent coordinate, or we can accommodate the constraint by means of a Lagrange multiplier. In the first approach, eliminating x_2 gives

$$T = \frac{1}{2}(m_1 + m_2)\dot{x}_1^2$$

$$V = -(m_1 - m_2) g x_1 - m_2 g l. \tag{3.177}$$

Neglecting any nonconservative forces, such as friction, we use the Lagrangian Eq. (3.173) to get

$$\ddot{x}_1 = \left(\frac{m_1 - m_2}{m_1 + m_2} \right) g. \tag{3.178}$$

In the second approach we utilize a Lagrange multiplier λ for the constraint (3.176). The Lagrangian Eqs. (3.173) then yield

$$m_1 \ddot{x}_1 - m_1 g = \lambda$$

$$m_2 \ddot{x}_2 - m_2 g = \lambda. \tag{3.179}$$

The constraint relation (3.176) yields $\ddot{x}_1 = -\ddot{x}_2$. Elimination of λ between Eqs. (3.179) then yields the result (3.178), as before. In this case, however, we can now solve for λ, the negative of the tension in the rope:

$$\lambda = -\frac{2m_1 m_2 g}{m_1 + m_2}, \tag{3.180}$$

which the first method does not yield. ■

Example 3.14 (*Atwood's Machine with a Climbing Monkey*) Consider Atwood's machine, as in the previous example, with a monkey of mass m_3 climbing up the cord above mass m_2 with a speed v_0 relative to mass m_2 (see Fig. 3.10). What is the subsequent motion of the system?

The constraints for this system are

$$\begin{aligned} x_1 + x_2 &= l \\ \dot{x}_2 - \dot{x}_3 &= v_0. \end{aligned} \tag{3.181}$$

The kinetic energy and the potential energy for the system are

$$\begin{aligned} T &= \frac{m_1}{2}\dot{x}_1^2 + \frac{m_2}{2}\dot{x}_2^2 + \frac{m_3}{2}\dot{x}_3^2 \\ V &= -m_1 g x_1 - m_2 g x_2 - m_3 g x_3. \end{aligned} \tag{3.182}$$

The Lagrangian equations of motion are

$$\begin{aligned} m_1\ddot{x}_1 - m_1 g &= \lambda_1 \\ m_2\ddot{x}_2 - m_2 g &= \lambda_1 + \lambda_2 \\ m_3\ddot{x}_3 - m_3 g &= -\lambda_2. \end{aligned} \tag{3.183}$$

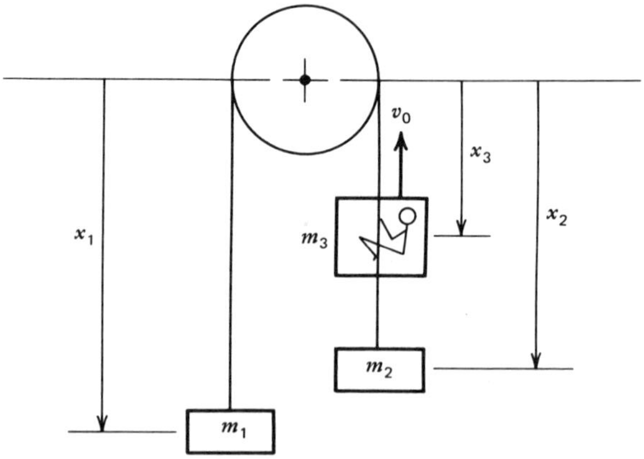

Figure 3.10 Atwood's machine with a climbing monkey.

The solution to this problem is straightforward, and we find that

$$\ddot{x}_1 = -\ddot{x}_2 = -\ddot{x}_3 = \left(\frac{m_1 - m_2 - m_3}{m_1 + m_2 + m_3} \right) g$$

$$\lambda_1 = -\frac{2m_1(m_2 + m_3)g}{m_1 + m_2 + m_3}$$

$$\lambda_2 = \frac{2m_1 m_3 g}{m_1 + m_2 + m_3}. \tag{3.184}$$

The Lagrange multiplier λ_1 is the negative of the tension in the cord attached to m_1. The tension exerted on the cord by the climbing monkey is λ_2. The tension in the cord attached to m_2 is given by $\lambda_1 + \lambda_2$. ■

3.4.10 Rotational Motion of Rigid Bodies

A rigid body can be defined as a system of mass points subject to the holonomic constraints that the distances between all pairs of mass points are fixed. These constraints are somewhat of an idealization since all solid bodies are deformable to some degree, however slight. Nevertheless the idealization of a rigid body is extremely useful in a great number of applications.

In general N particles can have as many as $3N$ degrees of freedom. The constraints of a rigid body greatly reduce the number of degrees of freedom when N is large. We first fix our attention on only three noncollinear reference points, denoted by the position vectors $\mathbf{r}_1$, $\mathbf{r}_2$, and $\mathbf{r}_3$, as shown in Fig. 3.11. Fixing the distance between these reference points provides three holonomic

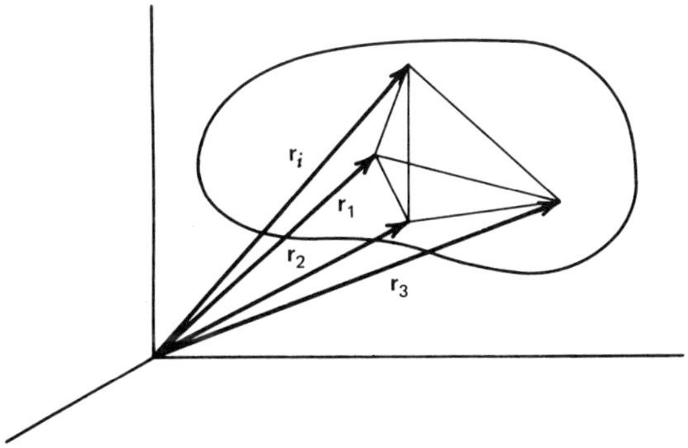

Figure 3.11 Reference points in a rigid body.

constraints:

$$|\mathbf{r}_1-\mathbf{r}_2|=C_{12}=\text{constant}$$
$$|\mathbf{r}_1-\mathbf{r}_3|=C_{13}=\text{constant}$$
$$|\mathbf{r}_2-\mathbf{r}_3|=C_{23}=\text{constant}. \tag{3.185}$$

Now let $\mathbf{r}_i$ denote the position of any other particle in the system except the three reference particles; there are $N-3$ such particles. For a rigid body the distances from $\mathbf{r}_i$ to any of the three reference points are fixed:

$$|\mathbf{r}_i-\mathbf{r}_1|=C_{i1}=\text{constant}, \qquad i=4,5,\ldots,N$$
$$|\mathbf{r}_i-\mathbf{r}_2|=C_{i2}=\text{constant}, \qquad i=4,5,\ldots,N$$
$$|\mathbf{r}_i-\mathbf{r}_3|=C_{i3}=\text{constant}, \qquad i=4,5,\ldots,N. \tag{3.186}$$

Equations (3.186) constitute $3(N-3)$ holonomic constraints, which together with the three constraints (3.185) for the reference points provide a total of $3N-6$ constraints. It thus follows that the rigid-body constraints reduce the original $3N$ degrees of freedom to only six. Consequently not more than six generalized coordinates are required to describe the motion of a rigid body. Of course if there are further constraints, then the number of generalized coordinates is reduced accordingly.

Now consider a possible description of the six generalized coordinates for a rigid body. The sum of the individual equations of motion (3.106) produces the result

$$\frac{d^2}{dt^2}\sum_{i=1}^{N} m_i\mathbf{r}_i=\sum_{i=1}^{N}\mathbf{F}_i. \tag{3.187}$$

This equation suggests the definition of the mass average position vector $\mathbf{R}$, called the *center of mass vector*:

$$\mathbf{R}\equiv\frac{\sum_{i=1}^{N} m_i\mathbf{r}_i}{M} \tag{3.188}$$

where $M\equiv\sum_{i=1}^{N}m_i$ is the total mass of the system. Equation (3.187) can thus be written

$$M\ddot{\mathbf{R}}=\sum_{i=1}^{N}\mathbf{F}_i. \tag{3.189}$$

The sum of all the forces acting on the particles thus appears as if it were acting on the total mass concentrated at the center of mass of the system.

It is now useful to reckon the position of each particle with respect to the center of mass:

$$\mathbf{R}_i \equiv \mathbf{r}_i - \mathbf{R}. \tag{3.190}$$

Clearly, it follows that

$$\sum_{i=1}^{N} m_i \mathbf{R}_i = 0 \tag{3.191}$$

and hence the positions of the particles with respect to the center of mass are not all independent. The kinetic energy of the system can now be expressed as follows:

$$\begin{aligned} T &\equiv \frac{1}{2} \sum_{i=1}^{N} m_i (\dot{\mathbf{r}}_i)^2 \\ &= \frac{1}{2} M(\dot{\mathbf{R}})^2 + \frac{1}{2} \sum_{i=1}^{N} m_i (\dot{\mathbf{R}}_i)^2 \end{aligned} \tag{3.192}$$

where use was made of Eq. (3.191). The total kinetic energy is thus found to be the *sum of the kinetic energy of the concentrated total mass of the system moving with the velocity of the center of mass plus the sum of the kinetic energies of the particles moving relative to the center of mass.*

For a rigid body the position vectors $\mathbf{R}_i$ relative to the center of mass of the system have a fixed length. Furthermore the relative velocity $\dot{\mathbf{R}}_i$ can be expressed as

$$\dot{\mathbf{R}}_i = \boldsymbol{\omega} \times \mathbf{R}_i \tag{3.193}$$

where $\boldsymbol{\omega}$ is the angular velocity vector of the rotational motion about the center of mass. As in Section 1.5.1, we can now show that the kinetic energy of the rotational motion about the center of mass can be written for rigid bodies as

$$\frac{1}{2} \sum_{i=1}^{N} m_i (\dot{\mathbf{R}}_i)^2 \equiv \frac{1}{2} \boldsymbol{\omega} \cdot \overleftrightarrow{\mathfrak{J}} \cdot \boldsymbol{\omega} \tag{3.194}$$

where

$$\overleftrightarrow{\mathfrak{J}} \equiv \sum_{i=1}^{N} m_i \left(R_i^2 \overleftrightarrow{\mathbf{I}} - \mathbf{R}_i \mathbf{R}_i \right) \tag{3.195}$$

is the *moment-of-inertia tensor* and $\overleftrightarrow{\mathbf{I}}$ is the idemfactor or unit tensor. (The extension to a continuous distribution of infinitesimal mass particles involves replacing m_i by $\rho\, dV$, where ρ is the mass density and dV the differential volume element, and replacing the summation by an integral over the whole body.) The total kinetic energy of the rigid body thus can be expressed as

$$T = \frac{1}{2} M(\dot{\mathbf{R}})^2 + \frac{1}{2} \boldsymbol{\omega} \cdot \overleftrightarrow{\mathcal{I}} \cdot \boldsymbol{\omega}. \tag{3.196}$$

The six generalized coordinates describe the center-of-mass position vector $\mathbf{R}$ and the angular rate of rotation vector $\boldsymbol{\omega}$.

3.4.11 Euler Angles

A useful choice of generalized coordinates for the description of the angular velocity vector is the set of three angles ϕ, θ, ψ, called the *Euler angles*, which are described in Exercise 1.5.6 and shown in Fig. 3.12. The Euler angles describe the orientation of a Cartesian body-fixed coordinate system with respect to a Cartesian inertial coordinate system.

In body-fixed coordinates the angular velocity vector is given by

$$\boldsymbol{\omega} = \omega_x \hat{\mathbf{e}}_x + \omega_y \hat{\mathbf{e}}_y + \omega_z \hat{\mathbf{e}}_z \tag{3.197}$$

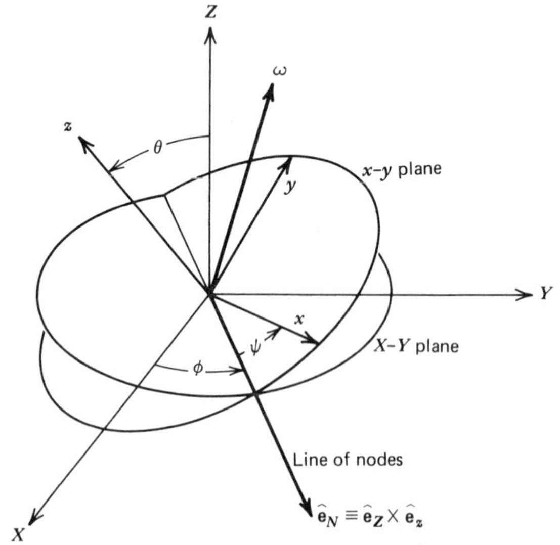

Figure 3.12 Euler angles.

where

$$\omega_x = \dot{\phi} \sin\theta \sin\psi + \dot{\theta} \cos\psi$$

$$\omega_y = \dot{\phi} \sin\theta \cos\psi - \dot{\theta} \sin\psi$$

$$\omega_z = \dot{\phi} \cos\theta + \dot{\psi}. \tag{3.198}$$

When the body-fixed coordinates are aligned with the principal axes of the rigid body, the moment-of-inertial tensor is given by

$$\overleftrightarrow{\mathcal{I}} = I_{xx}\hat{\mathbf{e}}_x\hat{\mathbf{e}}_x + I_{yy}\hat{\mathbf{e}}_y\hat{\mathbf{e}}_y + I_{zz}\hat{\mathbf{e}}_z\hat{\mathbf{e}}_z \tag{3.199}$$

and the kinetic energy of rotation is then expressed as

$$\begin{aligned} T_{\text{rot}} &\equiv \frac{1}{2}\omega \cdot \overleftrightarrow{\mathcal{I}} \cdot \omega \\ &= \frac{1}{2}\left[I_{xx}\omega_x^2 + I_{yy}\omega_y^2 + I_{zz}\omega_z^2\right], \end{aligned} \tag{3.200}$$

which is given in terms of the Euler angles by virtue of Eqs. (3.198).

3.4.12 Lagrangian Equations for Rotational Motion

Since the kinetic energy associated with the motion of the center of mass is independent of the Euler angles, we can write

$$\begin{aligned} T &= T(\dot{\mathbf{R}}, \phi, \theta, \psi, \dot{\phi}, \dot{\theta}, \dot{\psi}) \\ &= T_{CM}(\dot{\mathbf{R}}) + T_{\text{rot}}(\phi, \theta, \psi, \dot{\phi}, \dot{\theta}, \dot{\psi}). \end{aligned} \tag{3.201}$$

The Lagrangian equations for rotational motion are thus (for no constraints):

$$\frac{d}{dt}\frac{\partial T_{\text{rot}}}{\partial \dot{\phi}} - \frac{\partial T_{\text{rot}}}{\partial \phi} = Q_\phi \tag{3.202a}$$

$$\frac{d}{dt}\frac{\partial T_{\text{rot}}}{\partial \dot{\theta}} - \frac{\partial T_{\text{rot}}}{\partial \theta} = Q_\theta \tag{3.202b}$$

$$\frac{d}{dt}\frac{\partial T_{\text{rot}}}{\partial \dot{\psi}} - \frac{\partial T_{\text{rot}}}{\partial \psi} = Q_\psi. \tag{3.202c}$$

It is interesting to compare these equations with the vector Euler equation for rotational motion:

$$\overleftrightarrow{\mathcal{I}} \cdot \dot{\omega} + \omega \times (\overleftrightarrow{\mathcal{I}} \cdot \omega) = \mathbf{M}. \tag{3.203}$$

It is easy to show that Eq. (3.202c) for the generalized coordinate ψ can be rewritten as

$$I_{zz}\dot{\omega}_z + \omega_x\omega_y(I_{yy} - I_{zz}) = Q_\psi. \tag{3.204}$$

This equation is simply the z component of Eq. (3.203), such that $Q_\psi \equiv \mathbf{M}\cdot\hat{\mathbf{e}}_z$, that is, the generalized force Q_ψ is the component of the moment about the center of gravity in the body-fixed z direction. This is to be expected since the Euler angle ψ is associated with a rotation about the body-fixed z axis.

The Lagrangian Eqs. (3.202a) and (3.202b) for the generalized coordinates ϕ and θ do not take forms similar to Eq. (3.204). However, since ϕ and θ are associated with rotations around the inertial Z axis and the line of nodes, the Lagrangian Eqs. (3.202a) and (3.202b) can be rewritten in the forms

$$\hat{\mathbf{e}}_Z\cdot\left[\overleftrightarrow{\mathcal{I}}\cdot\dot{\boldsymbol{\omega}} + \boldsymbol{\omega}\times(\overleftrightarrow{\mathcal{I}}\cdot\boldsymbol{\omega})\right] = \mathbf{M}\cdot\hat{\mathbf{e}}_Z \equiv Q_\phi \tag{3.205a}$$

$$\hat{\mathbf{e}}_N\cdot\left[\overleftrightarrow{\mathcal{I}}\cdot\dot{\boldsymbol{\omega}} + \boldsymbol{\omega}\times(\overleftrightarrow{\mathcal{I}}\cdot\boldsymbol{\omega})\right] = \mathbf{M}\cdot\hat{\mathbf{e}}_N \equiv Q_\theta. \tag{3.205b}$$

The generalized forces Q_ϕ and Q_θ are the components of the moment about the center of mass in the inertial Z direction and in the direction of the line of nodes.

Example 3.15 (*Trajectory of a Curving Baseball*) Consider a homogeneous spherical ball (a baseball, for instance) traveling through the air with density ρ, at a velocity $\dot{\mathbf{R}} \equiv \mathbf{V}(t)$ and spinning with the angular velocity $\boldsymbol{\omega}(t)$. Let the radius of the ball be $r = a$, such that the cross-sectional area is $S = \pi a^2$. The moment of inertia is given by $\overleftrightarrow{\mathcal{I}} = I_0\overleftrightarrow{\mathbf{I}}$, where $I_0 \equiv 2ma^2/5$ and m is the mass of the ball. The forces acting on the ball are gravity, drag, and the Magnus force arising because of spin, which is perpendicular to both $\boldsymbol{\omega}$ and $\mathbf{V}$:

$$\mathbf{F} = m\mathbf{g} - \frac{1}{2}\rho S C_D V\mathbf{V} + \frac{1}{2}\rho S C_M a\boldsymbol{\omega}\times\mathbf{V}. \tag{3.206}$$

Here C_D is the dimensionless drag coefficient and C_M the dimensionless Magnus coefficient, both assumed constant. The moment on the ball is assumed to be

$$\mathbf{M} = -\frac{1}{2}\rho S C_S a^2 \frac{\mathbf{V}\times(\boldsymbol{\omega}\times\mathbf{V})}{V} \tag{3.207}$$

where C_S is the dimensionless spin coefficient, also assumed constant. For a homogeneous sphere the vector equations of motion are

$$m\frac{d\mathbf{V}}{dt} = \mathbf{F} \tag{3.208a}$$

$$I_0\frac{d\boldsymbol{\omega}}{dt} = \mathbf{M}. \tag{3.208b}$$

These two equations constitute two vector equations for $\mathbf{V}$ and $\boldsymbol{\omega}$, that is, for the generalized velocities. The generalized coordinates need not actually appear insofar as solving the equations is concerned. However, if the actual trajectory of the ball is desired, the center-of-mass position is obtained from

$$\mathbf{R}(t)=\mathbf{R}_0+\int_0^t \mathbf{V}(t)\,dt. \tag{3.209}$$

On the other hand, when the governing equations are established by means of the Lagrangian equations, the set of equations appears considerably more complicated because it involves the generalized coordinates, that is, the coordinates of the center of mass of the ball and the Euler angles, and the generalized forces have to be resolved into their respective component forms by means of Eqs. (3.206) and (3.207). Thus for this problem the Lagrangian formulation does not offer any particular advantage over the vector formulation (3.208), and indeed it actually may be more cumbersome. Frequently the advantage obtained from the Lagrangian method in formulating the governing equations comes from constraints that are involved with the problem. ■

3.4.13 Constraints for Rolling Motion

Consider a rigid body rotating such that its outer surface contacts the fixed surface of another solid body at a single point, called the point of contact. Let $\mathbf{r}_p$ denote the position vector to the point of contact on the rotating rigid body, and let $\mathbf{R}$ denote the position vector to the center of gravity of the rotating rigid body (see Fig. 3.13). The velocity of the point of contact of the rotating rigid body relative to an inertial reference frame is

$$\dot{\mathbf{r}}_p=\dot{\mathbf{R}}+\boldsymbol{\omega}\times(\mathbf{r}_p-\mathbf{R}). \tag{3.210}$$

If there is no relative motion between the surface of the rotating body and the fixed surface at the contact point $\mathbf{r}_p$, then since the fixed surface is stationary, we have $\dot{\mathbf{r}}_p=0$. The constraint condition for pure rolling without slipping is

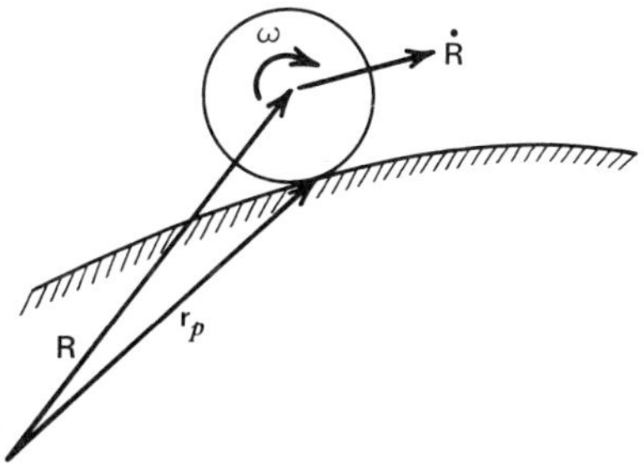

Figure 3.13 Configuration for a rolling body.

thus, from Eq. (3.210),

$$\dot{\mathbf{R}} = \boldsymbol{\omega} \times (\mathbf{R} - \mathbf{r}_p). \tag{3.211}$$

In general this constraint is nonholonomic.

Example 3.16 (*Trajectory of a Bowling Ball*) Consider a homogeneous spherical ball of radius $r = a$ (a bowling ball, for instance) rolling without slipping on a flat solid surface. Let the Z axis in a Cartesian inertial reference frame be measured upward opposite to gravity, such that $\mathbf{g} = -g\hat{\mathbf{e}}_Z$. The position of the center of gravity of the ball is

$$\mathbf{R} = X\hat{\mathbf{e}}_X + Y\hat{\mathbf{e}}_Y + Z\hat{\mathbf{e}}_Z. \tag{3.212}$$

The angular velocity is given in inertial components by

$$\boldsymbol{\omega} = \omega_X\hat{\mathbf{e}}_X + \omega_Y\hat{\mathbf{e}}_Y + \omega_Z\hat{\mathbf{e}}_Z, \tag{3.213}$$

where (see Exercise 1.5.6)

$$\begin{aligned} \omega_X &= \dot{\psi} \sin\theta \sin\phi + \dot{\theta} \cos\phi \\ \omega_Y &= -\dot{\psi} \sin\theta \cos\phi + \dot{\theta} \sin\phi \\ \omega_Z &= \dot{\psi} \cos\theta + \dot{\phi}. \end{aligned} \tag{3.214}$$

One constraint is $Z = a$, that is, the center of gravity of the ball is a fixed height above the flat surface. It follows that $\dot{Z} = 0$. Since $\mathbf{R} - \mathbf{r}_p = a\hat{\mathbf{e}}_Z$, the component form of the nonholonomic constraint Eq. (3.211) is

$$\begin{aligned} \dot{X} &= a\omega_Y \\ \dot{Y} &= -a\omega_X \end{aligned} \tag{3.215a}$$

or

$$\begin{aligned} &\dot{X} + a(\dot{\psi} \sin\theta \cos\phi - \dot{\theta} \sin\phi) = 0 \\ &\dot{Y} + a(\dot{\psi} \sin\theta \sin\phi + \dot{\theta} \cos\phi) = 0. \end{aligned} \tag{3.215b}$$

Because there are three constraints, there are only three degrees of freedom. Nevertheless since two constraints are nonholonomic, five generalized coordinates X, Y, ϕ, θ, and ψ are required to describe the motion.

When there are no forces other than gravity or the constraint forces, the Lagrangian equations of motion can be written with the Lagrange multipliers for the nonholonomic constraints (3.215b) as

$$m\ddot{X}=\lambda_1 \tag{3.216a}$$

$$m\ddot{Y}=\lambda_2 \tag{3.216b}$$

$$I_0\frac{d}{dt}[\dot{\phi}+\dot{\psi}\cos\theta]=0 \tag{3.216c}$$

$$I_0[\ddot{\theta}+\dot{\phi}\dot{\psi}\sin\theta]=a(-\lambda_1\sin\phi+\lambda_2\cos\phi) \tag{3.216d}$$

$$I_0\frac{d}{dt}[\dot{\phi}\cos\theta+\dot{\psi}]=a\sin\theta(\lambda_1\cos\phi+\lambda_2\sin\phi) \tag{3.216e}$$

where $I_0 \equiv 2ma^2/5$. These equations must be solved together with the constraint eqs. (3.215b).

Equation (3.216c) states that the angular velocity component ω_Z about the inertial Z axis is conserved. In fact, Eqs. (3.216 c, d, e) can be combined to the vector form

$$I_0\frac{d\boldsymbol{\omega}}{dt}=a\lambda_2\hat{\mathbf{e}}_X-a\lambda_1\hat{\mathbf{e}}_Y, \tag{3.217}$$

which together with Eqs. (3.216 a, b) shows how the Lagrange multipliers are identified with the forces and moments of constraint. Resolution of Eq. (3.217) into its inertial X and Y components and the use of Eqs. (3.216 a, b) and (3.215a) shows that $\lambda_1=\lambda_2=0$. Thus the forces of constraint to maintain the ball in pure rolling are zero. Moreover it is found that $\dot{X}=a\omega_Y=$constant and $\dot{Y}=-a\omega_X=$constant, and thus the ball rolls in a straight-line trajectory with a constant speed and a constant angular velocity.

Equation (3.216e) shows that the angular velocity component about the body-fixed z axis, $\omega_z \equiv \dot{\psi}+\dot{\phi}\cos\theta$, is a constant also. Thus since ω_Z and ω_z are both conserved, we can solve for $\dot{\phi}$ and $\dot{\psi}$ in terms of them and obtain

$$\dot{\phi}=\frac{\omega_Z-\omega_z\cos\theta}{\sin^2\theta} \tag{3.218a}$$

$$\dot{\psi}=\frac{\omega_z-\omega_Z\cos\theta}{\sin^2\theta}. \tag{3.218b}$$

When these are substituted into Eq. (3.216d), the governing equation for θ is obtained:

$$\ddot{\theta}+\frac{(\omega_z-\omega_Z\cos\theta)(\omega_Z-\omega_z\cos\theta)}{\sin^3\theta}=0. \tag{3.219}$$

After this equation has been solved, then θ and ψ can be obtained from Eqs. (3.218 a, b). ■

3.4.14 Hamilton's Principle for a Continuous Medium

Here we consider the variational formulation of the dynamics of continuous media. Following essentially the same procedure as that used for discrete systems, we can develop Hamilton's principle (3.114) or (3.116). Once again we begin with Newton's second law of motion, now applied to a continuous system,

$$m\mathbf{a}-\mathbf{F}=0. \tag{3.220}$$

A path that differs from the path of actual configuration of the body can be expressed as $\mathbf{u}+\delta\mathbf{u}$, where $\delta\mathbf{u}$ is the admissible variation (called *virtual displacement*) of the path. By admissible variation we mean that $\delta\mathbf{u}$ is consistent with the specified boundary conditions, and initial and final conditions $[\delta\mathbf{u}(t_1)=\delta\mathbf{u}(t_2)=0]$. The work done at time t by the inertia force and the resultant force in moving through the virtual displacement $\delta\mathbf{u}$ is given by

$$\int_\Omega \rho\frac{\partial^2\mathbf{u}}{\partial t^2}\cdot\delta\mathbf{u}\,d\tau+\int_\Omega \overleftrightarrow{\boldsymbol{\sigma}}:\delta\overleftrightarrow{\boldsymbol{\varepsilon}}\,d\tau-\int_\Omega \mathbf{f}\cdot\delta\mathbf{u}\,d\tau-\int_\Gamma \mathbf{t}\cdot\delta\mathbf{u}\,ds=0 \tag{3.221}$$

where ρ is the mass density of the medium, $\overleftrightarrow{\boldsymbol{\sigma}}$ and $\overleftrightarrow{\boldsymbol{\varepsilon}}$ are the stress and strain dyadics (see Section 1.5), $\mathbf{f}$ is the body force vector, and $\mathbf{t}$ is the specified surface stress vector on the surface Γ of the medium. The last two terms in the above expression can be represented by a virtual potential energy δV due to the applied conservation forces $\mathbf{f}$ and $\mathbf{t}$. We assume here the existence of a *strain energy density function* U_0 for the body such that

$$\sigma_{ij}=\partial U_0/\partial\varepsilon_{ij}. \tag{3.222}$$

We have from Eq. (3.221),

$$\int_\Omega \rho\frac{\partial^2\mathbf{u}}{\partial t^2}\cdot\delta\mathbf{u}\,d\tau+\delta\int_\Omega U_0\,d\tau+\delta V=0.$$

We now integrate the above equation with respect to time from t_1 to t_2 to obtain

$$\int_{t_1}^{t_2}\left[\int_\Omega \rho\frac{\partial^2\mathbf{u}}{\partial t^2}\cdot\delta\mathbf{u}\,d\tau+\delta(U+V)\right]dt=0 \tag{3.223}$$

where U is the strain energy,

$$U=\int_\Omega U_0\,d\tau. \tag{3.224}$$

Integrating the first term with respect to time by parts [see eq. (3.108)], we arrive at Hamilton's principle for a continuous body:

$$\delta \int_{t_1}^{t_2} [T-(U+V)]\, dt = 0 \tag{3.225}$$

where T is the kinetic energy,

$$T = \int_{\Omega} \frac{\rho}{2} \frac{\partial \mathbf{u}}{\partial t} \cdot \frac{\partial \mathbf{u}}{\partial t}\, d\tau. \tag{3.226}$$

The sum $U+V$ is known as the *total potential energy* of the body.

Example 3.17 Consider the tranverse motion of a membrane fixed at all points of the boundary Γ and subjected to distributed force f. The kinetic energy T, the strain energy U, and the potential energy V for the membrane (considering only small strains) are

$$T = \int_{\Omega} \frac{\rho}{2} \left(\frac{\partial u}{\partial t} \right)^2 dx\, dy$$

$$U = \int_{\Omega} \frac{T_0}{2} \left[\left(\frac{\partial u}{\partial x} \right)^2 + \left(\frac{\partial u}{\partial y} \right)^2 \right] dx\, dy$$

$$V = -\int_{\Omega} fu\, dx\, dy \tag{3.227}$$

where T_0 is the tension in the membrane. The Hamilton's principle gives

$$\delta \int_{t_1}^{t_2} L\, dt \equiv \int_{t_1}^{t_2} \int_{\Omega} \left[\rho \frac{\partial u}{\partial t} \frac{\partial \delta u}{\partial t} - T_0 \left(\frac{\partial u}{\partial x} \frac{\partial \delta u}{\partial x} + \frac{\partial u}{\partial y} \frac{\partial \delta u}{\partial y} \right) + f \delta u \right] dx\, dy\, dt = 0. \tag{3.228}$$

Using integration by parts on the first term, and the gradient theorem on the next two terms in the bracket, we have

$$0 = \int_{t_1}^{t_2} \int_{\Omega} \left[-\rho \frac{\partial^2 u}{\partial t^2} + T_0 \left(\frac{\partial^2 u}{\partial x^2} + \frac{\partial^2 u}{\partial y^2} \right) + f \right] \delta u\, dx\, dy\, dt$$

$$- \int_{t_1}^{t_2} \int_{\Gamma} T_0 \left(\frac{\partial u}{\partial x} n_x + \frac{\partial u}{\partial y} n_y \right) \delta u\, ds + \int_{\Omega} \left[\left(\rho \frac{\partial u}{\partial t} \delta u \right) \right]_{t_1}^{t_2} dx\, dy.$$

The Lagrange equation is given by

$$\rho\frac{\partial^2 u}{\partial t^2}=T_0\left(\frac{\partial^2 u}{\partial x^2}+\frac{\partial^2 u}{\partial y^2}\right)+f \text{ in } \Omega,\ t_1<t<t_2. \tag{3.229}$$

The boundary conditions are, for any $t_1<t<t_2$,

$$T_0\left(\frac{\partial u}{\partial x}n_x+\frac{\partial u}{\partial y}n_y\right)=0 \text{ on } \Gamma \quad \text{(natural)}, \quad \text{or} \quad u=\text{specified on } \Gamma \quad \text{(essential)} \tag{3.230}$$

The boundary term vanishes since $\delta u=0$ (since $u=0$) on Γ. ■

Exercises 3.3

Derive the Lagrangian function for the linear spring, mass, and damper systems in Exercises 1–10. (Assume that motion starts from rest, and the contact surfaces are free of friction.)

1

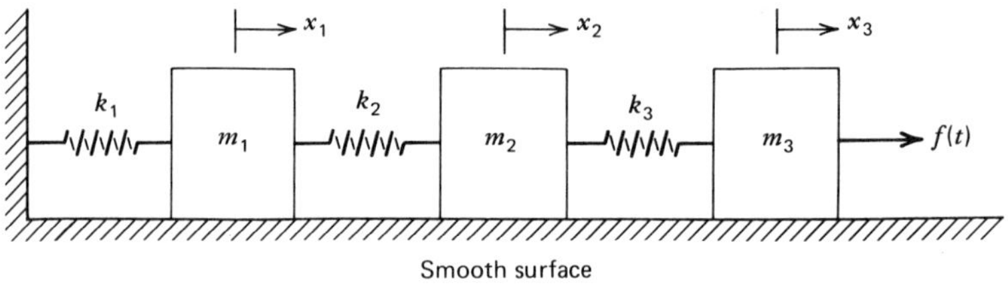

2

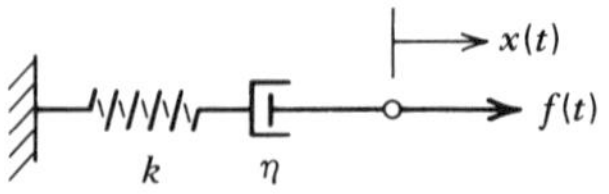

3

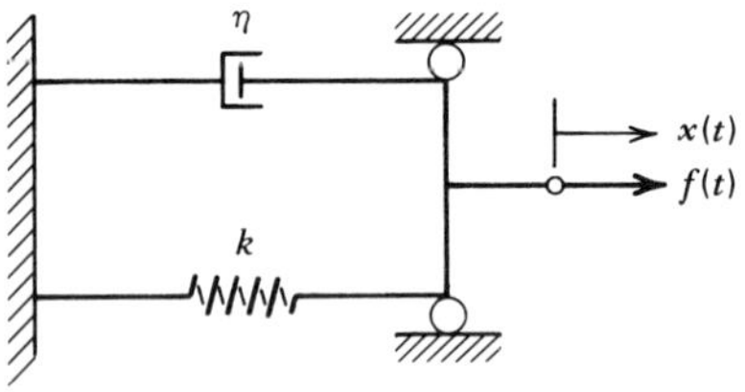

4

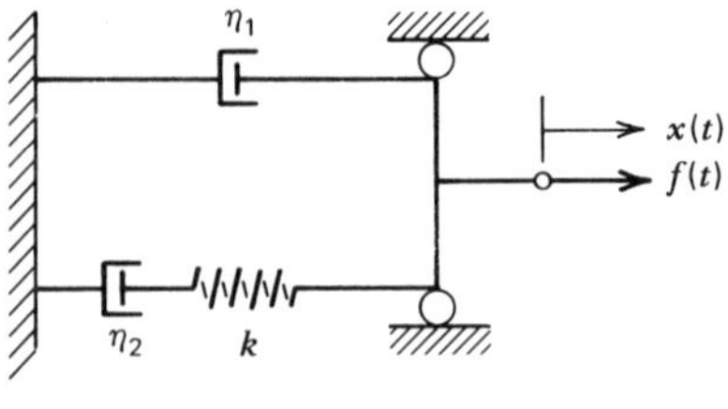

5

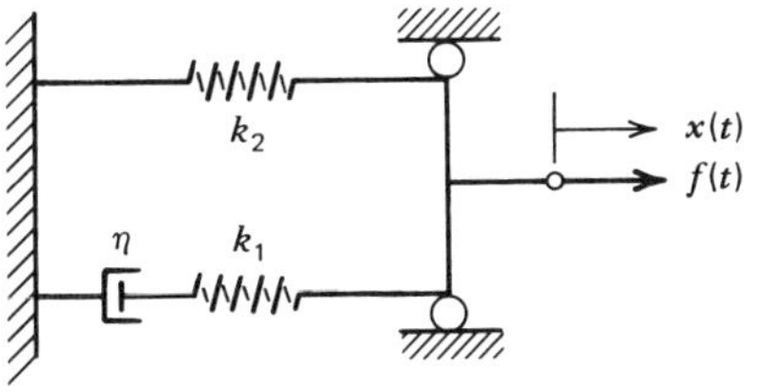

k2
η
k1
x(t)
f(t)

6

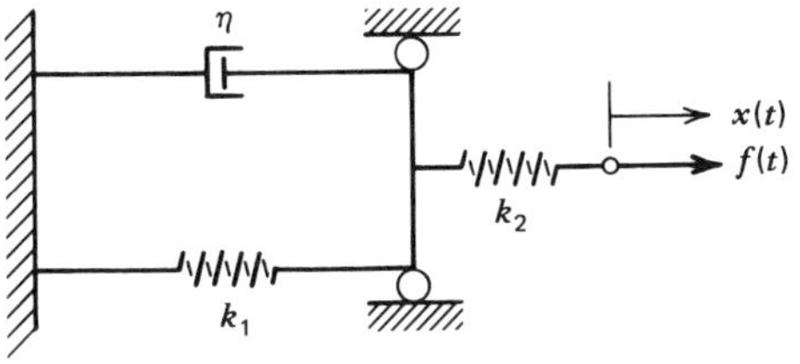

η
x(t)
f(t)
k2
k1

7

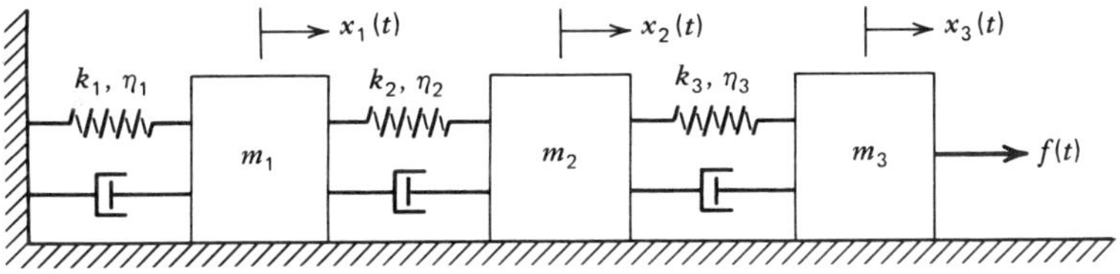

x1(t)
x2(t)
x3(t)
k1, η1
k2, η2
k3, η3
m1
m2
m3
f(t)

8

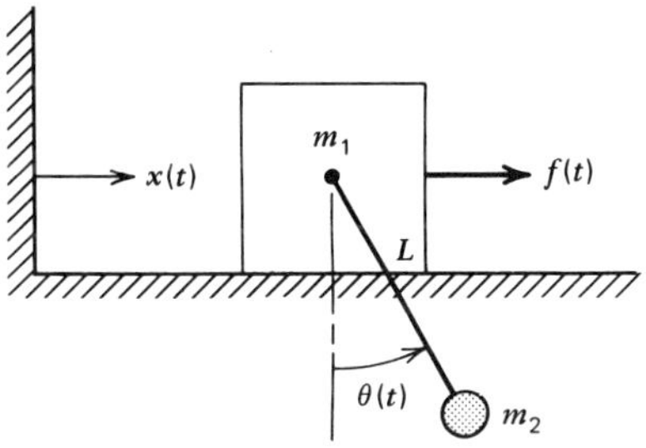

x(t)
m1
f(t)
L
θ(t)
m2

9

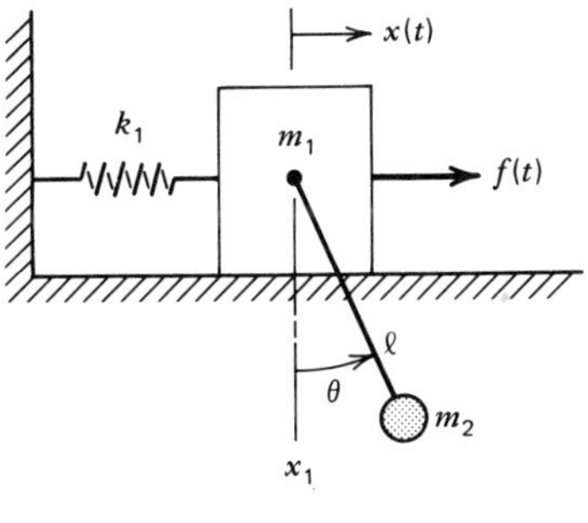

x(t)
k1
m1
f(t)
ℓ
θ
m2
x1

10

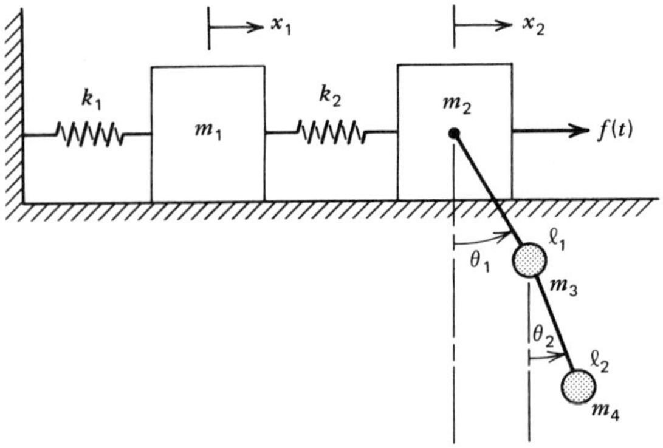

11 The governing differential equation for the motion of a simple pendulum in a plane was found to be the second-order nonlinear equation

$$\ddot{\theta}+\omega_0^2\sin\theta=0$$

where $\omega_0\equiv\sqrt{g/l}$. This sort of nonlinear equation appears frequently in dynamic problems. To obtain a solution, multiply the equation by $\dot{\theta}$ and write it in the form

$$\frac{d}{dt}\left[\frac{\dot{\theta}^2}{2}-\omega_0^2\cos\theta\right]=0.$$

(a) Show that this equation can be integrated once and placed in the form

$$\dot{\theta}^2=2\omega_0^2(\cos\theta-\cos\theta_m)$$

where θ_m is the maximum amplitude of the oscillation.

(b) Let $\theta=0$ when $t=0$, separate the variables, and show that the solution can be written in the form

$$\omega_0 t=\int_0^{\theta}\frac{d\bar{\theta}}{\sqrt{2(\cos\bar{\theta}-\cos\theta_m)}},$$

which holds until θ reaches its maximum value.

(c) By means of the substitutions $\sin\psi=\sin(\theta/2)/\sin(\theta_m/2)$ and $x=\sin\psi$, show that the solution can be written in the form

$$\omega_0 t=F(\psi;k)$$

where

$$F(\psi;k)\equiv\int_0^{\psi}\frac{d\bar{\psi}}{\sqrt{1-k^2\sin^2\bar{\psi}}}=\int_0^{\sin\psi}\frac{dx}{\sqrt{(1-x^2)(1-k^2x^2)}}$$

and $k \equiv \sin(\theta_m/2)$. The function $F(\psi; k)$ is called the *elliptic integral of the first kind with modulus k*.

(d) In the limit $k \to 0$, that is, in the limit where θ_m becomes vanishingly small, show that the solution is approximately

$$\sin\psi \simeq \sin\omega_0 t$$

or

$$\theta \simeq \theta_m \sin\omega_0 t,$$

which is the well-known result for small-amplitude oscillations.

When k is not small, but is finite, the above result suggests the definition of a new function $sn(\omega t; k)$ defined such that

$$\sin\psi \equiv sn(\omega_0 t; k).$$

As we have seen, when $k=0$, we have

$$sn(\omega_0 t; 0) = \sin\omega_0 t.$$

The function $sn(\omega_0 t; k)$ is called a *Jacobian elliptic function*, and it is a generalization of the circular sine function.

(e) Deduce that the period of the motion is given by

$$T = \frac{4F(\pi/2; k)}{\omega_0}.$$

By expanding the integral definition of $F(\pi/2; k)$ for small k, show that

$$T = \frac{4\pi}{\omega_0}\left[1 + \frac{1}{4}k^2 + \frac{9}{64}k^4 + \cdots\right], \qquad k \to 0.$$

12 Consider free undamped motion of a spherical pendulum, that is, a pendulum that is not restricted to motion in a plane. The kinetic energy is

$$T = \frac{m}{2}\left[(l\dot{\theta})^2 + (l\sin\theta\dot{\phi})^2\right],$$

and the potential energy is

$$V = mgl(1 - \cos\theta).$$

(a) Obtain the equations of motion by means of the Lagrangian equations.

(b) Find a single equation for $\theta(t)$ after two integrations in the form

$$\dot{\mu}^2 = 2\omega_0^2(\mu - \mu_m)(\mu_0 - \mu)\left[\mu + \frac{1 + \mu_0\mu_m}{\mu_0 + \mu_m}\right]$$

where $\mu \equiv \cos\theta$, θ_m and θ_0 are the maximum and minimum values of θ, and $\omega_0 = \sqrt{g/l}$.

(c) Find an expression for the period of the motion in terms of an integral, that is, in terms of a quadrature.

13 A mass $4m$ is attached to a string which passes over a frictionless pulley, as shown in the accompanying figure. The other end of the string is fastened to another frictionless pulley of mass m. Over this pulley passes a second string having masses m and $2m$ at its ends. Assume that the system starts from rest, and determine the motion of all the masses. What are the tensions in the strings?

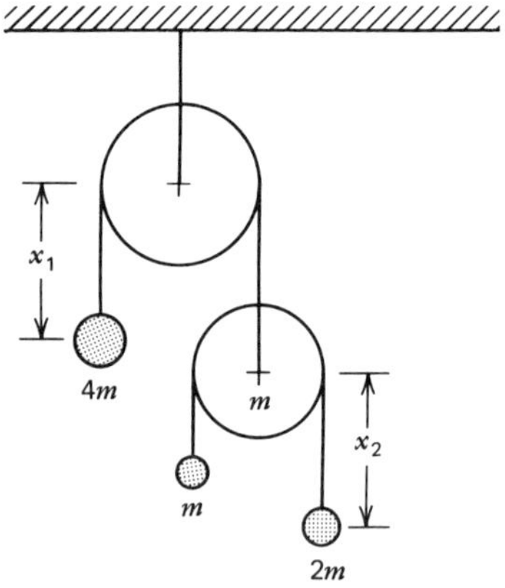

14 Consider a missile-shaped body rotating about its center of gravity which is traveling at a constant velocity $\mathbf{v}$ in the inertial Z direction. The body is axisymmetric about its body-fixed z axis, which is rotated from the inertial Z axis by the Euler angle θ, as shown in the accompanying figure. Because of axisymmetry, $I_{xx} \equiv I_{yy} \equiv I$.

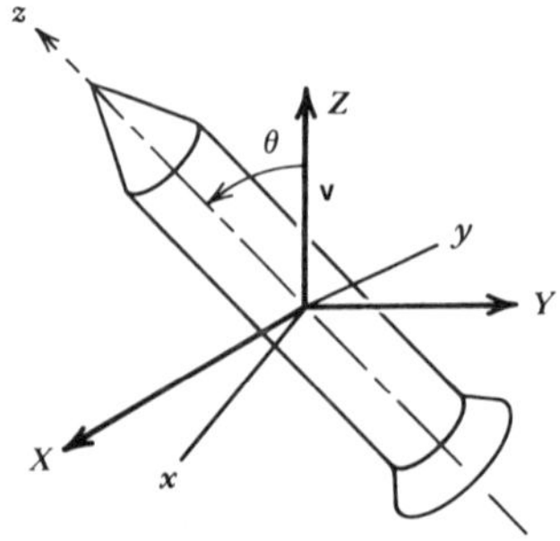

(a) Show that the rotational kinetic energy is given in terms of the Euler angles by

$$T=\frac{1}{2}\left[I(\dot{\phi}^2\sin^2\theta+\dot{\theta}^2)+I_{zz}(\dot{\phi}\cos\theta+\dot{\psi})^2\right]$$

(b) Assume that the aerodynamic moment on the body is conservative and given in terms of a potential energy by

$$V(\theta)=-k\cos\theta$$

where k is a constant. By means of the Lagrangian equations obtain the equations of rotational motion for the missile. Deduce that the angular-momentum components in the body-fixed z direction and inertial Z direction are constants, that is,

$$H_z\equiv I_{zz}(\dot{\phi}\cos\theta+\dot{\psi})=\text{constant}$$

$$H_Z\equiv I\dot{\phi}\sin^2\theta+I_{zz}(\dot{\phi}\cos\theta+\dot{\psi})\cos\theta=\text{constant}.$$

(c) Show that integration of the θ Lagrangian equation, after accounting for the constant angular momentum components, leads to the conservation of total energy:

$$E\equiv T+V=\frac{1}{2}\left[I\dot{\theta}^2+\frac{(H_Z-H_z\cos\theta)^2}{I\sin^2\theta}+\frac{H_z^2}{I_{zz}}\right]-k\cos\theta$$

$$=\text{constant}.$$

This is now a single differential equation for $\theta(t)$.

15 **(a)** Show for a homogeneous sphere that the kinetic energy is given in terms of the Euler angles by

$$T=\frac{1}{2}I_0\left[\dot{\theta}^2+\dot{\phi}^2+\dot{\psi}^2+2\dot{\phi}\dot{\psi}\cos\theta\right].$$

(b) Obtain the equations for the rotational motions of a curving baseball by Lagrangian methods. Discuss how these equations compare with those obtained by resolving the vector equation

$$I_0\frac{d\boldsymbol{\omega}}{dt}=\mathbf{M}$$

into its inertial Cartesian components.

16 Suppose for the trajectory of a bowling ball that, in addition to the constraint conditions for no slip at the contact point, there is a moment given by

$$\mathbf{M} = -\mu\omega$$

where μ is a constant. Determine the equations of motion for the bowling ball and describe its trajectory. How would you determine the instantaneous orientation of the finger holes in the bowling ball?

17 For certain systems the nonconservative parts of the generalized forces are damping forces that can be described, for n degrees of freedom, by

$$Q_i^* = -\sum_{j=1}^{n} C_{ij}\dot{q}_j$$

where the coefficients C_{ij} are real and symmetric in i and j and may depend upon the q_i's and t, but not on the $\dot{q}_i$'s. Rayleigh's dissipation function is defined such that

$$F(\dot{q}, q, t) \equiv \frac{1}{2}\sum_{i=1}^{n}\sum_{j=1}^{n} C_{ij}\dot{q}_i\dot{q}_j.$$

Show that for holonomic systems with n degrees of freedom the Lagrangian equations with Rayleigh dissipation can be written as

$$\frac{d}{dt}\left(\frac{\partial L}{\partial \dot{q}_i}\right) - \frac{\partial L}{\partial q_i} + \frac{\partial F}{\partial \dot{q}_i} = 0, \qquad i = 1, 2, 3, \ldots, n.$$

Given the Lagrange functions for continuous systems in Exercises 18–22, obtain the Euler-Lagrange equations. (In each case u represents deflection at time t, and f represents distributed force.)

18 (*Transverse Motion of a Cable of Mass Density* ρ *and Tension* T)

$$L = \int_0^a \left[\frac{\rho}{2}\left(\frac{\partial u}{\partial t}\right)^2 - \frac{T}{2}\left(\frac{\partial u}{\partial x}\right)^2 + fu\right] dx.$$

19 (*Transverse Motion of a Beam of Flexural Rigidity EI and Area A*)

$$L = \int_0^a \left[\frac{\rho A}{2}\left(\frac{\partial u}{\partial t}\right)^2 - \frac{EI}{2}\left(\frac{\partial^2 u}{\partial x^2}\right)^2 + fu\right] dx.$$

20 (*Transverse Motion of a Beam Including Rotatory Inertia*)

$$L = \int_0^a \left[\frac{\rho A}{2}\left(\frac{\partial u}{\partial t}\right)^2 + \frac{\rho I}{2}\left(\frac{\partial^2 u}{\partial x \partial t}\right)^2 - \frac{EI}{2}\left(\frac{\partial^2 u}{\partial x^2}\right)^2 + fu\right] dx.$$

21 (*Transverse Motion of a Timoshenko Beam*)

$$L=\int_0^a\left[\frac{\rho I}{2}\left(\frac{\partial\psi}{\partial t}\right)^2+\rho A\left(\frac{\partial u}{\partial t}\right)^2-\frac{EI}{2}\left(\frac{\partial\psi}{\partial x}\right)^2-\frac{kGA}{2}\left(\frac{\partial u}{\partial x}-\psi\right)^2\right]dx$$

where ψ and u are the dependent variables (ψ is the shear rotation), GA is shear rigidity, and k is a shear correction coefficient.

22 (*Transverse Motion of Reissner-Mindlin Plate*)

$$L=\frac{1}{2}\int_\Omega\left\{\frac{\rho h^3}{12}\left[\left(\frac{\partial\psi_x}{\partial t}\right)^2+\left(\frac{\partial\psi_y}{\partial t}\right)^2\right]+\rho h\left(\frac{\partial u}{\partial t}\right)^2\right.$$

$$-D\left[\left(\frac{\partial\psi_x}{\partial x}\right)^2+\left(\frac{\partial\psi_y}{\partial y}\right)^2+2\nu\left(\frac{\partial\psi_x}{\partial x}\right)\left(\frac{\partial\psi_y}{\partial y}\right)\right]$$

$$\left.-\frac{Gh^3}{12}\left(\frac{\partial\psi_y}{\partial x}+\frac{\partial\psi_x}{\partial y}\right)^2-kGh\left[\left(\frac{\partial w}{\partial x}-\psi_x\right)^2+\left(\frac{\partial w}{\partial y}-\psi_y\right)^2\right]+2fu\right\}dx\,dy$$

where h is the thickness of the plate, D is the flexural rigidity, $D=Eh^3/[12(1-\nu^2)]$, and ν is the Poisson's ratio.

3.5 CONSTRUCTION OF FUNCTIONALS FROM GOVERNING EQUATIONS—THE INVERSE PROBLEM

In Section 3.3 we were concerned with the necessary condition for an extremum of a given functional. The necessary condition that the first variation be zero leads us to the governing equations (or Euler equations) of the problem. That is, if we know the functional associated with a problem, we can obtain the associated governing equations including boundary and initial conditions. However, we are often faced with the inverse problem of finding the functional associated with a given set of equations in order to use variational methods of approximation. This section is devoted to the *inverse problem*: given a set of differential equations and associated boundary conditions, construct its variational problem.

3.5.1 Variational Formulation by the Inverse Procedure

The construction of a functional for a set of differential equations follows the variational procedure inverse to that used in arriving at the Euler equation from a functional. Here we use an intuitive and informal procedure to illustrate, by means of examples, the development of variational principles.

Consider the special case of one dependent variable and two independent variables. Suppose that we are required to find the functional associated with the equations

$$\frac{\partial F}{\partial u}-\frac{\partial}{\partial x}\left(\frac{\partial F}{\partial u_x}\right)-\frac{\partial}{\partial y}\left(\frac{\partial F}{\partial u_y}\right)=0 \qquad \text{in } \Omega$$

$$B(u)\equiv\frac{\partial F}{\partial u_x}n_x+\frac{\partial F}{\partial u_y}n_y=\hat{g} \qquad \text{on } \Gamma_1, \qquad u=\hat{u} \qquad \text{on } \Gamma_2 \tag{3.231}$$

where $F=F(x, y, u_x, u_y)$ and Γ_1 and Γ_2 are disjoint sets whose union is the total boundary $\Gamma=\Gamma_1\cup\Gamma_2$. Here u is assumed to be in $\mathcal{C}$,

$$\mathcal{C}=\{u:\ \ u\in H^2(\Omega),\ \ B(u)=\hat{g} \text{ on } \Gamma_1 \text{ and } u=\hat{u} \text{ on } \Gamma_2\}. \tag{3.232}$$

The first step in the construction of a functional involves multiplying the operator (Euler) equation with the variation δu of the dependent variable and integrating over the domain Ω of the problem:

$$\int_\Omega\left[\frac{\partial F}{\partial u}-\frac{\partial}{\partial x}\left(\frac{\partial F}{\partial u_x}\right)-\frac{\partial}{\partial y}\left(\frac{\partial F}{\partial u_y}\right)\right]\delta u\,dx\,dy=0. \tag{3.233}$$

The next step in the procedure involves the use of the divergence theorem to transfer the derivatives to δu:

$$0=\int_\Omega\left[\frac{\partial F}{\partial u}+\frac{\partial F}{\partial u_x}\frac{\partial \delta u}{\partial x}+\frac{\partial F}{\partial u_y}\frac{\partial \delta u}{\partial y}\right]dx\,dy-\int_\Gamma\left(\frac{\partial F}{\partial u_x}n_x+\frac{\partial F}{\partial u_y}n_y\right)\delta u\,ds. \tag{3.234}$$

The third step in the development involves expressing the boundary integral in terms of the specified values on the boundary:

$$\int_\Gamma\left(\frac{\partial F}{\partial u_x}n_x+\frac{\partial F}{\partial u_y}n_y\right)\delta u\,ds=\int_{\Gamma_1}\left(\frac{\partial F}{\partial u_x}n_x+\frac{\partial F}{\partial u_y}n_y\right)\delta u\,ds-\int_{\Gamma_1}\hat{g}\,\delta u\,ds \tag{3.235}$$

where the boundary integral over Γ_2 is zero since δu is zero there (because u is specified on Γ_2). Note that only the natural boundary conditions of the problem are included in the functional, while the essential boundary conditions are included in the description of the space of competing functions.

The fourth and last step in the development consists of bringing the variational operator δ outside the integrals by recognizing the integrands as variations of quadratic and/or product terms. In the present case the expression under the integral over Ω is δF and that under the line integral is

$\delta \int_{\Gamma_1} \hat{g}u\, ds$. Hence we get

$$\delta I(u) = \delta\left[\int_{\Omega} F(x, y, u, u_x, u_y)\, dx\, dy - \int_{\Gamma_1} \hat{g}u\, ds\right] = 0. \tag{3.236}$$

Thus the functional to be minimized is

$$I(u) = \int_{\Omega} F(x, y, u, u_x, u_y)\, dx\, dy - \int_{\Gamma_1} \hat{g}u\, ds. \tag{3.237}$$

It is clear from the above procedure that the natural and essential boundary conditions in Eq. (3.82) cannot be arbitrary but should be compatible with the Euler equation(s).

Example 3.18 Consider the steady heat conduction in a unit square. The normalized equation is given by

$$-\left(\frac{\partial^2 T}{\partial x^2} + \frac{\partial^2 T}{\partial y^2}\right) = Q(x, y), \qquad 0 < x, y < 1 \tag{3.238}$$

with the boundary conditions

$$\frac{\partial T}{\partial x} = \hat{g}_1(y) \quad \text{on } x = 0, \qquad \frac{\partial T}{\partial y} = \hat{g}_2(x) \quad \text{on } y = 0$$

$$T = \hat{T}_1 \quad \text{on } y = 1, \qquad T = \hat{T}_2 \quad \text{on } x = 1. \tag{3.239}$$

Using the first three basic steps described above for the development of a functional, we obtain

$$\begin{aligned}
\delta I(T) &= \int_0^1 \int_0^1 \left[-\left(\frac{\partial^2 T}{\partial x^2} + \frac{\partial^2 T}{\partial y^2}\right) - Q\right] \delta T\, dx\, dy \\
&= \int_0^1 \int_0^1 \left[\frac{\partial T}{\partial x}\frac{\partial \delta T}{\partial x} + \frac{\partial T}{\partial y}\frac{\partial \delta T}{\partial y} - Q\,\delta T\right] dx\, dy - \int_0^1 \left[\delta T \frac{\partial T}{\partial x}\Big|_{x=0}^{x=1} dy + \delta T \frac{\partial T}{\partial y}\Big|_{y=0}^{y=1} dx\right] \\
&= \int_0^1 \int_0^1 \left[\frac{\partial T}{\partial x}\frac{\partial \delta T}{\partial x} + \frac{\partial T}{\partial y}\frac{\partial \delta T}{\partial y} - Q\,\delta T\right] dx\, dy + \int_0^1 \left[(\delta T \hat{g}_1)|_{x=0}\, dy + (\delta T \hat{g}_2)|_{y=0}\, dx\right] \\
&= \frac{\delta}{2} \int_0^1 \int_0^1 \left[\left(\frac{\partial T}{\partial x}\right)^2 + \left(\frac{\partial T}{\partial y}\right)^2 - 2QT\right] dx\, dy + \delta \int_0^1 \hat{g}_1(y) T(0, y)\, dy \\
&\quad + \delta \int_0^1 \hat{g}_2(x) T(x, 0)\, dx.
\end{aligned} \tag{3.240}$$

From the third line in Eq. (3.240) we have the bilinear form B: $\mathcal{C}\times\mathcal{C}\to\mathbf{R}$ and the linear form l: $\mathcal{C}\to\mathbf{R}$,

$$B(T,\bar{T})\equiv\int_0^1\int_0^1\left(\frac{\partial T}{\partial x}\frac{\partial \bar{T}}{\partial x}+\frac{\partial T}{\partial y}\frac{\partial \bar{T}}{\partial y}\right)dx\,dy \tag{3.241}$$

$$l(\bar{T})\equiv\int_0^1\int_0^1 Q\bar{T}\,dx\,dy-\int_0^1\left[\hat{g}_1\bar{T}(0,y)\,dy+\hat{g}_2\bar{T}(x,0)\,dx\right]. \tag{3.242}$$

Then the weak (or variational) problem associated with Eq. (3.238) is to determine $T\in\mathcal{C}$ such that

$$B(T,\bar{T})=l(\bar{T}),\qquad \text{for all } \bar{T}\in\mathcal{C}. \tag{3.243}$$

Also, note that the functional (or quadratic form) I is related to $B(\cdot,\cdot)$ and $l(\cdot)$ by

$$\begin{aligned}I(T)&=\tfrac{1}{2}B(T,T)-l(T)\\&=\tfrac{1}{2}\int_0^1\int_0^1\left[\left(\frac{\partial T}{\partial x}\right)^2+\left(\frac{\partial T}{\partial y}\right)^2-2QT\right]dx\,dy+\int_0^1\hat{g}_1(y)T(0,y)\,dy\\&\quad+\int_0^1\hat{g}_2(x)T(x,0)\,dx. \quad\blacksquare\end{aligned} \tag{3.244}$$

Example 3.19 Construct the functional associated with the equations governing slow steady two-dimensional flow of an incompressible fluid,

$$\left.\begin{aligned}&\left[2\mu\frac{\partial^2 u}{\partial x^2}+\mu\frac{\partial}{\partial y}\left(\frac{\partial u}{\partial y}+\frac{\partial v}{\partial x}\right)\right]-\frac{\partial P}{\partial x}=f_1\\&\left[\mu\frac{\partial}{\partial x}\left(\frac{\partial u}{\partial y}+\frac{\partial v}{\partial x}\right)+2\mu\frac{\partial^2 v}{\partial y^2}\right]-\frac{\partial P}{\partial y}=f_2\end{aligned}\right\}\quad \text{in } R \tag{3.245}$$

$$\begin{aligned}&\left(2\mu\frac{\partial u}{\partial x}-P\right)n_x+\mu\left(\frac{\partial u}{\partial y}+\frac{\partial v}{\partial x}\right)n_y=\hat{t}_1 \quad \text{on } C_{1t}\\&\mu\left(\frac{\partial u}{\partial y}+\frac{\partial v}{\partial x}\right)n_x+\left(2\mu\frac{\partial v}{\partial y}-P\right)n_y=\hat{t}_2 \quad \text{on } C_{2t}\\&u=\hat{u}\quad \text{on } C_{1d},\qquad v=\hat{v}\quad \text{on } C_{2d}\end{aligned} \tag{3.246}$$

where the boundary segments C_{1t}, C_{2t}, etc. are as defined in Example 3.6. Note that there are three dependent variables: two velocity components (u, v) and the pressure P. Since there are three equations and three dependent variables, there is a question as to which variation must multiply which equation. In this regard it is important to know the origin of equations in Eq. (3.245). The first two equations are obtained from conservation of linear momentum and the third one is obtained from conservation of mass for an incompressible fluid. Tracing back to the origin of the equations, one observes that scalar multiplication of the momentum equation by the velocity is associated with the change of kinetic energy. Similarly, multiplication of the mass conservation equation by the pressure is associated with work done by the pressure in changing the volume of the fluid (which is zero for incompressible flow). Such a physical interpretation of the equations is essential in the inverse problem. Guided by these thoughts, we now proceed to construct the functional. Let $\Lambda=(u, v, P)$ and $\mathcal{C}=\{\Lambda\colon H^2(R)\times H^2(R)\times H^1(R),\ \Lambda$ satisfies the boundary conditions in Eq. (3.246)$\}$. Then we get

$$\begin{aligned}
\delta I(\Lambda)=&\int_R\left\{\left[2\mu\frac{\partial^2 u}{\partial x^2}+\mu\frac{\partial}{\partial y}\left(\frac{\partial u}{\partial y}+\frac{\partial v}{\partial x}\right)\right]-\frac{\partial P}{\partial x}-f_1\right\}\delta u\,dx\,dy\\
&+\int_R\left\{\left[\mu\frac{\partial}{\partial x}\left(\frac{\partial u}{\partial y}+\frac{\partial v}{\partial x}\right)+2\mu\frac{\partial^2 v}{\partial y^2}\right]-\frac{\partial P}{\partial y}-f_2\right\}\delta v\,dx\,dy\\
&+\int_R\left(\frac{\partial u}{\partial x}+\frac{\partial v}{\partial y}\right)\delta P\,dx\,dy\\
=&\int_R\left\{-\left[2\mu\left(\frac{\partial u}{\partial x}\frac{\partial \delta u}{\partial x}+\frac{\partial v}{\partial y}\frac{\partial \delta v}{\partial y}\right)+\mu\left(\frac{\partial u}{\partial y}+\frac{\partial v}{\partial x}\right)\left(\frac{\partial \delta u}{\partial y}+\frac{\partial \delta v}{\partial x}\right)\right]\right.\\
&\left.+P\left(\frac{\partial \delta u}{\partial x}+\frac{\partial \delta v}{\partial y}\right)+\delta P\left(\frac{\partial u}{\partial x}+\frac{\partial v}{\partial y}\right)-f_1\delta u-f_2\delta v\right\}dx\,dy\\
&+\oint_C\left[2\mu\left(\frac{\partial u}{\partial x}\delta u\,n_x+\frac{\partial v}{\partial y}\delta v\,n_y\right)+\mu\left(\frac{\partial u}{\partial y}+\frac{\partial v}{\partial x}\right)(\delta u\,n_y+\delta v\,n_x)\right.\\
&\left.-P(\delta u\,n_x+\delta v\,n_y)\right]ds. \qquad (3.247)
\end{aligned}$$

The variational problem thus becomes

$$B(\Lambda,\bar{\Lambda})=l(\bar{\Lambda}) \qquad (3.248)$$

where

$$B(\Lambda,\bar{\Lambda})=-\int_R\left\{\mu\left[2\frac{\partial u}{\partial x}\frac{\partial \bar{u}}{\partial x}+\left(\frac{\partial u}{\partial y}+\frac{\partial v}{\partial x}\right)\left(\frac{\partial \bar{u}}{\partial y}+\frac{\partial \bar{v}}{\partial x}\right)+2\frac{\partial v}{\partial y}\frac{\partial \bar{v}}{\partial y}\right]\right.$$

$$\left.-P\left(\frac{\partial \bar{u}}{\partial x}+\frac{\partial \bar{v}}{\partial y}\right)-\bar{P}\left(\frac{\partial u}{\partial x}+\frac{\partial v}{\partial y}\right)\right\}dx\,dy$$

$$l(\bar{\Lambda})=\int_R(f_1\bar{u}+f_2\bar{v})\,dx\,dy-\int_{C_{1t}}\hat{t}_1\bar{u}\,ds-\int_{C_{2t}}\hat{t}_2\bar{v}\,ds. \tag{3.249}$$

We have

$$I(\Lambda)=\tfrac{1}{2}B(\Lambda,\Lambda)-l(\Lambda).$$

or

$$I(u,v,P)=\int_R\left\{-\mu\left[\left(\frac{\partial u}{\partial x}\right)^2+\left(\frac{\partial v}{\partial y}\right)^2+\frac{1}{2}\left(\frac{\partial u}{\partial y}+\frac{\partial v}{\partial x}\right)^2\right]\right.$$

$$\left.+P\left(\frac{\partial u}{\partial x}+\frac{\partial v}{\partial y}\right)-f_1u-f_2v\right\}dx\,dy+\int_{C_{1t}}\hat{t}_1u\,ds+\int_{C_{2t}}\hat{t}_2v\,ds. \tag{3.250}$$

Again, if the form of the boundary conditions in Eq. (3.245) did not conform to the terms in the boundary integral in Eq. (3.246), we would not have been able to combine them as we did in arriving at the functional in Eq. (3.250). Indeed, in all physical problems of interest the boundary conditions conform to the governing equations in the above sense. ■

The next example is concerned with the construction of variational principles associated with the static equilibrium of an elastic body undergoing large displacement gradients (i.e., the strain-displacement equations are nonlinear). Nonlinear elasticity is one of the fields that admits variational formulation of the nonlinear field equations.

Example 3.20 Consider the equations of nonlinear elasticity governing the static equilibrium of a two-dimensional elastic body.

1 (*Strain-Displacement Equations*)

$$\gamma_1=\frac{\partial u}{\partial x}+\frac{1}{2}\left[\left(\frac{\partial u}{\partial x}\right)^2+\left(\frac{\partial v}{\partial x}\right)^2\right]$$

$$\gamma_2=\frac{\partial v}{\partial y}+\frac{1}{2}\left[\left(\frac{\partial u}{\partial y}\right)^2+\left(\frac{\partial v}{\partial y}\right)^2\right] \tag{3.251}$$

$$\gamma_3=\frac{1}{2}\left[\frac{\partial u}{\partial y}+\frac{\partial v}{\partial x}+\frac{\partial u}{\partial x}\frac{\partial u}{\partial y}+\frac{\partial v}{\partial x}\frac{\partial v}{\partial y}\right].$$

2 (*Equilibrium Equations*)

$$L_1(\sigma_i, u, v) \equiv \frac{\partial}{\partial x}\left[\sigma_1\left(1+\frac{\partial u}{\partial x}\right)+\sigma_3\frac{\partial u}{\partial y}\right] + \frac{\partial}{\partial y}\left[\sigma_3\left(1+\frac{\partial u}{\partial x}\right)+\sigma_2\frac{\partial u}{\partial y}\right] = -f_1$$

$$L_2(\sigma_i, u, v) \equiv \frac{\partial}{\partial x}\left[\sigma_1\frac{\partial v}{\partial x}+\sigma_3\left(1+\frac{\partial v}{\partial y}\right)\right] + \frac{\partial}{\partial y}\left[\sigma_3\frac{\partial v}{\partial x}+\sigma_2\left(1+\frac{\partial v}{\partial y}\right)\right] = -f_2. \tag{3.252}$$

3 (*Stress-Strain Relations, Linear Elastic Material*)

$$\sigma_1 = C_{11}\gamma_1 + C_{12}\gamma_2 + C_{13}\gamma_3$$

$$\sigma_2 = C_{21}\gamma_1 + C_{22}\gamma_2 + C_{23}\gamma_3 \quad \left(\text{or } \sigma_i = \sum_{j=1}^{3} C_{ij}\gamma_j\right) \tag{3.253}$$

$$\sigma_3 = C_{31}\gamma_1 + C_{32}\gamma_2 + C_{33}\gamma_3.$$

4 (*Boundary Conditions*)

$$\left.\begin{aligned}\left[\sigma_1\left(1+\frac{\partial u}{\partial x}\right)+\sigma_3\frac{\partial u}{\partial y}\right]n_x+\left[\sigma_3\left(1+\frac{\partial u}{\partial x}\right)+\sigma_2\frac{\partial u}{\partial y}\right]n_y=\hat{t}_1\\ \left[\sigma_1\frac{\partial v}{\partial x}+\sigma_3\left(1+\frac{\partial v}{\partial y}\right)\right]n_x+\left[\sigma_3\frac{\partial v}{\partial x}+\sigma_2\left(1+\frac{\partial v}{\partial y}\right)\right]n_y=\hat{t}_2.\end{aligned}\right\} \text{ on } C_t \tag{3.254}$$

$$u = \hat{u}, \qquad v = \hat{v} \qquad \text{on } C_d.$$

$$C_t \cap C_d = \text{empty}, \qquad C_t \cup C_d = \text{total boundary, } C. \tag{3.255}$$

Here γ_i and σ_i, $i=1,2,3$, are the strains and stresses ($\gamma_1=\gamma_{xx}, \gamma_2=\gamma_{yy},\ldots$), u and v are the displacements along x and y directions, respectively, and C_{ij} are material constants which may depend on the position (x, y). The body forces f_α and surface tractions $\hat{t}_\alpha$, $\alpha=1,2$, are given functions of position. The boundary C of the region R occupied by the body is divided into two portions: portion C_t on which the tractions are specified and portion C_d on which the displacements are specified. We wish to construct the functional associated with the nonlinear Eqs. (3.251)–(3.255). Let $\Lambda=(u, v, \gamma_1, \gamma_2, \gamma_3, \sigma_1, \sigma_2, \sigma_3)$. We write

$$\delta I(\Lambda)=\int_R\Bigg[\left\{\gamma_1-\left[\frac{\partial u}{\partial x}+\frac{1}{2}\left(\frac{\partial u}{\partial x}\right)^2+\frac{1}{2}\left(\frac{\partial v}{\partial x}\right)^2\right]\right\}\delta\sigma_1$$

$$+\left\{\gamma_2-\left[\frac{\partial v}{\partial y}+\frac{1}{2}\left(\frac{\partial u}{\partial y}\right)^2+\frac{1}{2}\left(\frac{\partial v}{\partial y}\right)^2\right]\right\}\delta\sigma_2$$

$$+\left\{\gamma_3-\frac{1}{2}\left[\frac{\partial u}{\partial y}+\frac{\partial v}{\partial x}+\frac{\partial u}{\partial x}\frac{\partial u}{\partial y}+\frac{\partial v}{\partial x}\frac{\partial v}{\partial y}\right]\right\}\delta\sigma_3$$

$$+\left[L_1(\sigma_i,u,v)+f_1\right]\delta u+\left[L_2(\sigma_i,u,v)+f_2\right]\delta v$$

$$+\sum_{j=1}^{3}\left(\sigma_j-\sum_{i=1}^{3}C_{ji}\gamma_i\right)\delta\gamma_j\Bigg]\,dx\,dy$$

$$=\int_R\Bigg\{\sum_{j=1}^{3}(\gamma_j\delta\sigma_j+\sigma_j\delta\gamma_j)-\sum_{i,j=1}^{3}C_{ij}\gamma_i\delta\gamma_j+f_1\delta u+f_2\delta v$$

$$+L_1(\sigma_i,u,v)\,\delta u+L_2(\sigma_i,u,v)\,\delta v-\left[\frac{\partial u}{\partial x}+\frac{1}{2}\left(\frac{\partial u}{\partial x}\right)^2+\frac{1}{2}\left(\frac{\partial v}{\partial x}\right)^2\right]\delta\sigma_1$$

$$-\left[\frac{\partial v}{\partial y}+\frac{1}{2}\left(\frac{\partial u}{\partial y}\right)^2+\frac{1}{2}\left(\frac{\partial v}{\partial y}\right)^2\right]\delta\sigma_2$$

$$-\frac{1}{2}\left[\frac{\partial u}{\partial y}+\frac{\partial v}{\partial x}+\frac{\partial u}{\partial x}\frac{\partial u}{\partial y}+\frac{\partial v}{\partial x}\frac{\partial v}{\partial y}\right]\delta\sigma_3\Bigg\}\,dx\,dy. \tag{3.256}$$

By means of the divergence theorem, the operator expressions in Eq. (3.256) can be written as

$$\int_R\left[L_1(\sigma_1,u,v)\,\delta u+L_2(\sigma_i,u,v)\,\delta v\right]dx\,dy$$

$$=-\int_R\Bigg\{\left[\sigma_1\left(1+\frac{\partial u}{\partial x}\right)+\sigma_3\frac{\partial u}{\partial y}\right]\frac{\partial\delta u}{\partial x}+\left[\sigma_3\left(1+\frac{\partial u}{\partial x}\right)+\sigma_2\frac{\partial u}{\partial y}\right]\frac{\partial\delta u}{\partial y}$$

$$+\left[\sigma_1\frac{\partial v}{\partial x}+\sigma_3\left(1+\frac{\partial v}{\partial y}\right)\right]\frac{\partial\delta v}{\partial x}+\left[\sigma_3\frac{\partial v}{\partial x}+\sigma_2\left(1+\frac{\partial v}{\partial y}\right)\right]\frac{\partial\delta v}{\partial y}\Bigg\}\,dx\,dy$$

$$+\oint_C\Bigg\{\left[\sigma_1\left(1+\frac{\partial u}{\partial x}\right)+\sigma_3\frac{\partial u}{\partial y}\right]n_x\delta u+\left[\sigma_3\left(1+\frac{\partial u}{\partial x}\right)+\sigma_2\frac{\partial u}{\partial y}\right]n_y\delta u$$

$$+\left[\sigma_1\frac{\partial v}{\partial x}+\sigma_3\left(1+\frac{\partial v}{\partial y}\right)\right]n_x\delta v+\left[\sigma_3\frac{\partial v}{\partial x}+\sigma_2\left(1+\frac{\partial v}{\partial y}\right)\right]n_y\delta v\Bigg\}\,ds$$

$$= -\int_R \left\{ \left(\frac{\partial \delta u}{\partial x} + \frac{\partial u}{\partial x}\frac{\partial \delta u}{\partial x} + \frac{\partial v}{\partial x}\frac{\partial \delta v}{\partial x} \right)\sigma_1 + \left(\frac{\partial \delta v}{\partial y} + \frac{\partial v}{\partial y}\frac{\partial \delta v}{\partial y} + \frac{\partial u}{\partial y}\frac{\partial \delta u}{\partial y} \right)\sigma_2 \right.$$

$$\left. + \left(\frac{\partial u}{\partial y}\frac{\partial \delta u}{\partial x} + \frac{\partial \delta u}{\partial y} + \frac{\partial u}{\partial x}\frac{\partial \delta u}{\partial y} + \frac{\partial \delta v}{\partial x} + \frac{\partial v}{\partial y}\frac{\partial \delta v}{\partial x} + \frac{\partial v}{\partial x}\frac{\partial \delta v}{\partial y} \right)\sigma_3 \right\} dx\,dy$$

$$+ \int_{C_t} (\hat{t}_1 \delta u + \hat{t}_2 \delta v)\, ds \tag{3.257}$$

where we collected the coefficients of σ_1, σ_2, and σ_3 in R and used the natural boundary conditions in Eq. (3.254) and the essential boundary conditions in Eq. (3.255) in arriving at the last expression. Using Eq. (3.257) in Eq. (3.256) and noting that all of the terms are of the form $F_2 \delta F_1 + F_1 \delta F_2 = \delta(F_1 F_2)$, we obtain

$$I(\Lambda) = \int_R \left[\left\{ \gamma_1 - \left[\frac{\partial u}{\partial x} + \frac{1}{2}\left(\frac{\partial u}{\partial x} \right)^2 + \frac{1}{2}\left(\frac{\partial v}{\partial x} \right)^2 \right] \right\} \sigma_1 \right.$$

$$+ \left\{ \gamma_2 - \left[\frac{\partial v}{\partial y} + \frac{1}{2}\left(\frac{\partial u}{\partial y} \right)^2 + \frac{1}{2}\left(\frac{\partial v}{\partial y} \right)^2 \right] \right\} \sigma_2$$

$$+ \left\{ \gamma_3 - \frac{1}{2}\left[\frac{\partial u}{\partial y} + \frac{\partial v}{\partial x} + \frac{\partial u}{\partial x}\frac{\partial u}{\partial y} + \frac{\partial v}{\partial x}\frac{\partial v}{\partial y} \right] \right\} \sigma_3$$

$$\left. - \frac{1}{2} \sum_{i,j=1}^{3} C_{ij}\gamma_i\gamma_j + f_1 u + f_2 v \right] dx\,dy + \int_{C_t} (\hat{t}_1 u + \hat{t}_2 v)\, ds \tag{3.258}$$

which is the functional associated with Eqs. (3.251)–(3.254). Again, the essential boundary conditions in Eq. (3.255) are not included in the functional, but are implied in the definition of the space of competing functions in which the solution lies. ■

We close this section with a comment that not all problems permit the construction of a (quadratic) functional. Certain linear problems and most nonlinear problems do not permit the functional formulation. As an example, consider the Navier-Stokes equations governing the steady two-dimensional flow of an incompressible fluid (see Example 3.19 for the Stokes problem):

$$\underline{\left(u\frac{\partial u}{\partial x} + v\frac{\partial u}{\partial y} \right)} + \frac{\partial P}{\partial x} - \mu\left[2\frac{\partial^2 u}{\partial x^2} + \frac{\partial}{\partial y}\left(\frac{\partial u}{\partial y} + \frac{\partial v}{\partial x} \right) \right] + f_1 = 0$$

$$\underline{\left(u\frac{\partial v}{\partial x} + v\frac{\partial v}{\partial y} \right)} + \frac{\partial P}{\partial y} - \mu\left[\frac{\partial}{\partial x}\left(\frac{\partial u}{\partial y} + \frac{\partial v}{\partial x} \right) + 2\frac{\partial^2 v}{\partial y^2} \right] + f_2 = 0 \tag{3.259}$$

$$\frac{\partial u}{\partial x} + \frac{\partial v}{\partial y} = 0.$$

The boundary conditions associated with Eq. (3.259) are the same as those in Eq. (3.246). The underlined terms in Eqs. (3.259) are nonlinear in u and v and do not permit the construction of a functional.

Without the nonlinear (underlined) terms, Eqs. (3.259) are the same as Eqs. (3.245) and permit us to construct a functional, as was done in Example 3.19. Let us consider the following integral over R of the underlined terms only. We have

$$\int_R \left[\left(u\frac{\partial u}{\partial x}+v\frac{\partial u}{\partial y}\right)\delta u+\left(u\frac{\partial v}{\partial x}+v\frac{\partial v}{\partial y}\right)\delta v\right]dx\,dy$$

$$\equiv\int_R\left(\frac{\partial u}{\partial x}u\,\delta u+\frac{\partial u}{\partial y}v\,\delta u+\frac{\partial v}{\partial x}u\,\delta v+\frac{\partial v}{\partial y}v\,\delta v\right)dx\,dy. \tag{3.260}$$

Note that one cannot bring the variational operator outside the integral, and therefore it is not possible for us to identify a quadratic functional for the Navier-Stokes equations. Also, note that both the nonlinearity as well as the odd-order derivatives prevented the construction of a functional. For example, even if we linearize the expression (3.260), we cannot bring the variational operation outside the integral.

3.5.2 Construction of Functionals Using the Vainberg Theorem

Here we present an alternate method for the construction of the functional associated with an operator equation. Recall from the previous section that the gradient of a functional I at u is defined by

$$\frac{\delta I(u;\eta)}{\alpha}\equiv\lim_{\alpha\to 0}\frac{\partial I(u+\alpha\eta)}{\partial\alpha}=(T(u),\eta) \tag{3.261}$$

where $T(u)$ is the gradient of I at u. When there exists a functional I such that Eq. (3.261) holds, the gradient operator $T(u)$ is said to be *potential*. More formally, an operator T: $\mathfrak{D}(T)\subset H\to H'$, where H is a Hilbert space, is said to be *potential* on $\mathfrak{D}(T)$ if and only if there exists a differentiable functional $I(u)$ on $\mathfrak{D}(T)$ such that $T(u)=\text{grad}\, I(u)$. Thus if T in

$$T(u)=0 \tag{3.262}$$

is potential, solutions of the weak problem

$$(T(u),\eta)=0,\qquad \text{for all } \eta\in H', \tag{3.263}$$

will be the critical points of some functional $I(u)$.

From the above discussion it is clear that one should have a means to determine whether a given operator is potential, and then a formula to determine the associated functional. The following two results provide the required tools for the construction of a functional associated with Eq. (3.262). For the proofs, the reader is referred to Vainberg (Ref. 31).

Theorem 3.2 Let T be a continuous operator from H into H' which has a derivative [in the sense of Eq. (3.27)] at every point $u \in \mathfrak{D}(T) \subset H$, where $\mathfrak{D}(T)$ is a convex subset of H. Then a necessary and sufficient condition that T be potential on $\mathfrak{D}(T)$ is that

$$(\delta T(u;\eta),\xi)=(\delta T(u;\xi),\eta). \tag{3.264}$$

In other words, the bilinear (in η and ξ) functional $(\delta T(u;\eta),\xi)$ must be symmetric in η and ξ for each $u \in \mathfrak{D}(T)$. ■

Theorem 3.3 Let T: $H \to H'$ be a continuous operator such that the conditions of Theorem 3.2 are met (i.e., T is potential). Then there exists a functional $I(u)$ (within a constant) whose gradient is the operator T,

$$I(u)=\int_0^1 (T(tu),u)\,dt \tag{3.265}$$

where t is a real parameter of integration. ■

As an example of the application of Eq. (3.265), we consider the nonlinear operator equation [see Eq. (2.128)]

$$T(u)\equiv Au-f=0 \tag{3.266}$$

where A: $H \to H'$ is a linear operator mapping a Hilbert space H into its dual H'. The derivative of $T(u)$ is given by

$$\begin{aligned}\frac{\delta T(u;\eta)}{\alpha} &= \lim_{\alpha\to 0}\frac{\partial T(u+\alpha\eta)}{\partial\alpha}\\ &= \lim_{\alpha\to 0}\frac{\partial}{\partial\alpha}[Au+\alpha A\eta-f]\\ &= A\eta.\end{aligned}$$

Note that in the present case $\delta T(u;\eta)$ does not depend on u. We have

$$(\delta T(u;\eta),\xi)=(A\eta,\xi),$$

which implies that T is potential if and only if A is symmetric.

Assuming that A is symmetric (so that T is a potential operator), we introduce $T(u)=Au-f$ into Eq. (3.265) to obtain

$$I(u)=\int_0^1 (A(tu)-f,u)\,dt=\int_0^1 [t(A(u),u)-(f,u)]\,dt$$

$$=\tfrac{1}{2}(Au,u)-(f,u). \tag{3.267}$$

Hence $I(u)$ is the (quadratic) functional whose gradient is equal to $T(u)$. Indeed we have

$$\delta I(u;\eta)=\tfrac{1}{2}(Au,\eta)+\tfrac{1}{2}(A\eta,u)-(f,\eta)$$

$$=(Au,\eta)-(f,\eta)=(Au-f,\eta)=(T(u),\eta).$$

We now illustrate the application of Eq. (3.265) to a couple of linear and nonlinear operator equations. For additional examples, see Ref. 19.

Example 3.21 Consider the nonlinear differential equation (see Example 3.3)

$$Au\equiv-\frac{d}{dx}\left[a(x)\left(\frac{du}{dx}\right)^3\right]+\frac{d^2}{dx^2}\left[b(x)\frac{d^2u}{dx^2}\right]+c(x)u=f, \qquad 0<x<1$$

where $u\in\mathfrak{D}(A)=\{u\in H^4(0,1):\quad u=du/dx=0 \text{ at } x=0,1\}$, and $H=L_2(0,1)$. The derivative of $T(u)\equiv Au-f$ is given by

$$\frac{\delta T(u;\eta)}{\alpha}=-\frac{d}{dx}\left[3a\left(\frac{du}{dx}\right)^2\frac{d\eta}{dx}\right]+\frac{d^2}{dx^2}\left[b\frac{d^2u}{dx^2}\frac{d^2\eta}{dx^2}\right]+c\eta.$$

We now check for the symmetry of $(\delta T(u;\eta),\xi)$; for any $\xi\in\mathfrak{D}(A)$,

$$\frac{1}{\alpha}\int_0^1 \delta T(u;\eta)\xi\,dx=\int_0^1\left[3a\left(\frac{du}{dx}\right)^2\frac{d\eta}{dx}\frac{d\xi}{dx}+b\frac{d^2u}{dx^2}\frac{d^2\eta}{dx^2}\frac{d^2\xi}{dx^2}+c\eta\xi\right]dx$$

$$-\left\{\left[3a\left(\frac{du}{dx}\right)^2\frac{d\eta}{dx}-\frac{d}{dx}\left(b\frac{d^2u}{dx^2}\frac{d^2\eta}{dx^2}\right)\right]\xi+b\frac{d^2u}{dx^2}\frac{d^2\eta}{dx^2}\frac{d\xi}{dx}\right\}_0^1.$$

The boundary terms vanish since $\xi=d\xi/dx=0$ at $x=0,1$. Since the integrand is symmetric in η and ξ, it follows that $(\delta T(u;\eta),\xi)$ is symmetric in η and ξ. Hence $T(u)=A(u)-f$ is potential.

Substituting for $T(tu)$ into Eq. (3.265), we get

$$I(u)=\int_0^1\left(-\frac{d}{dx}\left[at^3\left(\frac{du}{dx}\right)^3\right]+\frac{d^2}{dx^2}\left[bt\frac{d^2u}{dx^2}\right]+ctu-f,u\right)dt$$

$$=\left(-\frac{d}{dx}\left[\frac{a}{4}\left(\frac{du}{dx}\right)^3\right]+\frac{d^2}{dx^2}\left[\frac{b}{2}\frac{d^2u}{dx^2}\right]+\frac{c}{2}u-f,u\right)$$

$$=\int_0^1\left\{-\frac{d}{dx}\left[\frac{a}{4}\left(\frac{du}{dx}\right)^3\right]+\frac{d^2}{dx^2}\left[\frac{b}{2}\frac{d^2u}{dx^2}\right]+\frac{c}{2}u-f\right\}u\,dx$$

$$=\int_0^1\left[\frac{a}{4}\left(\frac{du}{dx}\right)^4+\frac{b}{2}\left(\frac{d^2u}{dx^2}\right)^2+\frac{c}{2}u^2-fu\right]dx. \tag{3.268}$$

In arriving at the last line, integration by parts is employed once on the first term and twice on the second term: The boundary terms vanish due to the fact that $u\in\mathfrak{D}(A)$. Compare the functional in Eq. (3.268) with that in Example 3.3. ■

Example 3.22 Consider the pair of linear equations

$$\left.\begin{aligned}&\operatorname{grad}u-\frac{\mathbf{v}}{k}=\mathbf{0}\\&-\operatorname{div}\mathbf{v}-Q=0\end{aligned}\right\}\quad \text{in } \Omega\subset\mathbb{R}^2 \tag{3.269}$$

$$u=\hat{u}\quad\text{on }\Gamma_1\qquad\text{and}\qquad\hat{\mathbf{n}}\cdot\mathbf{v}+h(u-u_0)=q\quad\text{on }\Gamma_2. \tag{3.270}$$

Here u and $\mathbf{v}$ are the dependent variables (u the temperature and $\mathbf{v}$ the heat flow), k, h, u_0, $\hat{u}$, Q, and q are given functions. The solution $(u,\mathbf{v})$ is assumed to be in the set

$$\mathfrak{D}=\begin{aligned}&\{(u,\mathbf{v}):\ \ u\in H^1(\Omega),\,\mathbf{v}\in H^1(\Omega)\times H^1(\Omega),\\&\delta u=0\ \text{ on }\ \Gamma_2,\ \text{ and }\ \delta(\hat{\mathbf{n}}\cdot\mathbf{v})=0\ \text{ on }\ \Gamma_1\}.\end{aligned} \tag{3.271}$$

The operator T is a matrix operator in the present case. The elements of the matrix T must be arranged in such a way that $(T(u,\mathbf{v}),(u,\mathbf{v}))$ makes sense (a task that requires experience and physical intuition). We have

$$\Gamma(\Lambda)\equiv\begin{bmatrix}0&-\operatorname{div}&0&0\\\operatorname{grad}&-1/k&0&0\\0&0&hI_2&I_2\\0&0&I_1&0\end{bmatrix}\begin{Bmatrix}u_\Omega\\\mathbf{v}_\Omega\\u_\Gamma\\\hat{\mathbf{n}}\cdot\mathbf{v}_\Gamma\end{Bmatrix}-\begin{Bmatrix}Q\\\mathbf{0}\\(hu_0+q)_{\Gamma_2}\\\hat{u}_{\Gamma_1}\end{Bmatrix}=\begin{Bmatrix}0\\\mathbf{0}\\0\\0\end{Bmatrix} \tag{3.272}$$

where $\Lambda = (u_\Omega, \mathbf{v}_\Omega, u_\Gamma, \hat{\mathbf{n}}\cdot\mathbf{v}_\Gamma)$, and I_1 and I_2 the identity restriction operators,

$$I_1 u \equiv u_{\Gamma_1}, \qquad I_2 \mathbf{v} \equiv \mathbf{v}_{\Gamma_2}.$$

We define the bilinear pairing $(\Lambda, \overline{\Lambda})$ by $\int_\Omega \{\Lambda\}^T\{\overline{\Lambda}\}\, dx\, dy$, where T denotes the transpose.

It can be verified that $T(u)$ is potential. Substituting $T(\Lambda)$ into Eq. (3.265), we obtain

$$I(\Lambda) = \int_0^1 (T(t\Lambda), \Lambda)\, dt = \int_\Omega \left[\int_0^1 ([T]\{t\Lambda\})^T \{\Lambda\}\, dt \right] d\Omega.$$

The inner product $(\Lambda, \overline{\Lambda})$ is defined by

$$(\Lambda, \overline{\Lambda}) = \int_\Omega \{\Lambda\}^T\{\overline{\Lambda}\}\, d\Omega = \int_\Omega (\Lambda_1 \overline{\Lambda}_1 + \boldsymbol{\Lambda}_2 \cdot \overline{\boldsymbol{\Lambda}}_2)\, dx\, dy$$

$$+ \int_{\Gamma_1} \Lambda_3 \overline{\Lambda}_3\, ds + \int_{\Gamma_2} \Lambda_4 \overline{\Lambda}_4\, ds.$$

Carrying out the integration with respect to t, we obtain

$$I(u, \mathbf{v}) = \int_\Omega \left\{ \frac{1}{2}(-\operatorname{div}\mathbf{v})u - Qu + \frac{1}{2}\operatorname{grad} u \cdot \mathbf{v} - \frac{1}{2k}\mathbf{v}\cdot\mathbf{v} \right\} dx\, dy$$

$$+ \int_{\Gamma_1} \left[\frac{hu}{2} + \frac{\hat{\mathbf{n}}\cdot\mathbf{v}}{2} - (hu_0 + q) \right] u\, ds + \int_{\Gamma_2} \left(\frac{u}{2} - \hat{u} \right) \hat{\mathbf{n}}\cdot\mathbf{v}\, ds$$

$$= \int_\Omega \left(\operatorname{grad} u \cdot \mathbf{v} - \frac{1}{2k}\mathbf{v}\cdot\mathbf{v} - Qu \right) dx\, dy + \int_{\Gamma_1} \left(\frac{h}{2}u - hu_0 - q \right) u\, ds - \int_{\Gamma_2} \hat{u}(\hat{\mathbf{n}}\cdot\mathbf{v})\, ds. \tag{3.273}$$

It can be verified by setting the gradient of I to zero that the Euler equations of the functional in Eq. (3.273) are

$$\delta u: \quad -\operatorname{div}\mathbf{v} - Q = 0 \quad \text{in } \Omega$$

$$\delta \mathbf{v}: \quad \operatorname{grad} u - \frac{\mathbf{v}}{k} = \mathbf{0} \quad \text{in } \Omega \tag{3.274}$$

$$\delta u: \quad \hat{\mathbf{n}}\cdot\mathbf{v} + h(u - u_0) - q = 0 \quad \text{on } \Gamma_1$$

$$\delta(\hat{\mathbf{n}}\cdot\mathbf{v}): \quad u - \hat{u} = 0 \quad \text{on } \Gamma_2. \tag{3.275}$$

It is important to note that all of the boundary conditions of the problem are included in the functional. ■

Exercises 3.4

Construct a functional associated with each set of equations in Exercises 1–10.

1

$$-\frac{d}{dx}\left[p(x)\frac{du}{dx}\right]+q(x)u=f, \qquad a<x<b$$

$$\left(\frac{du}{dx}-\alpha u\right)\Big|_{x=a}=0, \qquad \left(\frac{du}{dx}+\beta u\right)\Big|_{x=b}=0$$

where u is the dependent variable, $p(x)$, $q(x)$, and $f(x)$ are functions of x alone, and α and β are constants.

2

$$-\nabla\cdot(k\nabla u)+Q=0 \qquad \text{in } R\subset\mathsf{R}^2$$

$$k\frac{\partial T}{\partial n}-\hat{q}=0 \qquad \text{on } C_q$$

$$k\frac{\partial T}{\partial n}-h(T-T_\infty)=0 \qquad \text{on } C_h$$

$$T=\hat{T} \qquad \text{on } C_T$$

where k and Q are known functions of x and y, and where $\hat{q}$, h, T_∞, and $\hat{T}$ are specified values and C_q, C_h, and C_T are disjoint subsets whose union is the total boundary C.

3

$$D\nabla^2(\nabla^2 w)+P=0 \qquad \text{in } R\subset\mathsf{R}^2$$

$$\frac{\partial}{\partial n}(D\nabla^2 w)=\hat{V} \qquad \text{on } C_v, \qquad w=\hat{w} \qquad \text{on } C_w$$

$$\frac{\partial w}{\partial n}=\hat{\theta} \qquad \text{on } C_\theta, \qquad D\nabla^2 w=\hat{M} \qquad \text{on } C_m$$

where the boundary subsets C_v, C_w, C_θ, and C_m are such that

$$C_v+C_w=C_\theta+C_m=C, \qquad C_v\cap C_w=C_\theta\cap C_m=\varnothing.$$

Assume that D is a constant and $P=P(x,y)$.

4

$$\left.\begin{aligned}\delta P&: \quad \nabla\phi+\mathbf{v}=\mathbf{0}\\ \delta\mathbf{v}&: \quad \mathbf{P}-\rho_0\mathbf{v}=\mathbf{0}\\ \delta\phi&: \quad -\nabla\cdot\mathbf{P}+f=0\end{aligned}\right\} \qquad \text{in } R\subset\mathsf{R}^3$$

$$\phi=\phi_0 \qquad \text{on } S_1, \qquad \hat{\mathbf{n}}\cdot\mathbf{P}=P_0 \qquad \text{on } S_2$$

where ρ_0, ϕ_0, and P_0 are constants. S_1 and S_2 are disjoint subsets whose union is the total boundary S.

5 δu: $$-\frac{d}{dx}\left\{\alpha\left[\frac{du}{dx}+\frac{1}{2}\left(\frac{dw}{dx}\right)^2\right]\right\}=0, \qquad 0<x<L$$

δw: $$\frac{d^2}{dx^2}\left(\beta\frac{d^2w}{dx^2}\right)-\frac{d}{dx}\left\{\alpha\frac{dw}{dx}\left[\frac{du}{dx}+\frac{1}{2}\left(\frac{dw}{dx}\right)^2\right]\right\}+q=0, \qquad 0<x<L$$

$$\alpha\left[\frac{du}{dx}+\frac{1}{2}\left(\frac{dw}{dx}\right)^2\right]=-P \qquad \text{at } x=L$$

$$-\alpha\frac{dw}{dx}\left[\frac{du}{dx}+\frac{1}{2}\left(\frac{dw}{dx}\right)^2\right]+\beta\frac{d^3w}{dx^3}=0 \qquad \text{at } x=L$$

$$\beta\frac{d^2w}{dx^2}=0 \qquad \text{at } x=L, \qquad \frac{dw}{dx}=0 \qquad \text{at } x=0$$

$$u=w=0 \qquad \text{at } x=0.$$

Here α, β, and P are constants, and q is a function of x.

6 (*Bending of Isotropic Plates*)

$$\left.\begin{aligned} \delta M_1:&\quad \alpha(M_1-M_2\lambda)-\frac{\partial^2 w}{\partial x^2}=0\\ \delta M_2:&\quad \alpha(M_2-\lambda M_1)-\frac{\partial^2 w}{\partial y^2}=0\\ \delta w:&\quad 4\beta\frac{\partial^4 w}{\partial x^2\partial y^2}-\frac{\partial^2 M_1}{\partial x^2}-\frac{\partial^2 M_2}{\partial y^2}+P=0 \end{aligned}\right\} \quad \text{in } \Omega\subset\mathbf{R}^2$$

$$\left.\begin{aligned} &\left(\frac{\partial^2 w}{\partial x^2}-\hat{M}_1\right)n_x=0, \qquad \left(\frac{\partial^2 w}{\partial y^2}-\hat{M}_2\right)n_y=0\\ &\left(4\beta\frac{\partial^3 w}{\partial x\partial y^2}-\frac{\partial M_1}{\partial x}\right)n_x+\frac{\partial M_2}{\partial y}n_y-\hat{V}=0\\ &\frac{\partial^2 w}{\partial x\partial y}n_y=0 \end{aligned}\right\} \quad \text{on } \Gamma.$$

7 $$-\left(\frac{\partial^2 u}{\partial x^2}+2\frac{\partial^2 u}{\partial x\partial y}+\frac{\partial^2 u}{\partial y^2}\right)=f \qquad \text{in } \Omega\subset\mathbf{R}^2$$

$$\left(\frac{\partial u}{\partial x}+\frac{\partial u}{\partial y}\right)(n_x+n_y)=\hat{t} \quad \text{on } \Gamma_1, \qquad u=\hat{u} \quad \text{on } \Gamma_2.$$

8 (*Equations of plane Orthotropic Medium*)

$$\sigma_1 = C_{11}\frac{\partial u}{\partial x} + C_{12}\frac{\partial v}{\partial y} + C_{13}\left(\frac{\partial u}{\partial y} + \frac{\partial v}{\partial x}\right)$$

$$\sigma_2 = C_{12}\frac{\partial u}{\partial x} + C_{22}\frac{\partial v}{\partial y} + C_{23}\left(\frac{\partial u}{\partial y} + \frac{\partial v}{\partial x}\right)$$

$$\sigma_3 = C_{13}\frac{\partial u}{\partial x} + C_{23}\frac{\partial v}{\partial y} + C_{33}\left(\frac{\partial u}{\partial y} + \frac{\partial v}{\partial x}\right)$$

$$\frac{\partial \sigma_1}{\partial x} + \frac{\partial \sigma_3}{\partial y} + f_1 = 0$$

$$\frac{\partial \sigma_3}{\partial x} + \frac{\partial \sigma_2}{\partial y} + f_2 = 0, \qquad \text{in } \Omega$$

$$\sigma_1 n_x + \sigma_3 n_y = t_1, \qquad \sigma_3 n_x + \sigma_2 n_y = t_2 \qquad \text{on } \Gamma_1$$

$$u = \hat{u} \quad \text{and} \quad v = \hat{v} \qquad \text{on } \Gamma_2.$$

Assume that C_{ij}, $i, j = 1, 2, 3$, are functions of x and y, and

$$\Gamma_1 \cap \Gamma_2 = \varnothing, \qquad \Gamma_1 \cup \Gamma_2 = \Gamma$$

9

$$C_{11}\frac{\partial^2 u}{\partial x^2} + 2C_{13}\frac{\partial^2 u}{\partial x \partial y} + C_{33}\frac{\partial^2 u}{\partial y^2} + C_{13}\frac{\partial^2 v}{\partial x^2}$$

$$+ (C_{12} + C_{33})\frac{\partial^2 v}{\partial x \partial y} + C_{23}\frac{\partial^2 v}{\partial y^2} + f_1 = 0 \qquad \text{in } \Omega$$

$$C_{13}\frac{\partial^2 u}{\partial x^2} + (C_{12} + C_{33})\frac{\partial^2 u}{\partial x \partial y} + C_{23}\frac{\partial^2 u}{\partial y^2} + C_{33}\frac{\partial^2 v}{\partial x^2}$$

$$+ 2C_{23}\frac{\partial^2 v}{\partial x \partial y} + C_{22}\frac{\partial^2 v}{\partial y^2} + f_2 = 0 \qquad \text{in } \Omega$$

$$\left[C_{11}\frac{\partial u}{\partial x} + C_{12}\frac{\partial v}{\partial y} + C_{13}\left(\frac{\partial u}{\partial y} + \frac{\partial v}{\partial x}\right)\right]n_x$$

$$+ \left[C_{13}\frac{\partial u}{\partial x} + C_{23}\frac{\partial v}{\partial y} + C_{33}\left(\frac{\partial u}{\partial y} + \frac{\partial v}{\partial x}\right)\right]n_y = \hat{t}_1 \qquad \text{on } \Gamma_1$$

$$\left[C_{13}\frac{\partial u}{\partial x} + C_{23}\frac{\partial v}{\partial y} + C_{33}\left(\frac{\partial u}{\partial y} + \frac{\partial v}{\partial x}\right)\right]n_x$$

$$+ \left[C_{12}\frac{\partial u}{\partial x} + C_{22}\frac{\partial v}{\partial y} + C_{23}\left(\frac{\partial u}{\partial y} + \frac{\partial v}{\partial x}\right)\right]n_y = \hat{t}_2 \qquad \text{on } \Gamma_1$$

$$u = \bar{u} \text{ and } v = \bar{v} \qquad \text{on } \Gamma_2.$$

10 $$\left(x\frac{d^4w}{dx^4}+2\frac{d^3w}{dx^3}-\frac{1}{x}\frac{d^2w}{dx^2}+\frac{1}{x^2}\frac{dw}{dx}\right)-gx=0, \qquad 0<x<a.$$

Assume that the natural boundary conditions are specified at $x=0$ and $x=a$.

11–20 Identify the linear and bilinear forms and associated vector spaces for Exercises 1–10.

21 Identify the product space of the solution Λ of the plane elasticity problem discussed in Example 3.20.

Construct variational statements of the nonlinear problems in Exercises 22–24.

22 (*Poisson-Boltzmann Equation*)

$$-\frac{d^2u}{dx^2}+e^u-e^{-u}=0, \qquad 0<x<1, \qquad u(0)=u(1)=0.$$

23 (*Thomas-Fermi Equation*)

$$-\frac{d^2u}{dx^2}+\sqrt{\frac{u^3}{x}}=0, \qquad 0\le x<\infty, \qquad u(0)=1,\ x\frac{du}{dx}\to 0 \qquad \text{as } x\to\infty.$$

24 (*Föppl-Hencky Equation*)

$$-\frac{d}{dx}\left(x^3\frac{du}{dx}\right)-\frac{2x^3}{u^2}=0, \qquad 0\le x\le 1, \qquad u(1)=u_0,\ \frac{du}{dx}(0)=0.$$

25 (*Buckling of a Hinged Flexible Column*) Construct the functional associated with the differential equation

$$\frac{d^4w}{dx^4}-\lambda\left\{\frac{dw}{dx}+\left(x-\frac{1}{2}\right)\frac{d^2w}{dx^2}\right\}=0, \qquad 0<x<1$$

$$w(0)=w(1)=0$$

$$\frac{d^2w}{dx^2}(0)=\frac{d^2w}{dx^2}(1)=0.$$

26 Use the Vainberg theorem to construct the functional associated with the nonlinear differential equation

$$-\frac{d}{dx}\left[\frac{1}{12}\left(\frac{du}{dx}\right)^3+2\frac{du}{dx}\right]+6u^5=6, \qquad 0<x<1$$

$$u(0)=1, \qquad u(1)=0.$$

27 Use the Vainberg theorem to construct the functional associated with Eqs. (3.245) and (3.246).

28 Using the Vainberg theorem, construct the functional associated with the following nonlinear equations of (i.e., large displacement gradients) elastic body Ω (see Example 3.20 for notation):

(a) (*Equilibrium Equations*)

$$-\left[-\sigma_{ij}\left(\delta_{mj}+u_{m,j}\right)\right]_{,i}-f_m=0.$$

(b) (*Stress-Strain Relations*)

$$E_{ijkl}\varepsilon_{kl}-\sigma_{ij}=0.$$

(c) (*Strain-Displacement Equations*)

$$\tfrac{1}{2}\left(u_{i,j}+u_{j,i}+u_{m,i}u_{m,j}\right)-\varepsilon_{ij}=0.$$

(d) (*Boundary Conditions*)

$$n_i\sigma_{ij}\left(\delta_{mj}+u_{m,j}\right)-\hat{T}_m=0 \qquad \text{on } \Gamma_1, \qquad u_i=\hat{u}_i \qquad \text{on } \Gamma_2.$$

In the above equations the summation convention on repeated indices is implied, and a comma followed by an index denotes differentiation with respect to the independent variable ($u_{i,j}\equiv\partial u_i/\partial x_j$, etc.). Define the appropriate inner product to employ Eq. (3.265).

29 (*Tonti-Reddy Variational Formulation of Time-Dependent Problems*) Tonti (Ref. 30) pointed out that the first-order (more generally, odd-order) differential operator $\partial/\partial t$ is self-adjoint with respect to the *convolution bilinear form*:

$$((u,v))=\int_\Omega\int_0^{t_0}u(x,t)v(x,t-t_0)\,dt\,dx.$$

Following Tonti's observation, Reddy (Ref. 23) presented variational principles for boundary initial value problems using the theory of potential operators. The procedure is illustrated here through the following steps:

(a) Consider the following set of operator equations:

$$Tu=v$$

$$Ev=w$$

$$T^*w=f-C_1\frac{\partial u}{\partial t}-C_2\frac{\partial^2 u}{\partial t^2}$$

or

$$Eu = E^{-1}w$$

$$T^*w = f - C_1\frac{\partial u}{\partial t} - C_2\frac{\partial^2 u}{\partial t^2} \tag{i}$$

in $\Omega \subset \mathsf{R}^2$. Here C_1 and C_2 are given nonnegative functions $\mathbf{x} = (x_1, x_2) \in \Omega$, T is a linear (differential) operator from a Hilbert space H_1 into a Hilbert space H_0, E is isomorphism mapping H_0 onto its dual H_0', and T^* is the adjoint of T, mapping H_0' into H_1'. The operator T and its adjoint T^* are related by

$$((w, Tu))_0 = ((T^*w, u))_1, \qquad w \in H_0' \tag{ii}$$

where $((\cdot,\cdot))_0$ and $((\cdot,\cdot))_1$ denote duality pairings on $H_0' \times H_0$ and $H_1' \times H_1$, respectively. Assume the following form of boundary and initial conditions:

$$Bu = g_1 \quad \text{on } \partial\Omega_1, \qquad B^*Ev = B^*w = g_2 \quad \text{on } \partial\Omega_2, \qquad t \in (0, t_0) \tag{iii}$$

$$C_2 u = h_0, \qquad C_1 u + C_2\frac{\partial u}{\partial t} = h_1 \qquad \text{in } \overline{\Omega}, \qquad t = 0. \tag{iv}$$

Here $\partial\Omega_1$ and $\partial\Omega_2$ are disjoint sets whose union is $\partial\Omega$, and B and B^* are boundary operators which depend on T through the formal adjoint definition (or abstract Green's formula)

$$((w, Tu))_0 = ((T^*w, u))_1 + ((B^*w, u))_{\partial\Omega_2} - ((w, Bu))_{\partial\Omega_1}. \tag{v}$$

(b) Write Eqs. (i), (iii), and (iv) in operator form,

$$\mathcal{P}(\Lambda) = 0 \tag{vi}$$

and define the bilinear product by

$$\langle \Lambda_1, \Lambda_2 \rangle = ((u_1, u_2))_0 + ((v_1, v_2))_1 + ((u_1, u_2))_{\partial\Omega_2}$$
$$+ ((v_1, v_2))_{\partial\Omega_1} + \langle u_1, u_2 \rangle_0 + \langle \dot{u}_1, \dot{u}_2 \rangle_0 \tag{vii}$$

where

$$\langle u, v \rangle_0 \equiv \int_\Omega u(\mathbf{x}, 0)\, v(\mathbf{x}, t_0)\, d\mathbf{x}. \tag{viii}$$

(c) Using the Vainberg theorem, show that the functional associated with the operator equation (vi) [or, equivalently, Eqs. (i), (iii), and (iv)] is given by

$$I_1(u,v,w)=(w,Tu-v)_0+\frac{1}{2}(Ev,v)_0+\frac{1}{2}\left(C_1\frac{\partial u}{\partial t}+C_2\frac{\partial^2 u}{\partial t^2},u\right)_1$$

$$-(f,u)_1+(w,Bu-g_1)_{\partial\Omega_1}-(g_2,u)_{\partial\Omega_2}+\frac{1}{2}\left[u,C_1u+C_2\frac{\partial u}{\partial t}\right]_0$$

$$+\frac{1}{2}\left[\frac{\partial u}{\partial t},C_2u\right]_0-[u,h_1]_0-\left[\frac{\partial u}{\partial t},h_0\right]_0 \qquad \text{(ix)}$$

where

$$(w,v)_0=\int_0^{t_0}((w(\cdot,\tau),v(\cdot,t_0-\tau)))_0\,d\tau$$

$$(T^*w,u)_1=\int_0^{t_0}((T^*w(\mathbf{x},t),u(\mathbf{x},t_0)))_1\,dt \qquad \text{(x)}$$

$$[u_1,u_2]_0=((u_1(\mathbf{x},t_0),u_2(\mathbf{x},0)))_0.$$

(d) Set $v=E^{-1}w$ in Eq. (ix) and obtain the functional associated with Eq. (i):

$$I_2(u,w)=(w,Tu)_0-\frac{1}{2}(w,E^{-1}w)_0+\frac{1}{2}\left(C_1\frac{\partial u}{\partial t}+C_2\frac{\partial^2 u}{\partial t^2},u\right)_1$$

$$-(f,u)_1+(w,Bu-g_1)_{\partial\Omega_1}-(g_2,u)_{\partial\Omega_2}+\frac{1}{2}\left[u,C_1u+C_2\frac{\partial u}{\partial t}\right]_0$$

$$+\frac{1}{2}\left[\frac{\partial u}{\partial t},C_2u\right]_0-[u,h_1]_0-\left[\frac{\partial u}{\partial t},h_0\right]_0. \qquad \text{(xi)}$$

30 Specialize the functional in (xi) to the following two cases:

(a) (*Heat Conduction*) Consider the heat conduction problem described by the equation

$$g(\mathbf{x},t)=k(\mathbf{x})\nabla\theta(\mathbf{x},t) \qquad \text{in } \Omega$$

$$C(\mathbf{x})\frac{\partial\theta(\mathbf{x},t)}{\theta t}-\nabla\cdot\mathbf{g}=f(\mathbf{x},t) \qquad \text{in } \Omega \qquad \text{(xii)}$$

where $\theta(\mathbf{x},t)$ is the temperature, $f(\mathbf{x},t)$ is a given function (source term) of position $\mathbf{x}$ and time t, $C(\mathbf{x})$ is the heat capacity, and k is the thermal conductivity. Clearly, Eq. (xii) is of the general form in (i) with $C_2=0$, $T=$gradient, and $T^*=-$divergence. The boundary and initial conditions are

$$\theta(\mathbf{x}, t)=\hat{\theta}(\mathbf{x}, t), \qquad \mathbf{x}\in\partial\Omega_\theta$$

$$k(\mathbf{x})\nabla\theta(\mathbf{x}, t)\cdot\hat{\mathbf{n}}\equiv h=\hat{h}, \qquad \mathbf{x}\in\partial\Omega_q, \quad t>0 \tag{xiii}$$

$$\theta(\mathbf{x},0)=\theta_0(\mathbf{x}), \qquad \mathbf{x}\in\Omega$$

where $\partial\Omega_\theta$ and $\partial\Omega_q$ denote the disjoint sets whose union is $\partial\Omega$, $\hat{\theta}$ and $\hat{h}$ are prescribed functions on the respective boundaries, $\hat{\mathbf{n}}$ is the unit vector normal to $\partial\Omega$ at $\mathbf{x}$, and θ_0 is the prescribed initial value of the temperature. Write Eq. (xi) in explicit form for Eqs. (xii) and (xiii).

(b) (*The Wave Equation*) Consider the wave equation

$$\mathbf{v}(\mathbf{x}, t)=k(\mathbf{x})\operatorname{grad} u(\mathbf{x}, t) \quad \text{in } \Omega, \qquad u(\mathbf{x}, t)=\hat{u}(\mathbf{x}, t) \quad \text{on } \partial\Omega_1$$

$$\frac{\partial^2 u(\mathbf{x}, t)}{\partial t^2}-\nabla\cdot\mathbf{v}(\mathbf{x}, t)=f(\mathbf{x}, t) \quad \text{in } \Omega, \qquad \mathbf{n}\cdot\mathbf{v}=\hat{v}(\mathbf{x}, t) \quad \text{on } \partial\Omega_2 \tag{xiv}$$

with the boundary conditions

$$u(\mathbf{x}, t)=\hat{u}(\mathbf{x}, t) \qquad \text{on } \partial\Omega_1$$

$$k\,\mathbf{n}\cdot\nabla u(\mathbf{x}, t)=\hat{v}(\mathbf{x}, t) \qquad \text{on } \partial\Omega_2 \tag{xv}$$

and initial conditions

$$u(\mathbf{x},0)=u_0(\mathbf{x}), \qquad \frac{\partial u(\mathbf{x},0)}{\partial t}=d_0(\mathbf{x}), \qquad \mathbf{x}\in\Omega. \tag{xvi}$$

Again, Eq. (xiv) is of the general form (i) with $C_1=0$, $C_2=1$, $T=$gradient, and $T^*=-$divergence.

3.6 VARIATIONAL METHODS OF APPROXIMATION

3.6.1 Introduction

While the calculus of variations provides an alternate method to the derivation of governing equations of a problem, its use as a powerful means of constructing sequences of functions that converge to desired solutions of the governing (Euler) equations is also appealing. This section is devoted to the discussion of several methods of numerical approximation that are based on the variational formulations discussed in the previous section. Before embarking on various methods of approximation, we review briefly the essence of previous sections of this chapter.

Consider equations of the form

$$\mathcal{P}(u) \equiv A(u) - f = 0 \tag{3.276}$$

where A is a certain operator (most frequently a differential operator), f is a given function, and u is the desired solution. The domain $\mathcal{D}_A$ of A is assumed to be a linear set dense in a Hilbert space H, and A is a positive-definite operator mapping $\mathcal{D}_A$ into H. For a given f in H, the problem is to find u in $\mathcal{D}_A$ such that Eq. (3.276) is satisfied. In the previous section we established through concrete examples that equations of the form (3.276) (in which A is a positive operator) can be cast equivalently as an equation minimizing a quadratic functional. The following theorem summarizes the results of Section 3.5.

Theorem 3.4 Let A be a (linear) positive operator in $\mathcal{D}_A$ and $f \in H$. Suppose that Eq. (3.276) has a solution u_0 in $\mathcal{D}_A$. Then the solution is unique. Further, the quadratic functional

$$I(u) = \tfrac{1}{2}(Au, u) - (f, u) \tag{3.277}$$

assumes its minimum in $\mathcal{D}_A$ for only $u = u_0$.

Proof

Uniqueness Let $u_1, u_2 \in \mathcal{D}_A$ be two solutions to Eq. (3.276). Then we have

$$A(u_1 - u_2) = 0.$$

Taking the inner product of the above equation with $u_1 - u_2$, we get

$$(A(u_1 - u_2), u_1 - u_2) = 0.$$

Since A is positive, it follows that $u_1 - u_2 = 0$. Hence the solution is unique.

Minimum of a Quadratic Functional First note that $I(u)$ is well defined for all u in H. From previous discussions we know that I assumes a minimum (or maximum) at $u = u_0$ if $\delta I(u_0; \eta) = 0$. That is,

$$0 = \frac{dI}{d\alpha}(u_0 + \alpha\eta)\Big|_{\alpha=0}$$

$$= \frac{d}{d\alpha}\left[(A(u_0 + \alpha\eta), u_0 + \alpha\eta) - 2(f, u_0 + \alpha\eta)\right]\Big|_{\alpha=0}$$

$$= (Au_0, \eta) + (A\eta, u_0) - 2(f, \eta).$$

Since A is symmetric, it follows that (from Lemma 2.2)

$$2(Au_0 - f, \eta) = 0 \rightarrow Au_0 - f = 0.$$

Thus, Eq. (3.276) is the Euler equation of the functional I (or, equivalently, u_0 makes I stationary). To establish that u_0 indeed makes I a minimum, consider

$$
\begin{aligned}
I(u)-I(u_0)&=(Au,u)-(Au_0,u_0)-2(f,u-u_0)\\
&=(A(u-u_0),u-u_0)+(Au_0,u)+(Au,u_0)-2(Au_0,u_0)-2(f,u-u_0)\\
&=(A(u-u_0),u-u_0)+2(Au_0-f,u-u_0)\\
&=(A(u-u_0),u-u_0)
\end{aligned}
$$

where the linearity and symmetry of A is used. Since A is positive, we have

$$I(u)-I(u_0)\geq 0 \qquad \text{or} \qquad I(u)\geq I(u_0).$$

Thus u_0 is the critical point which makes I a minimum. ■

Remarks

1 Note that when A is a linear symmetric operator, the functional associated with the operator Eq. (3:276) is given by Eq. (3.277). Functional I in Eq. (3.277) can be arrived at by using the inverse procedure described in Section 3.5. The weak form $(Au_0-f,\eta)=0$ can also be written as

$$B(u_0,\eta)=l(\eta) \tag{3.278}$$

where

$$B(u_0,\eta)=(Au_0,\eta),\quad l(\eta)=(f,\eta). \tag{3.279}$$

2 From Eq. (3.278) it follows that

$$|B(u_0,u)|=|l(u)|\leq\|f\|_H\|u\|_H\leq c\|f\|_H\|u\|_A$$

for every u in $\mathfrak{D}_A$. For $u=u_0$, we have (since A is positive)

$$\|u_0\|_A\leq c\|f\|_H \tag{3.280}$$

which shows the *continuous dependence* of the generalized solution u_0 on the *data* f.

Example 3.23 Consider the equation governing the transverse deflection of a beam of length L, modulus of elasticity $E(x)>0$, and moment of inertia $I(x)>0$ in $[0, L]$, and subjected to continuously distributed load $q(x)$:

$$\frac{d^2}{dx^2}\left(EI\frac{d^2w}{dx^2}\right)=-q \qquad \text{in } \Omega=(0,L). \tag{3.281}$$

Suppose that the ends of the beam are fixed (i.e., clamped),

$$w(0)=\frac{dw}{dx}(0)=0, \qquad w(L)=\frac{dw}{dx}(L)=0. \tag{3.282}$$

Let $H=L_2(\Omega)$, and let $\mathfrak{D}_A$ be the linear set of functions which are continuous with their derivatives up to the fourth order and satisfy the boundary conditions (recall that all of them are of the essential type) in Eq. (3.282). The linear set $\mathfrak{D}_A$ is dense in $L_2(\Omega)$ [i.e., every function u in $L_2(\Omega)$ can be approximated to arbitrary accuracy by functions from the linear set]. The operator A is defined by the left-hand side of Eq. (3.281).

We prove that A is positive. First we prove that A is symmetric in $\mathfrak{D}_A$. For w, v in $\mathfrak{D}_A$ we have

$$\begin{aligned}(Aw,v)&=\int_0^L \frac{d^2}{dx^2}\left(EI\frac{d^2w}{dx^2}\right)v\,dx\\&=\int_0^L EI\frac{d^2w}{dx^2}\frac{d^2v}{dx^2}dx+\left[\frac{d}{dx}\left(EI\frac{d^2w}{dx^2}\right)v-EI\frac{d^2w}{dx^2}\frac{dv}{dx}\right]_0^L\\&=\int_0^L EI\frac{d^2w}{dx^2}\frac{d^2v}{dx^2}dx.\end{aligned} \tag{3.283}$$

The boundary terms are zero in view of Eq. (3.282). Since the right-hand side is symmetric in w and v (and w, v are in $\mathfrak{D}_A$), it follows that A is symmetric in $\mathfrak{D}_A$. Next set $v=w$ in Eq. (3.283),

$$(Aw,w)=\int_0^L EI\left(\frac{d^2w}{dx^2}\right)^2 dx. \tag{3.284}$$

Since by assumption $E(x)>0$, $I(x)>0$ (physically, these assumptions are justified for the modulus and the moments of inertia cannot be negative functions), we have $(Aw,w)\geq 0$ for every w in $\mathfrak{D}_A$. Furthermore, if $(Aw,w)=0$, it follows that $d^2w/dx^2=0$ in Ω. This implies that w is of the form $w(x)=a+bx$. However, in view of the boundary conditions (3.282), $a=b=0$, and therefore $w(x)=0$. Thus we established that A is positive.

The functional associated with Eqs. (3.281) and (3.282) is given by [cf. Eq. (3.277)]

$$I(w)=\frac{1}{2}\int_0^L EI\left(\frac{d^2w}{dx^2}\right)^2 dx+\int_0^L qw\,dx, \tag{3.285}$$

which represents the elastic total potential energy of the beam. The weak form is given by

$$\int_0^L EI\frac{d^2w}{dx^2}\frac{d^2\eta}{dx^2}\,dx = -\int_0^L q\eta\,dx. \quad \blacksquare \tag{3.286}$$

The Energy Space Let A be a linear operator which maps $\mathfrak{D}_A$, a dense subset of a Hilbert space H. Suppose that A is symmetric and positive definite on $\mathfrak{D}_A$, that is, there exists a constant $c>0$ such that

$$\begin{aligned}(Au,u)&\geq c\|u\|^2, &&\text{for all } u\in\mathfrak{D}_A\\ (Au,u)&>0, &&\text{for all } u\neq 0.\end{aligned} \tag{3.287}$$

Since A is linear and symmetric, it follows that (Au, v) is bilinear and symmetric. Therefore we can define a new inner product by

$$(u,v)_A \equiv (Au,v), \qquad \text{for all } u,v\in\mathfrak{D}_A. \tag{3.288}$$

The norm and metric associated with the inner product are given by

$$\|u\|_A=\sqrt{(u,u)_A}, \qquad d_A(u,v)=\|u-v\|_A. \tag{3.289}$$

Then $\mathfrak{D}_A$ constitutes a linear inner product space with respect to the inner product $(\cdot,\cdot)_A$. It is not necessarily complete. Of course by the "addition" of the limit points of all Cauchy sequences in $\mathfrak{D}_A$ to $\mathfrak{D}_A$ one can complete the space. Let us denote the completion of $\mathfrak{D}_A$ by H_A. The space H_A, called the *energy space*, is a Hilbert space.

3.6.2 The Ritz Method

Consider the problem of solving the operator Eq. (3.276) where A is a linear positive-definite operator. Then the quadratic functional $B(u, u)$ is positive on H_A, the energy space defined above. From Theorem 3.4 it follows that the solution u_0 of Eq. (3.276) minimizes the quadratic functional

$$I(u)=\tfrac{1}{2}B(u,u)-(f,u)=\tfrac{1}{2}\|u\|_A^2-l(u). \tag{3.290}$$

In other words, the problem of determining solutions to Eq. (3.276) is equivalent to finding the minimum of $I(u)$ on H_A. The Ritz method seeks to minimize the functional on a finite-dimensional subspace of H_A.

In the following we shall assume that the functional $l(u)$ is bounded in H_A, and that H_A is separable (i.e., there exists a countable linearly independent set whose finite linear combinations can be used to approximate any element in H_A to a specified accuracy). In order to approximate an element u in H_A, we choose a sequence of elements $\{\phi_i\}_i^n \subset H_A$, which satisfy the following conditions:

1 $\{\phi_n\} \subset H_A$.
2 For any n, the set $\{\phi_n\}$ is linearly independent. (3.291)
3 $\{\phi_n\}$ is complete in H_A.

The functions ϕ_n are usually called *coordinate functions*. The requirement that $\{\phi_n\} \subset H_A$ implies two properties: (a) ϕ_i should be such that $I(u_n)$ is not zero; in other words, u_n should be continuous as required by the functional $I(u)$. (b) The ϕ_i satisfy the boundary conditions (only the essential boundary conditions) of the problem. Condition 2 ensures the invertibility of the resulting matrix. Clearly, H_A contains all linear combinations of $\{\phi_n\}$.

In the Ritz method the approximate representation of the solution u to Eq. (3.276) is sought in the form

$$u_n = \sum_{i=1}^{n} c_i \phi_i \tag{3.292}$$

where the constants c_i, called *the Ritz coefficients*, are chosen so that

$$I(u_n) = \min. \tag{3.293}$$

Upon substituting Eq. (3.292) for u_n into Eq. (3.293), one obtains I as an ordinary function of $c_1, c_2, \ldots, c_n$. Then the necessary condition for I to attain a minimum (or maximum) is that the first partial derivatives of I with respect to each c_k, $k = 1, 2, \ldots, n$, be zero:

$$\frac{\partial I}{\partial c_k}(c_1, c_2, \ldots, c_n) = 0, \qquad k = 1, 2, \ldots, n. \tag{3.294}$$

We have

$$\frac{\partial}{\partial c_k}\left[\left(A\left(\sum_{i=1}^{n} c_i \phi_i\right), \sum_{j}^{n} c_j \phi_j\right) - 2\left(f, \sum_{j=1}^{n} c_j \phi_j\right)\right] = 0$$

$$\frac{1}{2}\sum_{j=1}^{n}\left[\left(A(\phi_k), \phi_j\right)c_j + \left(A(\phi_j), \phi_k\right)c_j\right] - (f, \phi_k) = 0$$

or

$$\sum_{j=1}^{n} B(\phi_k, \phi_j) c_j \equiv \sum_{j=1}^{n} b_{ij} c_j = l(\phi_i). \tag{3.295}$$

Equation (3.295) consists of a system of n equations for the n constants. Since, by assumption, $\phi_1, \phi_2, \ldots, \phi_n$ are linearly independent, the determinant of the matrix $[B]=[b_{ij}]$, $b_{ij}=B(\phi_i, \phi_j)$, is different from zero. Hence the system (3.295) is uniquely solvable provided the data f is compatible (see Section 2.8.3).

The following theorem summarizes the results of the above discussion.

Theorem 3.5 Let A be a positive operator on a linear set $\mathfrak{D}_A$ which is dense in a separable Hilbert space H, and let $f \in H$. Let H_A be the completion of $\mathfrak{D}_A$ with respect to the inner product induced by the bilinear form $B(\cdot, \cdot)$. Further, let $\phi_1, \phi_2, \ldots, \phi_n$ be a basis in H_A. Then the Ritz sequence $\{u_n\}$, with the constants c_i uniquely determined for each fixed n by Eq. (3.295), converges in H_A to the generalized solution of Eq. (3.276). Moreover, if $n > m$, then

$$\|u_n - u_0\|_A \leq \|u_m - u_0\|_A \tag{3.296}$$

where $\|\cdot\|_A$ is the norm in H_A.

Proof First we note that u_n given by Eq. (3.292) is in H_A. By Theorem 3.4, the problem of determining the solution to Eq. (3.276) is equivalent to finding the stationary values of the functional I of Eq. (3.277) on H_A. The problem of minimizing the quadratic functional I on H_A has a unique solution—the constants c_i are uniquely determined by the system of Eq. (3.295). On the other hand the true solution u_0 is given by the Fourier series theorem, Theorem 2.16.

$$u_0 = \sum_{i=1}^{\infty} \alpha_i \hat{\phi}_i \tag{3.297}$$

where $\{\hat{\phi}_i\}$ is an orthonormal set in H_A. Since $\{\phi_i\}_{i=1}^{n}$ can be orthonormalized in H_A using the Gram-Schmidt process, the sequence (3.292) is the sequence of partial sums of the series (3.297). Since H_A is complete, the series (3.292) converges in H_A to u_0.

In terms of the orthonormal functions $\hat{\phi}_i$, sequence (3.292) can be expressed in the form

$$u_n = \sum_{i=1}^{n} \alpha_i \hat{\phi}_i. \tag{3.298}$$

From Eqs. (3.297) and (3.298) it follows that

$$\|u_0 - u_n\|_A^2 = \sum_{i=n+1}^{\infty} \alpha_i^2. \tag{3.299}$$

Consequently for increasing n the norm $\|u_0 - u_n\|_A$ decreases (or, at least, it does not increase), implying that Eq. (3.296) holds. This completes the proof of the theorem. ■

The following remarks can be made concerning the Ritz method:

1 (*Method of Orthonormal Series*) Recall from Section 2.7.4 that a given element u of a Hilbert space H can be represented by the sum of orthogonal projections along the coordinate functions $\hat{\phi}_i$. The error introduced by approximating u by a finite sum $\Sigma_i^n \alpha_i \hat{\phi}_i$ was minimized in the least-squares sense. If the coordinate functions $\hat{\phi}_i$ lie in some subspace S of H (i.e., the ϕ_i are orthonormal with respect to the norm defined in S), then the constants α_i that minimize the error $\|u - \Sigma_i^n \alpha_i \hat{\phi}_i\|_S$ are given by

$$\alpha_i = (\hat{\phi}_i, u)_S. \tag{3.300}$$

Thus for $S = H_A$, α_i are given by

$$\alpha_i = (\hat{\phi}_i, u)_A = (Au, \hat{\phi}_i)_H = (f, \hat{\phi}_i)_H.$$

Since the $\hat{\phi}_i$ are orthonormal in $S = H_A$, it follows that $B(\hat{\phi}_i, \hat{\phi}_j) = \delta_{ij}$, and the Ritz method also gives the same result for α_i. In other words, the problem of best approximation discussed in Section 2.7.4, also known as the *method of orthonormal series*, is a special case of the Ritz method. Conversely, u_n given by Eq. (3.292) is the best approximation to u_0 with respect to the metric in H_A and $\{\phi_i\}$.

2 It also follows that when A is a linear positive symmetric operator, the Ritz approximation and the Galerkin approximation given in Eq. (2.105) are the same. However, the Galerkin method can also be applied, as will be seen in the next section, to nonlinear operator equations.

3 The data f should be such that the linear form $l(\phi_i) = (f, \phi_i)$ is continuous on H_A, and that the solvability condition (2.130) is satisfied.

4 Since the natural boundary conditions are included in the functional, it is not necessary to pick the basis (i.e., approximation functions) that satisfies all of the boundary conditions; it is sufficient to pick a basis that is complete and satisfies only the essential boundary conditions.

5 When the boundary conditions are nonhomogeneous, the generalized solution is approximated by expressions of the form

$$u_n = \phi_0 + \sum_{i=1}^{n} c_i \phi_i \tag{3.301}$$

where $\phi_0(x)$ is a function that satisfies the nonhomogeneous boundary conditions.

6 The completeness of the sequence $\{\phi_i\}_{i=1}^n$ is not always a strict requirement. Acceptable solutions can be obtained, depending on the problem, with an incomplete set of linearly independent functions. We shall see, via specific examples, how important is the choice of the sequence in applying the Ritz method.

7 As pointed out in Theorem 3.5, the Ritz approximation gives improved accuracy (in the norm of H_A) with increasing n. For increasing values of n, the previously generated elements of the matrix in Eq. (3.295) remain unchanged, and to the matrix of this system one must add newly computed rows and columns. This property of the Ritz method is very attractive from a numerical point of view.

8 (*Error Estimate*) The n-parameter Ritz solution $u_n \in \mathfrak{D}_A$ of the generalized solution u_0 can be viewed as the exact solution of the operator Eq. (3.276) for the data $f_n \equiv Au_n$. This interpretation enables us to estimate the error $(u_n - u_0)$ in the norm of H_A. Suppose that $B(u_0, u) = (f, u)$ and $B(u_n, u) = (f_n, u)$ for all $u \in H_A$. Then $z_0 = u_n - u_0$ is the generalized solution of the equation

$$B(z_0, u) = (f_n - f, u), \qquad \text{for all } u \in H_A.$$

Using Eq. (3.280), we have the error estimate

$$\|u_n - u_0\|_A \leq C\|f_n - f\|_H = C\|Au_n - f\|_H \tag{3.302}$$

where $C = C(\Omega)$ is a constant which is independent of u but depends on the domain Ω of the problem (see Exercise 2.4.6).

We now consider several examples of the Ritz method.

Example 3.24

1 Consider the differential equation

$$-\frac{d^2u}{dx^2} = \cos \pi x, \qquad 0 < x < 1, \qquad \left(A = -\frac{d^2}{dx^2}\right) \tag{3.303a}$$

subjected to the (*Dirichlet*) boundary conditions

$$u(0) = u(1) = 0. \tag{3.303b}$$

The domain $\mathfrak{D}_A$ of the operator A consists of twice differentiable functions in $(0, 1)$ which satisfy the boundary conditions (3.303b). The operator A is positive on $\mathfrak{D}_A \subset H = L_2(0, 1)$. The operator is also symmetric on $\mathfrak{D}_A$. The null space $\mathfrak{N}(A^*)$ contains elements of the form $v(x) = a + bx$. Since $v \in \mathfrak{D}_A$, we have $v(0) = v(1) = 0$, hence $\mathfrak{N}(A^*)$ is trivial. Consequently the solvability condition (2.130) is identically satisfied. The operator Eq. (3.303) has a unique solution. The exact solution is given by

$$u_0(x) = \frac{1}{\pi^2}(\cos \pi x + 2x - 1). \tag{3.304}$$

The quadratic functional is given by, $u \in \mathfrak{D}_A$,

$$I(u)=\frac{1}{2}(Au,u)_0-(f,u)_0$$

$$=\frac{1}{2}\int_0^1\left[-\frac{d^2u}{dx^2}u-2u\cos\pi x\right]dx$$

$$=\frac{1}{2}\int_0^1\left[\left(\frac{du}{dx}\right)^2-2u\cos\pi x\right]dx. \tag{3.305}$$

Next we wish to use the problem to illustrate several properties of the Ritz approximation. We choose the following basis (or approximation functions) for the n-parameter Ritz approximation:

$$u_n=\sum_{i=1}^{n}c_i\sin i\pi x. \tag{3.306}$$

The set $\{\phi_i\}=\{\sin i\pi x\}$ clearly satisfies the conditions (3.291). It is twice differentiable in $[0,1]$, satisfies the essential boundary conditions of the problem (both of the given boundary conditions in the present case are of the essential type), and forms a basis in H_A which consists of all differentiable functions that vanish at $x=0,1$. The weak form of the equations is given by setting δI to zero:

$$2\int_0^1\left(\frac{du}{dx}\frac{d\delta u}{dx}-\cos\pi x\,\delta u\right)dx=0$$

or, equivalently,

$$B(u,\delta u)-(\cos\pi x,\delta u)=0.$$

Substituting Eq. (3.292) into the above expression, we get

$$\sum_{j=1}^{n}B(\phi_i,\phi_j)c_j-(\cos\pi x,\phi_i)=0 \tag{3.307a}$$

where

$$b_{ij}=B(\phi_i,\phi_j)=\int_0^1(i\pi)^2\sin i\pi x\sin j\pi x\,dx=\begin{cases}0, & \text{if } j\neq i\\ \dfrac{(i\pi)^2}{2}, & \text{if } j=i\end{cases} \tag{3.307b}$$

$$l(\phi_i)=\int_0^1 \cos\pi x \sin i\pi x\, dx=\frac{1}{2}\int_0^1 [\sin\pi(i+1)x+\sin\pi(i-1)x]\, dx$$

$$=\frac{1}{2}\left[\frac{(-1)^{i+1}-1}{\pi(i+1)}+\frac{(-1)^{i-1}-1}{\pi(i-1)}\right]$$

$$=\begin{cases} 0, & \text{if } i \text{ is odd} \\ \dfrac{2i}{\pi(i^2-1)}, & \text{if } i \text{ is even.} \end{cases} \tag{3.307c}$$

Thus $[B]=[b_{ij}]$ is a diagonal matrix. Consequently we can solve for c_i very easily,

$$c_i=\frac{4}{\pi^3}\frac{1}{i(i^2-1)}, \qquad \text{for } i \text{ even.} \tag{3.308}$$

The solution is given by

$$u_n(\cdot x)=\frac{4}{\pi^3}\sum_{i=1}^{n}\frac{\sin i\pi x}{(i^2-1)i}, \qquad i \text{ even}$$

$$=\frac{2}{\pi^3}\sum_{j=1}^{n}\frac{\sin 2j\pi x}{(4j^2-1)j}. \tag{3.309}$$

It is not difficult to show that the series is uniformly convergent in the interval $[0,1]$ to u_0 given by Eq. (3.304). The error estimate (3.302) for the problem at hand becomes (with $C=1$)

$$\|u_n-u_0\|_A \le \|Au_n-f\|_0$$

$$=\left\|\sum_{j=1}^{n}\frac{8j^2\sin 2j\pi x}{\pi j(4j^2-1)}-\cos\pi x\right\|_0$$

$$=\left[\frac{\pi}{2}-\frac{32}{\pi}\sum_{j=1}^{n}\frac{j^2}{(4j^2-1)^2}\right]. \tag{3.310}$$

For $n=1$, 2, and 3, this error estimate is

$$\|u_n-u_0\|_A \le \begin{cases} 0.439, & n=1 \\ 0.258, & n=2 \\ 0.199, & n=3. \end{cases}$$

In fact the actual error $\|u_0-u_n\|_A$ is smaller than the estimate (3.310).

2 Next we consider the (mixed boundary-value) problem

$$-\frac{du^2}{dx^2}=\cos\pi x, \qquad 0<x<1$$

$$u(0)=0, \qquad u'(1)=0. \tag{3.311}$$

We note that the boundary condition $u(0)=0$ is of the essential type and $u'(1)=0$ is of the natural type. The operator A is self-adjoint on $\mathfrak{D}_A$, which consists of twice differentiable functions that satisfy the boundary conditions (3.311). The null space $\mathfrak{N}(A^*)$ consists of the zero element. Hence the solvability condition is trivially satisfied. Thus the solution exists and is unique.

Following the remark made earlier, we select a basis that satisfies only the essential boundary condition, $u(0)=0$. Obviously $\{\sin i\pi x\}$ meets the boundary condition $u(0)=0$. However, H_A in the present case consists of all differentiable functions which vanish at $x=0$. Therefore $\{\sin i\pi x\}$, being not complete, cannot form a basis for H_A. The set is incomplete because it cannot generate the functions in H_A that are nonzero at $x=1$. If we overlook the completeness and use $\{\sin i\pi x\}$ for computing the Ritz coefficients c_i, we obtain the same Ritz solution as in part 1, whereas the exact solution to Eq. (3.311) is given by

$$u_0=\frac{1}{\pi^2}(\cos\pi x-1). \tag{3.312}$$

Hence the Ritz solution does not converge to the exact solution [it converges to Eq. (3.304)].

To remedy this situation, we must select a complete set $\{\phi_i\}$ in H_A. If we add a function $\phi_0=x$, then the set $\{x,\sin\pi x,\sin 2\pi x,\dots\}$ will be complete in H_A. Note that $\phi_0=1$ is not in H_A [because it does not satisfy $\phi_0(0)=0$]. Since we have already computed b_{ij} for $i,j=1,2,\dots,n$, we need to compute only b_{ij} for $i=0, j=0,1,2,\dots,n$, and $l(\phi_0)$. We obtain

$$\left.\begin{aligned} b_{0j}&=B(\phi_0,\phi_j)=\int_0^1\phi_j'\,dx=\begin{cases}0, & \text{if } j\neq 0\\ 1, & \text{if } j=0\end{cases}\\ l(\phi_0)&=\int_0^1 x\cos\pi x\,dx=-\frac{2}{\pi^2}.\end{aligned}\right\} \tag{3.313}$$

The solution $u_n=\sum_{i=0}^n c_i\phi_i$ is given by

$$u_n=-\frac{2x}{\pi^2}+\frac{2}{\pi^3}\sum_{j=1}^{n}\frac{\sin 2\pi jx}{(4j^2-1)j} \tag{3.314}$$

which converges to

$$\left[-\frac{2x}{\pi^2}+\frac{1}{\pi^2}(\cos\pi x+2x-1)\right]=\frac{1}{\pi^2}(\cos\pi x-1),$$

the exact solution.

3 We now consider the differential equation

$$-\frac{du^2}{dx^2}=\cos\pi x, \qquad 0<x<1 \tag{3.315a}$$

subjected to the (Neumann) boundary conditions

$$u'(0)=u'(1)=0. \tag{3.315b}$$

In the present case $\mathfrak{N}(A^*)$ consists of constant functions. Then the solvability condition becomes

$$(f,k)_0\equiv k\int_0^1\cos\pi x\,dx=0, \qquad k \text{ is a constant,} \tag{3.316}$$

which is clearly satisfied. Hence the solution exists. The variational problem can be shown to have a unique solution in H_A by proving that the bilinear form satisfies the conditions of the Lax-Milgram theorem, Theorem 2.19. However, the solution is not unique in H_A, because if $u(x)$ is a solution of Eq. (3.315), then $v(x)=u(x)+c$, where c is a constant, is also a solution of Eq. (3.315). To make the solution unique, we impose the additional condition

$$\int_0^1 u(x)\,dx=0. \tag{3.317}$$

Since both of the boundary conditions are of the natural type, H_A consists of all differentiable functions. Neither of the two sets $\{\sin i\pi x\}$ and $\{x,\sin\pi x,\sin 2\pi x,\dots\}$ is complete in H_A. We must add an element to the set $\{\sin i\pi x\}$ that does not vanish at $x=0$ and $x=1$, and at the same time $u_n(0)\neq u_n(1)$. If the last requirement is not satisfied by the new element, then the new set spans only the subspace of H_A that consists of functions $u(x)$ with the property $u(0)=u(1)$, and therefore not complete in H_A.

We select $\phi_0=x-c$, where c is a constant not equal to 1. We obtain b_{0j} and $l(\phi_0)$ as in Eq. (3.122). Hence the solution becomes

$$u_n(x)=-\frac{2}{\pi^2}(x-c)+\frac{2}{\pi^3}\sum_{j=1}^{n}\frac{\sin 2\pi jx}{(4j^2-1)}. \tag{3.318}$$

The constant c must be selected such that the condition (3.317) is satisfied. This gives $c=\frac{1}{2}$. Then the Ritz approximation $u_n(x)$ converges to $-(2/\pi^2)(x-\frac{1}{2})+(1/\pi^2)(\cos \pi x+2x-1)=\cos \pi x/\pi^2$, which is the exact solution to Eq. (3.315).

4 Here we consider the Ritz approximation of Eq. (3.303) using algebraic polynomials. We select $\{\phi_i\}_{i=1}^n=\{x^i(1-x)\}\subset H_A$. Since the ϕ_i are not orthogonal, the matrix $[b_{ij}]$ is not diagonal. One must invert the matrix $[b_{ij}]$ to compute the Ritz coefficients c_i. We have

$$b_{ij}=\int_0^1\left[ix^{i-1}-(i+1)x^i\right]\left[jx^{j-1}-(j+1)x^j\right]dx$$

$$=\frac{2ij}{(i+j)\left[(i+j)^2-1\right]},\qquad i,j=1,2,\ldots \tag{3.319a}$$

$$l(\phi_i)=\int_0^1(x^i-x^{i+1})\cos \pi x\,dx$$

$$=\sum_{k=0}^{[(i-1)/2]}(-1)^k\frac{i!}{(i-2k-1)!}\frac{1}{\pi^{2(k+1)}}\left[\cos \pi x\, x^{i-2k-1}\right]_0^1$$

$$-\sum_{S=0}^{[i/2]}(-1)^S\frac{(i+1)!}{(i-2S)!}\frac{1}{\pi^{2(S+1)}}\left[\cos \pi x x^{i-2S}\right]_0^1 \tag{3.319b}$$

where $S=[i/2]$=greatest integer less than or equal to $i/2$. For example, if $n=1$, we have

$$b_{11}=\frac{1}{3},\qquad l(\phi_1)=\left[\frac{1}{\pi^2}(-2)\right]-\left[\frac{2}{\pi^2}(-1)\right]=0$$

and hence $c_1=0$. For $n=2$, we get

$$b_{11}=\frac{1}{3},\qquad b_{12}=\frac{1}{6},\qquad b_{22}=\frac{2}{15},\qquad l(\phi_1)=0,\qquad l(\phi_2)=\frac{2}{\pi^2}-\frac{12}{\pi^4}.$$

This yields $c_2=-2c_1=40(\pi^2-6)/\pi_4$. The two-parameter Ritz solution is not very accurate. Indeed the error estimate (3.302) yields

$$\|u_2-u_0\|_A\leq-\frac{480}{\pi^4}+\frac{240}{\pi^2}-\frac{5760}{\pi^6}+\frac{1}{2}\doteq 13.898.$$

The solution improves with increasing n, but the convergence is slow.

5 We now consider the question of finding the eigenvalues λ of Eq. (3.303) using the Ritz approximation (3.306). The Ritz system associated with the

eigenvalue problem is given by

$$\sum_{j}^{N}\left[B(\phi_i,\phi_j)-\lambda(\phi_i,\phi_j)_0\right]c_j=0. \tag{3.320}$$

We need to compute the coefficients m_{ij} (called the mass coefficients),

$$\begin{aligned} m_{ij}&=(\phi_i,\phi_j)_0\\ &=\int_0^1 \sin i\pi x \sin j\pi x\,dx=\begin{cases}0, & \text{if } i\neq j,\\ \tfrac{1}{2}, & \text{if } i=j.\end{cases}\end{aligned}$$

The eigenvalue problem involves solving for λ_n, $n=1,2,\ldots,N$, such that Eq. (3.320) holds for any nontrivial c_j. Recall from Chapter 1 that this requirement amounts to finding the roots of the polynomial (in λ) obtained by setting the determinant of Eq. (3.320) to zero:

$$\det(b_{ij}-\lambda m_{ij})=0.$$

Since both b_{ij} and m_{ij} are diagonal in the present case, the polynomial is given by

$$\prod_{i}^{N}\left[(i\pi)^2-\lambda\right]=0 \qquad \text{or} \qquad \lambda_i=\pi^2 i^2. \quad ■$$

Example 3.25 Consider the so-called Sturm-Louiville differential equation

$$-p(x)\frac{d^2u}{dx^2}-q(x)\frac{du}{dx}+r(x)u=\hat{f}(x), \qquad 0<x<1. \tag{3.321}$$

The differential equation permits a quadratic functional only if (why?) $q(x)=dp/dx$. We can then write Eq. (3.321) in the form

$$-\frac{d}{dx}\left(p(x)\frac{du}{dx}\right)+r(x)u=\hat{f}(x), \qquad 0<x<1. \tag{3.322}$$

We wish to solve the equation for $p(x)=1+x$, $r(x)=\hat{f}(x)=0$, and with the nonhomogeneous boundary conditions

$$u(0)=0, \qquad \left[p(x)\frac{du}{dx}\right]_{x=1}=1. \tag{3.323}$$

If we define the domain $\mathfrak{D}_A$ of the operator

$$A\equiv-\frac{d}{dx}\left[(1+x)\frac{d}{dx}\right]$$

as the set of twice differentiable functions which satisfy the boundary conditions, it will not be a linear space. To avoid this difficulty, we reformulate the differential equation by introducing the function $w = u - v$, with v such that it satisfies the conditions $(p(x)dv/dx)|_{x=1} = 1$, and $v(0) = 0$. Then we have

$$-\frac{d}{dx}\left(p(x)\frac{dw}{dx}\right) = f(x), \qquad 0 < x < 1$$

$$w(0) = 0, \qquad \left(p(x)\frac{dw}{dx}\right)\bigg|_{x=1} = 0 \tag{3.324}$$

where

$$f(x) = \frac{d}{dx}\left(p(x)\frac{dv}{dx}\right).$$

Clearly, $v = x/2$ satisfies the requirements. Then $f(x)$ becomes $f = \frac{1}{2}$. Now we define $\mathfrak{D}_A$ to be the space of twice differentiable functions which satisfy the homogeneous boundary conditions (3.324). Then the operator A can be shown to be positive definite and self-adjoint on $\mathfrak{D}_A$, and $\mathfrak{N}(A^*) = \{0\}$. We now turn to the Ritz approximation of the problem.

The variational formulation of Eq. (3.324) is given by

$$B(u, v) = l(v) \tag{3.325a}$$

where

$$B(u, v) = \int_0^1 \left[(1+x)\frac{du}{dx}\frac{dv}{dx}\right] dx, \qquad l(v) = \int_0^1 f(x)v\,dx. \tag{3.325b}$$

We construct the Ritz approximation in the form

$$w_n(x) = \sum_i^n c_i\phi_i, \qquad \phi_i = x^i. \tag{3.326}$$

Note that the $\phi_i \in H_A$ (consists of all differentiable functions that vanish at $x = 0$) satisfy only the essential boundary condition $w(0) = 0$. We have

$$b_{ij} = \int_0^1 ij(1+x)x^{i-1}x^{j-1}\,dx = \int_0^1 ij[x^{i+j-2} + x^{i+j-1}]\,dx$$

$$= ij\left[\frac{1}{i+j-1} + \frac{1}{i+j}\right], \qquad i, j = 1, 2, \ldots$$

$$l(\phi_i) = \int_0^1 \frac{1}{2}x^i\,dx = \frac{1}{2(i+1)}. \tag{3.327}$$

For $n=1$ we get

$$b_{11}=\tfrac{3}{2}, \qquad l(\phi_1)=\tfrac{1}{4}, \qquad c_1=\tfrac{1}{6}.$$

For $n=2$ we obtain

$$\begin{bmatrix} \frac{3}{2} & \frac{5}{3} \\ \frac{5}{3} & \frac{7}{3} \end{bmatrix}\begin{Bmatrix} c_1 \\ c_2 \end{Bmatrix}=\begin{Bmatrix} \frac{1}{4} \\ \frac{1}{6} \end{Bmatrix} \Rightarrow c_1=\tfrac{11}{26}, \qquad c_2=-\tfrac{3}{13}.$$

The one-parameter and two-parameter solutions are given by

$$u_1=\frac{x}{2}+\frac{1}{6}x=\frac{2}{3}x, \qquad u_2=\frac{x}{2}+\frac{11}{26}x-\frac{3}{13}x^2=\frac{12}{13}x-\frac{3}{13}x^2. \tag{3.328}$$

The exact solution to the original equation is given by

$$u_0=\log_e(1+x)=x-\frac{x^2}{2}+\frac{x^3}{3}-\frac{x^4}{4}+\cdots. \tag{3.329}$$

Table 3.2 shows the values of the Ritz coefficients for values of $n=1,2,\ldots,8$. Clearly, the Ritz solution is converging to the exact solution (3.329). ■

Example 3.26 Consider the cantilever beam shown in Fig. 3.14. We wish to find the transverse deflection of the beam under applied transverse uniform loading of intensity q per unit length and end moment M_0.

$$\frac{d^2}{dx^2}\left(EI\frac{d^2w}{dx^2}\right)+q=0, \qquad 0<x<L, \qquad EI>0$$

$$w(0)=\frac{dw}{dx}(0)=0, \qquad \left(EI\frac{d^2w}{dx^2}\right)\bigg|_{x=L}=M_0, \qquad \left[\frac{d}{dx}\left(EI\frac{d^2w}{dx^2}\right)\right]\bigg|_{x=L}=0. \tag{3.330}$$

Here we have

$$A\equiv\frac{d^2}{dx^2}\left(EI\frac{d^2}{dx^2}\right)$$

and $\mathfrak{D}_A=\{u\in H^4(0,L):$ u satisfies the homogeneous boundary conditions associated with Eq. (3.330)$\}$. In view of the definition of $\mathfrak{D}_A$, the given boundary-value problem must be transformed into

$$\frac{d^2}{dx^2}\left(EI\frac{d^2u}{dx^2}\right)+f=0, \qquad 0<x<L \tag{3.331}$$

Table 3.2 Ritz Coefficients in Eq. (3.326) for Eq. (3.324)

n	c_1	c_2	c_3	c_4	c_5	c_6	c_7	c_8
1	0.166667							
2	0.423077	−0.230769						
3	0.484127	−0.396825	0.105820					
4	0.496885	−0.467290	0.218069	−0.054517				
5	0.499406	−0.490790	0.285601	−0.131016	0.029946			
6	0.499889	−0.497608	0.316164	−0.190399	0.082234	−0.017132		
7	0.499979	−0.499414	0.327762	−0.224233	0.131894	−0.052921	0.010080	
8	0.499996	−0.499863	0.331661	−0.240053	0.165963	−0.093099	0.034595	−0.006054
Exact	0.500000	−0.500000	0.333333	−0.250000	0.200000	−0.166667	0.142857	−0.125000

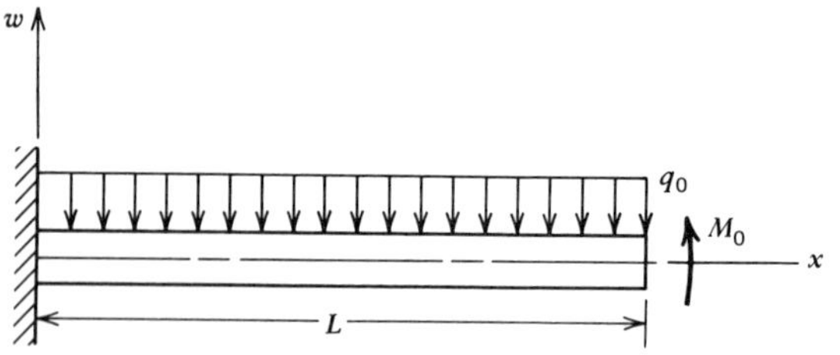

Figure 3.14 Bending of an elastic (cantilever) beam.

where

$$u = w - w_0, \qquad f = q - \frac{d^2}{dx^2}\left(EI\frac{d^2 w_0}{dx^2}\right)$$

and w_0 satisfies the actual (nonhomogeneous) boundary conditions (3.330). We can choose w_0 to be

$$w_0 = \frac{M_0 x^2}{2EI},$$

which satisfies the required boundary conditions. Then f becomes $f = q$. Once we find the solution u to Eq. (3.331), the actual solution w is given by $w = u + w_0$.

The operator A is self-adjoint and positive on $\mathfrak{D}_A$. We have

$$(Au, v)_0 = \int_0^L \frac{d^2}{dx^2}\left(EI\frac{d^2u}{dx^2}\right) v\,dx$$

$$= \int_0^L EI\frac{d^2u}{dx^2}\frac{d^2v}{dx^2}dx + \left[\frac{d}{dx}\left(EI\frac{d^2u}{dx^2}\right)v - EI\frac{d^2u}{dx^2}\frac{dv}{dx}\right]_0^L$$

Since $v \in \mathfrak{D}_A$, the boundary terms vanish, and we have a symmetric bilinear form in u and v:

$$\|Au\|_0^2 = EI\int_0^L \left(\frac{d^2u}{dx^2}\right)^2 dx > 0, \qquad \text{for } u \neq 0.$$

Also, $Au = 0$ implies that $d^2u/dx^2 = 0$ or $u = a_0 + a_1 x$. Since $u \in \mathfrak{D}_A$, we must have $u(0) = (du/dx)(0) = 0$. This gives $a_0 = a_1 = 0$. Thus A is positive definite on $\mathfrak{D}_A$. Further, the null space $\mathfrak{N}(A^*) = \mathfrak{N}(A)$ is trivial, and hence the solvability condition is satisfied. Thus the solution exists and is unique.

We now return to the original boundary-value problem (3.330) to demonstrate the application of the Ritz method. Everything said below will also apply to Eq. (3.331) with $M_0=0$. In both cases the space H_A is given by

$$H_A=\left\{u:\quad u\in H^2(0,L),\, u(0)=\frac{du}{dx}(0)=0\right\}. \tag{3.332}$$

Next we construct the variational form and associated functional for the differential Eq. (3.330). Following the procedure of Section 3.5, we write

$$\begin{aligned}\delta I(w)&=\int_0^L\left[\frac{d^2}{dx^2}\left(EI\frac{d^2w}{dx^2}\right)+q\right]\delta w\,dx\\&=\int_0^L\left(EI\frac{d^2w}{dx^2}\frac{d^2\delta w}{dx^2}+q\delta w\right)dx+\left[\frac{d}{dx}\left(EI\frac{d^2w}{dx^2}\right)\delta w-EI\frac{d^2w}{dx^2}\frac{d\delta w}{dx}\right]_0^L\end{aligned} \tag{3.333}$$

It follows from the boundary terms that the essential and natural conditions involve the specification of the following quantities at $x=0$ and L:

$$\begin{aligned}&w \text{ and } \frac{dw}{dx}, &&\text{essential (those associated with } \delta)\\&\frac{d}{dx}\left(EI\frac{d^2w}{dx^2}\right),\quad EI\frac{d^2w}{dx^2}, &&\text{natural (the coefficients of \textit{varied} quantities)}\end{aligned} \tag{3.334}$$

The first derivative of the transverse deflection w is called the *slope*, $EI\,d^2w/dx^2$ is the bending moment, and $(d/dx)(EI\,d^2w/dx^2)$ is the shear force. In view of the boundary conditions $(3.330)_2$ of the problem, the boundary terms in Eq. (3.333) can be simplified, and the weak form becomes

$$\delta I(w)=\int_0^L\left(EI\frac{d^2w}{dx^2}\frac{d^2\delta w}{dx^2}+q\delta w\right)dx-M_0\frac{d\delta w}{dx}(L) \tag{3.335}$$

or

$$I(w)=\int_0^L\left[\frac{EI}{2}\left(\frac{d^2w}{dx^2}\right)^2+qw\right]dx-M_0\frac{dw}{dx}(L) \tag{3.336}$$

where the first term under the integral represents the strain energy due to bending, and the second term represents the work done by the distributed load, and the last term represents the work done by the concentrated moment M_0, in rotating the end point through a slope $\theta_0=(dw/dx)(\mathrm{L})$.

We now construct the n-parameter Ritz solution using the weak form (3.335):

$$B(u,v)=\int_0^L EI\frac{d^2u}{dx^2}\frac{d^2v}{dx^2}dx,\ l(v)=\int_0^L -qv\,dx+M_0\frac{dv}{dx}(L). \quad (3.337)$$

We select algebraic coordinate functions ϕ_i that satisfy the essential boundary conditions $\phi_i(0)=\phi_i'(0)=0$. We represent w by

$$w_n(x)=\sum_i^n c_i\phi_i,\qquad \phi_i=x^{i+1}. \quad (3.338)$$

Substituting Eq. (3.338) into Eq. (3.337), we obtain

$$b_{ij}=\int_0^L EI(i+1)ix^{i-1}j(j+1)x^{j-1}dx$$

$$=\frac{EIij(i+1)(j+1)(L)^{i+j-1}}{(i+j-1)}, \quad (3.339a)$$

$$l(\phi_i)=\frac{-q(L)^{i+2}}{i+2}+M_0(i+1)(L)^i. \quad (3.339b)$$

Considering the two-parameter family of approximation ($n=2$), we have

$$EI(4Lc_1+6L^2c_2)=-\frac{q_0L^3}{3}+2M_0L$$

$$EI(6L^2c_1+12L^3c_2)=-\frac{q_0L^4}{4}+3M_0L^2. \quad (3.340)$$

Solving for c_1 and c_2, we obtain

$$c_1=-\frac{5q_0L^2-12M_0}{24EI},\qquad c_2=\frac{q_0L}{12EI}.$$

The two-parameter solution is given by

$$w_2(x)=\frac{(12M_0-5q_0L^2)}{24EI}x^2+\frac{q_0L}{12EI}x^3. \quad (3.341)$$

By rewriting the solution w_2 in the form

$$w_2(x)=\frac{x^2Lq_0}{24EI}(2x-5L)+\frac{M_0x^2}{2EI}\equiv\tilde{w}_2+w_0 \quad (3.342)$$

we note that $\tilde{w}_2$ is the two-parameter Ritz solution to Eq. (3.331). This can be easily seen by setting M_0 to zero in Eq. (3.342).

The three-parameter ($n=3$) solution gives the matrix equations

$$\begin{bmatrix} 4 & 6L & L^2 \\ 6L & 12L^2 & 18L^3 \\ 8L^2 & 18L^3 & \frac{144}{5}L^4 \end{bmatrix} \begin{Bmatrix} c_1 \\ c_2 \\ c_3 \end{Bmatrix} = \begin{Bmatrix} -\frac{q_0 L^2}{3} + 2M_0 \\ -\frac{q_0 L^3}{4} + 3M_0 L \\ -\frac{q_0 L^4}{5} + 4M_0 L^2 \end{Bmatrix} \tag{3.343}$$

which yields the *exact* solution

$$w(x) = \left(\frac{2M_0 - q_0 L^2}{4EI} \right) x^2 + \frac{q_0 L}{6EI} x^3 - \frac{q_0}{24EI} x^4. \tag{3.344}$$

If the three-parameter Ritz solution is exact, then one suspects that the four-parameter (or, for any $n > 3$) solution should also give the exact solution. This implies that all c_i, $i = 4, 5, \ldots$, should be identically zero. If one solves the system $\Sigma_j^n b_{ij} c_j = l(\phi_i)$ on a digital computer, the c_i, $i = 1, 2, 3$, computed will remain unchanged for any $n \geq 3$, whereas c_i, $i > 3$, will be of the order 10^{-10} (due to the round-off errors in the computer) instead of zero. ■

The next example is concerned with the application of the Ritz method to the Lagrange multiplier and mixed formulations of beam bending.

Example 3.27 Consider the problem of minimizing the functional

$$I(\theta, w) = \int_0^L \left[\frac{EI}{2} \left(\frac{d\theta}{dx} \right)^2 + qw \right] dx - M_0 \theta(L), \tag{3.345}$$

subject to the constraint

$$\theta - \frac{dw}{dx} = 0 \tag{3.346}$$

where w is the transverse displacement of the beam. Note that if Eq. (3.346) is used in Eq. (3.345), we obtain the functional (3.336) associated with the bending of a cantilever beam subjected to transversely distributed load q and end moment M_0.

1 (*Lagrange Multiplier Formulation*) The modified functional is given by (see Section 3.3)

$$I_L(\theta, w, \lambda) = \int_0^L \left[\frac{EI}{2} \left(\frac{d\theta}{dx} \right)^2 + qw \right] dx + \int_0^L \lambda \left(\theta - \frac{dw}{dx} \right) dx - M_0 \theta(L) \tag{3.347}$$

where λ is the Lagrange multiplier, which in the present case represents the shear force. In order to identify the essential boundary conditions, we find the Euler equations associated with the functional I_L.

$$\delta I_L = \int_0^L \left(EI \frac{d\theta}{dx}\frac{d\delta\theta}{dx} + q\delta w \right) dx + \int_0^L \left[\delta\lambda \left(\theta - \frac{dw}{dx} \right) + \lambda \left(\delta\theta - \frac{d\delta w}{dx} \right) \right] dx$$

$$- M_0 \delta\theta(L)$$

$$= \int_0^L \left\{ \left[-\frac{d}{dx}\left(EI \frac{d\theta}{dx} \right) + \lambda \right] \delta\theta + \left(q + \frac{d\lambda}{dx} \right) \delta w + \left(\theta - \frac{dw}{dx} \right) \delta\lambda \right\} dx$$

$$+ \left(EI \frac{d\theta}{dx}(L) - M_0 \right) \delta\theta(L) - EI \frac{d\theta}{dx}(0)\, \delta\theta(0) - (\delta w \lambda)\big|_{x=0}^{L} = 0. \tag{3.348}$$

Thus the *specified essential* boundary conditions for the problem are $\theta(0) = w(0) = 0$.

To solve the problem by the Ritz method, the approximating functions for θ and w can be linear (as opposed to quadratic in Example 3.26), and a constant for λ. Write Eq. (3.348) in the alternate form,

$$\delta I_L = \int_0^L \left(EI \frac{d\theta}{dx}\frac{d\delta\theta}{dx} + \lambda\delta\theta \right) dx - M_0 \delta\theta(L) + \int_0^L \left(-\lambda \frac{d\delta w}{dx} + q\delta w \right) dx$$

$$+ \int_0^L \delta\lambda \left(\theta - \frac{dw}{dx} \right) dx = 0. \tag{3.349}$$

Seeking first a one-parameter approximation of the form

$$\theta(x) = a_1 x, \qquad w(x) = b_1 x, \qquad \lambda(x) = c_1, \tag{3.350}$$

we obtain from Eq. (3.349)

$$\int_0^L (EIa_1 \delta a_1 + c_1 x \delta a_1)\, dx - M_0 L \delta a_1 = 0 \quad \text{or} \quad EILa_1 + \frac{c_1 L^2}{2} = M_0 L$$

$$\int_0^L (q\delta b_1 x - c_1 \delta b_1)\, dx = 0 \quad \text{or} \quad \frac{qL^2}{2} - c_1 L = 0, \quad c_1 = \frac{qL}{2}$$

$$\int_0^L (a_1 x - b_1)\, \delta c_1\, dx = 0 \quad \text{or} \quad \frac{a_1 L^2}{2} - b_1 L = 0 \quad b_1 = \frac{4M_0 L - qL^3}{8EI}.$$

Thus the one-parameter solution is

$$\theta(x)=\left(\frac{4M_0-qL^2}{2EI}\right)x,\qquad w(x)=\frac{L(4M_0-qL^2)}{8EI}x,$$

$$\lambda(x)=\frac{qL}{2}\tag{3.351}$$

whereas the *exact* solution is

$$\theta(x)=\frac{(2M_0-qL^2)x}{2EI}+\frac{qLx^2}{2EI}-\frac{qx^3}{6EI},\qquad \lambda=q(L-x)\tag{3.352}$$

and $w(x)$ is given by Eq. (3.344).

A two-parameter approximation of the form

$$\theta(x)=ax+a_2x^2,\qquad w(x)=b_1x+b_2x^2,\qquad \lambda(x)=c_1+c_2x\tag{3.353}$$

results in the following solution:

$$\theta(x)=\frac{12M_0-5qL^2}{12EI}x+\frac{qL}{4EI}x^2$$

$$w(x)=-\frac{qL^3}{24EI}x+\frac{6M_0-qL^2}{12EI}x^2$$

$$\lambda(x)=q(L-x).\tag{3.354}$$

The maximum slope θ and maximum deflection w (at $x=L$) given by Eq. (3.354) coincide with the exact solution.

It should be noted that approximations chosen for θ, w, and λ *cannot* be arbitrary (even after meeting the essential boundary conditions and continuity requirements). For example, the choice

$$\theta(x)=a_1x,\qquad w(x)=b_1x,\qquad \lambda(x)=c_1+c_2x\tag{3.355}$$

would result in an inconsistent set of equations whose solution is the trivial solution. This implies that there exists a relationship between the number of parameters used in the approximations of θ, w, and λ.

2 (*Mixed Method*) The mixed method involves rewriting a given higher order equation as a pair of lower order equations by introducing secondary dependent variables. This decomposition of a higher order equation into a pair of lower order equations enables one to employ a lower order polynomial for the approximation functions. To fix the ideas, let us consider the equation

$$\frac{d^2}{dx^2}\left(EI\frac{d^2w}{dx^2}\right)+q=0, \qquad 0<x<L.$$

Introducing the bending moment M as an additional variable, we write

$$\frac{M(x)}{EI}=\frac{d^2w}{dx^2}, \qquad \frac{d^2M(x)}{dx^2}+q=0, \qquad 0<x<L. \tag{3.356}$$

The functional corresponding to Eq. (3.356) is given by

$$I(w, M)=\int_0^L\left(\frac{dw}{dx}\frac{dM}{dx}+\frac{M^2}{2EI}-qw\right)dx. \tag{3.357}$$

The essential boundary conditions involve specifying w and M at $x=0, L$. In the present case the *specified* essential boundary conditions are [see Eq. (3.330)] $w(0)=0$, $M(L)=M_0$.

We seek a two-parameter approximation for w and M:

$$w(x)=\phi_0+a_1\phi_1+a_2\phi_2, \qquad M(x)=\psi_0+b_1\psi_1+b_2\psi_2 \tag{3.358}$$

with $\phi_0=0$, $\phi_1=x$, $\phi_2=x^2$, $\psi_0=M_0$, $\psi_1=x-L$, and $\psi_2=(x-L)^2$. Substituting into δI, we have

$$b_1=0, \qquad b_2=\frac{-q}{2}, \qquad a_1=-\frac{11qL^3}{120EI}, \qquad a_2=\frac{M_0}{2EI}-\frac{3qL^2}{40EI}. \tag{3.359}$$

The solution is

$$w(x)=-\frac{11qL^3}{120EI}x+\left(\frac{M_0}{2EI}-\frac{3qL^2}{40EI}\right)x^2$$

$$M(x)=-\left(\frac{q}{2}\right)(x-L)^2+M_0. \tag{3.360}$$

The Ritz solution for $M(x)$ coincides with the exact solution, whereas the Ritz solution and exact solutions for the deflection $w(x)$ at $x=L$ are

$$w(L)=\frac{3ML^2-qL^4}{6EI}, \qquad w_{\text{exact}}(L)=\frac{2ML^2-qL^4}{8EI}. \quad ■$$

Example 3.28 Consider the steady heat conduction with uniform energy generation Q in a square plate:

$$
\begin{aligned}
-\nabla^2 T &= Q && \text{in } R=\{(x,y):\ 0<x,y<1\}\\
T &= 0 && \text{on sides } x=1, \text{ and } y=1\\
\frac{\partial T}{\partial n} &= 0 && \text{on sides } x=0, \text{ and } y=0.
\end{aligned}
\tag{3.361}
$$

The operator $A=-\nabla^2$ is self-adjoint and positive on $\mathfrak{D}_A$, the linear space of functions from $H^2(\Omega)$ which satisfy the boundary conditions in Eq. (3.361). The null space $\mathfrak{N}(A^*)$ consists of functions of the form

$$v=a+bx+cy.$$

Since $v\in\mathfrak{D}_A$, we have

$$v(1,y)=v(x,1)=\frac{\partial v}{\partial x}(0,y)=\frac{\partial v}{\partial y}(x,0)=0.$$

These conditions imply $v=0$. Thus $\mathfrak{N}(A^*)$ is trivial, and therefore Eq. (3.361) has a unique solution. The same conclusion can be reached by showing that the bilinear form associated with the problem (see below) satisfies the conditions of the Lax-Milgram theorem, Theorem 2.19, on the space $H_A=\{u:\ u\in H^1(\Omega), u(1,y)=u(x,1)=0\}$.

We select the following Ritz approximation:

$$T_n=\sum_i^n c_i\phi_i,\qquad \phi_i=\cos\frac{(2i-1)\pi x}{2}\cos\frac{(2i-1)\pi y}{2},\qquad i=1,2,\ldots. \tag{3.362}$$

Incidentally, the ϕ_i also satisfy the natural boundary conditions of the problem. While the choice $\phi_i=\sin i\pi x\cdot\sin i\pi y$ meets the essential boundary conditions, it is not complete because it cannot be used to generate the solution that does not vanish on the sides $x=0$ and $y=0$.

The functional associated with Eq. (3.361) is given by

$$I(T)=\frac{1}{2}\int_0^1\int_0^1\left[\left(\frac{\partial T}{\partial x}\right)^2+\left(\frac{\partial T}{\partial y}\right)^2-2QT\right]dx\,dy. \tag{3.363}$$

The bilinear and linear forms are given by

$$
\begin{aligned}
B(u,v)&=\int_0^1\int_0^1\left(\frac{\partial u}{\partial x}\frac{\partial v}{\partial x}+\frac{\partial u}{\partial y}\frac{\partial v}{\partial y}\right)dx\,dy\\
l(v)&=\int_0^1\int_0^1 Qv\,dx\,dy.
\end{aligned}
\tag{3.364}
$$

The coefficient matrix b_{ij} and source vector l_i can be computed using ϕ_i in Eqs. (3.364):

$$b_{ij} = -\int_0^1 \int_0^1 \alpha_i \alpha_j (f_i f_j \cdot g_i g_j + p_i p_j \cdot q_i q_j)\, dx\, dy$$

where $f_i = \sin \alpha_i x$, $g_i = \cos \alpha_i y$, $p_i = \cos \alpha_i x$, $q_i = \sin \alpha_i y$, and $\alpha_i = [(2i-1)\pi/2]$. Carrying the integration we obtain

$$b_{ij} = \begin{cases} \dfrac{\alpha_i^2}{2}, & \text{if } i=j \\ \dfrac{\alpha_i^2 \alpha_j^2}{2(\alpha_i^2 - \alpha_j^2)^2}, & \text{if } i \neq j, \end{cases} \tag{3.365a}$$

$$l(\phi_i) = \frac{Q}{\alpha_i^2}. \tag{3.365b}$$

For the one-parameter approximation we get

$$b_{11} = \frac{\alpha_1^2}{2}, \qquad l(\phi_1) = \frac{Q}{\alpha_1^2}, \qquad c_1 = \frac{32Q}{\pi^4}. \tag{3.366}$$

Hence the solution is given by

$$T_1 = \frac{32Q}{\pi^4} \cos\left(\frac{\pi x}{2}\right) \cos\left(\frac{\pi y}{2}\right). \tag{3.367}$$

The exact solution to this problem is given by

$$T(x,y) = \frac{Q}{2}\left\{(1-y^2) + \frac{32}{\pi^3} \sum_{n=0}^{\infty} \frac{(-1)^n \cos[(2n+1)\pi y/2] \cosh[(2n+1)\pi x/2]}{(2n-1)^3 \cosh[(2n+1)\pi/2]}\right\}. \tag{3.368}$$

The one parameter Ritz solution at the center of the plate is 0.1643, whereas the exact solution is 0.1811 (9.3% error). The accuracy can be improved by selecting more parameters. For example, for $n=2$ we get

$$T_2 = Q\left[0.3283988 \cos\left(\frac{\pi x}{2}\right) \cos\left(\frac{\pi y}{2}\right) + 0.001976 \cos\left(\frac{3\pi x}{2}\right) \cos\left(\frac{3\pi y}{2}\right)\right],$$

which differs from the exact solution at the center of the plate only by 0.54%. Table 3.3 shows the Ritz coefficients for $n = 1, 2, \ldots, 6$.

Table 3.3 Ritz Coefficients in Eq. (3.362) for the Eq. (3.361)

n	c_1/Q	c_2/Q	c_3/Q	c_4/Q	c_5/Q	c_6/Q
1	0.32851					
2	0.328399	0.001976				
3	0.328395	0.001966	0.000267			
4	0.328394	0.001965	0.000264	0.000070		
5	0.328394	0.001965	0.000264	0.000069	0.000026	
6	0.328394	0.001965	0.000264	0.000069	0.000025	0.000012

If algebraic polynomials are to be used, one can choose $\phi_1=(1-x)(1-y)$ or $\phi_1=(1-x^2)(1-y^2)$, both of which satisfy the homogeneous essential boundary conditions. Obviously the choice $\phi_1=(1-x)(1-y)$ does not meet the natural boundary conditions of the problem.

For the one-parameter approximation by $\phi_1=(1-x^2)(1-y^2)$ we get

$$T_1(x,y)=\frac{5Q}{16}(1-x^2)(1-y^2). \tag{3.369}$$

This gives a value of $0.17578Q$ for T at the center of the plate. This is an error of 2.94% when compared to the exact solution.

For the one-parameter approximation by $\phi_1=(1-x)(1-y)$ we get

$$T_1(x,y)=\frac{3Q}{8}(1-x)(1-y). \tag{3.370}$$

When compared with the exact solution at the center of the plate, this is an error of 48.23%! Of course the solution would improve with a larger number of parameters of the type $\phi_i=(1-x^i)(1-y^i)$. ■

The next example is concerned with Dirichlet and Neumann boundary-value problems associated with the Poisson equation, which arises in many problems of engineering (see Exercise 3.5.15).

Example 3.29 Consider the Poisson equation

$$-\nabla^2 u=f \qquad \text{in } \Omega\subset \mathbf{R}^2 \tag{3.371}$$

where Ω is the unit square. We shall consider the Dirichlet and Neumann boundary conditions for the problem.

1 (*Dirichlet Boundary Conditions*) Suppose that

$$u=0 \qquad \text{on } \Gamma. \tag{3.372}$$

The domain of the operator, $A\equiv-\nabla^2$, is given by $\mathfrak{D}_A=H^2(\Omega)\cap H_0^1(\Omega)$.

The bilinear form for the operator is given by

$$B(u,v)=\int_0^1\int_0^1\left(\frac{\partial u}{\partial x}\frac{\partial v}{\partial x}+\frac{\partial u}{\partial y}\frac{\partial v}{\partial y}\right)dxdy.$$

The energy space H_A in the present case is $H_0^1(\Omega)$. The energy norm $\|u\|_A=\sqrt{B(u,u)}$ is equal to the seminorm $|u|_{1,\Omega}$ on $H_0^1(\Omega)$, which is equivalent to the norm $\|u\|_{1,\Omega}$ on $H_0^1(\Omega)$. The operator A is self-adjoint and positive definite on $\mathfrak{D}_A$ (and also in H_A). Hence $\mathfrak{N}(A^*)$ contains only the zero element, and therefore the solvability condition (2.130) is identically satisfied. Hence there exists a unique solution to the problem for any $f\in L_2(\Omega)$.

For the N-parameter Ritz approximation we choose

$$u_N=\sum_{n=1}^{N}\sum_{m=1}^{N}c_{mn}\phi_{mn},\qquad \phi_{mn}=\sin m\pi x\sin n\pi y. \tag{3.373}$$

The coefficient matrix $[b]\equiv[B(\phi_{ij},\phi_{mn})]$ can be computed to be

$$B(\phi_{ij},\phi_{mn})=\begin{cases}\dfrac{\pi^2}{4}(i^2+j^2), & \text{if } m=i \text{ and } n=j\\ 0, & \text{otherwise.}\end{cases} \tag{3.374a}$$

For $f=f_0=\text{constant}$, we get

$$l(\phi_{ij})=\begin{cases}\dfrac{4f_0}{\pi^2 ij}, & \text{if both } i \text{ and } j \text{ are odd}\\ 0, & \text{otherwise.}\end{cases} \tag{3.374b}$$

The solution becomes

$$u_N=\frac{16}{\pi^4}\sum_{m,n=1}^{N}\frac{\sin(2m-1)\pi x\sin(2n-1)\pi y}{(2m-1)(2n-1)\left[4(m^2+n^2)-2(m+n)+2\right]} \tag{3.375}$$

which coincides with the exact solution.

For $f=\cos\pi x$, we get

$$l(\phi_{ij})=\begin{cases}\dfrac{8}{\pi^2}\dfrac{1}{j(i^2-1)}, & \text{if } j \text{ is odd and } i \text{ is even}\\ 0, & \text{otherwise.}\end{cases} \tag{3.376}$$

The solution then becomes

$$u_N=\frac{32}{\pi^4}\sum_{j=1,3,\ldots}^{N}\;\sum_{i=2,4,\ldots}^{N}\frac{\sin i\pi x\sin j\pi y}{j(i^2-1)(i^2+j^2)}. \tag{3.377}$$

2 (*Neumann Boundary Conditions*) Now consider the boundary conditions

$$\frac{\partial u}{\partial n}=0\left(\frac{\partial u}{\partial x}=0 \text{ at } x=0,1;\ \frac{\partial u}{\partial y}=0 \text{ at } y=0,1\right). \tag{3.378}$$

The domain of the operator is $\mathfrak{D}_A$ contains functions from $H^2(\Omega)$ whose normal derivatives vanish on Γ, while the energy space is $H_A=H^1(\Omega)$. Then $\sqrt{B(u,u)}$ does not satisfy the axioms of a norm [because the null space $\mathfrak{N}(A^*)$ contains functions that are constant in Ω]. In order to eliminate the constants from the domain $\mathfrak{D}_A$, we define

$$\mathfrak{D}_A \equiv \hat{H}^2(\Omega)=\left\{u\in H^2(\Omega),\quad \int_\Omega u(x,y)\,dxdy=0,\ \frac{\partial u}{\partial n}=0 \text{ on } \Gamma\right\}. \tag{3.379}$$

Then

$$\int_\Omega\left|\frac{\partial u}{\partial x}\right|^2 dxdy=0,\qquad \int_\Omega\left|\frac{\partial u}{\partial y}\right|^2 dxdy=0\qquad \text{implies } u=\text{constant}$$

and since $u\in\hat{H}^2(\Omega)$, the constant must be zero. Consequently $\sqrt{B(u,u)}=\|u\|_A$ defines a norm on $\hat{H}^2(\Omega)$. Then the Neumann problem has a unique solution for all $f\in L_2(\Omega)$ such that (solvability condition)

$$\int_\Omega f(x,y)\,dxdy=0. \tag{3.380}$$

Clearly, in the present case f cannot be a nonzero constant; $f=\cos\pi x$ satisfies the solvability condition.

For the Ritz approximation we choose

$$u_N=\sum_{i=0}^{N}\sum_{j=0}^{N}c_{ij}\cos i\pi x\cos j\pi y. \tag{3.381}$$

Upon substituting into the variational problem $B(u,v)=l(v)$, we get the system of equations

$$\frac{\pi^2}{4}(a^2j^2+b^2i^2)c_{ij}=\int_0^1\int_0^1 f\cos i\pi x\cos j\pi y\,dxdy$$

$$=\begin{cases}\frac{1}{2}, & i=1,\ j=0\\ 0, & \text{otherwise.}\end{cases} \tag{3.382}$$

Thus we have

$$u_N=\frac{1}{\pi^2}\cos\pi x,\qquad 0<x<1 \tag{3.383}$$

which coincides with the classical solution of the Neumann problem for $f=\cos\pi x$. ■

The next section is devoted to the discussion of the Galerkin method, which is a generalization of the Ritz method. The Galerkin method can be applied to problems in which the underlying operator is not positive definite and even a nonlinear operator.

3.6.3 The Galerkin Method

Consider a separable Hilbert space H and let S be a dense linear subspace of H. From Lemma 2.2, if for some element $u \in H$

$$(u, v) = 0 \qquad \text{holds for every } v \in S,$$

then it follows that $u = 0$ in H. If $\{\phi_i\}$ is a basis in H, then

$$(u, \phi_k) = 0, \qquad \text{for all } k \Rightarrow u = 0 \text{ in } H. \tag{3.384}$$

Now consider the operator equation $A\colon \mathfrak{D}_A \subset H \to H$,

$$Au = f \qquad \text{in } \Omega. \tag{3.385}$$

If $u_0 \in \mathfrak{D}_A$ is such that

$$(Au_0 - f, \phi_k) = 0 \qquad \text{for every } k = 1, 2, \ldots, \tag{3.386}$$

then we have $Au_0 - f = 0$ in H, that is, u_0 is the solution of Eq. (3.385) in H. In other words, finding the solution of Eq. (3.386) is equivalent to finding the solution of Eq. (3.385). This equivalence forms the basis of the Galerkin method.

Consider the operator Eq. (3.385) and let $\mathfrak{D}_A$ be such that it contains all linear combinations of the form

$$u = \sum_i \alpha_i \phi_i.$$

In the Galerkin method we look for an approximate solution $u_n \in \mathfrak{D}_A$ of Eq. (3.385) in the form

$$u_N = \sum_i^N c_i \phi_i \tag{3.387}$$

where N is an arbitrary but fixed positive integer and c_i are constants to be determined using Eq. (3.386):

$$(Au_N - f, \phi_k) = 0, \qquad k = 1, 2, \ldots. \tag{3.388}$$

This gives N equations for the N unknown constants $c_1, c_2, \ldots, c_N$.

If the operator A is linear, then Eq. (3.388) becomes

$$\sum_{i}^{N}(A\phi_i,\phi_k)c_i=(f,\phi_k), \qquad k=1,2,\ldots. \tag{3.389}$$

Further, if A is symmetric, Eq. (3.389) becomes

$$\sum_{i}^{N}(\phi_i,A\phi_k)c_i=(f,\phi_k), \qquad k=1,2,\ldots, \tag{3.390}$$

which is the same as the system of Eqs. (3.295) obtained in the Ritz procedure.

It should be noted that the Galerkin method is applicable to a much larger class of operator equations than the positive-definite linear operator equations, which can be solved by the Ritz method. In the Galerkin method, the most general form of which is given by Eq. (3.388), the operator A is not restricted to being positive definite, symmetric, or even linear. Due to this general nature of the Galerkin procedure, the questions of solvability, existence, and uniqueness of solutions, as one might suspect, are more difficult, in general.

Although the Ritz and Galerkin methods lead to the same results in the case of linear positive-definite operators, we note that the basic ideas underlying these methods are distinctly different. The Ritz method is essentially based on the weak form resulting from the quadratic functional associated with a given equation, whereas the Galerkin method is based on the given equation itself. Use of the weak form relaxes the continuity requirement of the coordinate functions ($\phi_i \in H_A$), and imposes only the essential boundary conditions of the coordinate functions. On the other hand, the Galerkin method seeks solutions from $\mathfrak{D}_A$ (hence, ϕ_i must satisfy *all* of the boundary conditions of the problem).

Petrov-Galerkin Method In the Galerkin method the same basis was employed both for the coordinate functions and in the requirement (3.386). The modification of the Galerkin method which uses two different bases, one for the coordinate functions and another for ϕ_k in Eq. (3.386), is known as the Petrov-Galerkin method:

$$(Au_N-f,\psi_k)=0, \qquad k=1,2,\ldots,N \tag{3.391}$$

where u_N is given by Eq. (3.387).

For the Galerkin method and the Petrov-Galerkin method, the coordinate functions $\{\phi_i\}$ must satisfy conditions analogous to those of Eq. (3.291):

1 $\{\phi_i\}\subset \mathfrak{D}_A$.
2 For arbitrary N, the set $\{\phi_i\}_{i=1}^{N}$ is linearly independent. (3.392)
3 $\{\phi_i\}$ is complete in $\mathfrak{D}_A$.

One should note the difference between requirement 1 of Eq. (3.291) and that of Eq. (3.392). The difference in the spaces in which the solutions are sought indicates the basic difference between the two methods. The space $\mathfrak{D}_A$ used in the Galerkin method is more restrictive than the energy space H_A used in the Ritz method. For example, if A is a $2m$th-order differential operator, then $\mathfrak{D}_A \subset H^{2m}(\Omega)$, whereas $H_A \subset H^m(\Omega)$. Further, functions in $\mathfrak{D}_A$ satisfy all of the boundary conditions of the problem while functions in H_A satisfy only the essential boundary conditions of the problem. The restrictions on $\mathfrak{D}_A$ can be relaxed in most problems of interest by trading the differentiation between ϕ_k and ψ_k as equally as the problem permits. In most problems of interest, the highest derivatives are of even order. Therefore one can use the integration by parts to transfer half of the differentiation to ψ_k. The resulting weak form admits so-called *generalized solutions* (i.e., solutions in the variational sense) in H_A.

To fix the ideas, consider the differential operator (see Exercise 2.6.20)

$$A = \sum_{0 \leq |\boldsymbol{\beta}|, |\boldsymbol{\alpha}| \leq 2m} (-1)^{|\boldsymbol{\alpha}|} D^{\boldsymbol{\alpha}}\left[a^{\boldsymbol{\alpha\beta}}(\mathbf{x}) D^{\boldsymbol{\beta}}(\cdot)\right] \tag{3.393}$$

of order $2m$. We consider the problem of solving Eq. (3.385) [with A given by Eq. (3.393)], subjected to the homogeneous boundary conditions (recall that a linear operator equation with nonhomogeneous boundary conditions can be reformulated as an equivalent operator equation with homogeneous boundary conditions)

$$B_j u = 0, \qquad j = 0, 1, 2, \ldots, 2m-1. \tag{3.394}$$

Here the B_j are boundary operators compatible with the form of A. The first m conditions (i.e., $j = 0, 1, \ldots, m-1$) contains the derivatives of u up to and including order $m-1$ and constitute the essential boundary conditions. The remaining m conditions involve derivatives from m up to $2m-1$ (both inclusive), and constitute the natural boundary conditions for the problem. Then the spaces $\mathfrak{D}_A$ and H_A are defined by

$$\mathfrak{D}_A = \left\{u\colon\ u \in H^{2m}(\Omega),\, D^{\boldsymbol{\alpha}} u = 0 \text{ on } \Gamma, |\boldsymbol{\alpha}| \leq 2m-1\right\} = H_0^{2m}(\Omega)$$

$$H_A = \{u\colon\ u \in H^m(\Omega),\, D^{\boldsymbol{\alpha}} u = 0 \text{ on } \Gamma, |\boldsymbol{\alpha}| \leq m-1\} = H_0^m(\Omega). \tag{3.395}$$

The Galerkin and the Ritz methods applied to the problem at hand give the following systems of equations:

$$B_G(u_N, \phi_k) = (f, \phi_k)_0, \qquad \text{for } \phi_k \in H_0^{2m}(\Omega)$$

$$B_R(u_N, \phi_k) = (f, \phi_k)_0, \qquad \text{for } \phi_k \in H_0^m(\Omega) \tag{3.396}$$

for all $k=1,2,\ldots,N$, where

$$B_G(\phi_i,\phi_j)=\int_\Omega A(\phi_i)\phi_j\,dx\,dy$$

$$B_R(\phi_i,\phi_j)=\int_\Omega \sum_{0\le|\alpha|,|\beta|\le m} D^\alpha\phi_i D^\beta\phi_j\,dx\,dy. \tag{3.397}$$

It is clear that in the present case (because A is self-adjoint) we can integrate the expression in $B_G(.,.)$ m times by parts to reduce it to B_R, and thereby an approximate solution u_N can be sought in $H_0^m(\Omega)$.

Next consider the nonlinear operator equation of the form

$$\begin{aligned}\mathcal{P}(u)&\equiv N(u)+A(u)=f \qquad \text{in } \Omega\\ B(u)&=0 \qquad \text{on } \Gamma\end{aligned} \tag{3.398}$$

where A and B are operators defined in Eqs. (3.393) and (3.394), respectively, and N is the nonlinear operator of order m:

$$N(u)=u\sum_{|\alpha|\le m} D^\alpha u. \tag{3.399}$$

The variational problem associated with Eq. (3.398) involves finding $u\in H_0^m(\Omega)$ such that

$$B_N(u_N,\phi_k)+B_R(u_N,\phi_k)=(f,\phi_k)_0, \qquad \text{for any } k \tag{3.400}$$

where

$$B_N(\phi_i,\phi_k)=(N(u_N),\phi_k)_0. \tag{3.401}$$

One can interpret Eq. (3.400) as one obtained by applying the Galerkin method to Eq. (3.398), except that the resulting bilinear form $B_G(.,.)$ is converted to $B_R(.,.)$. To make a distinction between the usual Galerkin method, and the Galerkin method in which the bilinear form is defined on a larger (i.e., less restrictive) space, we will call the latter the *generalized Galerkin method*.

With these comments we now turn to some examples of application of the Galerkin method and its variants.

Example 3.30 Consider the nonlinear differential equation

$$\begin{aligned} u\frac{d^2u}{dx^2}+\left(\frac{du}{dx}\right)^2&=1, \qquad 0<x<1\\ u'(0)=0, \qquad u(1)&=\sqrt{2}.\end{aligned} \tag{3.402}$$

The domain of the operator (which is nonlinear) is the set $\mathfrak{D}$ of twice differentiable functions that satisfy the boundary conditions (3.402). The set is not a linear vector space. Let $u = w + \sqrt{2}$ in Eq. (3.402). We get

$$\left(w+\sqrt{2}\right)\frac{d^2w}{dx^2}+\left(\frac{dw}{dx}\right)^2=1, \qquad 0<x<1$$

$$w'(0)=0, \qquad w(1)=0. \tag{3.403}$$

The domain of the new operator [defined by Eq. (3.403)] is the set $\mathfrak{D}_A$ of twice differentiable functions which satisfy the homogeneous boundary conditions (3.403). The set constitutes a linear subspace of $H^2(0,1)$. In the present text we did not develop any theoretical results to establish that a given nonlinear problem has a solution, and if it has one, it may not be unique. Here we are primarily interested in the Ritz and Galerkin approximations to Eq. (3.403).

First we recast Eq. (3.403) in the more familiar form

$$-\frac{d}{dx}\left[\left(w+\sqrt{2}\right)\frac{dw}{dx}\right]+1=0. \tag{3.404}$$

1 (*Galerkin Method*) In the Galerkin method we must select the coordinate functions from $\mathfrak{D}_A$. Suppose that we select algebraic polynomials for the coordinates functions

$$w_N(x)=\sum_i^N c_i\phi_i, \qquad \phi_i=1-x^{i+1}. \tag{3.405}$$

We illustrate the steps involved in obtaining the Galerkin solution by taking $N=1$ (in the interest of algebraic simplicity). We have

$$A(w_1)=-\left[c_1(1-x^2)+\sqrt{2}\right](-2c_1)-(2xc_1)^2$$

$$=\sqrt{2}\,2c_1+2c_1{}^2-6x^2c_1{}^2$$

$$\left(A(w_1),\phi_1\right)_0=\int_0^1 A(w_1)(1-x^2)\,dx$$

$$=\tfrac{4}{3}\sqrt{2}\,c_1+\tfrac{8}{15}c_1{}^2$$

$$(f,\phi_1)_0=\int_0^1(-1)(1-x^2)\,dx=-\tfrac{2}{3}. \tag{3.406}$$

Thus we have a quadratic expression (as a result of the nonlinear equation) in c_1. Solving for c_1, we get

$$c_1^{(1)} = \frac{-5\sqrt{2}+\sqrt{30}}{4} \approx -0.39846$$

$$c_1^{(2)} = \frac{-5\sqrt{2}-\sqrt{30}}{4} \approx -3.13707. \tag{3.407}$$

Thus there are two solutions to the variational problem $(A(w_1), \phi_1) = (f, \phi_1)$. We must use some criteria to discard one of the two constants calculated above. Here we compute the L_2 norm of the residual, $R = A(w_1) - f$ using $c_1^{(1)}$ and $c_1^{(2)}$. This gives $R_1 < R_2$, where R_i is the residual obtained by using $c_1^{(i)}$. Consequently the one-parameter Galerkin solution to Eq. (3.389) is

$$u_1 \equiv w_1 + \sqrt{2} = 1.01575 + 0.39846x^2. \tag{3.408}$$

The exact solution of Eq. (3.389) is given by

$$u_0(x) = \sqrt{1+x^2} \approx 1 + \tfrac{1}{2}x^2 + \cdots. \tag{3.409}$$

The one-parameter solution is in fairly good agreement with the first two terms of the exact solution.

If we solve the differential Eq. (3.402) using the Ritz method, we must select the approximate solution u_N in the form

$$u_N = \phi_0 + \sum_i^N c_i \phi_i \tag{3.410}$$

where ϕ_0 must satisfy the specified boundary conditions. For the problem at hand, we must select $\phi_0 = \sqrt{2}\,x^2$ so that $\phi_0(1) = \sqrt{2}$ and $\phi_0'(0) = 0$. For $N = 1$, the parameters $\hat{c}_1^{(1)}$ and $\hat{c}_1^{(2)}$ are computed to be

$$\hat{c}_1^{(1)} = \frac{1}{2\sqrt{2}}\left(\sqrt{15}-1\right), \qquad \hat{c}_1^{(2)} = -\frac{1}{2\sqrt{2}}\left(1+\sqrt{15}\right). \tag{3.411}$$

Using the minimum residual criteria, we select $\hat{c}_1^{(1)}$, which gives

$$u = \sqrt{2}\,x^2 + 1.01575\,(1 - x^2),$$

which is *exactly* the same as Eq. (3.408).

2 (*Generalized Galerkin Method*) Although the differential Eq. (3.403) is nonlinear, the highest derivative in the equation is of even order. Therefore we can integrate the first term by parts to weaken the requirement on the coordinate functions. We have

$$(A(w_1), \phi_k)_0 = \int_0^1 \left[\left(w_1 + \sqrt{2}\right)\frac{dw_1}{dx}\right]\frac{d\phi_k}{dx}\,dx \tag{3.412}$$

where ϕ_k is such that it satisfies the essential boundary condition $\phi_k(1)=0$. The natural boundary condition, being homogeneous, is not seen in the weak form.

Since ϕ_i must satisfy only $\phi_i(1)=0$, we write

$$w_N(x) = \sum_i^N c_i\phi_i, \qquad \phi_i = 1 - x^i. \tag{3.413}$$

Substituting Eq. (3.413) in Eq. (3.412) and carrying the indicated integration, we obtain

$$c^2 + 2\sqrt{2}\,c + 1 = 0, \qquad c_1^{(1)} = 1 - \sqrt{2}\,, \qquad c_1^{(2)} = -1 - \sqrt{2}\,. \tag{3.414}$$

Using the same criterion as before (i.e., the minimum residual criterion in the L_2 sense), we select $c_1^{(1)}$ so that

$$u_1 \equiv w_1 + \sqrt{2} = 1 - 0.41421x. \tag{3.415}$$

The one-parameter solution is not good enough (especially for larger values of $x, 0 < x < 1$), but one can get better accuracy by taking more parameters in the approximation. When $N \geq 2$, the selection of the right parameters from $c_i^{(1)}$ and $c_i^{(2)}$ is not easy. One must use numerical methods, such as the Newton's method, to solve the nonlinear algebraic equations.

3 (*Petrov-Galerkin Method*) Let us consider the approximation in Eq. (3.405) with $N=1$. In the Petrov-Galerkin method we must select the "weight" function ψ_k in $(A(w_N) - f, \psi_k) = 0$ to be different from the coordinate functions ϕ_k. In the present case we select

$$\psi_1 = \cos\frac{\pi x}{2} \qquad \text{so that} \qquad \psi_1(1) = 0, \qquad \psi_1'(0) = 0. \tag{3.416}$$

We have

$$(A(w_1), \psi_1) = \frac{1}{\pi}\left[4\sqrt{2}\,c_1 + \left(-8 + \frac{96}{\pi^2}\right)c_1^2\right]$$

$$(f, \psi_1) = -\frac{2}{\pi}. \tag{3.417}$$

Solving for c_1, we get

$$c_1^{(1,2)} = \frac{-2\sqrt{2} \pm \sqrt{8-2(16/\pi^2-8)}}{(16/\pi^2-8)},$$

$$c_1^{(1)} = -0.40317, \qquad c_2^{(1)} = -2.87268. \tag{3.418}$$

Once again we use the "minimum residual" criterion and select $c_1^{(1)}$. The solution is given by

$$u_1 \equiv w_1 + \sqrt{2} = 1.01104 + 0.40317x^2, \tag{3.419}$$

which is a slight improvement on the Galerkin solution. It must be noted that we have not established that the Petrov-Galerkin method yields better accuracy than the Galerkin method or vice versa. ■

Example 3.31 Consider the boundary-value problem

$$-k\nabla^2 u = f \qquad \text{in } R$$

$$u = 0 \qquad \text{on } C, \tag{3.420}$$

which arises in heat conduction, fluid mechanics, and solid mechanics, among other areas of engineering (see Exercise 3.4.15). Let the domain R be an equilateral triangle as shown in Fig. 3.15.

We wish to construct a one-parameter Galerkin solution of the problem. Since $u=0$ on C, we choose $\phi_0 = 0$. We must choose ϕ_1 such that it is zero on C(i.e., on AB, BC, and CA) and differentiable twice with respect to x and y. In the process of constructing a ϕ_1 that meets the homogeneous boundary conditions, usually the continuity conditions are automatically met. If we choose ϕ_1 to be the equation of the boundary, then it automatically vanishes

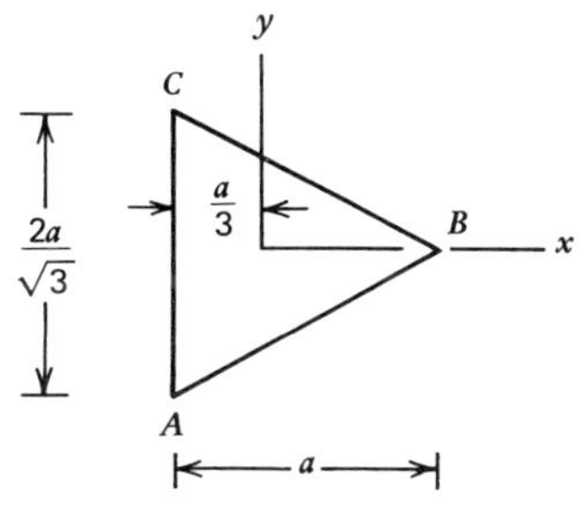

Figure 3.15 Poisson's equation on an equilateral triangle.

on the boundary. The equations of lines AB, BC, and CA are, respectively,

$$x-\sqrt{3}\,y-\tfrac{2}{3}a=0$$

$$x+\sqrt{3}\,y-\tfrac{2}{3}a=0$$

$$x+\tfrac{1}{3}a=0.$$

We choose

$$\phi_1=\left(x-\sqrt{3}\,y-\tfrac{2}{3}a\right)\left(x+\sqrt{3}\,y-\tfrac{2}{3}a\right)\left(x+\tfrac{1}{3}a\right)\left(-\frac{1}{2a}\right)$$

$$=\frac{1}{2}(x^2+y^2)-\frac{1}{2a}(x^3-3xy^2)-\frac{2}{27}a^2, \qquad (3.421)$$

which is zero on any of the three line segments but assumes a nonzero value inside the triangular domain.

We have

$$\nabla^2\phi_1=1+1-\frac{1}{2a}[6x-6x]=2.$$

Substituting into the Galerkin integral, we obtain

$$\int_R\left(-ck\nabla^2\phi_1-f\right)\phi_1\,dxdy=0$$

$$\int_R(-2ck-f)\phi_1\,dxdy=0 \qquad \text{or} \qquad c=-\frac{f}{2k}.$$

Hence the solution is

$$u=-\frac{f}{2k}\left[\frac{1}{2}(x^2+y^2)-\frac{1}{2a}(x^3-3xy^2)-\frac{2}{27}a^2\right]. \quad \blacksquare \qquad (3.422)$$

The next section is concerned with various special cases of the Petrov-Galerkin method (for various choices of the weight function).

3.6.4 Least-Squares, Collocation, Courant, and Weighted-Residuals Methods

Least-Squares Method Recall from Section 2.7.4 that the least-squares method consists in determining an approximate solution u_N of the equation $Au=f$ in the form

$$u_N=\sum_{i}^{N}c_i\phi_i \qquad (3.423)$$

where the constants c_i are to be determined from the condition

$$e^2 \equiv \| Au_N - f \|^2 = \text{minimum}. \tag{3.424}$$

Here the norm $\|\cdot\|$ is taken in the domain space $\mathfrak{D}_A$. It is assumed that the set $\{A\phi_i\}$ is also linearly independent (it will be apparent in the following discussion why this assumption is imposed). Substituting for u_N from Eq. (3.423) into Eq. (3.424) and using the necessary condition for the minimum of the function e, we get

$$\frac{\partial e^2}{\partial c_i} = 2(Au_N - f, A\phi_k)_0 = 0 \tag{3.425}$$

and if A is linear, we get

$$\sum_i^N (A\phi_k, A\phi_i)_0 c_i = (\phi_k, f)_0. \tag{3.426}$$

Since the $\{A\phi_i\}$ are linearly independent, the inverse of the coefficient matrix exits, and the system (3.426) is uniquely solvable.

It should be pointed out that the least-squares method yields a symmetric coefficient matrix for any linear operator (i.e., whether or not A is symmetric). For linear positive-definite operators A, the sequence $\{u_N\}$ converges in $\mathfrak{D}_A$ to the true solution u_0 [under the conditions (3.392) on ϕ_i]. The main advantage of the least-squares method lies (apart from the symmetric coefficient matrix) in the fact that not only u_N converges to u_0 in $\mathfrak{D}_A$, but also Au_N converges to f. The error in the approximate solution u_N can be estimated according to Eq. (3.302):

$$\| u_N - u_0 \|_A \leq \frac{1}{c_0} \| Au_N - f \|_0 \tag{3.427}$$

where $\| u \|_A^2 = (Au, Au)_0$, and c_0 is a constant that depends on the domain of the problem.

Collocation Method The collocation method seeks an approximate solution u_N to $Au - f = 0$ in the form (3.423) by requiring the residual in the equation $R \equiv Au_N - f$ to be identically zero at N selected points $\mathbf{x}^i$, $i = 1, 2, \ldots, N$, in the domain

$$R\big(u_N(\mathbf{x}^i)\big) = 0, \qquad i = 1, 2, \ldots, N. \tag{3.428}$$

The selection of the points $\mathbf{x}^i$ is crucial in obtaining a well-conditioned system of equations and ultimately in obtaining an accurate solution.

Courant Method The so-called Courant method combines the basic concepts of the Ritz and the least-squares methods (for linear operator equations). The method seeks an approximate solution u_N in the form (3.423) by requiring the modified functional

$$\tilde{I}(u)=I(u)+\tfrac{1}{2}\|Au-f\|^2, \qquad u\in\mathfrak{D}_A \tag{3.429}$$

where $I(u)$ is the quadratic functional associated with $Au=f$, to attain a minimum on $\mathfrak{D}_A$. Obviously the statement makes sense only for operator equations that admit functional formulation. While the method requires a laborious calculation, the method improves convergence (i.e., as $N\to\infty$).

One must immediately note that the Courant method is nothing but the same as the penalty method (with the penalty parameter $\varepsilon=1$) described earlier in this chapter. In the present case the condition $Au-f=0$ can be viewed as the constraint.

Weighted-Residuals Method (Petrov-Galerkin Method) Before looking at some examples of the application of various methods, it is informative to know how various variational methods discussed thus far can be viewed as special cases of the Petrov-Galerkin method (or better known as the *weighted-residuals method*).

By comparing Eq. (3.426) with Eq. (3.391) one can easily see that the least-squares method is a special case of the Petrov-Galerkin method in which the weight function ψ_k is equal to $A\phi_k$. Similarly, the collocation method can be shown (not as obviously as in the least-squares method) to be a special case of the Petrov-Galerkin method with $\psi_k=\delta(\mathbf{x}-\mathbf{x}^k)$, where $\delta(\mathbf{x})$ is the Dirac delta function,

$$\int_\Omega f(\mathbf{x})\delta(\mathbf{x}-\xi)\,d\mathbf{x}\equiv f(\xi). \tag{3.430}$$

We now consider several examples of the application of the least-squares and collocation methods.

Example 3.32 Consider the equilibrium of a flexible beam subjected to a distributed load while resting on a continuous elastic foundation, (see Fig. 3.16). The governing equation is

$$\frac{d^2}{dx^2}\left(EI\frac{d^2w}{dx^2}\right)+kw+q_0=0, \qquad 0<x<L. \tag{3.431}$$

The boundary conditions for the problem are given by

$$w(0)=w(L)=0, \qquad \frac{d^2w}{dx^2}(0)=\frac{d^2w}{dx^2}(L)=0. \tag{3.432}$$

From the previous discussions we know that the first two boundary conditions

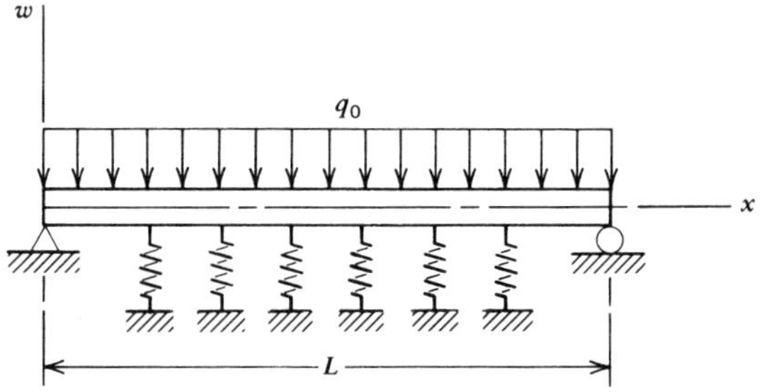

Figure 3.16 Beam on an elastic foundation.

in Eq. (3.432) are of the essential type, while the last two are of the natural type.

We now consider a two-parameter solution by various special cases of the Petrov-Galerkin method.

We assume an approximation of the form

$$\begin{aligned} w_2 &= c_1\phi_1 + c_2\phi_2 \\ &= c_1 \sin\left(\frac{\pi x}{L}\right) + c_2 \sin\left(\frac{3\pi x}{L}\right). \end{aligned} \tag{3.433}$$

Functions ϕ_1 and ϕ_2 satisfy all of the boundary conditions (w and d^2w/dx^2 are zero at $x=0, L$). The residual of the approximation is

$$\begin{aligned} R(c_1, c_2) &= \frac{d^2}{dx^2}\left(EI\frac{d^2 w_2}{dx^2}\right) + kw_2 + q_0 \\ &= \left[EI\left(\frac{\pi}{L}\right)^4 + k\right] c_1 \sin\left(\frac{\pi x}{L}\right) + \left[EI\left(\frac{3\pi}{L}\right)^4 + k\right] c_2 \sin\left(\frac{3\pi x}{L}\right) + q_0. \end{aligned}$$

1 (*Galerkin Method, or Ritz Method*)

$$\int_0^L R(c_1, c_2)\phi_1\, dx = L\left[EI\left(\frac{\pi}{L}\right)^4 + k\right]\frac{c_1}{2} + q_0\frac{2L}{\pi} = 0$$

$$\int_0^L R(c_1, c_2)\phi_2\, dx = L\left[EI\left(\frac{3\pi}{L}\right)^4 + k\right]\frac{c_2}{2} + q_0\frac{2L}{3\pi} = 0.$$

Solving for c_1 and c_2, we obtain

$$c_1 = -\frac{4L^4 q_0}{\pi(EI\pi^4 + kL^4)}, \qquad c_2 = -\frac{4L^4 q_0}{3\pi(81EI\pi^4 + kL^4)}. \tag{3.434}$$

2 (*Least-Squares Method*)

$$\int_0^L R\frac{\partial R}{\partial c_1}dx = \int_0^L R\frac{(EI\pi^4 + kL^4)}{L^4}\sin\left(\frac{\pi x}{L}\right)dx = 0$$

$$\int_0^L R\frac{\partial R}{\partial c_2}dx = \int_0^L R\frac{(81EI\pi^4 + kL^4)}{L^4}\sin\left(\frac{3\pi x}{L}\right)dx = 0. \quad (3.435)$$

These equations are just constant multiples of those obtained in the Galerkin method. Hence c_1 and c_2 are the same as above.

3 (*Collocation*) We select $x_1 = L/4$ and $x_2 = L/2$ as the collocation points. We get

$$x_1:\ 0 = \left[EI\left(\frac{\pi}{L}\right)^4 + k\right]c_1\sin\left(\frac{\pi}{4}\right) + \left[EI\left(\frac{3\pi}{L}\right)^4 + k\right]c_2\sin\left(\frac{3\pi}{4}\right) + q_0$$

$$x_2:\ 0 = \left[EI\left(\frac{\pi}{L}\right)^4 + k\right]c_1\sin\left(\frac{\pi}{2}\right) + \left[EI\left(\frac{3\pi}{L}\right)^4 + k\right]c_2\sin\left(\frac{3\pi}{2}\right) + q_0$$

$$(3.436)$$

Solving for c_1 and c_2, we obtain

$$c_1 = -\frac{(1+\sqrt{2})q_0L^4}{2(EI\pi^4 + kL^4)}, \qquad c_2 = -\frac{(\sqrt{2}-1)q_0L^4}{2(81EI\pi^4 + kL^4)}. \quad (3.437)$$

The exact solution of Eq. (3.431) is given by

$$w(x) = -\frac{q_0}{k}\left[1 - \frac{\cosh\alpha x\cos\alpha(x-L) + \cosh\alpha(x-L)\cos\alpha x}{\cosh\alpha L + \cos\alpha L}\right] \quad (3.438)$$

where $\alpha = \sqrt[4]{k/4EI}$. ■

Example 3.33 We consider the nonlinear boundary-value problem of Example 3.30. We wish to solve it by the least-squares method, and choose the same approximation as that used for the Galerkin method,

$$u = \sqrt{2}\,x^2 + c(1-x^2). \quad (3.439)$$

We construct the least-squares approximation of the problem,

$$\frac{d}{dc}\int_0^1\left[\frac{d}{dx}\left(u\frac{du}{dx}\right) - 1\right]^2 dx = 0$$

or

$$0=\int_0^1\left\{2\left[6x+\sqrt{2}\,c(1-6x^2)-c^2(1-3x^2)\right]-1\right\}$$
$$\times\left[2\sqrt{2}\,(1-6x^2)-4c(1-3x^2)\right]dx.$$

Carrying the indicated integration, we obtain

$$16c^3-48\sqrt{2}\,c^2+116c-47\sqrt{2}=0.$$

Among the three roots, we again select $c=1.0885$ on the basis of minimizing the error in the approximation. We have

$$u_1(x)=\sqrt{2}\,x^2+1.0885(1-x^2)=1.0885+0.3257x^2. \qquad (3.440)$$

The exact solution to Eq. (3.402) is given by Eq. (3.409):

$$u_{\text{exact}}=\sqrt{1+x^2}\,.$$

In Table 3.4 the Ritz, Galerkin, and least-squares solutions are compared for various values of x. ■

Example 3.34 Consider the steady-state heat conduction in a homogeneous isotropic semi-finite slab whose edges are maintained at temperature T_0. Along the short edge the temperature is specified to be

$$q(x)=T_0+T_1x(h-x).$$

We wish to find the temperature distribution in the slab. The equation to be solved is

$$-\nabla^2T=0 \qquad \text{in } R$$

subject to the boundary conditions shown in Fig. 3.17.

Table 3.4 Comparison of Approximate Solutions with the Exact Solution of the Nonlinear Eq. (3.402)

x	0	0.2	0.4	0.6	0.8	1.0
Exact	1.000	1.020	1.077	1.166	1.261	1.414
Galerkin	1.016	1.031	1.079	1.159	1.271	1.414
Ritz	1.000	1.083	1.166	1.248	1.331	1.414
Least-Squares	1.088	1.102	1.141	1.206	1.297	1.414

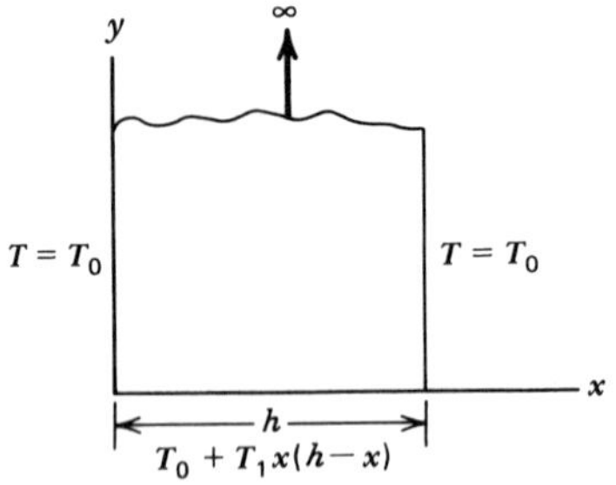

Figure 3.17 Heat conduction in a semi-finite slab.

A nondimensionalized form of the equation is convenient. Let

$$\theta=\frac{(T-T_0)}{(T_1-T_0)},\qquad \bar{x}=\frac{x}{h},\qquad \bar{y}=\frac{y}{h}.$$

Then we have (omitting the bars on x and y),

$$-\nabla^2\theta=0\qquad 0<x<1,\qquad 0<y<\infty$$

$$\theta(0,y)=\theta(1,y)=0,\qquad \text{for } y>0$$

$$\theta(x,0)=x(1-x),\qquad \text{and}\qquad \theta(x,\infty)=0,\qquad 0\le x\le 1. \tag{3.441}$$

We assume an approximate solution of the form

$$\theta_1(x,y)=c_1\phi_1=c_1(y)x(1-x) \tag{3.442}$$

where the parameter c_1 is assumed to be a function of y. Note that ϕ_1 satisfies the first two boundary conditions. The second part of boundary conditions will be satisfied if

$$c_1(0)=1\qquad \text{and}\qquad c_1(\infty)=0. \tag{3.443}$$

We solve the problem first in x dimension using the collocation method, and then we solve the resulting differential equation exactly using the boundary conditions in Eq. (3.443). Using collocation at $x=\frac{1}{3}$ (or $x=\frac{2}{3}$), we have

$$R=\frac{\partial^2\theta_1}{\partial x^2}+\frac{\partial^2\theta_1}{\partial y^2}=-2c_1(y)+x(1-x)\frac{d^2c_1}{dy^2}$$

$$0=-2c_1(y)+\frac{1}{3}\cdot\frac{2}{3}\cdot\frac{d^2c_1}{dy^2}$$

or

$$\frac{d^2c_1}{dy^2} - 9c_1 = 0.$$

The solution to this equation is given by

$$c_1(y) = Ae^{-3y} + Be^{3y}.$$

Using the boundary conditions in Eq. (3.443), we have $A=1$ and $B=0$. Hence the solution is

$$\theta(x, y) = x(1-x)e^{-3y}. \quad \blacksquare \tag{3.444}$$

The above example illustrates that the parameters c_i in the variational approximation can be taken as functions of a variable, depending on the convenience. Indeed, in time-dependent (unsteady) problems, the parameters c_i are assumed to be functions of time (only) while ϕ_i are assumed to depend on spatial coordinates x and y. This leads, as can be seen from the above example, to two stages of solution, both of which could be obtained by approximate methods. In the solution of time-dependent problems, usually the spatial approximation is considered first, and the temporal (or time) approximation next. Such a procedure is commonly known as *semidiscrete approximation* (in space). Semidiscrete variational approximation in space results in a set of ordinary differential equations (in time) as opposed to algebraic equations. Complete discretization is achieved by approximating the time derivatives, either by finite difference methods or by variational methods.

Example 3.35 Consider the dynamics (i.e., unsteady transverse motion) of a uniform beam of length L, flexural rigidity EI (constant), and mass per unit length ρ, and clamped at the ends. The equation of motion can be written as

$$\rho\frac{\partial^2 w}{\partial t^2} + \frac{\partial^2}{\partial x^2}\left(EI\frac{\partial^2 w}{\partial x^2}\right) = 0 \tag{3.445a}$$

where $w(x, t)$ is the transverse displacement and t is the time. The boundary conditions are

$$x=0: \quad w=0 \qquad \frac{\partial w}{\partial x}=0$$

$$x=L: \quad w=0, \qquad \frac{\partial w}{\partial x}=0, \qquad \text{for } t>0. \tag{3.445b}$$

We assume the initial configuration to be

$$w(x,0)=\frac{q_0L^4}{EI\pi^4}\left[\sin\frac{\pi x}{L}-\frac{\pi x}{L}\left(1-\frac{x}{L}\right)\right],\qquad \frac{\partial w}{\partial t}(x,0)=0. \quad (3.445c)$$

We wish to determine the motion of the beam after a sudden removal of the static load, q_0 = constant.

Using the following nondimensional variables,

$$\bar{x}=\frac{x}{L},\qquad \bar{t}=\frac{t}{L^2\sqrt{\rho/EI}},\qquad \bar{w}=\frac{w}{q_0L^4/EI\pi^4}$$

we obtain (after dropping the bars over the variables),

$$\frac{\partial^4 w}{\partial x^4}+\frac{\partial^2 w}{\partial t^2}=0,\qquad 0<x<1, t>0$$

$$w=\frac{\partial w}{\partial x}=0\quad \text{at}\quad x=0,1, t>0$$

$$w=\sin\pi x-\pi x(1-x),\qquad \frac{\partial w}{\partial t}=0\quad \text{at}\quad t=0,\qquad \text{for}\quad 0<x<1. \quad (3.446)$$

Note that all of the boundary conditions in the present problem are of the essential type. Hence the selection criteria for the approximation functions in the Galerkin and the Ritz methods become the same. The following approximation functions meet the boundary conditions:

$$\phi_1=1-\cos 2\pi x, \phi_2=1-\cos 4\pi x,\ldots, \phi_n=1-\cos 2n\pi x.$$

Using the semidiscrete Galerkin approximation,

$$\sum_{j=1}^{n}\int_0^1\left(\frac{d^2c_j}{dt^2}\phi_i\phi_j+c_j\frac{d^2\phi_i}{dx^2}\frac{d^2\phi_j}{dx^2}\right)dx=0,$$

we obtain

$$[M]\{\ddot{c}\}+[K]\{c\}=0 \quad (3.477a)$$

where

$$M_{ij}=\int_0^1\phi_i\phi_j\,dx,\qquad K_{ij}=\int_0^1\frac{d^2\phi_i}{dx^2}\frac{d^2\phi_j}{dx^2}dx. \quad (3.447b)$$

Suppose now that $n=1$. We get $M_{11}=\frac{3}{2}$, $K_{11}=(2\pi)^4/2$. Hence

$$\frac{d^2c_1}{dt^2}+\beta^2c_1=0, \qquad \beta=\frac{(2\pi)^2}{\sqrt{3}}\approx 22.7929. \tag{3.448}$$

The exact solution of Eq. (3.448) is given by

$$c_1=A\sin\beta t+B\cos\beta t \tag{3.449}$$

where A and B are constants to be determined using the initial conditions. The residuals in the initial values of w and $\partial w/\partial t$ are

$$R_1^0=w(x,0)=\sin\pi x-\pi x(1-x)=c_1(0)(1-\cos 2\pi x)-\sin\pi x-\pi x(1-x)$$

$$R_2^0=\frac{\partial w}{\partial t}(x,0)=0=\frac{dc_1}{dt}(0)(1-\cos 2\pi x). \tag{3.450}$$

Using the Galerkin method on these residuals with $\phi_1=1-\cos 2\pi x$, we get

$$c_1(0)=0.1107, \qquad \frac{dc_1}{dt}(0)=0. \tag{3.451}$$

Equation (3.449) becomes, in view of the initial conditions (3.451),

$$c_1=0.1107\cos\beta t$$

and

$$w(x,t)=0.1107(1-\cos 2\pi x)\cos 22.7989t. \quad \blacksquare \tag{3.452}$$

3.6.5 Concluding Remarks on Variational Methods of Approximation

We conclude this section with some observations and remarks on the variational methods of approximation. As pointed out earlier, physical and mathematical reasoning gives information about the form of the solution, which in turn is very helpful in the selection of appropriate trial functions. It is always desirable to pick admissible functions from the lowest order, and, if possible, to pick algebraic polynomials that are easy to work with. Without physical insight into the problem, certain choices of trial functions for some problems could result in either trivial or unacceptable solutions.

While the variational methods provide simple means of finding approximate solutions to most physical problems, their application to problems with irregular domains and complex boundary conditions becomes difficult due to the inability of selecting the trial functions that meet the requirements of the method used. Further, without a judicious choice of these functions, the resulting solution may be erroneous. In some problems, at the end of complex algebraic calculations one might end up with a trivial (or zero) solution for the undetermined parameters. The choice of the trial functions becomes even more difficult if the boundary of the domain is irregular. Moreover, there is no systematic method for the selection, and hence it cannot be automated on digital computers. In summary, the variational methods have the following shortcomings which make them less attractive when compared with other approximate methods:

1. There is no systematic method of selecting the approximation functions from the linear set on which the operator is defined.
2. Since the choice differs for each set of boundary conditions for the same type of problem, the selection process cannot be automized.
3. Without a judicious choice of the approximation functions, the resulting algebraic equations may be ill-conditioned, and/or may lead to trivial solutions.

Exercises 3.5

1 Consider the differential equation

$$-\frac{d^2u}{dx^2}=\frac{1}{1+x}, \qquad 0<x<1$$

subjected to the following three types of boundary conditions:

$$u(0)=u(1)=0 \tag{i}$$

$$u(0)=u'(1)=0 \tag{ii}$$

$$u'(0)=u'(1)=0. \tag{iii}$$

(a) Define the domain of the operator $A\equiv-d^2/dx^2$ for each of the boundary conditions.

(b) Give the variational formulation $B(u,v)=l(v)$ for each of the problems, and identify the domain H_A of the bilinear form in each case.

(c) Using an N-parameter Ritz approximation, set up the algebraic equations for each case. Solve the systems of equations for $N=2$, and compare the Ritz solution with the exact solution for (ii),

$$u_0(x)=x(1+\ln 2)-(1+x)\ln(1+x).$$

2 Consider the differential equation in Example 3.25 subject to the boundary conditions

$$u'(0)=0, \qquad u'(1)=0.$$

Obtain the n-parameter Ritz approximation by choosing $\phi_i = \cos i\pi x$ as the basis. Compute the actual error and the error estimate for the problem.

3 Consider the differential equation

$$-\frac{d}{dx}\left[a(x)\frac{du}{dx}\right]+c(x)u=f(x), \qquad 0<x<1$$

$$u(0)=1, \qquad u'(1)=3.$$

(a) Reformulate the differential equation such that the boundary conditions are homogeneous.

(b) Solve the differential equation when $a(x)=1+x^2, c=1$, and $f(x)=\sin\pi x$, using an N-parameter Ritz approximation.

4 Repeat Exercise 3 for the following boundary conditions;

$$u(0)=1, \qquad u(1)=0.$$

5 Consider the differential equation

$$\frac{d^2}{dx^2}\left[b(x)\frac{d^2w}{dx^2}\right]+c(x)w=q(x), \qquad 0<x<1$$

$$w(0)=w(1)=0, \qquad w''(0)=w''(1)=0$$

associated with a simply supported beam on an elastic foundation (see Example 3.32).

(a) Define the energy space H_A for the problem, and prove that the operator of the equation is positive definite on H_A.

(b) Choose $\phi_n(x)=\sin n\pi x$, and derive the N-parameter Ritz system when $b(x)=1+x, c=10^2$, and $q=10^3(x-x^2)$. Specialize the results for $N=3$.

6 For a beam on an elastic foundation as in Example 3.32, show that the n-parameter Ritz-Galerkin approximation of the form in Eq. (3.423) is given by

$$w_n=\sum_{i=1}^{n} a_n \sin\left(\frac{n\pi x}{L}\right)$$

where

$$a_n=-\frac{4q_0L^4}{n\pi(n^4\pi^4EI+kL^4)}, \qquad \text{for odd values of } n$$

$$a_n=0, \qquad \text{for even values of } n.$$

7 Given the differential equation and end conditions

$$-\frac{d}{dx}\left[(1+x)\frac{du}{dx}\right]-u=0, \qquad 0<x<1$$

$$u(1)=1, \qquad u(0)=0,$$

determine a one-parameter solution using the Ritz method.

8 For a clamped-hinged beam of length L, constant flexural rigidity EI, and mass per unit length m, the functional associated with its free vibration is given by

$$I(w)=\tfrac{1}{2}\int_0^L\left[EI(w'')^2-\lambda w^2\right]dx$$

where $\lambda=m\omega^2$, ω being the natural frequency of vibration. Using two-parameter Ritz approximation with algebraic functions, set up the eigenvalue problem and solve for the eigenvalues $\lambda_i (i=1,2)$, in terms of EI and L. The specified boundary conditions are

$$w=0 \quad \text{at } x=0, L$$

$$\frac{dw}{dx}=0 \quad \text{at } x=0 \quad \text{and} \quad \frac{d^2w}{dx^2}=0 \quad \text{at } x=L.$$

9 Given the functional

$$I(u,v)=\int_0^L\left\{\frac{du}{dx}\frac{dv}{dx}+\frac{v^2}{2\alpha}-qu\right\}dx$$

with the essential boundary conditions

$$u(0)=0, \qquad v(L)=\hat{v}$$

determine:

(a) A one-parameter Ritz solution.

(b) A one-parameter Galerkin solution.

10 Show that the Galerkin method as well as the least-squares method (why?) gives the same solution as that obtained by the Ritz method for the problem in part 1 of Example 3.24 when the same coordinate functions are used.

11 Use the Ritz and the Galerkin methods to solve the nonlinear differential equation

$$\beta\frac{d^4w}{dx^4}-\frac{\alpha}{2}\frac{d}{dx}\left(\frac{dw}{dx}\right)^3+1=0, \qquad 0<x<1$$

subject to the boundary conditions

$$w=\frac{dw}{dx}=0 \qquad \text{at } x=0$$

$$-\frac{\alpha}{2}\left(\frac{dw}{dx}\right)^3+\beta\frac{d^3w}{dx^3}=0 \qquad \text{and} \qquad \frac{d^2w}{dx^2}=0 \qquad \text{at } x=1.$$

Here α and β are constants.

12 Consider the problem of a vertically loaded clamped circular plate of radius a with a central hole of radius b. The governing differential equation is given by

$$\left(\frac{d^2}{dr^2}+\frac{1}{r}\frac{d}{dr}\right)\left(\frac{d^2w}{dr^2}+\frac{1}{r}\frac{dw}{dr}\right)=q, \qquad b<r<a$$

where w is the transverse deflection and q is the uniformly distributed loading. Obtain a two-parameter Galerkin solution by choosing

$$\phi_1=(r^2-b^2)^2(a^2-r^2)^2, \qquad \phi_2=r^2(r^2-b^2)^2(a^2-r^2)^2.$$

These functions satisfy the essential boundary conditions $w=\partial w/\partial n=0$ at the boundary.

13 Consider the two-point boundary-value problem

$$\frac{d^2u}{dx^2}-\lambda u=0$$

$$u(0)=0, \qquad u(1)+\frac{du}{dx}(1)=0.$$

Determine the first three characteristic (or eigen) values. Use:

(a) The Ritz method.

(b) The collocation at $x=\frac{1}{4}, \frac{1}{2}$, and $\frac{3}{4}$.

Compare your solution with the exact solution.

14 Use collocation at $x=\frac{1}{2}$ to solve the problem in Example 3.34.

15 Consider the partial differential equation

$$-k\nabla^2=f \qquad \text{in } R$$

$$u=0 \qquad \text{on } C$$

where R is a rectangular domain of dimensions $2a\times 2b$ and C is its boundary. The origin of the coordinate system is at the center of the domain. This equation arises in many fields of engineering science. A few areas are listed in the accompanying table, with the physical meaning of the variables.

Table 3.5 Some Examples of the Poisson Equation $-\nabla \cdot (k \nabla u) = Q$ *

Field	Variable u	Material Constant k	Source Variable Q	Secondary Variables $q, \partial u/\partial x, \partial u/\partial y$
Heat transfer	Temperature T	Conductivity k	Heat source q	Heat flow q [comes from conduction $k\partial u/\partial n$ and convection $h(u-u_\infty)$]
Irrotational flow of an ideal fluid	Stream function ψ	Density ρ	Mass production σ, normally zero	Velocities $\frac{\partial \psi}{\partial x} = -v, \frac{\partial \psi}{\partial y} = u$
	Velocity potential ϕ	Density ρ	Mass production, σ normally zero	$\frac{\partial \phi}{\partial x} = u, \frac{\partial \phi}{\partial y} = v$
Ground-water flow	Piezometric head ϕ	Permeability k	Recharge Q (or pumping $-Q$)	See page q $q = k\frac{\partial \phi}{\partial n}$ $u = -k\frac{\partial \phi}{\partial x}, \quad v = -k\frac{\partial \phi}{\partial y}$
Torsion of constant cross-section members	Stress function ϕ	Shear modulus G $k = \frac{1}{G}$	Angle of twist per unit length θ $Q = 2\theta$	Shear stresses τ_{zx}, τ_{zy} $\frac{\partial \phi}{\partial x} = -\tau_{zy}, \quad \frac{\partial \phi}{\partial y} = \tau_{zx}$
Electrostatics	Scalar potential ϕ	Dielectric constant ε	Charge density ρ	Displacement flux density D_n
Magnetostatics	Magnetic potential ϕ	Permeability μ	Charge density ρ	Magnetic flux density B_n

*Natural boundary condition: $k\frac{\partial u}{\partial n} + h(u - u_\infty) = q$; essential boundary condition: $u = \hat{u}$

Using a two-parameter approximation of the form (with $a=b$)

$$u_2=\phi_0+c_1\phi_1+c_2\phi_2$$

with $\phi_0=0, \phi_1=a^2(x^2-a^2)(y^2-a^2)$, and $\phi_2=(x^2+y^2)(x^2-a^2)(y^2-a^2)$, obtain the Ritz solution to the problem. What is the Galerkin solution to the problem?

16 Repeat Exercise 15 with

$$u_N=\sum_{m=1}^{N}\sum_{n=1}^{N} c_{mn}\sin\left(\frac{m\pi x}{a}\right)\sin\left(\frac{n\pi y}{b}\right)$$

and show that

$$c_{mn}=\frac{16f_0}{\pi^4}\left\{mn\left[\left(\frac{m}{a}\right)^2+\left(\frac{n}{b}\right)^2\right]\right\}^{-1}.$$

17 Consider the Dirichlet problem for the Poisson equation

$$-\nabla^2u=1 \qquad \text{in } R$$

$$u=0 \qquad \text{on } S$$

where S is the boundary of the two-dimensional domain, $R=\{(x,y): 0<x<a, 0<y<b\}$. Obtain a one-parameter Ritz approximation.

(a) Use algebraic basis functions.

(b) Use trigonometric basis functions.

18 A square plate of constant thickness is clamped (fixed) along all of its four sides and subjected to a uniformly distributed load P in the transverse (to the plane of the plate) direction. If the functional associated with the bending of a thin elastic plate is given by

$$I(w)=\frac{D}{2}\int\left[w_{xx}^2+w_{yy}^2+2\alpha w_{xx}w_{yy}+2(1-\alpha)w_{xy}^2\right]dxdy-\int_R Pw\,dxdy$$

$$w=\frac{\partial w}{\partial n}=0 \qquad \text{on the boundary,}$$

determine a one-parameter Rayleigh-Ritz solution using trigonometric approximation functions.

19 Consider the partial differential equation

$$Au\equiv-\nabla^2u=1 \qquad \text{in } \Omega\subset\mathbf{R}^2$$

$$u=0 \qquad \text{on } \Gamma$$

where the domain Ω is an elliptical region, that is, Γ is the arc of an ellipse

$$\frac{x^2}{a^2}+\frac{y^2}{b^2}=1.$$

Find a one-parameter Galerkin solution of the equation.

20 Find the first eigenvalue of the differential equation in Exercise 19 using the Galerkin method.

21 Repeat Exercises 19 and 20 when Ω is the positive quadrant of the ellipse.

22 Consider the mixed problem for the Poisson equation

$$-\nabla^2 u=f \quad \text{in } \Omega\subset\mathbf{R}^2, \qquad \Omega \text{ is the unit square}$$

$$u=0 \quad \text{along } x=0 \text{ and } x=1$$

$$\frac{\partial u}{\partial n}=0 \quad \text{along } y=0 \text{ and } y=1.$$

Find the N-parameter Galerkin (Ritz) solution in the form

$$u_N=\sum_{i=1}^{N}\sum_{j=0}^{N} c_{ij}\sin i\pi x\cos j\pi y.$$

23 Solve Part 2 of Example 3.24 by the least-squares method.

24 Determine a three-parameter solution that satisfies the boundary conditions

$$u(0)=1, \qquad u(1)=0$$

and minimizes the integral

$$I(u)=\int_0^1\left[\frac{1}{48}\left(\frac{du}{dx}\right)^4+\left(\frac{du}{dx}\right)^2+u^6-6u\right]dx.$$

The resulting nonlinear equations must be solved numerically using a numerical technique. The exact solution is given by $u=1-x^2$.

25 Solve the differential equation in Exercise 1, boundary condition (i), using the Petrov-Galerkin method.

26 Solve the differential equation in part 3 of Example 3.24 using the Petrov-Galerkin method with $\psi_i=\cos i\pi x$.

27 Solve the problem in Exercise 24 by the least-squares method.

28 Use the Ritz/Galerkin method to find the eigenvalues of the problems in Exercise 1.

REFERENCES AND ADDITIONAL READING

1 Arthurs, A. M., *Complementary Variational Principles*, Oxford University Press, London, 1967.

2 Becker, M., *The Principles and Applications of Variational Methods*, M.I.T. Press, Cambridge, MA 1964.

3 Biot, M. A., *Variational Principles in Heat Transfer*, Clarendon, London, 1972.

4 Courant, R., "Variational methods for the solution of equilibrium and vibrations," *Bulletin of the American Mathematical Society*, Vol. 49, pp. 1–23, 1943.

5 Denn, M. M., *Optimization by Variational Methods*, McGraw-Hill, New York, 1969.

6 Dym, C. L., and Shames, I. H., *Solid Mechanics: A Variational Approach*, McGraw-Hill, New York, 1973.

7 Ekeland, I., and R. Temam, *Convex Analysis and Variational Problems*, North-Holland, New York, 1976.

8 Finlayson, B. A., *The Method of Weighted Residuals and Variational Principles*, Academic, New York, 1972.

9 Forray, M. J., *Variational Calculus in Science and Engineering*, McGraw-Hill, New York, 1968.

10 Hestenes, M. R., *Optimization Theory: The Finite Dimensional Case*, Wiley, New York, 1975.

11 Hildebrand, F. B., *Methods of Applied Mathematics*, 2nd ed., Prentice-Hall, Englewood Cliffs, NJ, 1965.

12 Lanczos, C., *The Variational Principles of Mechanics*, The University of Toronto Press, Toronto, 1964.

13 Leipholz, H., *Direct Variational Methods and Eigenvalue Problems in Engineering*, Noordhoff, Leyden, 1977.

14 Lippmann, H., *Extremum and Variational Principles in Mechanics*, Springer-Verlag, New York, 1972.

15 Lovelock, D., *Tensors, Differential Forms and Variational Principles*, Wiley, New York, 1975.

16 Mikhlin, S. G., *Variational Methods in Mathematical Physics*, Pergamon (distributed by Macmillan), New York, 1964.

17 Mikhlin, S. G., *Mathematical Physics, An Advanced Course*, North-Holland, Amsterdam, 1970.

18 Mikhlin, S. G., *The Numerical Performance of Variational Methods*, Wolters-Noordhoff, Groningen, Netherlands, 1971.

19 Oden, J. T., and J. N. Reddy, *Variational Methods in Theoretical Mechanics*, Springer-Verlag, New York, 1976.

20 Parkus, H., *Variational Principles in Thermo- and Magneto-Elasticity*, Lecture Notes, No. 58, Springer-Verlag, New York, 1972.

21 Petrov, I. P., *Variational Methods in Optimal Control Theory*, translated from the 1965 Russian edition, Academic, New York, 1968.

22 Prenter, P. M., *Splines and Variational Methods*, Wiley, New York, 1975.

23 Reddy, J. N., "A note on mixed variational principles for initial-value problems," *Quarterly Journal of Mechanics and Applied Mathematics*, Vol. 28, Part 1, pp. 123–132, 1975.

24 Rektorys, K., *Variational Methods in Mathematics, Science and Engineering*, Reidel, Boston, 1977.

25 Rustagi, J. S., *Variational Methods in Statistics*, Academic, New York, 1976.

26 Schechter, R. S., *The Variational Methods in Engineering*, McGraw-Hill, New York, 1967.

27 Smith, D. K., *Variational Methods in Optimization*, Prentice-Hall, Englewood Cliffs, NJ, 1974.

28 Stacey, W. M., *Variational Methods in Nuclear Reactor Physics*, Academic, New York, 1974.

29 Strieder, W., and R. Aris, *Variational Methods Applied to Problems of Diffusion and Reaction*, Springer, Berlin, 1973.

30 Tonti, E., "On the variational formulation of linear initial-value problems," *Anal. Mat. pura Appl.*, Vol. 95, pp. 331–359, 1973.

31 Vainberg, M. M., *Variational Method and Method of Monotone Operators in the Theory of Nonlinear Equations*, translated from Russian by A. Libin, Wiley, New York, 1973.

32 Washizu, K., *Variational Methods in Elasticity and Plasticity*, 2nd ed., Pergamon, New York, 1975.

33 Weinberger, H. F., *Variational Methods for Eigenvalue Problems*, University of Minnesota, Minneapolis, 1962.

34 Weinstock, R., *Calculus of Variations with Applications to Physics and Engineering*, McGraw-Hill, New York, 1952.

35 Yourgrau, W., and S. Mandelstam, *Variational Principles in Dynamics and Quantum Theory*, 3rd ed., W. B. Saunders, Philadelphia, PA, 1968.

ANSWERS TO SELECTED EXERCISES

CHAPTER ONE

Exercises 1.1

1 $\mathbf{r}=\mathbf{A}+\alpha\mathbf{B},\ -\infty<\alpha<\infty.$

3 $[(\mathbf{C}-\mathbf{A})\times(\mathbf{B}-\mathbf{A})]\cdot(\mathbf{r}-\mathbf{A}).$

11 $-5\mathbf{r}_1+2\mathbf{r}_2+\mathbf{r}_3=0.$

Exercises 1.2

1 **(e)**
$$\begin{aligned} A_iA_j\varepsilon_{ijk} &= \tfrac{1}{2}(A_iA_j+A_jA_i)\varepsilon_{ijk},\\ &= \tfrac{1}{2}(A_iA_j\varepsilon_{ijk}+A_pA_q\varepsilon_{qpk}),\\ &= \tfrac{1}{2}A_iA_j(\varepsilon_{ijk}-\varepsilon_{ijk})=0. \end{aligned}$$

5 Linearly dependent, $\mathbf{A}+0\cdot\mathbf{B}+\mathbf{C}-\mathbf{D}=\mathbf{0}.$

6
$$\begin{bmatrix} \frac{1}{2} & -\frac{\sqrt{3}}{2} & 0\\ \frac{\sqrt{3}}{2} & \frac{1}{2} & 0\\ 0 & 0 & 1 \end{bmatrix}.$$

7 $A=3,\ \cos\alpha=-\cos\beta=\frac{2}{3},\ \cos\gamma=\frac{1}{3}.$

8 **(a)** $\mathbf{A}\cdot\hat{\mathbf{e}}_B=\frac{2}{3}.$
(b) $\cos\theta=\hat{\mathbf{e}}_A\cdot\hat{\mathbf{e}}_B=\frac{2}{15}.$

9 $\mathbf{V}=\dfrac{\omega}{\sqrt{3}}(\hat{\mathbf{i}}-2\mathbf{j}+\hat{\mathbf{k}}),\quad v=\dfrac{6}{\sqrt{3}}\omega.$

10 (a) $\Delta \equiv [\mathbf{e}_1\mathbf{e}_2\mathbf{e}_3] = (3\sqrt{3}-1)/8,$

$\mathbf{e}^1 = (\tfrac{3}{2}\hat{\mathbf{i}} - \tfrac{1}{2}\hat{\mathbf{j}})/\Delta,$

$\mathbf{e}^2 = (-\hat{\mathbf{i}} + \sqrt{3}\,\hat{\mathbf{j}})/4\Delta,$

$\mathbf{e}^3 = \hat{\mathbf{k}}$

(b) $|\mathbf{e}_1| = \dfrac{1}{2},\ |\mathbf{e}_2| = \dfrac{\sqrt{10}}{2},\ |\mathbf{e}_3| = 1,$

$|\mathbf{e}^1| = \dfrac{\sqrt{10}}{2\Delta},\ |\mathbf{e}^2| = \dfrac{1}{\Delta},\ |\mathbf{e}^3| = 1.$

11 $A_1 = 1 + \dfrac{\sqrt{3}}{4},\ A_2 = \dfrac{43+\sqrt{3}}{8},\ A_3 = 3.$

13 $\mathbf{e}_1' = \dfrac{\sqrt{3}}{4}\hat{\mathbf{i}} + \dfrac{1}{4}\hat{\mathbf{j}},$

$\mathbf{e}_2' = \dfrac{1-3\sqrt{3}}{8}\hat{\mathbf{i}} + \dfrac{9-\sqrt{3}}{8}\hat{\mathbf{j}},$

$\mathbf{e}_3' = \hat{\mathbf{k}},\ \hat{\mathbf{e}}_i' = \dfrac{\mathbf{e}_i'}{|\mathbf{e}_i'|}.$

14 $\mathbf{e}_1' = \hat{\mathbf{e}}_1 + \hat{\mathbf{e}}_3,$

$\mathbf{e}_2' = \tfrac{1}{2}(3\hat{\mathbf{e}}_1 + 4\hat{\mathbf{e}}_2 - 3\hat{\mathbf{e}}_3),$

$\mathbf{e}_3' = \tfrac{5}{17}(2\hat{\mathbf{e}}_1 - 3\hat{\mathbf{e}}_2 - 2\hat{\mathbf{e}}_3),$

$\hat{\mathbf{e}}_i' = \dfrac{\mathbf{e}_i'}{|\mathbf{e}_i'|}.$

15 $\mathbf{e}_1' = 2\hat{\mathbf{e}}_1 + \hat{\mathbf{e}}_2 + \hat{\mathbf{e}}_3,$

$\mathbf{e}_2' = \tfrac{1}{6}(4\hat{\mathbf{e}}_1 - 13\hat{\mathbf{e}}_2 + 5\hat{\mathbf{e}}_3),$

$\mathbf{e}_3' = \tfrac{1}{105}(-108\hat{\mathbf{e}}_1 + 36\hat{\mathbf{e}}_2 + 180\hat{\mathbf{e}}_3),$

$\hat{\mathbf{e}}_i' = \dfrac{\mathbf{e}_i'}{|\mathbf{e}_i'|}.$

16 See Example 2.43.

Exercises 1.3

1 (a) $$\begin{bmatrix} 3 & 5 & 1 & -1 \\ -4 & 1 & -2 & 1 \\ 2 & 1 & 5 & 1 \\ 2 & -1 & 1 & 0 \end{bmatrix} \begin{Bmatrix} x_1 \\ x_2 \\ x_3 \\ x_4 \end{Bmatrix} = \begin{Bmatrix} 2 \\ -3 \\ 5 \\ 4 \end{Bmatrix}.$$

(b) $$\begin{bmatrix} 3 & 2 & 16 & 5 \\ 2 & 1 & 5 & 1 \\ 1 & -1 & -1 & -1 \\ -5 & 11 & 13 & 3 \end{bmatrix} \begin{Bmatrix} x_1 \\ x_2 \\ x_3 \\ x_4 \end{Bmatrix} = \begin{Bmatrix} 0 \\ 5 \\ -2 \\ 6 \end{Bmatrix}.$$

2 $$\begin{bmatrix} 0 & 1 & 0 \\ -\dfrac{1}{\sqrt{2}} & 0 & \dfrac{1}{\sqrt{2}} \\ \dfrac{1}{\sqrt{2}} & 0 & \dfrac{1}{\sqrt{2}} \end{bmatrix}.$$

3 **(a)** $$\begin{bmatrix} -9 & 16 & 20 \\ 4 & -6 & -7 \\ -7 & 12 & 14 \end{bmatrix}.$$

(b) $$\begin{bmatrix} 6 & 0 & 0 \\ 0 & 7 & 0 \\ 0 & 0 & 6 \end{bmatrix}.$$

(c) $$\begin{bmatrix} 190 & 102 \\ 100 & 64 \\ 92 & 7 \end{bmatrix}.$$

6 $f(A)=[I]$.

7 $[\hat{K}]=[K_{11}]-[K_{12}][K_{22}]^{-1}[K_{21}]$,

$\{\hat{F}\}=\{F_1\}-[K_{22}]^{-1}\{F_2\}$.

8 $$[A]=\begin{bmatrix} 1 & 1 & 2 \\ 1 & 3 & \frac{1}{2} \\ 2 & \frac{1}{2} & 7 \end{bmatrix}.$$

9 **(a)** $$\begin{bmatrix} 1 & -6 & 1 & 0 & 1 \\ 0 & 1 & 4 & 2 & 6 \\ 0 & 0 & 1 & \frac{3}{4} & \frac{11}{4} \\ 0 & 0 & 0 & 1 & \frac{239}{47} \end{bmatrix}.$$

(b) $$\begin{bmatrix} 1 & 1 & 11 & 4 & -5 \\ 0 & 1 & 17 & 7 & -15 \\ 0 & 0 & 1 & \frac{9}{22} & -\frac{27}{22} \\ 0 & 0 & 0 & 1 & -\frac{323}{44} \end{bmatrix}.$$

11 **(a)** 8.

(b) 252.

16 **(a)** $\frac{1}{44}\begin{bmatrix} 15 & -2 & -7 \\ -2 & 12 & -2 \\ -7 & -2 & 15 \end{bmatrix}$.

(b) det$=0$.

(c) $\frac{1}{12}\begin{bmatrix} -9 & 6 & 3 \\ 6 & -4 & 2 \\ 3 & 2 & -1 \end{bmatrix}$.

(d) $\frac{1}{27}\begin{bmatrix} 3 & -6 & -6 \\ -6 & 3 & -6 \\ 6 & 6 & -3 \end{bmatrix}$.

19 **(a)** $x_1=\frac{4}{3}$, $x_2=0$, $x_3=\frac{2}{3}$.

(b) $x_1=1$, $x_2=\frac{7}{13}$, $x_3=1$.

20 **(a)** $x_1=\frac{1}{9}$, $x_2=\frac{11}{9}$, $x_3=-\frac{5}{9}$, $x_4=\frac{2}{9}$.

(b) $x_1=\frac{1}{12}(27c_1-36c_2+15c_3)$,

$x_2=\frac{1}{12}(-36c_1+64c_2-30c_3)$,

$x_3=\frac{1}{12}(15c_1-30c_2+15c_3)$.

(c) $x_1=\frac{684}{84}$, $x_2=-\frac{344}{84}$, $x_3=-\frac{184}{84}$, $x_4=\frac{316}{84}$.

Exercises 1.4

2 **(a)** $\mathbf{e}_1=\theta^2\cos\theta^3\hat{\mathbf{i}}_1+\theta^2\sin\theta^3\hat{\mathbf{i}}_2+\theta^1\hat{\mathbf{i}}_3$,

$\mathbf{e}_2=\theta^1\cos\theta^3\hat{\mathbf{i}}_1+\theta^1\sin\theta^3\hat{\mathbf{i}}_2-\theta^2\hat{\mathbf{i}}_3$,

$\mathbf{e}_3=-\theta^1\theta^2\sin\theta^3\hat{\mathbf{i}}_1+\theta^1\theta^2\cos\theta^3\hat{\mathbf{i}}_2$.

(b) $g_{11}=(\theta^2)^2+(\theta^1)^2$, $g_{12}=0$, $g_{13}=0$,

$g_{22}=(\theta^1)^2+(\theta^2)^2$, $g_{23}=0$, $g_{33}=(\theta^1\theta^2)^2$.

(c) $\mathbf{e}^1=\dfrac{\mathbf{e}_1}{g_{11}}$, $\mathbf{e}^2=\dfrac{\mathbf{e}_2}{g_{22}}$, $\mathbf{e}^3=\dfrac{\mathbf{e}_3}{g_{33}}$.

(d) $ds=[(\theta^1)^2+(\theta^2)^2][(d\theta^1)^2+(d\theta^2)^2]+(\theta^1\theta^2d\theta^3)^2$.

(e) $\hat{A}^1=A^1\sqrt{(\theta^1)^2+(\theta^2)^2}$, $\hat{A}^2=\sqrt{A^2(\theta^1)^2+(\theta^2)^2}$,

$\hat{A}^3=A^3\theta^1\theta^2$.

4 $\cos\theta = \dfrac{A^iB_i}{|A||B|} = \dfrac{A^ig_{ij}B^j}{|A||B|} = \dfrac{A^iB^jg_{ij}}{\left(A^iA^jg_{ij}\right)\left(B^mB^ng_{mn}\right)}.$

6 $$\begin{Bmatrix} \mathbf{e}_1 \\ \mathbf{e}_2 \\ \mathbf{e}_3 \end{Bmatrix} = \begin{bmatrix} \theta^2\cos\theta^3 & \theta^2\sin\theta^3 & \theta^1 \\ \theta^1\cos\theta^3 & \theta^1\sin\theta^3 & -\theta^2 \\ -\theta^1\theta^2\sin\theta^3 & \theta^1\theta^2\cos\theta^3 & 0 \end{bmatrix} \begin{Bmatrix} \hat{\mathbf{e}}_x \\ \hat{\mathbf{e}}_y \\ \hat{\mathbf{e}}_z \end{Bmatrix}.$$

10 $I_1 = 3 + 2\gamma_{ii}$, $I_2 = 3 + 4\gamma_{ii} + 2(\gamma_{ii}\gamma_{jj} - \gamma_{ij}\gamma_{ij})$, $I_3 = \det(\delta_{ij} + 2\gamma_{ij})$.

Exercises 1.6

1 $5/\sqrt{11}$.

3 $\dfrac{\partial r}{\partial x_j} = \dfrac{x_j}{r}$.

Exercises 1.8

1 $\dfrac{1}{R}\dfrac{d}{dR}\left(R\dfrac{dE_{\phi\phi}}{dR} + E_{\phi\phi} - E_{RR}\right) = 0.$

9 $\operatorname{curl} D = \dfrac{1}{R}(\hat{\mathbf{e}}_z\hat{\mathbf{e}}_R + \hat{\mathbf{e}}_z\hat{\mathbf{e}}_z + \hat{\mathbf{e}}_R\hat{\mathbf{e}}_\phi).$

Exercises 1.9

1 $\hat{\bar{\mathbf{e}}}_1 = \dfrac{1}{\sqrt{3}}(\hat{\mathbf{e}}_1 - \hat{\mathbf{e}}_2 + \hat{\mathbf{e}}_3)$, $\hat{\bar{\mathbf{e}}}_2 = \dfrac{1}{\sqrt{14}}(2\hat{\mathbf{e}}_1 + 3\hat{\mathbf{e}}_2 + \hat{\mathbf{e}}_3)$,

$\hat{\bar{\mathbf{e}}}_3 = \dfrac{1}{\sqrt{42}}(-4\hat{\mathbf{e}}_1 + \hat{\mathbf{e}}_2 + 5\hat{\mathbf{e}}_3).$

$$[\beta] = \frac{1}{\sqrt{42}}\begin{bmatrix} \sqrt{14} & -\sqrt{14} & \sqrt{14} \\ 2\sqrt{3} & 3\sqrt{3} & \sqrt{3} \\ -4 & 1 & 5 \end{bmatrix}.$$

2 $$[\bar{T}] = \begin{bmatrix} 2 & -2.16 & 0 \\ -2.16 & 1.07 & -1.526 \\ 0 & -1.526 & 0.929 \end{bmatrix}.$$

3 Use $\det(\sigma_{ij} - \delta_{ij}) = 0$ and Exercise 1.3.13.

4 **(a)** $\lambda_1 = 2+\sqrt{20}$, $\lambda_2 = 3$, $\lambda_3 = 2-\sqrt{20}$,

$\hat{\mathbf{e}}_1 = \pm(-0.8506, 0.526, 0)$.

(b) $\lambda_1 = 5$, $\lambda_2 = 4$, $\lambda_3 = 1$,

$\hat{\mathbf{e}}_1 = \pm\frac{1}{\sqrt{2}}(1, -\sqrt{3}, 0)$.

(c) $\lambda_1 = 4$, $\lambda_2 = 2$, $\lambda_3 = 1$,

$\hat{\mathbf{e}}_1 = \pm\frac{1}{\sqrt{2}}(0, 1, -1)$, $\hat{\mathbf{e}}_2 = \pm\frac{1}{\sqrt{2}}(0, 1, 1)$, $\hat{\mathbf{e}}_3 = (1, 0, 0)$.

(d) $\lambda_1 = 3$, $\lambda_2 = 2$, $\lambda_3 = -1$,

$\hat{\mathbf{e}}_1 = \pm\frac{1}{\sqrt{2}}(1, 0, 1)$, $\hat{\mathbf{e}}_2 = \pm\frac{1}{\sqrt{3}}(-1, 1, 1)$, $\hat{\mathbf{e}}_3 = \pm\frac{1}{\sqrt{14}}(3, 2, 1)$.

(e) $\lambda_1 = 11.824$, $\lambda_2 = 1.285$, $\lambda_3 = -7.109$,

$\hat{\mathbf{e}}_1 = (0.5329, 0.2462, 0.4396)$.

(f) $\lambda_1 = 3.247$, $\lambda_2 = 1.555$, $\lambda_3 = 0.1981$.

6 **(i)** **(a)** $\mathbf{\Pi}(\hat{\mathbf{n}}) = 2\hat{\mathbf{e}}_1 + 14\hat{\mathbf{e}}_2 + 2\hat{\mathbf{e}}_3$ (kips)

(b) $\Pi = \sqrt{204}$ (kips), $\theta = 120.89°$.

(c) $\sigma_n = -\frac{22}{3}$ (kips), $\sigma_t = 12.26$ (kips).

(ii) **(a)** $\mathbf{\Pi}(\hat{\mathbf{n}}) = 10\hat{\mathbf{e}}_1 + \frac{50}{3}\hat{\mathbf{e}}_2 + \frac{40}{3}\hat{\mathbf{e}}_3$ (kips).

(b) $\Pi = \frac{10}{3}\sqrt{50}$ (kips), $\theta = 90°$.

(c) $\sigma_n = 0$, $\sigma_t = \frac{10}{3}\sqrt{50}$ (kips).

(iii) **(a)** $\mathbf{\Pi}(\hat{n}) = \left(\frac{8+\sqrt{2}}{3}\right)\hat{\mathbf{e}}_1 - \left(\frac{8+\sqrt{2}}{3}\right)\hat{\mathbf{e}}_2 + \frac{4}{3}(1+\sqrt{2})\,\hat{\mathbf{e}}_3$ (kips).

(b) $\Pi = 5.48$ (kips), $\theta = 16.4°$.

(c) $\sigma_n = 5.257$ (kips), $\sigma_t = 1.547$ (kips).

10 $\lambda_i = a_{ii}$ (no sum on i).

CHAPTER TWO

Exercises 2.1

1 **(a)** $\{a, b, c, d, e, f, g, h\}$.

(b) $\{b, f\}$.

(c) $\{a, b, d, f\}$.

(d) $\{c, e, g\}$.

2 **(a)** upper bound$=1$, supremum$=1$.
(b) lower bound$=$infimum$=$minimum$=2$.
(c) lower bound$=$infimum$=$minimum$=10^{-6}$.
upper bound$=$supremum$=$maximum$=10^{6}$.
(d) lower bound$=$infimum$=$minimum$=1$; upper bound$=2$, etc.

4 $A\times A=\{(1,1), (1,2), (1,4), (1,5), (2,1), (2,2), (2,4), (2,5), (4,1), (4,2), (4,4), (4,5), (5,1), (5,2), (5,4), (5,5)\}$.
$R=\{(2,1), (2,2), (4,1), (4,2), (4,4), (5,1), (5,5), (1,1)\}$.
R is reflexive and transitive, but not symmetric.

5 $R[2]=\{2,4,6,\ldots\}$.
$R[6]=\{6,7,\ldots\}$.
$R[8]=\{8,10,\ldots\}$.

6 **(a)** Reflexive and transitive, but not symmetric.
(b) An equivalence relation.
(c) An equivalence relation.

7 Onto.

9 $f(a)=\{9,4,1,9,16\}$: *not* one-to-one (of course, it is onto).

10 **(a)** $(f\cdot g)(x)=(x-1)^3+1$.
(b) $(g\cdot f)(x)=x^3$.
(c) $f^{-1}(y)=(y-1)^{1/3}$.

11 Onto mapping, not one-to-one.

13 **(a)** One-to-one, not onto.
(b) One-to-one, but not onto.
(c) One-to-one and onto: $f^{-1}(y)=\dfrac{(y-1)}{(y-2)}$.
(d) One-to-one and onto: $f^{-1}(y)=\dfrac{(y+3)}{5}$.
(e) Neither one-to-one nor onto.

Exercises 2.2

2 Yes.

4 Metric; pseudometric.

11 **(a)** and **(c)** use hit-and-miss metric defined in Exercise 10.

13 $(u,v)\in C_*^2(\Omega)\times C_+^2(\Omega)$, where

$$C_*^2(\Omega)=\{u\colon u\in C^2(\Omega),\ u=0 \text{ at } x=0,1, \text{ and } y=0;\ u=1 \text{ at } y=1\}$$

$$C_+^2(\Omega)=\{v\colon v\in C^2(\Omega),\ v=0 \text{ at } x=0,1 \text{ and } y=0,1\}.$$

16 It is a Cauchy sequence in the L_2-metric as well as in the sup-metric. The limit in the former case is zero, and in the latter case it is 1.

18 **(a)** $[1, \infty)$; limit $=1$.
(b) $[1, \infty)$; limit $=0$.
(c) $[1, \infty)$; limit $=0$.
(d) $[1, \infty)$; limit $=0$.
(e) $[1, \infty)$; limit $=1$.
(f) $[1, \infty)$; limit $=1/3$.
(g) $[1, \infty)$; limit $=1/3$.
(h) $[1, \infty)$; limit $=0$.
(i) $[1, \infty)$; limit $=0$.

Exercises 2.3

1 **(a)** Group.
(c) A vector space; $\boldsymbol{\theta}$ (identity element) $=(1,1)$, and $\mathbf{I}_a=\left(\frac{1}{a_1}, \frac{1}{a_2}\right)$.

2 **(a)** Addition is neither associative nor commutative.

3 Yes.

4 Yes.

7 **(a)** Not a subspace.
(b) A subspace.
(c) Not a subspace.
(d) A subspace.

8 **(a)** A subspace.
(b) Not a subspace.
(c) A subspace.
(d) A subspace.

10 **(a)** U is the line $x_1=x_2=x_3$, and V is the x_2x_3-plane. $U \cap V=\{0\}$, $U \cup V=\mathbf{R}^3$. Hence, it is a direct sum.
(b) U is the x_1x_2-plane, and V is the x_2x_3-plane. $U \cap V=\{\mathbf{x}=(0, x_2, 0)\}$; $U \cup V=\mathbf{R}^3$.
(c) $U \cap V=\{\mathbf{x}=(x_1, -2x_1, x_1)\}$, $U \cup V=\{\mathbf{x}=(x_1, x_2, x_1)\}=\mathbf{R}^2$
(d) U is the x_3-axis. $U \cap V=\{0\}$, $U \cup V=\mathbf{R}^3$, hence it is a direct sum.

11 **(a)** Linearly dependent: $4(-1,1,0)-5(-1,1,1)+(-2,-1,1)+3(1,1,1)=0$.
(b) Linearly independent.
(c) Linearly independent.

13 **(a)** Does not span $\mathbf{R}^3$.

(b) Spans $\mathbf{R}^3$; for example, one can write

$$(x, y, z) = (2x + 0.5y)(1,1,1) + x(-2,-1,2)$$
$$+ (z - 4x - 0.5y)(-1,1,1) + (y - z + 3x)(-1,1,0).$$

(c) Spans $\mathbf{R}^3$; for example one can write

$$(x, y, z) = (2x + y - z)(-1,1,0) + (z - 3x - 0.5y)(-1,1,1)$$
$$+ x(-2,-1,1) + (2x + 0.5y)(1,1,1).$$

(d) Spans $\mathbf{R}^3$; one can write

$$(x, y, z) = (x - 0.5y)(1,0,0) + (0.5y - z/3)(1,2,0)$$
$$+ (z/3)(1,2,3).$$

14 S_1 is the three-dimensional subspace defined by

$$S_1 = \{\mathbf{x} \in \mathbf{R}^4 \colon \mathbf{x} = (x_1, x_2, x_3, x_4), x_4 = x_1 + x_2 + x_3\}.$$

15 S_2 is the two-dimensional subspace defined by

$$S_2 = \{\mathbf{x} \in \mathbf{R}^4 \colon \mathbf{x} = (x_1, x_2, 3x_1 + 2x_2, 4x_1 + 3x_2)\}.$$

17 The set $(e^{-kx}, e^{kx}, e^{ikx}, e^{-ikx})$ is a basis of the linear space of solutions of the differential equation

$$\frac{d^4u}{dx^4} - k^4u = 0, \quad k > 0.$$

18 From Exercises 14 and 15, it follows that $S_1 + S_2$ is the three-dimensional subspace defined by

$$S_1 + S_2 = \{\mathbf{x} \in \mathbf{R}^4 \colon \mathbf{x} = (x_1, x_2, x_3, x_4), x_4 = x_1 + x_2 + x_3, \quad \text{or}$$
$$x_2 = 3x_1 + 2x_2 \quad \text{and} \quad x_4 = 4x_1 + 3x_2\}.$$

A basis for $S_1 + S_2$ is provided by

$$\{(-1,2,1,2), (1,2,3,6), (1,-1,1,1)\}.$$

20 The solution is of the form $(a, 7-2a, -2+a, 0)$, where a is an arbitrary real number. Note that the set of such elements, such as $\{(0,7,-2,0), (1,5,-1,0), (2,3,0,0), \dots\}$ does not constitute a linear vector space.

21 The dimension is equal to 3. $\{e^{-x}, e^{-2x}, e^{-3x}\}$ provides a basis.

22 The dimension is equal to 1. $\{(8,7,1)\}$ provides a basis.

23 The two-dimensional subspace is given by

$$S=\left\{\mathbf{x}:\left(x_1, x_2, \frac{-ax_1-bx_2}{c}\right)\right\},$$

which describes a plane. $\{(c,0,-a), (0,c,-b)\}$ provides a basis for the subspace.

Exercises 2.4

3 **(a)** $\|u\|_0=\sqrt{\frac{5}{6}-\frac{2}{\pi}}$, sup-norm$=1$.

(b) $\|u\|_0=\sqrt{\frac{3}{5}}$, sup-norm$=1$.

(c) $\|u\|_0=\sqrt{\frac{5}{6}-\frac{8}{\pi^2}}$, sup-norm$=0.207$.

(d) $\|u\|_0=\frac{2}{\sqrt{3}}$, sup-norm$=\sqrt{2}$.

(e) $\|u\|_0=\sqrt{50}$, sup-norm$=100$.

(f) $\|u\|_0=\sqrt{\frac{3}{8}}$, sup-norm$=1$.

12 **(a)** $(u,v)_0=\frac{4}{\pi^3}$, $(u,v)_1=\frac{4}{\pi^3}+\frac{4}{\pi}$.

(b) $(u,v)_0=0$, $(u,v)_1=0$.

(c) $(u,v)_0=0$, $(u,v)_1=0$.

(d) $(u,v)_0=\frac{1}{\pi}(2a+b+c)-\frac{4c}{\pi^3}$, $(u,v)_1=(u,v)_0-\frac{4c}{\pi}$.

(e) $(u,v)_0=\frac{1}{\pi}(-1+\frac{2}{\pi}+\frac{4}{\pi^2}+\frac{8}{\pi^3})$, $(u,v)_1=(u,v)_0+\frac{16}{\pi^2}$.

13 $(-a+ib)\sin \pi x$ provided $a=b$.

16 **(a)** Both sup-norm and L_2-norms do not exist.

(b) Sup-norm does not exist; L_2-norm exists ($\|u\|_0=\sqrt{3}$).

(c) Both sup-norm and L_2-norm exist; sup-norm$=1$; $\|u\|_0=\sqrt{\ln 2}$.

17 **(a)** $C=1/3$.

(b) $C=(-7\pm\sqrt{50})/2$.

18 **(a)** $S=\{\mathbf{x}\in\mathbf{R}^3: \mathbf{x}=(x_1,-x_1,x_3)\}$.

(b) $S=\{\mathbf{x}\in\mathbf{R}^3: \mathbf{x}=(x_1,x_2,x_1-2x_3\}$.

(c) $S=\{\mathbf{x}\in\mathbf{R}^3: \mathbf{x}=(-2x_2-3x_3,x_2,x_3)\}$.

19 $p(x)=107-548x+510x^2$.

22 $S=\{\mathbf{x}\in\mathbf{R}^3: \mathbf{x}=(x_1,x_2,x_1+\frac{3}{2}x_2)\}$,

$S^\perp=\{\mathbf{x}\in\mathbf{R}^3: \mathbf{x}=(-x_3,-\frac{3}{2}x_3,x_3)\}$.

25 The subspace of all even functions, $f(-x)=f(x)$.

26 **(b)** Zero matrices.

Exercises 2.5

1 **(a)** Nonlinear.

(b) Nonlinear.

(c) Linear.

2 Onto; not one-to-one.

3 $[T]=\begin{bmatrix}1 & 1 & 1\\ 0 & 1 & 1\\ 0 & 0 & 1\end{bmatrix}$.

4 One-to-one and onto.

5 **(a)** $\mathcal{N}=\{0\}$.

(b) Not a linear space.

(c) $\mathcal{N}=\{0\}$.

7 $\mathcal{N}=\{0\}$.

13 **(a)** $(T_1+T_2)(\mathbf{x})=(x_2-x_3,3x_1-x_2-x_3,x_1+x_2+x_3)$.

(b) $(T_1T_2)(\mathbf{x})=(0,x_2-x_3-x_1,3x_1-x_2)$.

(c) $(T_2T_1)(\mathbf{x})=(x_1-x_2-2x_3,x_3-x_1,0)$.

15 **(a)** $[T_1+T_2]=\begin{bmatrix}0 & 0 & -1\\ 3 & -1 & -1\\ 1 & 1 & 1\end{bmatrix}$.

(b) $[T_1T_2]=\begin{bmatrix}0 & 0 & 0\\ -1 & 1 & -1\\ 3 & -1 & 0\end{bmatrix}$.

(c) $[T_2T_1]=\begin{bmatrix}1 & -1 & -2\\ -1 & 0 & 1\\ 0 & 0 & 0\end{bmatrix}$.

16 **(a)** $(1,8,3)$.

(b) $(7,6,-5)$.

(c) $(-3,2,-5)$.

30 **(a)** $v(x)=x^2-\frac{1}{3}$.
(b) $v(x)=-x+\frac{1}{2}$.
(c) $v(x)=x(1-x)-\frac{1}{6}$.

32 $$[T]=\begin{bmatrix} 1 & 2 & 2 & 0 & 0 \\ 0 & 1 & 4 & 6 & 0 \\ 0 & 0 & 1 & 6 & 12 \\ 0 & 0 & 0 & 1 & 8 \\ 0 & 0 & 0 & 0 & 1 \end{bmatrix}.$$

33 $$[T]=\begin{bmatrix} 1 & 2 & 0 & 0 \\ 0 & 2 & 1 & 0 \\ 1 & 0 & 1 & 0 \end{bmatrix}.$$

34 $$[D]=\begin{bmatrix} 1 & 8 & -9 \\ 1 & 6 & -7 \\ 1 & 6 & -7 \end{bmatrix}.$$

36 $$[B]=\frac{\alpha}{L^2}\begin{bmatrix} 12 & -6L & -12 & -6L \\ -6L & 4L^2 & 6L & 2L^2 \\ -12 & 6L & 12 & 6L \\ -6L & 2L^2 & 6L & 4L^2 \end{bmatrix}, \quad \alpha=\text{constant}.$$

$$\{l\}=\frac{\beta L}{12}\{6 \quad -L \quad 6 \quad L\}^T, \quad \beta=\text{constant}.$$

41 **(a)** Represents a bilinear form.
(b) Represents a bilinear form.
(c) Represents a symmetric bilinear form.
(d) Does not represent a bilinear form.

42 **(a)** $$[B]=\begin{bmatrix} 1 & -3 & 2 \\ -1 & 1 & 0 \end{bmatrix}.$$

(b) $$[B]=\begin{bmatrix} 2 & 0 & -3 \\ 0 & 4 & 5 \\ 1 & 0 & 0 \end{bmatrix}.$$

(c) Not a bilinear form.

43 **(a)** $B\colon \mathbf{R}^3 \to \mathbf{R}^3$,
$$B(\mathbf{x},\mathbf{y})=x_1(y_1-y_2+2y_3)+x_2(y_1+y_2+y_3)+x_3(2y_1+y_3).$$

44 **(a)** $$[Q]=\begin{bmatrix} 1 & 1 & 2 \\ 1 & 3 & \frac{1}{2} \\ 2 & \frac{1}{2} & 6 \end{bmatrix}.$$

(c) $$[Q]=\begin{bmatrix} a & \dfrac{c}{2} \\ \dfrac{c}{2} & b \end{bmatrix}.$$

45 $[B]$ is the zero matrix.

Exercises 2.6

12 **(a)** $[T^*]=[T]^T=\begin{bmatrix} \cos\theta & \sin\theta \\ -\sin\theta & \cos\theta \end{bmatrix}.$

15 **(a)** $\mathfrak{D}(T)=H_0^4(0, L)$, $\mathfrak{R}(T)=L_2(0, L)$.

20 Self-adjoint operator.

23 **(a)** Elliptic but not strongly elliptic.
(b) Hyperbolic.

30 **(a)** $8\left(\sin x+\frac{\sin 3x}{3^3}+\cdots+\frac{\sin(2n-1)x}{(2n-1)^2}+\cdots\right).$

(b) $96\left(\cos x+\frac{\cos 3x}{3^4}+\cdots+\frac{\cos(2n-1)x}{(2n-1)^4}+\cdots\right).$

(c) $f(x)=\frac{2}{\pi}\sum_{n=1}^{\infty}\frac{\sin 2nx}{n(4n^2-1)}.$

33 The Galerkin solution:

$$u_N=\sum_{n=1}^{N}\alpha_n\phi_n(x),\ \phi_n(x)=\sqrt{\frac{2}{\pi}}\sin nx,\quad \text{and}$$

$$\alpha_n=\begin{cases}\sqrt{\dfrac{2}{\pi}}\dfrac{2}{n(n^2-1)}, & \text{if } n \text{ is even.}\\ 0, & \text{if } n \text{ is odd.}\end{cases}$$

34 Both the Galerkin and the least squares solution is $u=\frac{x}{2}(1-x)$, which coincides with the exact solution.

Exercises 2.7

1 **(a)** Yes; $x_1=\frac{5}{9}$, $x_2=-\frac{23}{9}$, $x_3=\frac{2}{3}$.
(b) No; $\beta_1=2\beta_2=-\frac{2}{3}\beta_3$; no solution.
(c) Yes; $x_1=-\frac{1}{4}$, $x_2=-\frac{5}{4}$, $x_3=\frac{3}{4}$.
(d) Yes; $x_1=3$, $x_2=-14$, $x_3=15$.
(e) No; $\beta_1=-3\beta_2$, $\beta_3=5\beta_2$; many solutions.
(f) Yes; $x_1=\frac{4}{3}$, $x_2=0$, $x_3=\frac{1}{3}$.

10 **(a)** $\mathfrak{D}(A)=\left\{u\colon u\in H^4(\Omega),\ u=\frac{\partial u}{\partial n}=0 \text{ on } \partial\Omega\right\}.$

12 $v(x)=y\sin\pi x.$

CHAPTER THREE

Exercises 3.1

1 $x=1$, $y=4$, $f(1,4)=-8$.

3 $x=2$, $y=0$, $f(2,0)=0$.

5 $x=y=0$, $z=2$, $f(0,0,2)=20$.

7 $x=y=z=0$; $f(0,0,0)=0$.

8 $x=\frac{3}{4}$, $y=\frac{1}{4}$.

9 $x=\sqrt{\dfrac{3}{2}}$, $y=0$, $z=\dfrac{3}{2}$.

10 $x=\dfrac{2}{\sqrt{5}}$, $y=-\dfrac{1}{\sqrt{5}}$, $\lambda=-3$, or

$x=-\dfrac{2}{\sqrt{5}}$, $y=\dfrac{1}{\sqrt{5}}$, $\lambda=-3$ for the minimum.

12 $(x,y,z,\lambda)=(0,0,5,-5)$, $(2,-2,1,-1)$ and $(-2,2,1,-1)$ are the solutions.

13 $x=-4$, $y=2$, $\lambda=10$.

Exercises 3.2

6 $$\delta I=\int_0^1\left[\frac{y'}{\sqrt{1+y'^2}}\right]\delta y'dx,\ \delta y(0)=\delta y(1)=0.$$

8 $$\delta I=\int_0^1\sqrt{\frac{y}{1+(y')^2}}\cdot\left[\frac{2yy'\delta y'-1+(y')^2\delta y}{y^2}\right]dx.$$

9 $$\delta I=\int_0^1\left(\frac{d\delta u}{dx}\frac{dv}{dx}+\frac{du}{dx}\frac{d\delta u}{dx}-v\delta v-q\delta u\right)dx.$$

10 $$\delta I=D\int_R\left\{\nabla^2 w\nabla^2\delta w+(1-\nu)\left[2\frac{\partial^2 w}{\partial x\partial y}\frac{\partial^2\delta w}{\partial x\partial y}-\frac{\partial^2 w}{\partial x^2}\frac{\partial^2\delta w}{\partial y^2}-\frac{\partial^2\delta w}{\partial x^2}\frac{\partial^2 w}{\partial y^2}\right]\right\}dx\,dy-\int_R q\delta w\,dx\,dy$$

11 $\mathcal{C}=\{u\in H^2(a,b),\ u(a)=u_a,\ u(b)=u_b\}$,

$\mathcal{H}=\{u\in H^2(a,b),\ u(a)=0,\ u(b)=0\}$.

13 $\mathcal{C}=\{u\in H^2(R),\ u=\hat{u}$ and $v=\hat{v}$ on $C\}$,

$\mathcal{H}=\{u\in H^2(R),\ u=v=0$ on $C\}$.

15 $$\frac{d^4u}{dx^4}-\frac{d^2u}{dx^2}+1=0 \text{ in } (a,b).$$

Natural *B.C.*: specify $(-u'''+u')$ and $(u''+u')$ at a and b. Essential *B.C.*: u and u' at a and b.

16 $$-\left(\frac{\partial^2u}{\partial x^2}+\frac{\partial^2v}{\partial x^2}+\frac{\partial^2v}{\partial y^2}-f\right)=0,$$
$$-\left(\frac{\partial^2v}{\partial y^2}+\frac{\partial^2u}{\partial x^2}+\frac{\partial^2u}{\partial y^2}-g\right)=0. \qquad \text{(in } R.)$$

NBC: specify, $n_x\left(\frac{\partial u}{\partial x}+\frac{\partial v}{\partial x}\right)+n_y\frac{\partial v}{\partial y}$, and, $n_x\frac{\partial u}{\partial x}+n_y\left(\frac{\partial u}{\partial y}+\frac{\partial v}{\partial y}\right)$ on C.

EBC: specify u and v on C.

17 $$\nabla\mathbf{u}-\mathbf{v}/k=0,$$
$$-\nabla\cdot\mathbf{v}+Q=0.$$

NBC: $\hat{n}\cdot\mathbf{v}-q=0$ on C_1 and specify $\hat{\mathbf{n}}\cdot\hat{\mathbf{v}}$ on C_2.

EBC: specify u on C_2.

18 $$-\frac{\mu}{2}(u_{i,j}+u_{j,i})_{,j}+\frac{\partial P}{\partial x_i}-f_i=0 \text{ in } R.$$

NBC: $\frac{\mu}{2}(u_{i,j}+u_{j,i})n_j-Pn_i+t_i=0$ on C.

EBC: specify u_i.

19 $$\frac{\partial}{\partial x}\left(\frac{\partial M_1}{\partial x}+\frac{\partial M_3}{\partial y}\right)+\frac{\partial}{\partial y}\left(\frac{\partial M_3}{\partial x}+\frac{\partial M_2}{\partial y}\right)+P=0 \text{ in } R,$$

$$S(M_1-\nu M_2)+\frac{\partial^2w}{\partial x^2}=0,$$

$$S(M_2-\nu M_1)+\frac{\partial^2w}{\partial y^2}=0,$$

$$2S(1+\nu)M_3+\frac{2\partial^2w}{\partial x\partial y}=0.$$

NBC: $$-\frac{\partial M^2}{\partial s}-\left(\frac{\partial M_1}{\partial x}+\frac{\partial M_3}{\partial y}\right)n_x-\left(\frac{\partial M_3}{\partial x}+\frac{\partial M_2}{\partial y}\right)n_y+\hat{V}=0 \text{ on } C_2,$$

$$\frac{\partial w}{\partial n}=\hat{\theta}_n \text{ on } C_3.$$

EBC: $w=\hat{w}$ on C_1 and $M_n=\hat{M}_n$ on C_4.

20 $\varepsilon_{ij}-\frac{1}{2}(u_{i,j}+u_{j,i}+u_{m,i}u_{m,j})=0,$

$-\sigma_{ij}+E_{ijkl}\varepsilon_{kl}=0,$

$-[(\delta_{mi}+u_{m,i})\sigma_{ij}]_{,j}-\rho f_m=0.$

EBC: $u_m=\hat{u}_m$ on C_u.

NBC: $(\delta_{mi}+u_{m,i})\sigma_{ij}n_j-\hat{T}_m=0$ on C_σ.

21
$$\frac{\partial F}{\partial u}-\frac{\partial}{\partial x}\left(\frac{\partial F}{\partial u_x}\right)-\frac{\partial}{\partial y}\left(\frac{\partial F}{\partial u_y}\right)+\frac{\partial^2}{\partial x^2}\left(\frac{\partial F}{\partial u_{xx}}\right)+\frac{\partial^2}{\partial x\partial y}\left(\frac{\partial F}{\partial u_{xy}}\right)+\frac{\partial^2}{\partial y^2}\left(\frac{\partial F}{\partial u_{yy}}\right)=0$$

NBC: $\dfrac{\partial F}{\partial u_x}-\dfrac{\partial}{\partial x}\left(\dfrac{\partial F}{\partial u_{xx}}\right)-\dfrac{\partial}{\partial y}\left(\dfrac{\partial F}{\partial u_{xy}}\right)=0$ at $x=x_1, x_2$.

$\dfrac{\partial F}{\partial u_y}-\dfrac{\partial}{\partial x}\left(\dfrac{\partial F}{\partial u_{xy}}\right)-\dfrac{\partial}{\partial y}\left(\dfrac{\partial F}{\partial u_{yy}}\right)=0$ at $y=y_1, y_2$.

$\dfrac{\partial F}{\partial u_{xx}}=0$ at $x=x_1$ and x_2; $\dfrac{\partial F}{\partial u_{yy}}=0$ at $y=y_1, y_2$.

EBC: $u=u_x=0$ at $x=x_1, x_2$,

$u=u_y=0$ at $y=y_1, y_2$.

24
$$-\frac{1}{4}\left[\left(\frac{du}{dx}\right)^2+8\right]\frac{d^2u}{dx^2}+6(u^5-1)=0.$$

25 Linear and bilinear forms for Example 3.5:

$$l(\bar{T})=\int_{C_q}\hat{q}\bar{T}\,ds-\int_{C_h}h(T-T_\infty)\bar{T}\,ds-\int_R Q\bar{T}\,dx\,dy,$$

$$B(T,\bar{T})=\int_R \text{grad}\,T\cdot\mathbf{k}\cdot\text{grad}\,\bar{T}\,dx\,dy.$$

26 Penalty method: The functional is given by

$$I(w,\psi)=\int_0^L\left[\frac{EI}{2}\left(\frac{d\psi}{dx}\right)^2+fw\right]dx+\frac{\gamma}{2}\int_0^L\left(\frac{dw}{dx}+\psi\right)^2dx.$$

The Euler equations are:

$$-\frac{d}{dx}\left(EI\frac{d\psi}{dx}\right)+\gamma\left(\frac{dw}{dx}+\psi\right)=0,$$

$$-\gamma\frac{d}{dx}\left(\frac{dw}{dx}+\psi\right)+f=0.$$

NBC: $EI\dfrac{d\psi}{dx}=0, \gamma\left(\dfrac{dw}{dx}+\psi\right)=0$ at $x=0, L$,

or

EBC: $\psi=0$ and $w=0$ at $x=0, L$.

27 The Lagrange multiplier method: The functional is given by

$$I_L(u,v,w,Q_x,Q_y)=I(u,v,w)$$
$$+\int_R\left[Q_x\left(\frac{\partial w}{\partial x}-u\right)+Q_y\left(\frac{\partial w}{\partial y}-v\right)\right]dx\,dy.$$

The Euler equations are:

$$-D\left[\frac{\partial}{\partial x}\left(\frac{\partial u}{\partial x}+\nu\frac{\partial v}{\partial y}\right)+(1-\nu)\frac{\partial}{\partial y}\left(\frac{\partial u}{\partial y}+\frac{\partial v}{\partial x}\right)\right]-Q_x=0,$$

$$-D\left[(1-\nu)\frac{\partial}{\partial x}\left(\frac{\partial u}{\partial y}+\frac{\partial v}{\partial x}\right)+\frac{\partial v}{\partial y}\left(\frac{\partial v}{\partial y}+\nu\frac{\partial u}{\partial x}\right)\right]-Q_y=0,$$

$$-\frac{\partial Q_x}{\partial x}-\frac{\partial Q_y}{\partial y}-q=0,\text{ in } R.$$

$$NBC\colon\ D\left[n_x\left(\frac{\partial u}{\partial x}+\nu\frac{\partial v}{\partial y}\right)+(1-\nu)n_y\left(\frac{\partial u}{\partial y}+\frac{\partial v}{\partial x}\right)\right]=0,$$

$$D\left[n_x(1-\nu)\left(\frac{\partial u}{\partial y}+\frac{\partial v}{\partial x}\right)+n_y\left(\frac{\partial v}{\partial y}+\nu\frac{\partial u}{\partial x}\right)\right]=0,$$

$$Q_xn_x+Q_yn_y=0,$$

or, *EBC*: $u=v=w=0$ on the boundary of R.

Exercises 3.3

1 $L=\frac{1}{2}[m_1\dot{x}_1^2+m_2\dot{x}_2^2+m_3\dot{x}_3^2-k_1x_1^2+k_2(x_2-x_1)^2+k_3(x_3-x_2)^2]$.

2 $\delta L=-(kx\delta x+\eta\dot{x}\delta x)$, $\dot{x}=\dfrac{\dot{f}}{k}+\dfrac{f}{\eta}$; $x=x_1=x_2$.

7 $\delta L_D=\delta L-(\eta_1\dot{x}_1\delta x+\eta_2\dot{x}_2\delta x+\eta_3\dot{x}_3\delta x_3)$, where L is given in the answer to Exercise 1.

9 $L=\frac{1}{2}[m_1\dot{x}^2+m_2(\dot{x}+l\dot{\theta})^2-\frac{1}{2}(k_1x^2+m_2ql\theta^2)$.

12 **(a)** $-\ddot{\theta}+\sin\theta\cos\theta\,\dot{\phi}+g\sin\theta=0$; $-\dfrac{d}{dt}(\sin\theta\dot{\phi})=0$.

13 $\ddot{x}_1=\dfrac{g}{23}$.

18 $-\dfrac{\partial}{\partial t}\left(\rho\dfrac{\partial u}{\partial t}\right)+\dfrac{\partial}{\partial x}\left(T\dfrac{\partial u}{\partial x}\right)+f=0.$

19 $-\dfrac{\partial}{\partial t}\left(\rho A\dfrac{\partial u}{\partial t}\right)-\dfrac{\partial^2}{\partial x^2}\left(EI\dfrac{\partial^2 u}{\partial x^2}\right)+f=0.$

20 $$-\frac{\partial}{\partial t}\left(\rho A\frac{\partial u}{\partial t}\right)+\frac{\partial^2}{\partial x\partial t}\left(\rho I\frac{\partial^2 u}{\partial x\partial t}\right)-\frac{\partial^2}{\partial x^2}\left(EI\frac{\partial^2 u}{\partial x^2}\right)+f=0.$$

21 $$-\frac{\partial}{\partial t}\left(\rho I\frac{\partial \psi}{\partial t}\right)+\frac{\partial}{\partial x}\left(EI\frac{\partial \psi}{\partial x}\right)+kGA\left(\frac{\partial u}{\partial x}-\psi\right)=0,$$

$$-\frac{\partial}{\partial t}\left(\rho A\frac{\partial u}{\partial x}\right)+\frac{\partial}{\partial x}\left[kGA\left(\frac{\partial u}{\partial x}-\psi\right)\right]=0.$$

Exercises 3.4

1 $$I(u)=\int_a^b\left[\frac{p}{2}\left(\frac{du}{dx}\right)^2+\frac{q}{2}u^2-fu\right]dx+\frac{1}{2}\left[\alpha p(a)u^2(a)+\beta p(b)u^2(b)\right].$$

2 $$I(T)=\int_R\left\{\frac{k}{2}\nabla T\cdot\nabla T+QT\right\}dx\,dy-\int_{C_q}\hat{q}T\,ds-h\int_{C_h}\left(\frac{T}{2}-T_\infty\right)T\,ds.$$

3 $$I(w)=\int_R\left[\frac{D}{2}(\nabla^2 w)^2+Pw\right]dx\,dy+\int_{C_v}\hat{V}w\,ds-\int_{C_m}\hat{M}\frac{\partial w}{\partial n}ds.$$

4 $$I(\phi,\mathbf{v},\mathbf{P})=\int_R\left[(\nabla\phi+\mathbf{v})\cdot\mathbf{P}-\frac{\rho_0}{2}\mathbf{v}\cdot\mathbf{v}\right]dx\,dy\,dz-\int_{S_2}P_0\phi\,ds.$$

6 See Exercise 3.2.19.

7 $$I(u)=\int_\Omega\left[\frac{1}{2}\left(\frac{\partial u}{\partial x}+\frac{\partial u}{\partial y}\right)^2-fu\right]dx\,dy-\int_{\Gamma_1}\hat{t}u\,ds.$$

8 Use inverse relations (relating $u_{i,j}$ with σ_{ij} through the compliances, S_{ij}).

9 $$I(u,v)=-\int_\Omega\left\{\frac{1}{2}\left[C_{11}\left(\frac{\partial u}{\partial x}\right)^2+2C_{12}\frac{\partial u}{\partial x}\frac{\partial v}{\partial y}+2C_{13}\frac{\partial u}{\partial x}\left(\frac{\partial u}{\partial y}+\frac{\partial v}{\partial x}\right)\right.\right.$$

$$\left.+C_{22}\left(\frac{\partial v}{\partial y}\right)^2+2C_{23}\frac{\partial v}{\partial y}\left(\frac{\partial u}{\partial y}+\frac{\partial v}{\partial x}\right)+C_{33}\left(\frac{\partial u}{\partial y}+\frac{\partial v}{\partial x}\right)^2\right]$$

$$\left.+f_1u+f_2v\right\}dx\,dy-\int_{\Gamma_1}(\hat{t}_1u+\hat{t}_2v)\,ds.$$

10 See Exercise 3.2.23.

14 $$B(\mu,\mathbf{v},\mathbf{P};\psi,\mathbf{u},Q)=\int_R\left[(\nabla\phi+\mathbf{v})\cdot\mathbf{Q}+(\nabla\psi+\mathbf{u})\cdot\mathbf{P}-\rho_0\mathbf{v}\cdot\mathbf{u}\right]dx\,dy,$$

$$l(\psi,u,Q)=\int_{S_2}P_0\psi\,ds.$$

17 $$B(u,v)=\int_{\Omega}\left(\frac{\partial u}{\partial x}+\frac{\partial u}{\partial y}\right)\left(\frac{\partial v}{\partial x}+\frac{\partial v}{\partial y}\right)dx\,dy,$$

$$l(v)=\int_{\Omega} fv\,dx\,dy+\int_{\Gamma_1}\hat{t}v\,ds.$$

22 $$I(u)=\int_0^1\left[\frac{1}{2}\left(\frac{du}{dx}\right)^2+2\cosh u\right]dx.$$

24 $$I(u)=\int_0^1\left[\frac{x^3}{2}\left(\frac{du}{dx}\right)^2+\frac{2x^3}{u}\right]dx.$$

25 $$I(w)=\frac{1}{2}\int_0^1\left[\left(\frac{d^2w}{dx^2}\right)^2+\lambda\left(x-\frac{1}{2}\right)\left(\frac{dw}{dx}\right)^2\right]dx.$$

26 $$I(u)=\int_0^1\left[\frac{1}{48}\left(\frac{du}{dx}\right)^4+\left(\frac{du}{dx}\right)^2+u^6-6u\right]dx.$$

28 See Exercise 3.2.20

Exercises 3.5

1 **(a)** **(i)** $\mathfrak{D}(A)=H^2(0,1)\cap H_0^1(0,1)$

(ii) $\mathfrak{D}(A)=\{u\colon u\in H^2(0,1), u(0)=u'(1)=0\}$

(b) **(i)** $H_A=H_0^1(0,1)$

(iii) $H_A=\{u\in H^1(0,1), \int_0^1 u(x)\,dx=0\}.$

(c) **(i)** $\phi_i=(1-x)x^i, \psi_0=0,$

$$B(\phi_i,\phi_j)=\frac{ij}{i+j-1}-\frac{2ij+i+j}{i+j}+\frac{(i+1)(j+1)}{i+j+1},$$

$$l(\phi_i)=2(-1)^i\left[\ln 2+\sum_r^i\frac{1}{r}(-1)^r\right]-\frac{1}{1+i},$$

$$u_{2R}=0.38x-0.458x^2+0.07786x^3,$$

$$u_0=2\times\ln 2-(1+x)\ln(1+x),$$

$$=0.3863x-0.5x^2+0.1667x^3-0.0833x^4+\cdots$$

(ii) $\phi_i = x^i$ (satisfy the essential boundary condition only),

$$B(\phi_i, \phi_j) = ij/(i+j-1),$$

$$l(\phi_i) = (-1)^i \left[\ln 2 + \sum_r^i \frac{1}{r}(-1)^r \right],$$

$$u_{2R} = 0.648x - 0.579x^2,$$

$$u_0 = 0.693x - 0.5x^2 + 0.1667x^3 - \cdots$$

3 **(a)** $$-\frac{d}{dx}\left(a\frac{dw}{dx}\right) + cw = f + 3\frac{da}{dx} - c(3x+1).$$

(b) $\phi_0 = 0, \phi_i = x^i$ (for w, not for u),

$$B(\phi_i, \phi_j) = \frac{ij}{i+j-1} + \frac{ij+1}{i+j+1},$$

$$l(\phi_i) = \sum_{r=0}^{[i/2]} (-1)^{r+1} \frac{i!}{(i-2r)!} \frac{1}{\pi^{2r+1}},$$

where $[s]$ means greatest integer $\leq s$;

$[\frac{1}{2}] = 0$, $[1\frac{1}{2}] = 1$, etc.

5 **(a)** $H_A = H^2(0,1) \cap H_0^1(0,1)$.

(b) $\phi_0 = 0$, $\phi_i = \sin i\pi x$ (or $\phi_0 = 0, \phi_i = (1-x)x^i$).

7 $\phi_0 = x, \phi_i = x^i(1-x)$,

$$B(\phi_i, \phi_j) = \frac{2ij}{(i+j)^2 - 1} - \frac{2(ij+i+j)}{(i+j)(i+j+2)} + \frac{2(i+j)+7}{(i+j+2)(i+j+3)},$$

$$l(\phi_i) = \frac{2}{(1+i)(3+i)},$$

$$u_{1R} = 1.5357x - 0.5357x^2.$$

8 $\phi_0 = 0, \phi_i = \left(\frac{x}{L}\right)^{i+1}\left(1 - \frac{x}{L}\right),$

$$M_{ij} \equiv \int_o^L \phi_i \phi_j \, dx = \frac{1}{i+j+3} + \frac{1}{i+j+5} - \frac{2}{i+j+4},$$

$$B(\phi_i, \phi_j) = \frac{EI}{L^3} \left\{ \frac{(1+i)ij}{i+j-1} - \frac{i(i+1)(j+1)(j+2)}{i+j} \right.$$

$$\left. - \frac{j(j+1)(i+1)(i+2)}{i+j} + \frac{(i+1)(j+1)(i+2)(j+2)}{i+j+1} \right\},$$

$|B_{ij} - \lambda M_{ij}| = 0,$

$\omega_1 = 15.45\alpha, \omega_2 = 75.33\alpha, \alpha = \sqrt{EI/mL^4}.$

(exact: $\omega_1 = 15.42\alpha, \omega_2 = 49.96\alpha$).

9 **(a)** $u = \phi_0 + c_1\phi_1, v = \psi_0 + D_1\psi_1,$

$\phi_0 = 0, \phi_1 = x, \psi_0 = \hat{v}, \psi_1 = (x - L),$

$$c_1 = -\frac{qL^3}{6\alpha} + \frac{\hat{v}L}{2\alpha}, D_1 = \frac{qL}{2}.$$

(b) $\phi_0 = 0, \phi_1 = x^2, \psi_0 = \hat{v}, \psi_1 = (x - L)^2.$

12 For $a = 2$, and $b = 1$ the Galerkin solution is given by $w_{2G} = 10^{-3}(0.9485 - 0.19r^2)(r^2 - 1)^2(4 - r^2)^2$;

The exact solution is given by

$$w_{\text{exact}} = -(0.266 + 0.2792r^2)\ln r - 0.257 + 0.2414r^2 + 0.01563r^4.$$

14 $\theta(x, y) = x(1 - x)e^{-2\sqrt{2}y}.$

15 $u_{2R} = u_{2G} = c_1\phi_1 + c_2\phi_2,$

$$c_1 = \frac{3885}{13296}\frac{f}{ka^4}, c_2 = \frac{1575}{26592}\frac{f}{ka^4}.$$

17 **(a)** See Exercise 15.

(b) See Exercise 16.

18 Use $\phi_0 = 0, \phi_1 = \left(1 - \cos\frac{2\pi x}{a}\right)\left(1 - \cos\frac{2\pi y}{b}\right).$

19 Use $\phi_0 = 0$, and $\phi_1 = \left(1 - \frac{x^2}{a^2} - \frac{y^2}{b^2}\right)$, and obtain $c_1 = \frac{a^2b^2}{2(a^2 + b^2)}.$

21 **(i)** Use $\phi_0 = 0$ and $\phi_1 = xy(1 - \frac{x^2}{a^2} - \frac{y^2}{b^2})m$ and compute

$B(\phi_1, \phi_1) = \frac{\pi}{64}ab(a^2 + b^2)$ and $l(\phi_1)$ to determine c_1.

(ii) For ϕ_1 given above, one gets

$$(\phi_1, \phi_1)_0 = \frac{\pi}{960}a^3b^3, \text{ and } \lambda_1 = 15\left(\frac{a^2 + b^2}{a^2b^2}\right).$$

24 Use $\phi_0 = 1 - x, \phi_i = \sin(2i - 1)\pi x$. The resulting nonlinear algebraic equations are:

$$0 = -3.1831 + 12.1033c_1 + 3.044c_1^3 + 54.7926c_1c_2^2$$
$$+ 152.2017c_1c_3^2 + 91.321c_1c_2c_3 + 0.1321c_1^2c_2$$
$$+ 136.9815c_2^2c_3,$$

$$0 = -1.061 + 3.044c_1^3 + 54.7926c_1^2c_2 + 45.6605c_1^2c_3$$

$$+ 100.9297c_2 + 246.5667c_2^3 + 273.9631c_2c_3$$

$$+ 1369.8153c_2c_3^2,$$

$$0 = -0.6366 + 136.9815c_1c_2^2 + 45.6605c_1^2c_2 + 152.2017c_1^2c_3$$

$$+ 1369.8153c_2^2c_3 + 278.5826c_3 + 1902.5213c_3^3.$$

INDEX